CONTENTS

TO THE STUDENT

□

This study guide has been prepared to help you learn biology. We have included the ways of studying and learning that successful students find helpful. If you study the textbook, use this study guide regularly, and follow the guidelines provided here, you will learn more about biology and, in the process, you are likely to earn a better grade. It is our sincere desire to provide you with enjoyable experiences during your excursion into the world of life.

Study guide chapters contain the following sections: *Chapter Overview, Reviewing Concepts*, *Building Words, Matching, Making Comparisons, Making Choices,* and *Visual Foundations.* Although these sections provide a fairly comprehensive review of the material in each chapter, they may not cover all of the material that your instructor expects you to know. The ways that you can use these sections are described below.

CHAPTER OVERVIEW

Each chapter in this study guide begins with an overview of the chapter. It focuses on the major concepts of the chapter, and, when appropriate, relates them contextually to what you have learned in previous chapters. We believe you will find it beneficial to read this section carefully *before* you undertake any of the exercises in the chapter, and again *after* you have completed them. Review this material any time you find yourself losing sight of "the big picture."

REVIEWING CONCEPTS

This section outlines the textbook chapter. It contains important concepts and fill-in questions. Be sure that you understand them. Examine the textbook material to be sure that you have correctly filled in all of the blank spaces. Once completed, check your work in the *Answers* section at the back of the study guide.

BUILDING WORDS

The language of biology is fundamental to an understanding of the subject. Many terms used in biology have common word roots. If you know word roots, you can frequently figure out the meaning of a new term. This section provides you with a list of some of the prefixes and suffixes (word roots) that are used in the chapter.

Learn the meanings of these prefixes and suffixes, then use them to construct terms in the blank spaces next to the definitions provided. Once you have filled in the blanks for all of the definitions, check your work in the *Answers* section at the back of the study guide.

Learning the language of biology also means that you can communicate in its language, and effective communication in any language means that you can read, write, and speak the language. Use the glossary in your textbook to learn how to pronounce new terms. Say the words out loud. You will discover that pronouncing words aloud will not only teach you verbal communication, it will also help you remember them.

MATCHING

This section tests your knowledge of key terms used in the chapter. Check you answers in the *Answers* section.

MAKING COMPARISONS

This section uses tables to test your ability to recognize relationships between key elements in the chapter. Check your answers in the *Answers* section

MAKING CHOICES

The items in this section were selected for two purposes. First, and perhaps most importantly, they help you develop study skills. They provide a means to re-examine, integrate, and analyze selected textbook material. To answer many of them, you must understand how material presented in one part of a chapter relates to material presented in other parts of the chapter. This section provides a means to evaluate your mastery of the subject matter and to identify material that needs additional study.

Once you have studied the textbook material and your notes, close your textbook, put away your notes, and try answering the multiple choice items. Check your answers in the *Answers* section. Don't be discouraged if you miss some answers. Take each item that you missed one at a time, go back to the textbook and your notes, and determine what you missed and why. Do not merely copy down the correct answers — that defeats the purpose. If you dig the answers out of the textbook, you will find that your understanding of the material will improve. Some points that you thought you already knew will become clearer, and you will develop a better understanding of the relationships between different aspects of biology.

VISUAL FOUNDATIONS

Selected visual elements from the textbook are presented in this section. Coloring the parts of biological illustrations causes one to focus on the subject more carefully. Coloring also helps to clearly delineate the parts and reveals the relationship of parts to one another and to the whole.

Begin by coloring the open boxes (❑). This will provide you with a quick reference guide to the illustration when you study it later. Close your textbook while you are doing this section. After completing the *Visual Foundations* section, use your textbook to check your work. Make notes about anything you missed for future reference. Come back to these figures periodically for review. If an item to be colored is listed in the singular form, and there is more than one of them in the illustration, be sure to color all of them. For example, if you are asked to color the "mitochondrion," be sure to color all of the mitochondria shown in the illustration.

OTHER STUDY SUGGESTIONS

When possible, read material in the textbook *before* it is covered in class. Don't be concerned if you don't understand all of the material at first, and don't try to memorize every detail in this first reading. Reading before class will contribute greatly to your understanding of the material as it is presented in class, and it will help you take better notes. As soon as possible after the lecture, recopy all of your class notes. You should do this while the classroom presentation is still fresh in your mind. If you are prompt and disciplined about it, you will be able to recall additional details and include them in your notes. Use material in this study guide and your textbook to fill in the gaps.

What is the best way for you to study? That is something you have to figure out for yourself. Reading all of the chapters assigned by your instructor is basic, but it is not all there is to studying.

A technique that many successful students have found beneficial is to organize a study group of 4-6 students. Meet periodically, especially before examinations, and go over the material together. Bring along your textbooks, study guides, and expanded notes. Have each member of the group ask every other member several questions about the material you are reviewing. Discuss the answers as a group until the principles and details are clear to everyone. Use the combined notes of all members to fill in any gaps in your notes. Consider the reasons for correct and incorrect answers to questions and items in the *Reviewing Concepts* and *Making Choices* sections. This study technique can do wonders. There is probably no better way to find out how well you know the material than to verbalize it in front of a critical audience. Just ask any biology instructor!

If you follow the guidelines presented here and carefully complete the work provided in this study guide, you will make giant strides toward understanding the principles of biology. If you decide to specialize in some area of biology, you will find the general background you obtained in this course to be invaluable.

CHAPTER 1

A View of Life

Biology is the science of life. Knowledge about biological principles is important if individuals are to make informed decisions concerning the role biology plays in their everyday lives including health care, food safety and the environment. Indeed, biology affects many personal, governmental, and societal decisions, and scientific endeavors have many ethical implications as scientific techniques and technology continue to advance.. Students studying biology cover many topics but the underlying principles include evolution, information transfer, and energy transfer. Living things share a common set of characteristics that distinguishes them from nonliving things. They are organized at several levels, each more complex than the previous, from the chemical level up through the biosphere. Organisms use chemical, electrical, and behavioral signals to transmit information within themselves and to other organisms in order to survive. They require a continuous input of energy to maintain the chemical transactions and cellular organization essential to their well-being. Organisms are studied by a system of observation, question, hypothesis, experiment, data analysis, and revised hypothesis known as the scientific method. Evolution shows the relationships that exist between diverse life forms on Earth and provides a framework for study. Historically, to make sense of the millions of organisms that have evolved on our planet, scientists developed a binomial and hierarchical system of nomenclature. More recently scientists have used newly developed technology in molecular biology to organize organisms into more closely related groups called clades. Ultimately, the tree of life indicates evolutionary relationships between organisms.

REVIEWING CONCEPTS

Fill in the blanks.

INTRODUCTION

1. Stem cells are (a)________________________ that have the capacity to (b)________________________ giving rise to (c)__________________ and to one or more (d) __________________. Stem cells also allow the body to (d)________________________ .
2. Biologist continue to discover new types of stem cells within the bodies of (a)________________________, (b)________________________, and (c)________________________.
3. The most versatile stem cells are called (a)________________________because they can give rise to (b)________________________ of the body.

4. Knowledge of biological concepts is a vital tool for understanding the challenges that confront modern society. Some of the more important challenges are:
   (a)____,
   (b)____,
   (c)____,
   (d)____,
   (e)____.

MAJOR THEMES OF BIOLOGY

5. The five forces that give life its unique characteristics are:
   (a)______________________,
   (b)______________________,
   (c)______________________,
   (d)______________________, and
   (e)______________________.

CHARACTERISTICS OF LIFE

6. The characteristics of life include:
   (a)______________________,
   (b)______________________,
   (c)______________________,
   (d)______________________,
   (e)______________________, and
   (f)______________________.

Organisms are composed of cells

7. Organisms can be (a)______________________ or (b)______________________.
8. The two major cell types are (a)______________________ and (b)______________________. These two cell types are distinguished by the presence or absence of (c)______________________.

Organisms grow and develop

9. Living things grow by increasing (a)______________________of cells, (b)______________________ of cells or both.
10. ______________________ is a term that encompasses all the changes that occur during the life of an organism.

Organisms regulate their metabolic processes

11. Metabolism can be described as the sum of all the ______________________ that take place in the organism.
12. (a) ______________________ describes the tendency of organisms to maintain an appropriate, balanced internal environment. The cellular mechanisms involved in this process are (b)______________________ control systems

Organisms respond to stimuli

13. Physical or chemical changes in an internal or external environment that evoke a response from all life forms are called ________________.
14. Animals that do no move from place to place are said to be ________.
15. Like animals, plants respond to (a) __________, (b)__________, (c) ____________, (d) ____________, and (e) ____________.

Organisms reproduce

16. We know that organisms come from previously existing organisms as a result of scientists like Francesco Redi in the 17th century and ________________ in the 19th century.
17. In general, living things reproduce in one of two ways: (a) ____________________ or (b) __________________.

Populations evolve and become adapted to the environment

18. Inherited characteristics that enhance an organism's ability to survive in a particular environment are called (a) __________________. These inherited characteristics may be at the (b)____________________, (c)____________________, (d)__________________, (e)_______________________, or be a combination of all four.

LEVELS OF BIOLOGICAL ORGANIZATION

19. Biologists learn about living things using ________________ to study their parts.
20. ____________________ also help biologists see what characteristics are new at a higher level of organization.

Organisms have several levels of organization

21. The smallest unit of a chemical element that retains the characteristic properties of that element is an (a) ______________ compared to the (b) ________________ which is the basic and functional unit of life.
22. Most animals have (a) ______________ tissue and (b)______________ tissue.

Several levels of ecological organization can be identified

23. All of the members of one species that live in the same geographic area make up a ____________________.
24. All of the ecosystems on Earth are collectively referred to as the ___________________________.
25. The study of how organisms relate to one another and to their physical environment is called _____________________.

INFORMATION TRANSFER

DNA transmits information from one generation to the next

26. _____________ are the units of hereditary material.

27. Each strand of DNA is made up of four types of individual chemical subunits called ____________
28. The two scientists credited with working out the structure of DNA are (a) ____________ and (b) ____________.

Information is transmitted by chemical and electrical signals

29. ____________ are large molecules important in determining the structure and function of cells and tissues.
30. Multicellular organisms use ____________ as one method of communication between cells.
31. ____________ are chemical compounds used as communication molecules by the nervous system

Organisms also communicate information to one another

32. Whole organisms may communicate with one by (a)____________, by emitting (b) ____________, or by using (c)____________

THE ENERGY OF LIFE

33. Organisms that produce their own food from raw materials are called (a)____________, or (b)____________
34. All the energy transformations and chemical processes that occur in an organism is referred to as ____________.
35. ____________ is the process by which molecular energy is released to do cellular work.
36. In addition to needing a continuous input of energy, an ecosystem contains three major categories of organisms: (a)____________, (b)____________, and (c)____________.

EVOLUTION: THE BASIC UNIFYING CONCEPT OF BIOLOGY

37. Populations change over time via ____________.

Biologists use a binomial system for naming organisms

38. The science of classifying and naming organisms is called ____________.
39. The category of classification with the least number of organisms is the ____________.
40. The scientific name (binomial) of an organism consists of the (a)____________ and (b)____________ names for that organism.

Taxonomic classification is hierarchical

41. Families are grouped into ____________.
42. The broadest of the taxonomic groups is the ____________.

Systematists classify organisms in three domains

43. A group of organisms with a common ancestor is called ____________.

44. The three domains of life are (a) ____________, (b) ____________, and (c) ____________, which includes plants, protists, animals, and fungi.

Species adapt in response to changes in their environment

45. Adaptations to environmental change occur due to ____________________.

Natural selection is an important mechanism by which evolution proceeds

46. (a) ____________________________ and (b)________________________ were the first scientists to bring the theory of evolution forward to the public and explain how natural selection is involved in the process.
47. Charles Darwin's famous book, "____," published in 1859, was focused on many of these ideas.
48. Darwin's theory of natural selection was based on the observations that individuals within species show (a)________________, and more individuals are produced than can possibly (b)______________, this causes (c)______________, and individuals with the most advantageous characteristics are the most likely to (d)____________________ and pass those adaptations on to their offspring.

Populations evolve as a result of selective pressures from changes in their environment

49. All the genes in a population make up its ____________________________.
50. Natural selection favors organisms with traits that enable them to effectively respond to ____.
51. Evolution may result in the formation of ____________________.

THE PROCESS OF SCIENCE

52. The ________________________________ is a system of observations, questions, a hypothesis, experiments, data analysis, and a revised hypothesis.

Science requires systematic thought processes

53. With __________________________ reasoning, one begins with supplied information or premises and draws conclusions based on those premises.
54. With __________________________ reasoning, one draws a conclusion from specific observations.

Scientists make careful observations and ask critical questions

55. In 1928, bacteriologist _______________________ discovered the antibiotic, penicillin, but had difficulty growing the antibiotic producing organism.

Chance often plays a role in scientific discovery

56. Significant discoveries are usually made by those who are in the habit of looking critically at nature and recognizing a ________________.

A hypothesis is a testable statement

57. The characteristics of a good hypothesis include being (a)________________, (b)________________, (c) ________________, and (d)________________.
58. A hypothesis cannot be ________________.
59. ________________ are becoming very useful in hypothesis development.
60. An experimental group differs from a control group only with respect to the ________________ being studied.

Researchers must avoid bias

61. In a ________________, neither the patient nor the physician knows who is getting the experimental drug and who is getting the placebo.

Scientists interpret the results of experiments and make conclusions

62. Sampling errors can lead to inaccurate conclusions because ________________.
63. ________________ are strengthened when others repeat a scientist's work.

A theory is supported by tested hypotheses

64. A ________________ is an integrated explanation of a number of hypotheses, each supported by consistent results from many observations or experiments.

Many hypotheses cannot be tested by direct experiment

65. Two examples of hypotheses that cannot be tested by direct experiment are (a)________________, and (b)________________.

Paradigm shifts accommodate new discoveries

66. A ________________ is a set of assumptions or concepts that constitute a way of thinking about reality.

Systems biology integrates different levels of information

67. ________________are biologists who study the simplest components of biological processes.
68. ________________ are biologists who integrate data from various levels of complexity with the goal of understanding the big picture.

Science has ethical dimensions

69. Scientists must be (a) ________________, and (b)________________. About the importance of their work.

Science, technology, and society interact

70. Scientists face many societal and political issues in many areas concerning their work including (a) ______________________, (b)______________________, (c) ______________________, (d)______________________, and (e)______________________.

BUILDING WORDS

Use combinations of prefixes and suffixes to build words for the definitions that follow.

Prefixes	The Meaning	Suffixes	The Meaning
a-	without, not, lacking	-logy	"study of"
auto-	self, same	-stasis	equilibrium caused by opposing forces
bio-	life		
eco-	home	-troph	nutrition, growth, "eat"
hetero-	different, other	-zoa	animal
homeo-	similar, "constant"		
hypo-	under		
photo-	light		
pro-	first, earliest form of		

Prefix	Suffix	Definition
________	________	1. The study of life.
________	________	2. The study of groups of organisms interacting with one another and their nonliving environment.
________	-sexual	3. Reproduction without sex.
________	________	4. Maintaining a constant internal environment.
________	-system	5. A community along with its nonliving environment.
________	-sphere	6. The planet earth including all living things.
________	________	7. An organism that produces its own food.
________	________	8. An organism that is dependent upon other organisms for food, energy, and oxygen.
________	-karyote	9. Simplest cell type.
________	-synthesis	10. The conversion of solar (light) energy into stored chemical energy by plants, blue-green algae, and certain bacteria.
________	-thesis	11. A tentative explanation.

MATCHING

Terms:

a.	Adaptation	g.	Flagella	m.	Species
b.	Cell	h.	Kingdom	n	Stimulus
c.	Cilia	i.	Order	o.	Stem cells
d.	Consumer	j.	Phylum	p.	Taxonomy
e.	Decomposer	k.	Pluripotent stem cells	q.	Organ
f.	Domain	l	Population	r.	Community

For each of these definitions, select the correct matching term from the list above.

____ 1. Can give rise to all tissues of the body
____ 2. Unspecialized cells with the capacity to divide
____ 3. Populations of different species that live in the same geographical area at the same time.
____ 4. The broadest category of classification used.
____ 5. The science of naming, describing, and classifying organisms.
____ 6. Short, hair like structures that project from the surface of some cells and are used by some organisms for locomotion.
____ 7. A physical or chemical change in the internal or external environment of an organism potentially capable of provoking a response.
____ 8. A group of tissues that work together to carry out specific functions.
____ 9. An evolutionary modification that improves the chances of survival and reproductive success in a given environment.
____ 10. A microorganism that breaks down organic material.
____ 11. The basic structural and functional unit of life, which consists of living material bounded by a membrane.
____ 12. A group of organisms with similar structural and functional characteristics that naturally breed only with each other and have a close common ancestry; a group of organisms with a common gene pool.

MAKING COMPARISONS

Fill in the blanks.

Level of Organization	Examples
#1	Group of bison living s in Yellowstone National Park
Community	#2
Ecosystem	#3
#4	All ecosystems on Earth
#5	Cells associate to form these
The most basic level of chemical organization	#6
Units of hereditary material	#7

MAKING CHOICES

Place your answer(s) in the space provided. Some questions may have more than one correct answer.

____ 1. In science, a tentative explanation for an observation or phenomenon is known as a/an
a. supposition
b. law
c. idea
d. theory
e. hypothesis

____ 2. Organisms that rely on other organisms for complex food molecules are known as
a. heterotrophs.
b. cyanobacteria.
c. autotrophs.
d. consumers.
e. producers.

____ 3. The kingdom(s) containing unicellular organisms that have a nucleus surrounded by a membrane ("true nucleus") is/are
a. bacteria.
b. protista.
c. fungi.
d. plantae.
e. animalia.

____ 4. A snake is a member of the kingdom/phylum
a. prokaryotae/vertebrate.
b. fungi/ascomycete.
c. protista/aves.
d. animalia/chordata.
e. plantae/angiosperm.

____ 5. The domain(s) containing organisms that are both single-celled and have a chromosome that is not separated from the cytoplasm by a membrane is/are
a. eukarya.
b. animalia.
c. bacteria.
d. plantae.
e. archae.

____ 6. Maintenance of a balanced internal environment within an organism is known as
a. homeostasis.
b. metabolism.
c. equilibrium.
d. adaptation.
e. regulation.

____ 7. Prokaryotic organisms
a. lack a nuclear membrane.
b. include bacteria.
c. contain membrane-bound organelles.
d. are simple plants.
e. comprise two domains.

____ 8. To comply with the scientific method, a hypothesis must
a. be true.
b. be testable.
c. be falsifiable.
d. withstand rigorous investigations.
e. eventually reach the status of a principle.

____ 9. The characteristic(s) common to all living things include(s)
a. reproduction.
b. presence of a true nucleus.
c. adaptation.
d. growth.
e. multicellularity.

____10. Deductive reasoning

a. is the opposite of inductive reasoning.
b. proceeds from general principles to specific conclusions.
c. begins with a conclusion being made about an observation.
d. establishes relationships between known facts
e. is based upon a premise

____11. The approach to biology that is the opposite of systems biology is

a. called organismal biology.
b. based upon paradigms.
c. called integrative biology.
d. based upon growth of the organism.
e. called reductionism.

The Federal Center for Disease Control tested 600 female prostitutes who regularly use condoms. They found that none of the women had the AIDS virus. Use this study and data to answer questions 10-12.

____12. In terms of determining a relationship between use of condoms by women and the kind of sexual behavior that might lead to contraction of AIDS, this experiment

a. could be called a theory.
b. is well-designed.
c. should be repeated with more subjects.
d. is of no value.
e. should include male subjects.

____13. The experiment

a. proves that condoms are effective in preventing AIDS.
b. indicates that AIDS is not transmitted through sexual activity.
c. suggests that condoms may be effective in preventing AIDS.
d. probably means that none of the partners of the female prostitutes had AIDS.
e. shows that women are less prone to contract AIDS than men.

____14. To determine the effectiveness of using condoms in preventing AIDS among women, this study was lacking in not having

a. enough subjects.
b. male subjects.
c. controls.
d. subjects who do not use condoms during sex.
e. women who are not prostitutes.

____15. The number of domains in the tree of life is

a. one.
b. two.
c. three.
d. four
e. none.

____16. Evolution by natural selection depends upon:

a. mutations.
b. adaptations.
c. differential reproduction
d. change in individuals.
e. environmental change.

____17. Which of the following is true concerning the levels of organization in multicellular organisms?

a. tissues are comprised of cells
b. organs are made up of tissues
c. a group of tissues and organs make up an organ system
d. molecules are comprised of atoms
e. cells are made up of atoms, molecules, and organelles.

VISUAL FOUNDATIONS

Questions 1-3 refer to the diagram below. Some questions may have more than one correct answer.

____ 1. The diagram represents:
a. developmental theory.
b. a controlled experiment.
c. deductive reasoning.
d. scientific method.
e. all of the above.

____ 2. Which of the following statements is/are supported by the diagram?
a. A theory and a hypothesis are identical.
b. Conclusions are based on results obtained from carefully designed experiments.
c. Using results from a single sample can result in sampling error.
d. A tested hypothesis may lead to new observations.
e. Science is infallible.

____ 3. Which of the following statements best describes the activity depicted by the box labeled #4 in the diagram?
a. Make a prediction that can be tested.
b. Generate a related hypothesis.
c. Form generalized inductive conclusions.
d. Develop a scientific principle.
e. Hypothesis not supported.

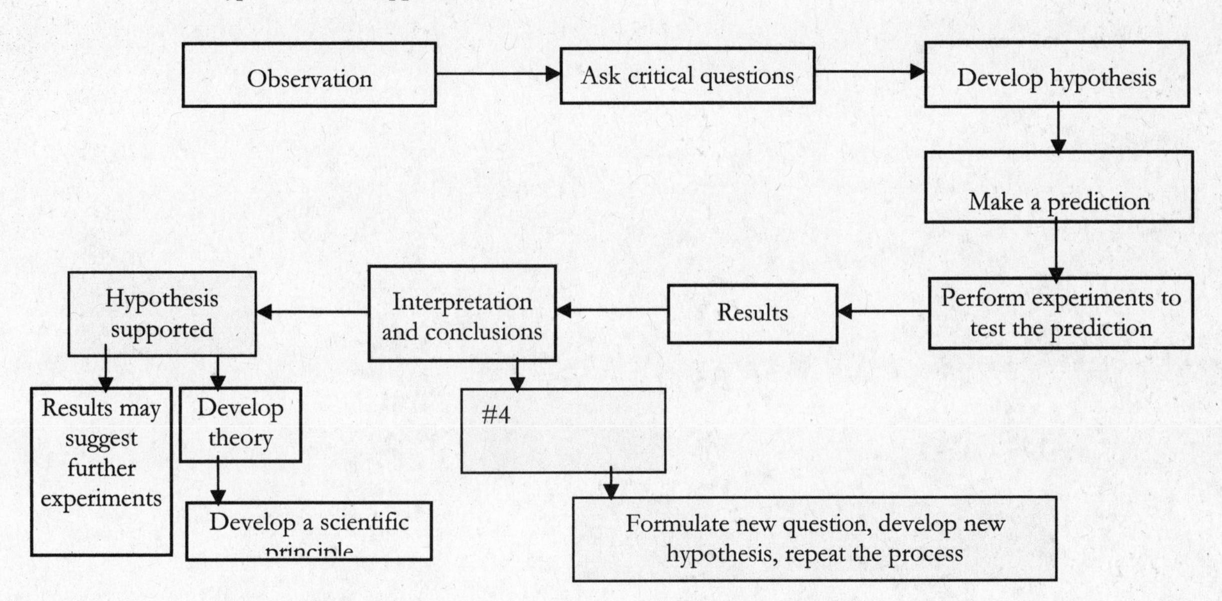

CHAPTER 2

❑

Atoms and Molecules: The Chemical Basis of Life

Living things, like everything else on our planet, are made of atoms and molecules. Therefore, to understand life you must first understand the interaction and activity of atoms and molecules. Living things share remarkably similar chemical compositions. About 96% of an organism's mass is made up of just four elements — oxygen, carbon, hydrogen, and nitrogen. The smallest portion of an element that retains its chemical properties is the atom. Atoms, in turn, consist of subatomic particles. Those most important to an understanding of life are the protons, neutrons, and electrons. The number and arrangement of electrons are most important biologically as they determine the chemical behavior of atoms. Electrons transferred during redox reactions are also very important to energy conversions. Molecules and compounds are formed by the joining of atoms. The interaction between atoms determines chemical bonds and attractive forces that produce compounds. The focus is on small, simple substances, known as inorganic compounds, which play critical roles in life's processes. One of these, water, is considered to have been essential to the origin of life, as well as to the continued survival and evolution of life on Earth. Acids, bases, and salts and how they generate electrolytes are also important for proper cellular function.

REVIEWING CONCEPTS

Fill in the blanks.

INTRODUCTION

1. The only two elements formed at the beginning of the universe were (a)____________________ and (b)____________________.
2. Among the biologically important groups of inorganic compounds are (a)____________________, (b)____________________, and (c)____________________.
3. Large complex carbon containing molecules are called ____________________.

ELEMENTS AND ATOMS

4. An element is a substance that cannot be separated into simpler parts by ____________________.
5. Over 96% of an organism's mass is made up of the four elements (a)____________________, (b)____________________, (c)____________________, and (d)____________________.

6. Elements required by the body in only very small amounts are called ______________________________.
7. The three main parts of an atom are the (a)______________, (b)______________, and (c)____________. The (d)______________________________ compose the atomic nucleus, and the (e)______________ move rapidly through the mostly empty space surrounding the nucleus.
8. (a)______________ have one unit of positive charge, (b)______________ are neutral, and (c)__________________ have a negative charge.

An atom is uniquely identified by its number of protons

9. An atom's atomic number is determined by the (a)________________________. To indicate this, Oxygen's atomic number is written as (b) ____________.

Protons plus neutrons determine atomic mass

10. An atomic mass unit (amu) is also called a ____________________.
11. The total number of protons and neutrons in an atom is referred to as the ________________________.

Isotopes of an element differ in number of neutrons

12. The (a) ______________________________ of isotopes vary, but the (b) __________________ remains constant.
13. Some isotopes are unstable and emit radiation when they decay. These isotopes are termed ____________________.
14. Radioactive decay darkens silver grains in film. The technique which uses this change in silver grains to detect the presence of radioactivity is called ____________________.

Electrons move in orbitals corresponding to energy levels

15. Electrons with similar energy levels comprise an ________________.
16. The electrons in the outermost energy level of an atom are called ______________ electrons.

CHEMICAL REACTIONS

17. The chemical behavior of an atom is determined by its ______________________________
18. All elements in the same vertical column (belonging to the same group) on the periodic table have similar ______________________.

Atoms form compounds and molecules

19. A chemical compound consists of ____________________________ ______________________________that are combined in a fixed ratio.
20. A ________________ forms when two or more atoms are joined very strongly.

Simplest, molecular, and structural chemical formulas give different information

21. A ________________________ formula simply uses chemical symbols to describe the chemical composition of a compound.
22. A ________________________ formula shows not only the types and numbers of atoms in a molecule but also their arrangement.

One mole of any substance contains the same number of units

23. The molecular mass of a compound is the sum of the ____.
24. The amount (mass) of a compound in grams that is equal to its atomic mass is called one __________.
25. A 1 molar solution contains one (a) ____________ of substance (b)______________in one (c) ___________.

Chemical equations describe chemical reactions

26. (a) __________________ are substances that participate in the reaction, and (b) _________________ are substances formed by the reaction.
27. Arrows are used to indicate the direction of a reaction. The reactants are placed to the (a) ________ of the arrows and the products are placed to the (b) _________ of the arrows.

CHEMICAL BONDS

28. (a) ______________ can hold atoms together and (b) __________________ determine the number of (c) ______________.

In covalent bonds electrons are shared

29. (a) ______________________ involve the sharing of electrons between atoms resulting in each atom having a (b)______________________.
30. Depending on the number of electrons shared covalent bonds can be (a)____________, (b) ______________, or (c) __________.

The function of a molecule is related to its shape

31. The functions of molecules in living cells are dictated largely by ______________________.
32. When covalent bonds are formed between atoms the orbitals in the valence shell may become rearranged in a process called ______________________

Covalent bonds can be nonpolar or polar

33. Covalently bonded atoms with similar electronegativities are bound together by ______________________________.
34. Covalently bonded atoms with differing electronegativities are bound together by ______________________________.

Ionic bonds form between cations and anions

35. Ionic compounds are comprised of positively charged ions called (a)________________, and negatively charged ions called (b)___________.

36. In the presence of a (a) ________________ an ionic compound will dissolve to produce a (b) ______________.
37. The process known as ___________________ has occurred when cations and anions of an ionic compound are surrounded by the charged ends of water molecules.

Hydrogen bonds are weak attractions

38. Hydrogen bonds tend to form between an (a)______________________________ and a hydrogen atom covalently bonded to either an (b)_______________ or a (c) _______________.
39. An individual hydrogen bonds is relatively (a)_____________ but, when present in large numbers, they are collectively (b)_______________.

Van der Waals interactions are weak forces

40. Van der Waals interactions are weak forces based on fluctuating ____________________________.

REDOX REACTIONS

41. Redox reactions are actually a combination of (a) _______________ and (b) _______________ reactions involving (c) _________________ transfers from one atom to another. The oxidizing agent is the (d)___________________ and the reducing agent is the (e)______________________.

WATER

42. In humans, water makes up about _______% of total body weight.

Hydrogen bonds form between water molecules

43. Cohesion of water molecules is due to (a)__________________________ between the molecules and contributes to (b) ____________________.
44. ______________ is the ability of water molecules to stick to other substances.
45. The tendency of water to move through narrow tubes as the result of adhesion and cohesion is known as ____________________________.

Water molecules interact with hydrophilic substances by hydrogen bonding

46. Substances that interact readily with water are said to be (a)________________, whereas those that don't are called (b)________________.
47. When nonpolar molecules interact with each other instead of water it is called an ________________________.

Water helps maintain a stable temperature

48. The total amount of energy kinetic energy in a substance is called __________.

49. ______________ is a measurement of the average kinetic energy of the particles in a substance.
50. Evaporative cooling depends on the high ____________________________ of water.
51. A ____________________ is a unit of heat energy.
52. Water has a high __________________________, which helps marine organisms maintain a relatively constant internal temperature.

ACIDS, BASES, AND SALTS

53. An acid is a proton donor that dissociates in solution to yield (a)______________________________. A base is a proton acceptor that generally dissociates in solution to yield (b)______________________________.

pH is a convenient measure of acidity

54. The pH scale covers the range (a) __________________ and is a (b)_________________scale. Neutrality is at pH (c) _____________.
55. A solution with a pH above 7 is a(an) (a) _________________. A solution with a pH below 7 is a(an) (b) _________________.

Buffers minimize pH change

56. When present in a solution, a buffer resists (a)___________________________. A buffer may be either a (b)_________________________ or a (c) _______________________.

An acid and a base react to form a salt

57. A compound in which the hydrogen ion is replaced by another cation is called a ____________.
58. Dissociated ions called ________________________ can conduct an electric current when a salt, an acid, or a base is dissolved in water.

BUILDING WORDS

Use combinations of prefixes and suffixes to build words for the definitions that follow.

Prefixes	The Meaning
co-	together
equi-	equal
hydr(o)-	water
iso-	alike
neutro-	neutral
non-	not
pol-	end of an axis
tetra-	four

Suffixes	The Meaning
-ar	pertaining to
-hedron	face
-philic	loving, friendly, lover
-phobic	fearing
-topos(tope)	place
-val-	strong

Prefix	Suffix	Definition
____________	-n	1. An uncharged subatomic particle.
____________	-librium	2. The condition of a chemical reaction when the forward and reverse reaction rates are equal.
____________	____________	3. The shape of a molecule when it appears as a three-dimensional pyramid.
____________	-ation	4. The process whereby ionic compounds combine chemically with water and dissolve.
____________	____________	5. Not attracted to water; insoluble in water.
____________	-electrolyte	6. A substance that does not form ions when dissolved in water, and therefore does not conduct an electric current.
____________	____________	7. Having a strong affinity for water; soluble in water.
____________	____________-s	8. Two atoms having the same number of electrons and protons, but a different number of neutrons
____________	____________-ent	9. Bonds formed by the sharing of electrons between atoms
____________	____________	10. Type of covalent bond formed between atoms that differ in electronegativity.

MATCHING

Terms:

a.	Anion	g.	Cation	m.	Ionic bond
b.	solvent	h.	Covalent bond	n.	Oxidation
c.	solute	i.	Electrolyte	o.	Proton
d.	Atom	j.	Electron	p.	Reduction
e.	Atomic mass	k.	Hydrogen bond	q.	Valence electrons
f.	Calorie	l.	Polar	r.	Nonpolar

For each of these definitions, select the correct matching term from the list above.

____ 1. The loss of electrons, or the loss of hydrogen atoms from a compound.
____ 2. The total number of protons and neutrons in the atomic nucleus.
____ 3. Substances dissolved in water that can conduct electrical current.
____ 4. The smallest subdivision of an element that retains its chemical properties.
____ 5. Chemical bond that involves one or more shared pairs of electrons.
____ 6. The name of the outermost electrons.
____ 7. An ion with a negative electric charge.
____ 8. A molecule with partial opposite charges.
____ 9. An ion with a positive electric charge.
____10. The amount of heat required to raise 1 gram of a substance 1 degree Celsius.
____11. Type of interaction key to water's unique properties.
____12. A covalent bond in which the atoms have similar electronegativities and the electrons are shared equally.
____13. A liquid in which a solid dissolves.

MAKING COMPARISONS

Fill in the blanks.

Name	Formula
Sodium ion	Na^+
Carbon dioxide	#1
Calcium	#2
#3	Fe_2O_3
Methane	#4
Chloride ion	#5
#6	K^+
Hydrogen ion	#7
#8	Mg
Ammonium ion	#9
Water	#10
Glucose	#11
Bicarbonate ion	#12
#13	H_2CO_3
#14	NaCl
#15	OH^-
Sodium hydroxide	#16

MAKING CHOICES

Place your answer(s) in the space provided. Some questions may have more than one correct answer.

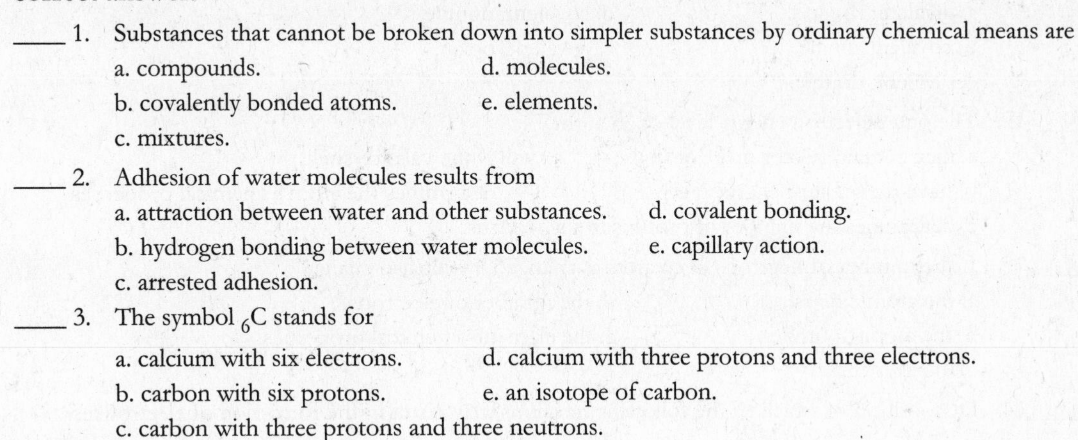

____ 1. Substances that cannot be broken down into simpler substances by ordinary chemical means are
a. compounds.
b. covalently bonded atoms.
c. mixtures.
d. molecules.
e. elements.

____ 2. Adhesion of water molecules results from
a. attraction between water and other substances.
b. hydrogen bonding between water molecules.
c. arrested adhesion.
d. covalent bonding.
e. capillary action.

____ 3. The symbol $_6C$ stands for
a. calcium with six electrons.
b. carbon with six protons.
c. carbon with three protons and three neutrons.
d. calcium with three protons and three electrons.
e. an isotope of carbon.

____ 4. Anions are

a. negative ions.
b. positive ions.
c. isotopes.
d. salts.
e. negative molecules.

Use the following chemical equation to answer questions 5-9.

$2\ H_2 + O_2 \rightarrow 2\ H_2O$ (water)

____ 5. The product has a molecular mass of

a. one.
b. twelve.
c. larger than either reactant alone.
d. sixteen.
e. eighteen.

____ 6. The reactant(s) in the formula is/are

a. hydrogen.
b. oxygen.
c. H_2O.
d. water.
e. the arrow.

____ 7. How many hydrogen atoms are involved in this reaction?

a. one
b. two
c. three
d. four
e. eight

____ 8. The product(s) is/are

a. hydrogen.
b. oxygen.
c. H_2O.
d. water.
e. the arrow.

____ 9. The lightest reactant has a mass number of

a. one.
b. two.
c. larger than the product.
d. eight.
e. ten.

____10. Reduction has occurred when

a. an atom gains an electron.
b. a molecule loses an electron.
c. hydrogen bonds break.
d. an atom loses an electron.
e. an ion gains an electron.

____11. When two atoms combine by sharing two pairs of their electrons, the bond is a

a. divalent, single.
b. covalent, single.
c. divalent, double.
d. covalent, double.
e. covalent, polar.

____12. The outer electron energy level of an atom

a. may contain two or more orbitals.
b. have the greatest energy level.
c. determines the number of protons in the nucleus.
d. is the valance shell.
e. determines the atom's chemical properties.

____13. If the number of neutrons in an atom is changed, it will also change

a. the atomic mass.
b. the atomic number.
c. the element.
d. the number of electrons.
e. the elements chemical properties.

____14. Dissociation of which of the following into ions may result in the formation of electrolytes?

a. salts
b. acids
c. isotopes
d. bases
e. alcohols

____15. A solution containing more hydrogen (H^+) ions than hydroxyl (OH^-) ions would be

a. acidic with a pH below 7.
b. alkaline with a pH below 7.
c. acidic with a pH above 7.
d. alkaline with a pH above 7.
e. neutral, with a pH of 7.

____16. H–O–H is

a. a structural formula.
b. water.
c. a chemical formula.
d. a molecule.
e. a structure with a molecular mass of 18.

____17. The atomic nucleus contains

a. protons.
b. neutrons.
c. electrons.
d. positrons.
e. isotopes.

____18. Inorganic compounds include

a. simple Bohr elements.
b. simple salts.
c. simple bases.
d. water.
e. simple acids.

____19. Matter

a. makes up atoms.
b. can be an electron.
c. has mass.
d. takes up space.
e. can be a proton.

____20. A chemical compound

a. consists of atoms.
b. lacks mass.
c. has an electrical charge.
d. is made with at least two different elements.
e. is made of molecules.

VISUAL FOUNDATIONS

Color the parts of the illustration below as indicated.

RED ❑ electron
YELLOW ❑ neutron
BLUE ❑ proton

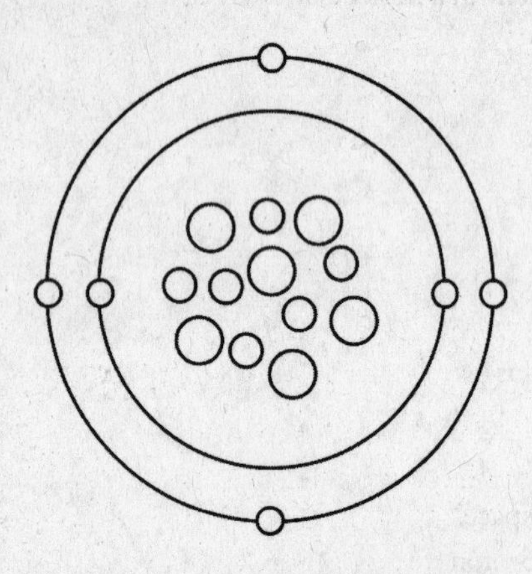

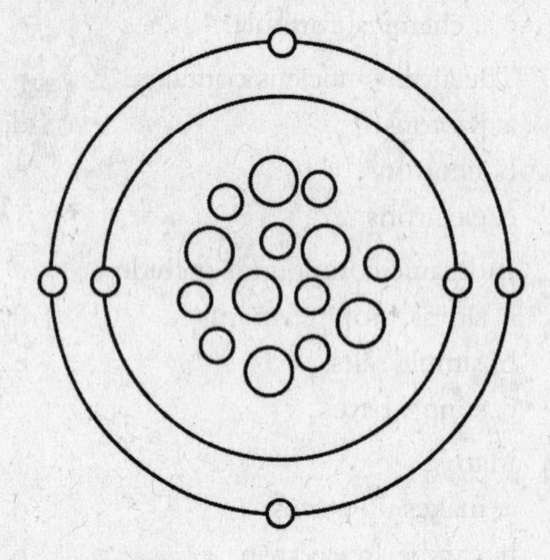

Questions 1-2 refer to the illustration above. Two atoms are represented in this illustration. Some questions may have more than one correct answer.

1. The atom on the left is called (a) ______________. The atom on the right is called (b) ________________.

2. These atoms:
   a. are ions.
   b. have the same atomic mass.
   c. have the same mass.
   d. are from different elements.
   e. have the same atomic number.

3. Water is an example of a polar covalent bond. Draw a figure of a water molecule showing the polar nature of the oxygen and hydrogen bond.

4. In one sentence explain why this is a polar bond.

CHAPTER 3

❑

The Chemistry of Life: Organic Compounds

In the previous chapter, the focus was on the smaller, simpler substances known as inorganic compounds and the bonds that produce them. The focus in this chapter is on the larger and more complex organic compounds. Most of the chemical compounds in organisms are organic compounds. Carbon is the central component of organic compounds, probably because it forms bonds with a greater number of different elements than any other type of atom. Adding functional groups allows large organic compound to interact better with each other. The major groups of organic compounds are the carbohydrates, lipids, proteins, and nucleic acids. Carbohydrates, built from monosaccharides, serve as energy sources for cells, and as structural components of fungi, plants, and some animals. Lipids, which are not characterized by structure, function as energy storage molecules, as structural components of cell membranes, and as hormones. Proteins, made from amino acids, serve a multitude of functions within the cell important to structure and basic cellular function. They also exhibit several levels of organization, allowing the production of a fully functional protein. Nucleic acids (made from nucleotides) transmit hereditary information and determine what proteins a cell manufactures.

REVIEWING CONCEPTS

Fill in the blanks.

INTRODUCTION

1. In organic compounds, carbon is covalently bonded to ____________.
2. In carbon containing inorganic molecules carbon is bonded to something other than (a)____________, or (b) ____________.
3. The properties of an organic molecule are influenced by the type of ____________ bound to the molecule.
4. The four major groups of organic compounds found in living systems are (a) ____________, (b)____________, (c)____________, and (d)____________.

CARBON ATOMS AND MOLECULES

5. Carbon has ____________ (#?) valence electrons.
6. Individual carbon atoms can form (a) ________, (b) ________, (c) ________, or (d) ________ (#?) bonds with a large variety of other atoms.
7. Organic compounds consisting of only carbon and hydrogen are called (a) ____________. These molecules may exist as

(b)____________________, (c) ____________________, or
(d)____________________.

Isomers have the same molecular formula but different structures

8. ____________________ isomers are compounds that differ in the covalent arrangements of their atoms.
9. ____________________ isomers are compounds that are different in the spatial arrangement of atoms or groups of atoms.
10. ____________________ are isomers that are mirror images of one another.

Functional groups change the properties of organic molecules

11. R — OH is the symbol for a (a) ____________________ group, and its chemical nature is (b) __________.
12. R — COOH is the symbol for a (a) ____________________ group, and its chemical nature is (b) __________.
13. R-NH_2 is the symbol for a (a) ____________________group, and its chemical nature is (b) __________.

Many biological molecules are polymers

14. Polymers are macromolecules produced by linking ____________________ together.
15. Polymers are built by (a) ____________________. They are degraded into their subunit components by (b) ____________________.

CARBOHYDRATES

16. The three groups of carbohydrates are (a) ____________________, (b)____________________, and (c) ____________________.
17. Carbohydrates contain the elements (a) ____________________, (b)____________________, and (c) ____________________ in a ratio of about 1:2:1.

Monosaccharides are simple sugars

18. The simple sugar (a) ____________________ is an abundant fuel molecule in most organisms. In cells its structure is usually in a (b)__________from.
19. Glucose is a structural isomer of ____________________.

Disaccharides consist of two monosaccharide units

20. Disaccharides form when two monosaccharides are bonded by a ____________________ linkage.
21. Two examples would be common table sugar, (a)__________, and the sugar in milk, (b)____________________.

Polysaccharides can store energy or provide structure

22. Long chains of ____________________ link together to form a polysaccharide.
23. ____________________ is the main carbohydrate used for energy storage in plants.

24. Starch occurs in two forms. The simpler form, _______________, is unbranched.
25. Plant cells store starch in organelles called _______________.
26. _______________ is the main storage carbohydrate of animals.
27. (a) _______________ is the most abundant carbohydrate on earth and is used to make plant (b) _______________.

Some modified and complex carbohydrates have special roles

28. (a) _______________ is a complex carbohydrate that forms the external skeleton of arthropods and the (b) _______________ of fungi..
29. A carbohydrate combined with a protein forms (a)_______________; carbohydrates combined with lipids are called (b) _______________.

LIPIDS

30. Lipids are not defined by (a) _______________ but by being (b)_______________ in water.
31. Lipids are composed of the elements (a) _______________, (b)_______________, and (c) _______________.

Triacylglycerol is formed from glycerol and three fatty acids

32. These molecules are commonly called (a) _________ and are used for energy (b) _______________.
33. Glycerol is a three carbon (a) _______________and each fatty acid that binds to it is a long (b) _______________ with a (c)_______________ on the end.

Saturated and unsaturated fatty acids differ in physical properties

34. Saturated fatty acids are solid due to _______________.
35. Unsaturated fats are (a) _______________ due to the presence of (b)_______________ which cause the bend in the hydrocarbon chain.
36. Trans fats are produced by (a) _______________ and mimic (b) _______________.

Phospholipids are components of cell membranes

37. A phospholipid molecule assumes a distinctive configuration in water because of its _______________ property, which means that one end of the molecule is hydrophilic and the other end is hydrophobic.
38. Phospholipids are made up of a (a) _______________ attached at one end to two (b) _______________ and at the other end to a (c)_______________ group attached to an (d) _______________compound.

Carotenoids and many other pigments are derived from isoprene units

39. Carotenoids consist of five-carbon hydrocarbon monomers known as (a) _______________. Animals convert these pigments to (b) _______________.

Steroids contain four rings of carbon atoms

40. A steroid consists of (a) ________________ atoms arranged in four attached rings, three of which contain (b) __________ carbon atoms and the fourth contains (c) __________.
41. ______________is an important example in animal cell membranes.

Some chemical mediators are lipids

42. Some chemical mediators are produced by the modification of ________________ that have been removed from membrane phospholipids.

PROTEINS

43. Proteins are (a) ________________ composed of (b) ____________ and they are the most (c) ______________ of cell components.

Amino acids are the subunits of proteins

44. All amino acids contain (a) ______________, and (b)____________ groups bonded to the (c) ______________. They vary in the (d) ________________ group.
45. ______________amino acids are ones animals cannot synthesize.

Peptide bonds join amino acids

46. The bonds that join two amino acids together are called ____________ bonds.
47. A chain of amino acids bound together is called ________________.
48. Amino acids combine chemically during (a) ______________ reactions when a bond is formed between the (b) ____________ group of one amino acid and the (c) __________ group of the other.

Proteins have four levels of organization

49. The (a) ______________of a protein is its 3-D shape, which is key to its (b) ____________.
50. The levels of organization distinguishable in protein molecules are (a)________________, (b) ________________, (c)________________, and (d) ____________.
51. The amino acid sequence is the primary structure of a ________________ and is what other levels are derived from.
52. Secondary structure results from (a) ____________ bonding involving the backbone. The two most common types are (b)______________, and (c) ______________.
53. Tertiary structure depends on interactions among (a)____________ and generates the (b) __________of the overall molecule.
54. Quaternary structure results from interactions among ____________ to generate a functional protein.

The amino acid sequence of a protein determines its conformation

55. ____________________________ help proteins fold.
56. The conformation of a protein determines its ________________.
57. When a protein loses its shape and function it is said to be ____________________.
58. Three examples of human disease caused by misfolded proteins are (a)__________________, (b) ________________________, and (c)____________________.

NUCLEIC ACIDS

59. Nucleic acids are polymers of subunits called ____________________.
60. Nucleotides are composed of a five carbon sugar which is either (a)________________, or (b) ________________, a nitrogenous base, which is either (c) ________________, or (d) ________________, and an inorganic (e) ________________ group.
61. DNA contains the purines (a) ________________, and (b) and the pyrimidines (c) ________________, and (d) ______________.
62. Nucelotides are connected by ______________________________.

Some nucleotides are important in energy transfers and other cell functions

63. Energy for life functions is supplied mainly by the nucleotide ______________.

BUILDING WORDS

Use combinations of prefixes and suffixes to build words for the definitions that follow.

Prefixes	The Meaning
amphi-	two, both, both sides
amylo-	starch
di-	two, twice, double
glycol-	sweet
hex-	six
hydr(o)-	water, also hydrogen
iso-	equal, "same"
macro-	large, long, great, excessive
mono-	alone, single, one
nucle(o)	nucleus,kernel
poly-	much, many
tri-	three
pent-	five

Suffixes	The Meaning
-ide	pertaining to
-lysis	breaking down, decomposition
-mer	part
-plast	formed, molded, "body"
-saccharide	sugar

Prefix	Suffix	Definition
____________	____________	1. A compound that has the same molecular formula as another compound but a different structure and different properties.
____________	-molecule	2. A large molecule consisting of thousands of atoms.

___________	___________	3. A single organic compound that links with similar compounds in the formation of a polymer.
___________	___________	4. The degradation of a compound by the addition of water.
___________	___________	5. A simple sugar.
___________	-ose	6. A sugar that consists of six carbons.
___________	___________	7. Two monosaccharides covalently bonded to one another.
___________	___________	8. A macromolecule consisting of repeating units of simple sugars.
___________	___________	9. A starch-forming granule.
___________	-acylglycerol	10. A compound formed of one fatty acid and one glycerol molecule.
___________	- acylglycerol	11. A compound formed of two fatty acids and one glycerol molecule.
___________	- acylglycerol	12. A compound formed of three fatty acids and one glycerol molecule.
___________	-pathic	13. Having one hydrophilic end and one hydrophobic end.
___________	-peptide	14. A compound consisting of two amino acids.
___________	-peptide	15. A compound consisting of a long chain of amino acids.
___________	-ose	16. A sugar that consists of five carbons.
___________	-carbon	17. An organic compound consisting only of carbon and hydrogen.
___________	-sidic	18. The linkage formed in a dissacharide bond.
___________	___________	19. .Individual monomer in nucleic acids.

MATCHING

Terms:

a.	Amino acid	g.	Glycogen	m.	Phospholipid
b.	Amino group	h.	Glucose	n.	Protein
c.	Carotenoid	i.	Hydrocarbon	o.	Starch
d.	Condensation	j.	Lipid	p.	Steroid
e.	Cellulose	k.	Nucleotide	q.	Tertiary structure
f.	Domain	l.	Phosphate group	r	Triglycerol

For each of these definitions, select the correct matching term from the list above.

____ 1. A molecule composed of one or more phosphate groups, a 5-carbon sugar, and a nitrogenous base.

____ 2. The most abundant type of lipid found in living organisms.

____ 3. A large complex organic compound composed of chemically linked amino acid subunits.

____ 4. Complex lipid molecules containing carbon atoms arranged in four interlocking rings.

____ 5. An organic compound containing an amino group and a carboxyl group.

____ 6. The three-dimensional structure of a protein molecule.

____ 7. An organic compound that is made up of only carbon and hydrogen atoms.

____ 8. Fatlike substance found in cell membranes.
____ 9. Polysaccharide used by plants to store energy.
____10. The principle carbohydrate stored in animal cells.
____11. Insoluble polysaccharide composed of joined glucose molecules.
____12. A distinct structural region in proteins.

MAKING COMPARISONS

Fill in the blanks.

Functional Group	Abbreviated Formula	Class of Compounds Characterized by Group	Description
Carboxyl	R — COOH	Carboxylic acids (organic acids)	Weakly acidic; can release a H^+ ion
#1	R — PO_4H_2	#2	Weakly acidic; one or two H^+ ions can be released
Methyl	#3	Component of many organic compounds	#4
#5	R---NH_2	#6	Weakly basic; can accept an H^+ ion
Sulfhydryl	R — SH	#7	Helps stabilize internal structure of proteins
#8	R---OH	#9	#10

MAKING CHOICES

Place your answer(s) in the space provided. Some questions may have more than one answer.

____ 1. The primary structure of a protein results from
a. ionic bonds.
b. hydrogen bonds.
c. covalent bonds.
d. polymerization.
e. peptide bonds.

____ 2. Condensation reactions result in
a. smaller molecules.
b. larger molecules.
c. reduced ambient water.
d. increase in polymers.
e. increase in monomers.

____ 3. An isomer can be described as one of a group of molecules with the same molecular formulas but different
a. elements.
b. compounds.
c. structures.
d. physical properties.
e. chemical properties.

____ 4. The complex carbohydrate storage molecule in plants is
a. glycogen.
b. starch.
c. usually in an extracellular position.
d. composed of simple glucose subunits.
e. more soluble than the animal storage molecule.

____ 5. Nucleotides
a. are polypeptide polymers.
b. are polymers of simple carbohydrates.
c. contain a base, phosphate, and sugar.
d. comprise complex proteins.
e. contain genetic information.

____ 6. Proteins contain
a. glycerol.
b. polypeptides.
c. peptide bonds.
d. amino acids.
e. nucleotides.

____ 7. Glycerol
a. is an alcohol.
b. is three carbons long.
c. forms an ester linkage with fatty acids.
d. is one component of phospholipids.
e. is hydrolyzed to produce fatty acids.

____ 8. Compared to saturated fatty acids, unsaturated fatty acids
a. contain more hydrogens.
b. contain more double bonds.
c. are more like oils.
d. are more prone to be solids.
e. contain more carbohydrates.

____ 9. Glycogen is
a. a molecule in animals.
b. a molecule in plants.
c. stored mainly in the liver and muscle of vertebrates.
d. composed of simple carbohydrate subunits.
e. also called animal starch.

____10. A monomer is to a polymer as a
a. chain is to a link.
b. link is to a chain.
c. monosaccharide is to a polysaccharide.
d. necklace is to a bead.
e. bead is to a necklace.

____11. $C_6H_{12}O_6$ is the formula for
a. glucose.
b. a type of monosaccharide.
c. cellulose.
d. fructose.
e. a type of disaccharide.

____12. Proteins
a. are a major source of energy.
b. contain carbon, hydrogen, and oxygen.
c. can combine with carbohydrates.
d. are found in plants and animals.
e. are polypeptide polymers.

____13. Phospholipids
a. are found in cell membranes.
b. are in the amphipathic lipid group.
c. contain glycerol, fatty acids, and phosphate groups.
d. are polar.
e. are polymers.

____14. Polymers are
a. made from monomers.
b. formed by condensation reactions.
c. formed by linking small organic compounds.
d. macromolecules.
e. biological molecules.

____15. –COOH

a. is a carboxyl group.
b. is a carbonyl group.
c. becomes ionized –COO^{-}.after donating a proton
d. coverts to –COO^{+} after accepting a proton.
e. is a functional group

____16. Carbon has ______ valance electrons

a. 1
b.2
c. 3
d. 4
e. 6

____17. Isomers that are mirror images of one another are called

a. geometric isomers
b. concentric isomers
c. trans isomers
d. structural isomers
e. enantiomers

____18. R-SH

a. is a sulfhydryl group.
b.is a polar molecule.
c. forms monomers when placed in solution.
d. is found in thiols.
e. contributes to protein structure,

____19. Which sugar is a hexose?

a. glucose
b. fructose
c. galactose
d. ribose
e. deoxyribose

____20. Glycoproteins

a. are composed of glycogen and proteins
b. are involved in protection
c. are involved in cell to cell adherence
d. are present on the outer surface of cells
e. are the majority of secreted proteins

____21. Two unsaturated fatty acids that must be obtained from food are

a. palmitic and linoleic acids.
b. palmitic and arachidonic acids
c. lauric and stearic acids.
d. linoleic and arachidonic acids.
e. myristic and stearic acids.

____22. Carotinoids are

a. plant pigments
b. insoluble in water
c. synthesized from cholesterol.
d. orange and green plant pigments
e. made up of six-carbon monomers.

____23. ___________ are chemical mediators derived from fatty acids.

a. retinals
b. isoprenes
c. prostaglandins
d. steroids
e. juvenile hormone

____24. Amino acids

a. have a carboxyl group.
b .are identified by their side chain
c. are mainly dipolar ions at neutral pH.
d. are important biological buffers.
e. exist mainly in ionized form at pH 7.

____25. .Hydrogen bonding contributes to the __________structure of proteins.

a. primary
b. secondary
c. tertiary
d. beta pleated sheet
e. quaternary

____26. Misfolded proteins called ____________ lead to some serious human diseases.
a. helical bodies
b. polar bodies
c. globular bodies
d. chaperones
e. prions

VISUAL FOUNDATIONS

Color the parts of the illustration below as indicated.

RED ❑ saturated fatty acid
GREEN ❑ glycerol
YELLOW ❑ phosphate group
BLUE ❑ choline
ORANGE ❑ unsaturated fatty acid

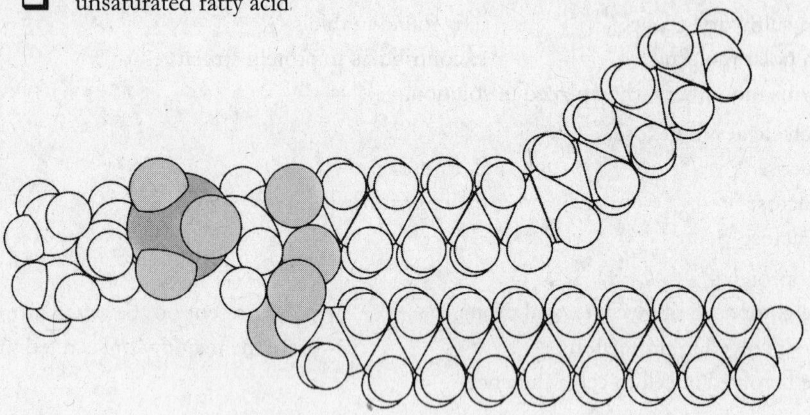

Questions 1-4 refer to the illustration above. Some questions may have more than one correct answer.

____ 1. The illustration represents
a. a polypeptide.
b. a phospholipid.
c. glycogen.
d. lecithin.
e. starch.

____ 2. The hydrophobic end of this molecule is composed of
a. amino acids.
b. fatty acids.
c. glucose.
d. phosphate groups.
e. steroids.

____ 3. This kind of molecule contributes to the formation of
a. neutral fat.
b. cell membranes.
c. RNA.
d. cellulose.
e. complex proteins.

____ 4. This molecule is amphipathic, meaning that
a. it contains nonpolar groups.
b. it is very large.
c. it contains hydrophobic and hydrophilic regions.
d. it is found in all life forms.
e. it can act as an acid.

CHAPTER 4

Organization of the Cell

Chapters 2 and 3 introduced the inorganic and organic materials critical to an understanding of the cell, the basic unit of life. In this chapter and those that follow, you will see how cells utilize these chemical materials. Because all cells come from preexisting cells, they have similar needs and therefore share many fundamental features. Most cells are microscopically small because of limitations on the movement of materials across the plasma membrane as well as the functions they perform. Cells are studied by a combination of methods. These include various types of light and electron microscopy and cell fractionation. Prokaryotic cells do not have membrane-bound organelles, but they do share several similarities with eukaryotes. Eukaryotic cells are more complex, containing a variety of membranous organelles. Although each organelle has its own particular structure and function, all of the organelles of a cell work together in an integrated fashion, many as part of the endomembrane system. The cytoskeleton provides mechanical support to the cell and functions in cell movement, the transport of materials within the cell, and cell division. Most cells are surrounded by some type of covering, which could be a glycocalyx, an extracellular matrix, or a cell wall.

REVIEWING CONCEPTS

Fill in the blanks.

INTRODUCTION

1. (a) ______________ are the building blocks of complex multicellular organisms, but an organism can also be a (b)__________cell.
2. The cytoskeleton of a cell are composed of two proteins (a)________________, which comprises the microfilaments, and (b)________________, which comprise the intermediate filaments.

THE CELL: BASIC UNIT OF LIFE

The cell theory is a unifying concept in biology

3. The two components of the cell theory are (a) ____________________ ____, and (b) ____.
4. Evidence that all living cells have a common origin is supported by (a)____, and (b)____.

The organization and basic functions of all cells are similar

5. Cellular organization allows cells to maintain ________________in a constantly changing environment.
6. All cells are enclosed by a ______________________________.

7. The internal structures of cells that carry out cell activities are called ________________________________.
8. Cell membranes also serve as ______________ surfaces for interacting proteins involved in certain types of biochemical reactions.

Cell size is limited

9. A micrometer is 1/1,000,000 of a meter and therefore ____________ of a millimeter.
10. Most cellular components are measured in ____________________, a unit that is actually 1 billionth of a meter.
11. In birds, frogs and humans, the __________ is the largest single cell in the organism.
12. In general, the smaller the cell, the larger the ________________________________.

Cell size and shape are adapted to function

13. The (a) ______________, and (b) ______________ of cells are adapted to the functions they perform.

METHODS FOR STUDYING CELLS

14. The most important tool scientists have for studying cells is the ____________________.
15. (a) __________________________ was the first person to describe cells, and (b) ______________________ was the first person to observe living cells.

Light microscopes are used to study stained or living cells

16. (a) ______________________ is the ratio of microscopic size to actual size of an object, and (b) ______________________ is the capacity to distinguish fine detail in an image.
17. Light microscopes have limited resolution due to the (a)________________________ of (b) ________________ used.
18. Lack of contrast can be remedied by ______________.
19. In (a) ______________________ microscopy, an image is formed by transmitting light through a cell and in (b) ____________________ rays of light are directed from the side of the specimen resulting in a bright image against a dark background.
20. Scientists use ______________________ microscopy to study internal structures that are constantly changing shape and location inside a cell.

Electron microscopes provide a high-resolution image that can be greatly magnified

21. Powerful electron microscopes are used to study the ________________ of cells.
22. In SEM, the beam of (a) __________________ does not (b) _______ through the sample while in (c) ________ it does.

Biologists us biochemical and genetic methods to connect cell structures with their functions

23. Cell components are separated by a method known as ___________________________.

PROKARYOTIC AND EUKARYOTIC CELLS

Organelles of prokaryotic cells are not surrounded by membranes

24. (a) _____________________ cells are relatively lacking in complexity, and their genetic material is not enclosed by membranes but is located in a (b) __________.
25. Outside of their cell membrane, most prokaryotic cells have a (a)_______________________ and many have (b)_______________________ to aid in movement.
26. The bacterial (a) ______________, used for polypeptide synthesis, is (b)______________ than the eukaryotic version.

Membranes divide the eukaryotic cell into compartments

27. Eukaryote cells are relatively complex and possess both (a)_________________________, and (b) ______________________.
28. _______________ is the fluid of the cytoplasm.
29. When they are present in the cytosol, ribosomes produce (a)____________________ proteins and when they are bound to membranes they produce (b) ______________________ proteins.

The unique properties of biological membranes allow eukaryotic cells to carry on many diverse functions

30. Compartmentalization allows cellular activity to be (a) ____________ and occur (b) _______________________ inside the cell.
31. Many chemical reactions are carried out by _______________ bound to the cell membrane.
32. The difference in both electrical charge and ion concentration across a cell membrane results in an __________________________.

THE CELL NUCLEUS

33. The (a) ____________________________ consists of two concentric membranes that separate the nuclear contents from the surrounding cytoplasm and contains (b) _______________that regulate material passage.
34. The _________________ provides support and helps organize nuclear contents.
35. DNA is associated with proteins and RNA, forming a complex known as ___________________.
36. The (a) __________________ helps to make (b) _______ found in ribosomes.

RIBOSOMES MANUFACTURE PROTEINS IN THE CYTOPLASM

37. Ribosomes are tiny particles found free in the cytoplasm or attached to certain membranes; they consist of (a) _______, and (b) __________.

MEMBRANOUS ORGANELLES IN THE CYTOPLASM

38. Most cytoplasmic organelles are components of the (a)_____________________ system. The organelles exchange materials using (b) ____________________.

The endoplasmic reticulum is a multifunctional network of membranes

39. The ______________________ is a complex of intracytoplasmic membranes with different domains that have distinct structures and functions.
40. The smooth ER synthesizes lipids including (a) ______________, and (b) ______________ used in making cell membranes. In fat tissue the smooth ER has (c) ______________ that are used in the formation of triglyerides. In liver the smooth ER is important in breaking down (d) ______________.
41. Rough RER is studded with ______________ that are involved in protein synthesis.
42. Some proteins constructed on RER are transported by (a)____________ for secretion to the outside or insertion in other membranes. Misfolded proteins will be degraded by the (b) ______.

The ER is the primary site of membrane assembly for components of the endomembrane system

43. The phospholipids found in membranes are synthesized on the ______________ surface of the smooth ER.

The Golgi complex processes, sorts, and routes proteins from the ER to different parts of the endomembrane system

44. In many cells the Golgi consists of flattened membranous sacs called ______________.
45. Cells that make large quantities of ________________ will have many Golgi stacks.

Lysosomes are compartments for digestion

46. Lysosomes are small sacs containing ______________ that can break down (lyse) complex molecules, foreign substances, and "dead" organelles.
47. Primary lysosomes are formed by budding from the (a)______________ and they are filled with enzymes synthesized in the (b) ____________________.

Vacuoles are large, fluid-filled sacs with a variety of functions

48. The membrane of a vacuole is called the ________________.
49. In plants vacuoles contain a high concentration of solutes and are responsible for the (a) ______________________ that provides mechanical strength to the cells.. In seeds store (b) ____________.

50. Protozoans have contractile vacuoles which remove ____________.

Peroxisomes metabolize small organic compounds

51. Peroxisomes get their name from the fact that they produce ______________________________ during oxidation reactions.
52. Peroxisomes are involved in the (a) ________________, (b)________________, and (c) ________________ of lipids.
53. Peroxisomes synthesize certain ____________________ that are components of the myelin sheath covering nerve cells.

Mitochondria and chloroplasts are energy-converting organelles

54. These organelles are thought to have evolved from (a)____________________ as a result of a (b)________________ relationship with eukaryotic cells..

Mitochondria make ATP through aerobic respiration.

55. Mitochondria have a smooth outer membrane which allows the passage of (a) ________________ into the intermembranous space and an inner membrane that has many (b) ________________ called (c)____________. The inner membrane (d) ____________________ the types of molecules entering the mitochondrial matrix.
56. Mitochondria are also involved in (a) ____________________, which is programmed cell death and different from (b) ______________ which is uncontrolled cells death.
57. Mitochondria negatively affect health and aging through the release of (a)____________________which form highly reactive compounds called (b) ____________________.

Chloroplasts convert light energy to chemical energy through photosynthesis

58. (a) ________________________ are organelles that contain a green pigment (b) ______________ that absorb light energy for photosynthesis. These organelles also contain light-absorbing yellow and orange pigments called (c) ______________.
59. ____________________ membranes contain chlorophyll that traps sunlight energy and converts it to chemical energy in ATP.
60. The matrix within chloroplasts, where carbohydrates are synthesized, is called the (a) ____________________, while the ATP that supplies energy to drive the process is synthesized on membranes called (b)____________________________________.

THE CYTOSKELETON

61. The cytoskeleton is a dense network of protein fibers responsible for the (a) ____________________, (b) ____________________, and (c)________________________ of cells.
62. (a) ________________________ and (b)________________________ are both made up of globular protein subunits that can rapidly assemble and disassemble.

Microtubules are hollow cylinders

63. Microtubules consist of the two proteins (a) ______________ and (b) ______________.
64. Structural (a) ______________ help regulate microtubule assembly and (b) ______________ use ATP to produce energy.
65. One motor protein, ______________, moves organelles toward the plus end of a microtubule.
66. One motor protein, ______________, transports organelles toward the minus end of a microtubule.

Centrosomes and centrioles function in cell division

67. In animal cells, the (a) ______________ is the location of the (b) ______________ where minus ends of microtubules are anchored.
68. In animal cells, (a) ______________ are located at right angles to one another within the (b) ______________ and are important in (c) ______________.

Cilia and flagella are composed of microtubules

69. The cellular appendages that are long and few in number are called (a) ______________ while numerous, short ones are called (b) ______________. Both are used for (c) ______________.
70. These structures are anchored in the cell by the ______________.
71. Almost every vertebrate cell has a ______________ on the cell surface that serves as a cellular antenna.

Microfilaments consist of intertwined strings of actin

72. Microfilaments themselves cannot contract but they do generate movement by (a) ______________ or by (b) ______________.
73. In muscle cells actin is associated with the protein ______________ to form fibers associated with muscle contraction.
74. Cells that move along a surface change shape. In these cells, actin filaments push the plasma membrane outward forming ______________.

Intermediate filaments help stabilize cell shape

75. Keratin and neurofilaments are examples of tough, flexible fibers called ______________ that stabilize cell shape and provide mechanical strength.

CELL COVERINGS

76. In most eukaryotic cells polysaccharide side chains of proteins and lipids form a ______________, or cell coat, that is part of the plasma membrane.
77. The animal cell extracellular matrix is composed of (a) ______________, and (b) ______________.
78. Plant cell walls are composed primarily of (a) ______________ which is a (b) ______________.

BUILDING WORDS

Use combinations of prefixes and suffixes to build words for the definitions that follow.

Prefixes	The Meaning
chloro-	green
chromo-	color
cyto-	cell
eu-	good, well, "true"
leuko-	white (without color)
lyso-	loosening, decomposition
micro-	small
myo-	muscle
pro-	"before"

Suffixes	The Meaning
-karyo(te)	nucleus
-plast(id)	formed, molded, "body"
-some	body

Prefix	Suffix	Definition
____________	-phyll	1. A green pigment that traps light for photosynthesis.
____________	____________	2. Organelles containing pigments that give fruits and flowers their characteristic colors.
____________	-plasm	3. Cell contents exclusive of the nucleus.
____________	-skeleton	4. A complex network of protein filaments within the cell.
glyoxy-	____________	5. A microbody containing enzymes used to convert stored fats in plant seeds to sugars.
____________	____________	6. An organelle that is not pigmented and is found primarily in roots and tubers, where it is used to store starch.
____________	____________	7 An organelle containing digestive enzymes.
____________	-filaments	8. Small, solid filaments, made of actin,that make up part of the cytoskeleton of eukaryotic cells.
____________	-tubules	9. Small, hollow filaments, made of tubulin that make up part of the cytoskeleton of eukaryotic cells.
____________	-villi	10. Small, finger-like projections from cell surfaces that increase surface area.
____________	-sin	11. A muscle protein that, together with actin, is responsible for muscle contraction.
peroxi-	____________	12. An organelle containing enzymes that split hydrogen peroxide, rendering it harmless.
____________	____________	13. Precursor organelles.
____________	-karyotes	14. Organisms that evolved before organisms with nuclei.
____________	-sol	15. The fluid component of the cytoplasm.
____________	____________	16. An organism with a distinct nucleus surrounded by nuclear membranes.

MATCHING

Terms:

a.	Actin	f.	Mitochondrion	k.	Secretory vesicle
b.	Centriole	g.	Nuclear envelope	l.	Stroma
c.	Chloroplast	h.	Nucleolus	m.	Vacuole
d.	Endoplasmic reticulum	i.	Plasma membrane	n.	Vesicle
e.	Granum	j.	Ribosome	o.	Cristae

For each of these definitions, select the correct matching term from the list above.

____ 1. One of a pair of small, cylindrical organelles lying at right angles to each other near the nucleus.

____ 2. Site of ribosome synthesis.

____ 3. An intracellular organelle that is the site of aerobic respiration.

____ 4. A fluid-filled, membrane-bounded sac found within the cytoplasm; may function in storage, digestion, or water elimination.

____ 5. Any small sac, especially a small spherical membrane-bounded compartment, within the cytoplasm.

____ 6. A chlorophyll containing intracellular organelle of some plant cells.

____ 7. A stack of thylakoids within a chloroplast.

____ 8. The fluid region of the chloroplast.

____ 9. An interconnected network of intracellular membranes.

____10. An organelle that is part of the protein synthesis machinery.

____11. Folds of the innermembrane in the mitochondria.

MAKING COMPARISONS

Fill in the blanks.

Cell Structure	Location of Structure	Function of Structure	Kind of Cell Containing This Structure
Nuclear area containing a single strand of circular DNA	Cytoplasm	Inheritance, control center	Prokaryotic
#1	Cytoplasm and RER	Synthesis of polypeptides	#2
Golgi complex	#3	#4	Most eukaryotic
Mitochondria	#5	#6	#7
Complex chromosomes	#8	Inheritance	#9
#10	Centrosome	#11	#12
#13	#14	Protection, support	Prokaryotic and eukaryotic

Cell Structure	Location of Structure	Function of Structure	Kind of Cell Containing This Structure
#15	#16	Encloses cellular contents, regulates passage of materials into and out of cell	#17
Chloroplasts	#18	#19	Eukaryotic (primarily plants)

MAKING CHOICES

Place your answer(s) in the space provided. Some questions may have more than one correct answer.

____ 1. A micrometer (μm) is
a. one billionth of a meter.
b. one millionth of a centimeter.
c. one thousandth of a millimeter.
d. one millionth of a micrometer.
e. one thousandth of a micrometer.

____ 2. Which of the following is in the nucleolus, but not normally found in the rest of chromatin?
a. DNA
b. protein
c. chromosomes
d. RNA
e. ribosomes

____ 3. Cell membrane functions include
a. isolation of chemical reactions.
b. selective permeability.
c. moderating between a cell's internal and external environments.
d. maintaining cell shape.
e. concentration of reactants.

____ 4. The low resolution attained by the light microscope is attributed to its use of
a. fixed specimens.
b. lenses.
c. visible wavelengths of light..
d. differentiating stains.
e. electrons instead of light.

____ 5. As cells become larger
a. the plasma membrane becomes limiting.
b. they take on a spherical shape.
c. the volume increases at a greater rate. proportionately than the surface area.
d. chemical reactions slow down.
e. surface area and volume increase

____ 6. Inner membranous folds in mitochondrion
a. are called cristae.
b. increase membrane surface area.
c. extend into the intermembrane space.
d. contain enzymes.
e. contain structural proteins.

____ 7. The Golgi complex functions to
a. modify proteins.
b. process proteins.
c. packages glycoproteins.
d. sort molecules.
e. break down large carbohydrates.

____ 8. The membranes that comprise the endomembrane system include
a. Golgi complex.
b. lysosomes.
c. endoplasmic reticulum.
d. transport vesicles.
e. plasma membrane.

____ 9. Lysosomes
a. contain digestive enzymes.
b. contain nucleic acids.
c. possess a membrane.
d. break down complex molecules.
e. break down organelles.

____10. Which of the following cells contain plastids?
a. animal
b. plant
c. some eukaryotic
d. some prokaryotic
e. algae

____11. The "cytoskeleton" of eukaryotic cells
a. changes constantly.
b. includes microfilaments.
c. includes some DNA.
d. functions in cell movements.
e. includes protein.

____12. The cell theory states that
a. new cells come from preexisting cells.
b. all cells are descended from ancient cells.
c. cells divide.
d. living things are composed of cells.
e. cells contain genetic material.

____13. Chloroplasts and mitochondria both
a. are found in plant cells.
b. have two membranes.
c. contain DNA.
d. are found in animal cells.
e. contain a matrix.

____14. Ribosomal RNA is synthesized in the
a. cytoplasm.
b. nuclear matrix.
c. cristae.
d. mitochondria.
e. nucleolus.

____15. Actin, myosin, and tubulin are
a. proteins.
b. in chromatin.
c. primarily in neurons.
d. components of filaments.
e. components of the plasma membrane.

____16. Fluorescence microscopy
a. uses stains bound to antibodies.
b. uses stains that absorb visible light.
c. uses special filters.
d. can detect location of specific molecules.
e. provide better resolution than electron microscopes.

____17. Transmission electron microscopy (TEM)
a. uses antibodies bound to gold molecules.
b. relies on the emission of secondary electrons.
c. provides information on cell contents.
d. allows electrons to pass through the specimen.
e. provides information about a cells external surface.

____18. Ribosmes
a. are smallest in prokaryotic cells.
b. are made up of protein and RNA molecules.
c. contain enzyme activity to form peptide bonds.
d. have three subunits
e. are connected to the Golgi complex.

____19. The endoplasmic reticulum
a. consists of a number of different domains.
b. has unique enzymes in its lumen.
c. when stained can be seen with light microscopy.
d. connects to other organelles in the cell.
e. encircles the nucleus.

____20. The nuclear envelope
a. has ribosomes on the outer surface.
b. has channels connecting the nucleoplasm and cytoplasm.
c. is a functional domain of the endoplasmic reticulum.
d. has transporters for newly synthesized RNA.
e. consists of two membranes.

____21. Vacuoles
a. are found in plants and animals.
b. breakdown unneeded organelles.
c. .carry out functions performed by lysosomes.
d. have no internal structure.
e. store plant defense compounds.

____22. Chloroplasts
a. produce ATP.
b. have a fluid filled inner space, the stroma.
c. are filled with stacks of amyloplasts.
d. form when plastids are exposed to light.
e. are one type of leukoplasts.

____23. Microtubules
a. are attached to actin.
b. are exported from the ER.
c. have ATPase activity.
d. are composed of tubulin.
e. are polarized.

____24. Microfilamenst
a. are flexible fibers.
b. are contractile fibers.
c. are located in the cell medulla.
d. are solid fibers.
e. generate movement.

VISUAL FOUNDATIONS

Color the parts of the illustration below as indicated.

RED ❑ plasma membrane
GREEN ❑ nuclear area
YELLOW ❑ cell wall
PINK ❑ capsule
BLUE ❑ flagellum
ORANGE ❑ pili

Questions 1 and 2 pertain to the illustration above.

1. Is this a prokaryotic cell or a eukaryotic cell? ____________________
2. Which of the colored parts is/are not unique to this kind of cell? ____________________

Color the parts of the illustration below as indicated.

RED ❑ plasma membrane

ORANGE ❑ nucleus

YELLOW ❑ cell wall

BLUE ❑ prominent vacuole

GREEN ❑ chloroplast

BROWN ❑ mitochondrion

TAN ❑ endomembrane system

(Note: The plasma membrane and the outer portion of the nuclear envelop were assigned other colors. Use tan to identify the remaining structures in this system.)

Questions 3-5 pertain to the illustration below.

3. What kind of cell is illustrated by this generalized diagram? ______________________________

4. Which of the colored parts is/are considered components of the cytoplasm of this cell?

____

5. Which of the colored parts is/are characteristic of this kind of cell and not generally associated with other types of cells?

____

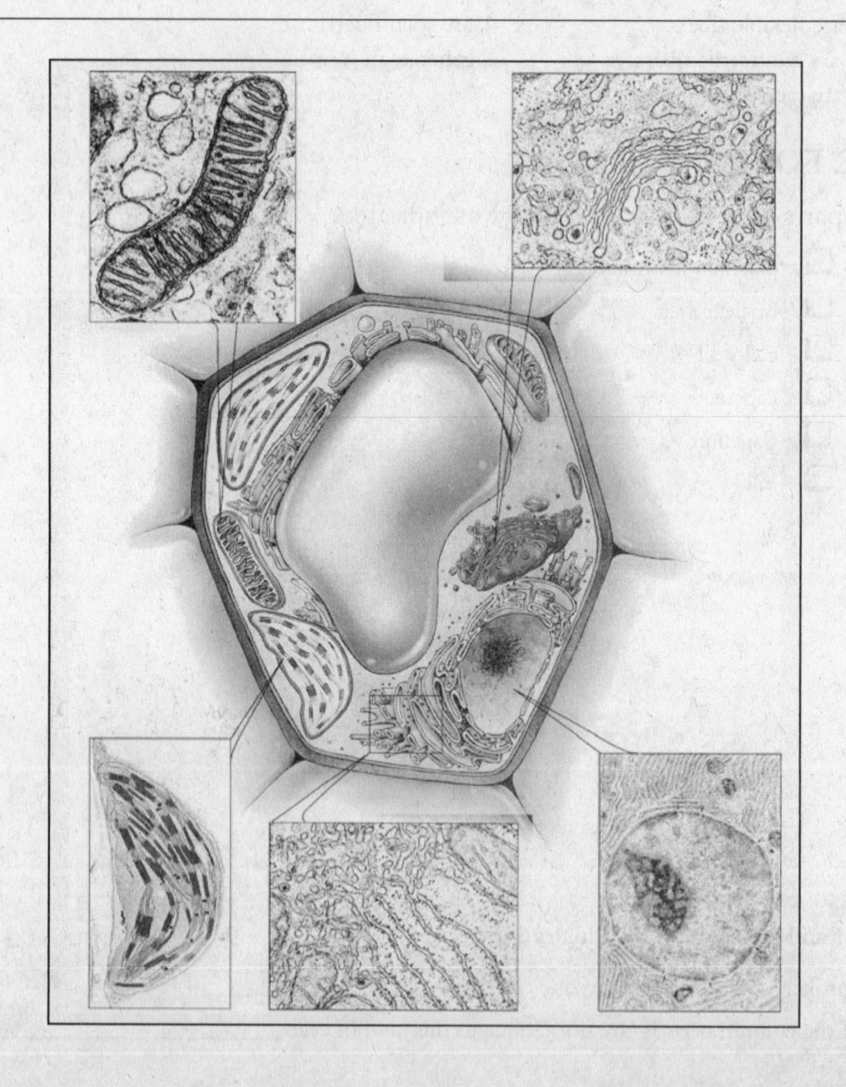

Color the parts of the illustration below as indicated.

RED ❑ plasma membrane

ORANGE ❑ nucleus

YELLOW ❑ centriole

BROWN ❑ mitochondrion

BROWN ❑ endomembrane system

(Note: The plasma membrane and the outer portion of the nuclear envelope have already been assigned colors. Use tan to identify the remaining structures in this system.)

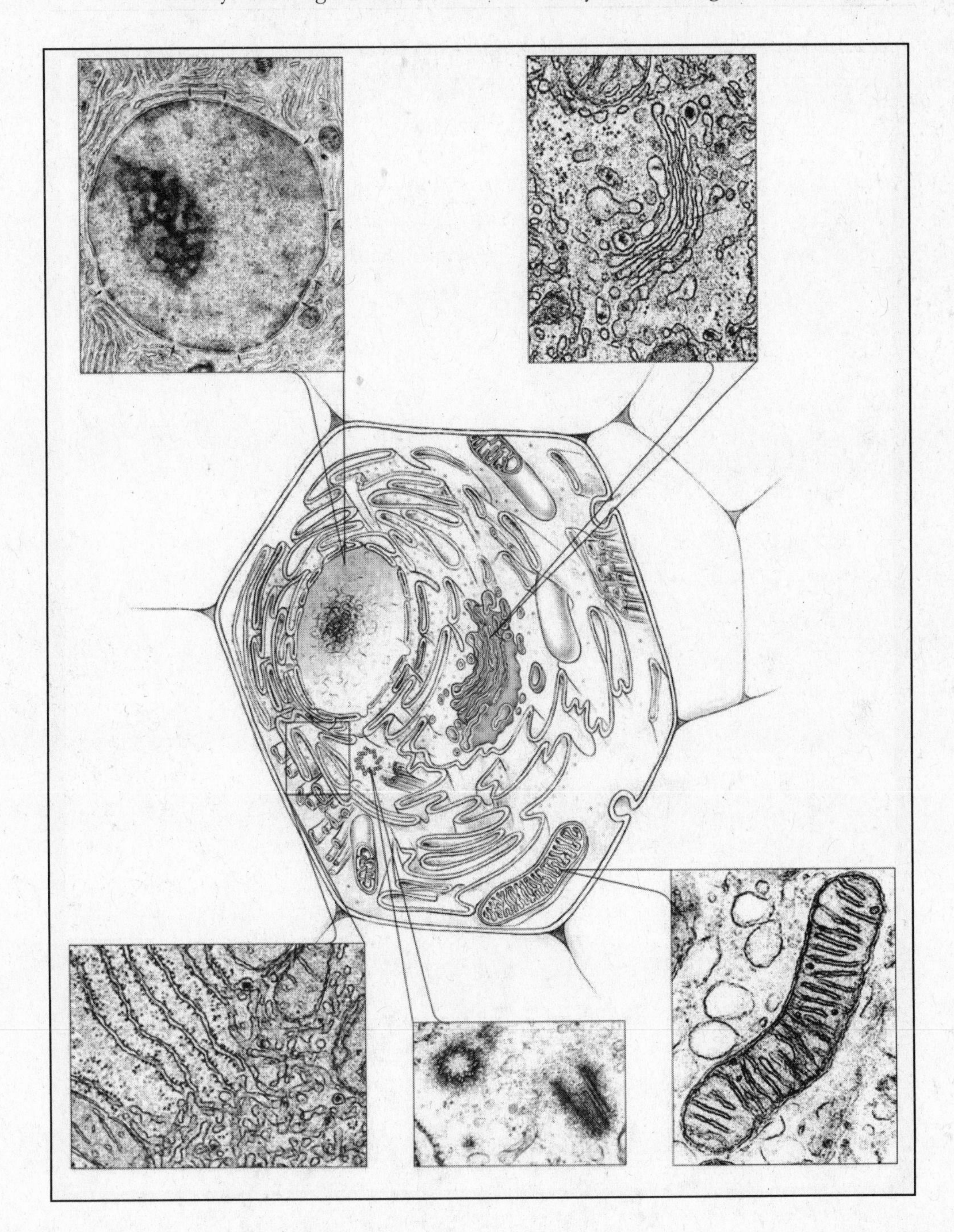

Questions 6-8 pertain to the illustration above.

6. What kind of cell is illustrated by this generalized diagram?

____

7. Which of the colored parts is/are considered components of the cytoplasm of this cell?

____

____

8. Which of the colored parts is/are characteristic of this kind of cell and not generally associated with other types of cells?

____

CHAPTER 5

❑

Biological Membranes

Biological membranes serve to enclose cells. They separate the interior world of the cell from the exterior and are an essential step in the evolution of life. In eukaryotes, biological membranes also enclose a variety of internal structures and form extensive internal membrane systems. Membranes are phospholipid bilayers with both integral and peripheral proteins associated. The fluid mosaic model explains our current understanding of their structure. This structure allows membranes to have several important physical properties. Membranes and their associated proteins have varied functions including; providing work surfaces for many chemical reactions, regulating movement of materials in and out of the cell, transmitting signals and information, holding cells together, and being an essential part of energy transfer and storage systems. Molecules pass through membranes in a variety of ways because of their selective permeability. In multicellular organisms, cell membranes have specialized junctions associated with them that allow neighboring cells to form strong connections, prevent material passage, or to establish rapid communication.

REVIEWING CONCEPTS

Fill in the blanks.

INTRODUCTION

1. All cells are physically separated from the external environment by a ____________________.
2. Membrane proteins known as ____________________ are responsible for connecting cells in multicellular sheets to form tissues.

THE STRUCTURE OF BIOLOGICAL MEMBRANES

3. Biological membranes are composed of lipids and proteins that are in ____________________.

Phospholipids form bilayers in water

4. As a result of their chemical structure, phospholipids are ____________________ molecules.
5. The headgroups of phospholipid molecules are said to be (a)____________________ because they readily associate with water. On the other hand, the (b)____________________ ends turn away from water and associate with each other.
6. The ____________________ of the membrane bilayer prevents many small, hydrophilic molecules from entering cells.
7. The constant motion of the hydrocarbon chains in the membrane gives the bilayer the property of a ____________________.

The fluid mosaic model explains membrane structure

8. The theory of membrane structure known as the (a)____________________ model holds that membranes consist of a dynamic, fluid (b)____________________ in which (c)________________are embedded or otherwise associated.

Biological membranes are two-dimensional fluids

9. The crystal-like properties of many phospholipid bilayers is due to the orderly arrangement of (a)____________________ on the outside and (b)____________________ on the inside.
10. The fluidity of the membrane bilayer is significantly affected by the properties of ____________________________.
11. The presence of __________________helps to prevent the membrane from solidifying.
12. ____________________functions as a "fluidity buffer" in animal cells. Plant cells have similar molecule types.

Biological membranes fuse and form closed vesicles

13. Products are secreted from cells by the fusion of vesicles with the ____________________________.

Membrane proteins include integral and peripheral proteins

14. (a)__________________________________ have hydrophobic regions that interact with the fatty acid tails of the membrane phospholipids. One type (b)____________________ pass through the entire membrane.
15. __________________________________ are not embedded in the lipid bilayer.

Proteins are oriented asymmetrically across the bilayer

16. Proteins destined for the outer cell surface are manufactured on (a)____________________ and then modified in the ER lumen to become (b)______________________. They then move through the (c)____________________.

OVERVIEW OF MEMBRANE PROTEIN FUNCTIONS

17. The large variety of activities associated with the plasma membrane is made possible by the diversity of ____________________ molecules in membranes.
18. The proteins called (a)__________________help anchor the cell to the extracellular matrix. Channel and pump proteins are important for (b)__________________. (c)________________proteins transmit signals by (d)____________________________.
19. Antigens are important in (a)__________________and other membrane proteins form (b)________________between cells.

CELL MEMBRANE STRUCTURE AND PERMEABILITY

20. Membranes are ______________________because they let some, but not all, substances pass.

Biological membranes present a barrier to polar molecules

21. Small ___________________ molecules easily permeate plasma membranes.
22. The membrane is relatively impermeable to (a)_________ and most (b)_______polar molecules.

Transport proteins transfer molecules across membranes

23. (a)______________bind the molecule being transported and undergo a (b)__________________.
24. (a)________________form tunnels through the membrane and are often (b)_______________.
25. _________________________ are gated channel proteins that facilitate the transport of water through the plasma membrane in response to osmotic gradients.

PASSIVE TRANSPORT

26. Passive transport does not require the cell to spend metabolic _____________.
27. Two types of diffusion are (a)______________________, and (b)_________________________

Diffusion occurs down a concentration gradient

28. In diffusion, there is a net movement of particles from (a)_________to (b)___________areas of concentration.
29. Diffusion rate is affected by temperature and the (a)____________________, (b)______________, and (c)____________________of the moving particles.

Osmosis is diffusion of water across a selectively permeable membrane

30. The movement of water through a selectively permeable membrane from a region of higher concentration to a region of lower concentration is called ____________________________.
31. The __________________________ of a solution is determined by the amount of dissolved substances in the solution.
32. When there is no net movement of water, a cell is in an ___________ solution.
33. If a solution has a high solute concentration, it is (a)_______________. If a cell is placed in this solution there is net movement of water (b)_________of the cell, causing the cell to (c)________.
34. A (a)_______________solution has a low solute concentration and a cell placed in this type of solution will (b)________as the net movement of water is (c)_______the cell.

35. (a)________________is generated in cells that have a cell wall when placed in a (b)____________solution. This same cell will undergo (c)______________when placed in a (d)____________solution.

Facilitated diffusion occurs down a concentration gradient

36. In facilitated diffusion, the net movement is always from a region of (a)________________ solute concentration to a region of (b)________________ solute concentration.
37. In this type of diffusion, a (a)__________protein makes the membrane permeable. These proteins are either (b)______________ or (c)____________.
38. After binding one or more molecules it is going to transport, a carrier protein undergoes ____________________________.

ACTIVE TRANSPORT

39. The energy for diffusion comes from the (a)________________, while the energy for active transport usually comes from (b)________.

Active transport systems "pump" substances against their concentration gradients

40. The sodium-potassium pump helps to maintain the (a)________________________________ which occurs as a result of (b)________________ (fewer or more) potassium ions being located inside the cell relative to the sodium ions outside
41. The distribution of sodium and potassium across cell membranes causes the inside of the cell to be ________________ charged relative to the outside.
42. The unequal ion distribution creates an (a)________________. When both a charge and concentration difference are present an (b)________________ will be established.

Carrier proteins can transport one or two solutes

43. (a)________________ transport one type of substance in one direction and (b)____________________ move two types of substances in one direction, while (c)________________move two substances in opposite directions.

Co-transport systems indirectly provide energy for active transport

44. The process whereby the movement of one solute down its concentration gradient provides the energy to transport another solute up its concentration gradient is known as ______________________________.

EXOCYTOSIS AND ENDOCYTOSIS

45. These mechanisms are similar to active transport because the cell must expend ______________.

In exocytosis, vesicles export large molecules

46. Exocytosis is the exportation of waste materials or specific products of secretion by the fusion of a _________________ with the plasma membrane of a cell.

In endocytosis, the cell imports materials

47. In (a)_______________________ a cell ingests large, solid particles. In (b)_______________________ a cell takes in dissolved materials.

48. _______________________________is how cells take in macromolecules like cholesterol.

CELL JUNCTIONS

49. Three types of cell junctions in animal cells, are (a)_______________________, (b)_______________________, and (c)_______________________, while (d)_______________________ connect plant cells.

Anchoring junctions connect cells of an epithelial sheet

50. The two common types of anchoring junctions are (a)_______________________ that are anchored to intermediate filaments in the cell, and (b)____________________ that connect with microfilaments of the cytoskeleton.

Tight junctions seal off intercellular spaces between some animal cells

51. The _______________________, composed of tight junctions, prevents many substances in the blood from passing into the brain.

Gap junctions allow the transfer of small molecules and ions

52. Gap junctions are (a)____________________junctions between cells. Gap junctions consist of a cluster of (b)________________, an integral membrane protein, molecules that form (c)____________________ that connect the cytoplasm of adjacent cells to allow communication.

Plasmodesmata allow certain molecules and ions to move between plant cells

53. Plasmodesmata in plant cells are functionally equivalent to ________________ in animal cells.

54. Most plasmodesmata contain a cylinder-shaped structure called the ______________________.

BUILDING WORDS

Use combinations of prefixes and suffixes to build words for the definitions that follow.

Prefixes	The Meaning
aqua-	water
desm(o)-	bond
endo-	within
equ-	equal
exo-	outside, outer, external
facility(ated)	easy
glyc(o)-	sweet, sugar

Suffixes	The Meaning
-able	able to
-cyto(sis)	cell
-desm(a)	bond
-some	body

hyper-	over
hypo-	under
iso-	equal, "same"
lipo-	fat
peripher-	outer surface
perm(e)-	to pass through
phago-	eat, devour
pino-	drink

Prefix	Suffix	Definition
________	________	1. A process whereby materials are taken into the cell.
________	________	2. The process whereby waste or secretion products are ejected from a cell by fusion of a vesicle with the plasma membrane.
________	-tonic (-osmotic)	3. Having an osmotic pressure or solute concentration that is greater than a standard solution.
________	-tonic (-osmotic)	4. Having an osmotic pressure or solute concentration that is less than a standard solution.
________	-tonic (-osmotic)	5. Having an osmotic pressure or solute concentration that is the same as a standard solution.
________	________	6. The ingestion of large solid particles, such as bacteria and food, by a cell.
________	________	7. A type of endocytosis whereby fluid is engulfed by vesicles originating at the cell surface.
________	________	8. A button-like plaque (body) present on two opposing cell surfaces, that holds (bonds) the cells together by means of protein filaments that span the intercellular space.
plasmo-	________	9. A cytoplasmic channel connecting (bonding) adjacent plant cells and allowing for the movement of small molecules and ions between cells.
________	-al	10. A protein not embedded in the lipid bilayer.
________	-protein	11. A protein with an added sugar group
________	________	12. Membranes that allow substances to pass through.
________	-porins	13. Gated water channels.
________	-ilibrium	14. Uniform distribution of particles across a membrane.
________	-diffussion	15. Uses a transport protein to move molecules across a membrane.
________	________	16. Artificial vesicle enclosed by phospholipid bilayers.

MATCHING

Terms:

a. Active transport
b. Concentration gradient
c. Co-transport
d. Dialysis
e. Diffusion
f. Facilitated diffusion
g. Fluid-mosaic model
h. Plasmodesmata
i. Osmosis
j. Signal transduction
k. Selectively permeable membrane
l. Tight junction
m. Turgor pressure

For each of these definitions, select the correct matching term from the list above.

____ 1. The modern picture of membranes in which protein molecules float in a phospholipid bilayer.

____ 2. The diffusion of water across a selectively permeable membrane.

____ 3. The transport of ions or molecules across a membrane and down a concentration gradient by a specific carrier protein.

____ 4. Regions in a system of differing concentration, such as exist in a cell and its environment, that cause molecules to move from areas of higher concentration to lower concentration.

____ 5. Energy-requiring transport of a molecule across a membrane from a region of low concentration to a region of high concentration.

____ 6. A membrane that allows some substances to cross it more easily than others.

____ 7. The random movement of molecules from a region of higher concentration to one of lower concentration of that substance.

____ 8. The internal pressure in a plant cell caused by the diffusion of water into the cell.

____ 9. A specialized structure between some animal cells, producing a tight seal that prevents materials from passing through the spaces between the cells.

____ 10. Structure allowing passage of certain small molecules and ions between plant cells.

MAKING COMPARISONS

Fill in the blanks.

Transport Mechanism	Description of the Transport Mechanism
Diffusion	Net movement of a substance from an area of high concentration to an area of low concentration
Phagocytosis	#1
#2	The passive transport of solutes down a concentration gradient aided by a specific protein in a membrane
#3	The active transport of substances into the cell by the formation of invaginated regions or "folds" of the plasma membrane that pinch off and become cytoplasmic vesicles
#4	Transfer of solutes by proteins located within the membrane
Desmosomes	#5
Exocytosis	#6
Osmosis	#7
#8	Transport of substances across a membrane requiring the expenditure of energy by the cell
Pinocytosis	#9
#10	Points of attachment between cells that hold cells together

MAKING CHOICES

Place your answer(s) in the space provided. Some questions may have more than one correct answer.

____ 1. Simple diffusion
a. moves molecules with a gradient.
b. does not occur in prokaryotes.
c. requires use of ATP.
d. involves protein channels.
e. moves substances both into and out of cells.

____ 2. Active transport
a. can move molecules against a gradient.
b. does not occur in prokaryotes.
c. requires use of ATP.
d. occurs in animal cells, not in plant cells.
e. moves substances both into and out of cells.

____ 3. A molecule is called amphipathic when it
a. prevents free passage of substances.
b. is in a membrane.
c. has hydrophobic and hydrophilic regions.
d. is a lipid.
e. is embedded in a bilipid layer.

____ 4. If membranes were not fluid and dynamic, which of the following might still occur normally?
a. active transport
b. facilitated diffusion
c. simple diffusion
d. endo- and exocytosis
e. osmosis

____ 5. Membrane fusion enables
a. diversity of proteins.
b. exocytosis.
c. endocytosis.
d. fusion of vesicles and plasma membrane.
e. formation of Golgi complexes.

Plants are placed in three beakers containing the following solutions: beaker A, distilled water; beaker B, isotonic solution; beaker C, 13% salt solution. Use this information to answer questions 7-9.

____ 6. The cells in beaker C
a. shriveled.
b. swelled.
c. plasmolyzed.
d. were unaffected.
e. probably eventually burst.

____ 7. The cells in beaker A
a. shriveled.
b. swelled.
c. plasmolyzed.
d. were unaffected.
e. probably eventually burst.

____ 8. The cells in beaker B
a. shriveled.
b. swelled.
c. plasmolyzed.
d. were unaffected.
e. probably eventually burst.

____ 9. Plasma membranes of eukaryotic cells have a large amount of
a. cholesterols.
b. phospholipid.
c. rigidity.
d. fluidity.
e. protein.

____ 10. Diffusion rate depends on
a. the flow of water.
b. concentration gradient.
c. energy from the cell.
d. the plasma membrane.
e. kinetic energy.

___11. Endocytosis may include

a. secretion vacuoles.
b. pinocytosis.
c. phagocytosis.
d. a combination of inbound particles with proteins.
e. receptor-mediation.

___12. Plasmodesmata

a. are channels in cytoplasm.
b. connect plant cells.
c. are plant cell structures equivalent to animal cell desmosomes.
d. are the same as several plasmodesma.
e. connect ER of adjacent cells.

___13. The principal cell adhesion molecules in vertebrates are known as

a. cadherins.
b. integral proteins.
c. glycoproteins.
d. plasmodesma.
e. β-pleated molecules.

___14. Sugars may be added to proteins

a. and porins.
b. to form gated channels.
c. resulting in a glycoprotein.
d. in the lumen of ER.
e. in the extracellular space.

___15. Membrane lipids

a. have a shape that is favorable for association into a bilayer.
b. can move sideways in the membrane leaflet.
c. can be moved from one leaflet of the membrane to the other by flippases.
d. organize in membranes with the hydrophobic head facing outwards.
e. form an ordered array in the membrane.

___16. In order to maintain optimal membrane fluidity a cell may

a. change the fatty acid content of the membrane.
b. adjust the ratio of saturated and unsaturated lipids.
c. add steroids to the membrane.
d. add cholesterol to the membrane
e. add lipids to increase the number of van der Walls interactions.

____17. Cotransport systems

a. transport two solutes at the same time.
b. moves two solute molecules down their concentration gradient.
c. use carrier proteins called uniporters.
d. uses ATP in the transport process.
e. move substances across a membrane by indirect transport.

____18. Endocytosis

a. moves material out of a cell.
b. moves material into a cell.
c. .results in the formation of a vacuole.
d. occurs by a receptor-mediated process.
e. occurs by pinocytosis.

____19. Gap junctions

a. function like desmosomes.
b. are an integral part of the membrane.
c. connect some nerve cells.
d. regulate the passage of substances between cells.
e. connect ER of adjacent cells.

VISUAL FOUNDATIONS

Color the parts of the illustration below as indicated. Label the interior and exterior of the cell, the carbohydrate chains, and the lipid bilayer.

RED	❑	alpha helix
GREEN	❑	hydrophilic region of transmembrane proteins
YELLOW	❑	hydrophobic region of transmembrane proteins
BLUE	❑	glycolipid
ORANGE	❑	glycoprotein
BROWN	❑	cholesterol
TAN	❑	peripheral protein

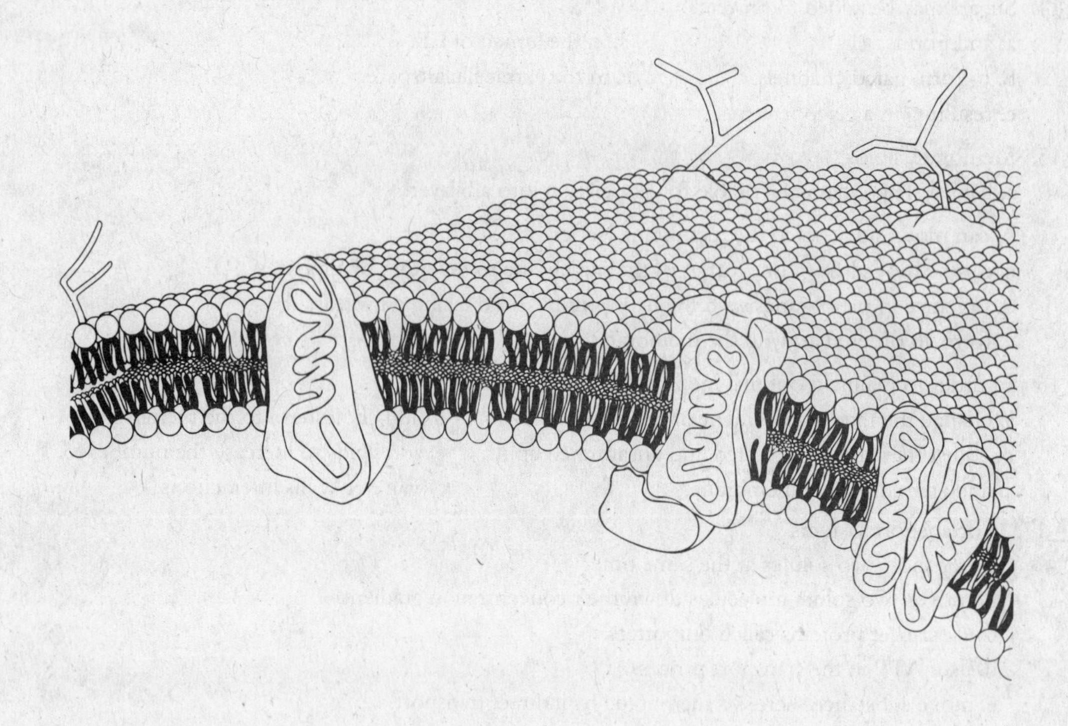

1. What is the name of the model for membrane structure illustrated above?
____

2. Which of the colored parts is most important in forming junctions between adjacent cells?
____

3. Which of the colored parts is important in maintaining fluidity in animal cells? _______________

CHAPTER 6

Cell Communication

The process of cell communication is vital to maintaining homeostasis in all types of cells. In multicellular organisms, cells must also communicate with one another, sometimes directly and sometimes indirectly. Organisms of different species, and even different kingdoms and domains, communicate with one another. This process involves cells sending and receiving signals, information crossing the plasma membrane and being transmitted through the cell, and finally, cells responding to the signals they receive. These signals come from the extracellular environment, many times from other cells and are typically chemical in nature. The signals are received by several known receptor proteins. Once received the signal is transmitted through a signaling pathway, which moves the signal inside the cell and amplifies it. Problems in cell communication can cause or contribute to a variety of diseases, including diabetes and cancer. Over billions of years, elaborate systems of cell communication have evolved, but many molecules involved are highly conserved across species. Developing a better understanding of how cells communicate may provide new insight and strategies for preventing and treating diseases.

REVIEWING CONCEPTS

Fill in the blanks.

INTRODUCTION

1. To maintain homeostasis, cells must continuously ______________________ with each other.
2. Prokaryotes, protists, fungi, plants, and animals all communicate with other members of their species by ______________________.
3. A community of microorganisms attached to a solid surface is known as a ______________________.

CELL COMMUNICATION: AN OVERVIEW

4. Like animals, plants send ______________________ to one another.
5. Organisms must receive and respond to signals from the environment in order to ______________________.
6. In plants and animals ______________________ are important signaling molecules for communication between cells.
7. The term ______________________ refers to the mechanisms by which cells communicate with one another.
8. The cells that can respond to a signaling molecule are called ______________________.

9. ______________________ is the process by which a cell converts an extracellular signal into an intracellular signal that results in a response.

SENDING SIGNALS

10. Cells communicate in several ways including (a)______________________, (b)______________________, (c)______________________, (d)______________________.
11. Most neurons signal one another by releasing chemical compounds called ______________________.
12. In animals specialized structures called ______________________ secrete hormones.
13. ______________________ is a process in which a signaling molecule diffuses through the interstitial fluid and acts on nearby cells.
14. ______________________ are paracrine regulators that modify cAMP levels and interact with other signaling molecules to regulate metabolic activities.

RECEPTION

15. A cell is programmed to receive signals based on the ______________________ it can synthesize.
16. A signaling molecule that binds to a specific receptor is called a (a)______________. Most are (b)______________ but some are (c)______________________ and can pass through the membrane.
17. A ligand is a molecule other than an ______________________.
18. A receptor generally has at least three (a)______________. The external one is for (b)______________________ of the ligand, while the internal (c)______________________ transmits the signal.
19. Pigments in plants, some algae, and some animals that absorb blue light and play a role in biological rhythms are called ______________________.

Cells regulate reception

20. Receptor down-regulation occurs in response to a (a)______________________ hormone concentration and involves (b)______________________ the number of receptors.
21. Receptor up-regulation occurs in response to a (a)______________________ hormone concentration and involves (b)______________________ the number of receptors.

Three types of receptors occur on the cell surface

22. Ion channel-linked receptors are also called (a) ______________ receptors. These receptors convert (b) ______________ signals to (c) ______________ ones.

23. (a) ______________________ receptors couple signaling molecules to (b) ______________________ pathways and are major targets for pharmaceutical development.
24. (a) ______________________________ are transmembrane proteins with a binding site for a signaling molecule outside the cell and an enzyme component inside the cell. A (b) ________________ is an example that performs phosphorylation.

Some receptors are located inside the cell

25. Intracellular receptors that regulate the expression of specific genes are called (a) ______________________ and usually bind (b)______________________ molecules.

SIGNAL TRANSDUCTION

26. Many regulatory molecules transmit information to the cell's interior without physically crossing the ______________________________.
27. A chain of signaling molecules is called a (a)______________________________ and its job is to relay and (b)______________________________ the original signal.

Signaling molecules can act as molecular switches

28. (a)______________________________ typically activates kinases while (b)______________________________ typically inactivates them.

Ion channel-linked receptors open or close channels

29. Gamma-aminobutyric acid is a neurotransmitter that binds to ______________________________ in neurons, thereby inhibiting neural signaling.

G-protein-linked receptors initiate signal transduction

30. G proteins are a group of regulatory proteins important in many ______________________________ pathways.
31. Some G proteins regulate ion channels in the (a) ________________ and others are involved in (b) ______________________________ and (c)______________________________.
32. G proteins usually relay a signal to a ______________________________.

Second messengers are intracellular signaling agents

33. Second messengers are usually (a)______________________________, or (b)______________________________.
34. In many signaling cascades in prokaryotic and animal cells, the second messenger is ______________________________.
35. Adenylyl cyclase is an enzyme on the ______________________ side of the plasma membrane.
36. When adenylyl cyclase is activated, it catalyzes the formation of (a)______________________________, which in turn activates (b)______________________________, leading to the phosphorylation of target (c)________________, resulting in some response in the cell.

37. The enzyme (a)______________________ splits PIP_2 into (b)________________ and (c)________________ which act as second messengers to affect (d)____________ activity and levels of (e)______________ ions.
38. Ion pumps in the plasma membrane normally maintain a ________________ calcium ion concentration in the cytosol compared to its concentration in the extracellular fluid.
39. Calcium ions are important in many cellular activities including (a)________________, (b)________________, (c)________________, and (d)________________.

Many activated intracellular receptors are transcription factors

40. Some hydrophobic signaling molecules diffuse across the membranes of target cells and bind with ________________ in the cytosol or in the nucleus.

Scaffold proteins increase efficiency

41. Scaffold proteins organize groups of (a)________________ into (b) ________________, ensuring that signals are relayed accurately, rapidly, and more efficiently.

Signals can be transmitted in more than one direction

42. Transmembrane proteins that connect the cell to the extracellular matrix are called (a) ________________ and transduce signals in (b) ________________.

RESPONSES TO SIGNALS

43. The three categories of responses are (a)________________, (b)________________, and (c)________________.

Ras pathways involve tyrosine kinase receptors and G proteins

44. (a) ________________ are a group of small G proteins active when bound to (b) ________________. Mutations in these proteins are associated with (c) ________________.
45. A well-studied ras pathway called (a) ________________ is the main signaling pathway for cell (b) ________________, and (c) ________________.

The response to a signal is amplified

46. The process of magnifying the strength of a signaling molecule is called ________________.

Signals must be terminated

47. Signal termination returns the receptor and each of the components of the (a) ________________ pathway to their (b)________________ states.

EVOLUTION OF CELL COMMUNICATION

48. Some signaling molecules are highly (a)________________________ such as, (b)__________________________, (c)________________________, and (d)________________________.
49. Evidence suggests that cell communication first evolved in ________________________ and continued to change over time as new types of organisms evolved.

BUILDING WORDS

Use combinations of prefixes and suffixes to build words for the definitions that follow.

Prefixes	The Meaning	Suffixes	The Meaning
di-	two, twice, double	-chrome	color
intra-	within	-crine	to secrete
inter	between		
neuro-	nerve		
para-	beside, near		
phyto-	plant		
substrat-	strewn under		
tri-	three		

Prefix	Suffix	Definition
____________	____________	1. Pertains to cell secretions (signaling molecules) that diffuse through the interstitial fluid and act on nearby cells.
____________	-transmitter	2. Substance used by neurons to transmit impulses across a synapse.
____________	____________	3. A blue-green, proteinaceous pigment involved in photoperiodism and a number of other light-initiated physiological responses of plants.
____________	-cellular	4. Pertains to the inside of a cell.
____________	-phosphate	5. A chemical compound with three phosphate groups.
____________	-phosphate	6. A chemical compound with two phosphate groups.
____________	-acylglycerol	7. A compound formed of two fatty acids and one glycerol molecule.
____________	-cellular	8. Signals that are sent between cells.
____________	-stitial	9. The fluid surrounding body cells.
____________	-e	10. Substance upon which an enzyme works.

MATCHING

Terms:

a. Adenylyl cyclase
b. Calmodulin
c. Enzyme-linked receptor
d. G protein-linked receptor
e. Guanosine diphosphate
f. Hormone
g. Intracellular receptor
h. Ligand
i. Neurotransmitter
j. Nitric oxide
k. Ras protein
l. Signal amplification
m. Signal termination
n. Signal transduction
o. Transcription factor
p. Tyrosine kinase

For each of these definitions, select the correct matching term from the list above.

____ 1. A signaling molecule that binds to a specific receptor.

____ 2. The process in which a receptor converts an extracellular signal into an intracellular signal that causes some change in the cell.

____ 3. A gaseous signaling molecule that passes into target cells.

____ 4. A signaling molecule released by neurons to signal one another.

____ 5. A chemical messenger secreted by endocrine glands.

____ 6. A transmembrane protein composed of seven alpha helices connected by loops that extend into the cytosol or outside the cell.

____ 7. A transmembrane protein with a binding site for a signaling molecule outside the cell and a binding site for an enzyme inside the cell.

____ 8. A receptor located in the cytosol or in the nucleus.

____ 9. An enzyme-linked receptor that phosphorylates specific tyrosines in certain signaling proteins inside the cell.

____ 10. One of a group of small G proteins that, when activated, trigger a cascade of reactions.

____ 11. A receptor located in the nucleus that activates or represses specific genes.

____ 12. An enzyme on the cytoplasmic side of the plasma membrane that catalyzes the formation of cyclic AMP from ATP.

____ 13. A protein that when combined with calcium ions affects the activity of protein kinases and protein phosphatases.

____ 14. The process of enhancing signal strength as a signal is relayed through a signal transduction pathway.

____ 15. The process of inactivating the receptor and each component of the signal transduction pathway once they have done their jobs.

MAKING COMPARISONS

Refer to this illustration of signal transduction to fill in blanks in the table below.

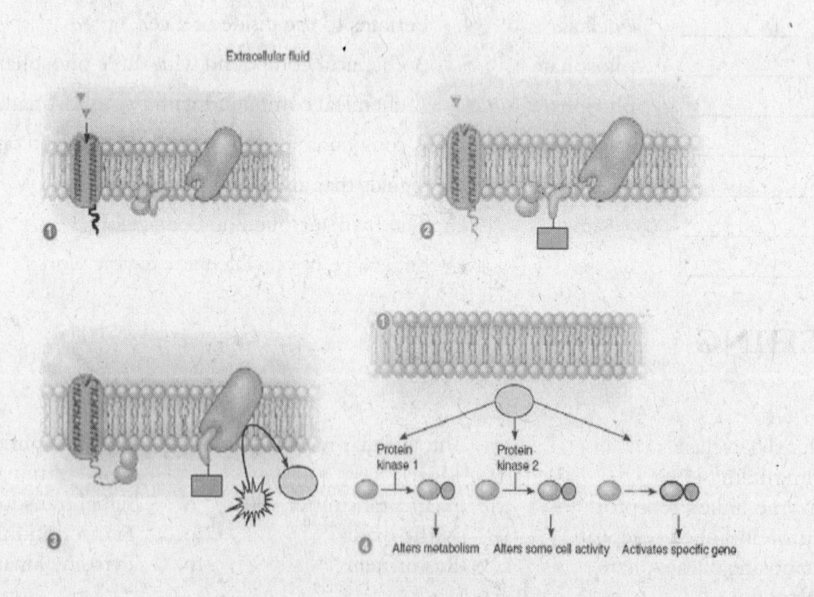

Identify the structures in the table and fill in the blanks.

Structure	Name of Structure	Structure	Name of Structure
entire structure	receptor molecule	at the arrow	#1
at the arrow	#2	at the arrow	#3
at the arrow	#4	entire structure	#5
entire structure	#6	entire structure	#7
at the arrow	#8		

MAKING CHOICES

Place your answer(s) in the space provided. Some questions may have more than one correct answer.

____ 1. A community of bacteria attached to a solid surface is a
a. clone.
b. quorum.
c. biofilm.
d. population.
e. colony.

____ 2. Cells in multicellular organisms that respond to a chemical signal from other cells in the same organism are called
a. transmitter cells.
b. signaling cells.
c. responders.
d. sensory cells.
e. target cells.

____ 3. The process in cells of converting an extracellular signal into an intracellular signal to which a response occurs is known as
a. optimization.
b. transduction.
c. internal regulation.
d. chemical transposition.
e. chemical signal reception.

____ 4. The process whereby a signaling molecule diffuses through interstitial fluid and affects nearby cells is
a. paracrine regulation.
b. cellular reception.
c. an example of hormonal regulation.
d. autocrine regulation.
e. endocrine regulation.

____ 5. The signaling chemicals transferred across the gap between neurons are called
a. neuroreceptors.
b. synaptic transmitters.
c. neurotransmitters.
d. neuronal enhancers.
e. neuron communicators.

____ 6. Endocrine glands
a. have no ducts.
b. secrete hormones.
c. communicate directly with neurons.
d. secrete chemical messengers into interstitial fluid.
e. secrete hormones into capillaries.

____ 7. Proteins and glycoproteins that bind with ligands are called
a. enzymes.
b. messengers.
c. signaling molecules.
d. bonding elements.
e. receptors.

____ 8. Receptor down-regulation may involve
a. destruction of receptors by lysosomes.
b. decreased numbers of receptors.
c. regulation of blood glucose levels.
d. increased intake of insulin by cells.
e. degradation of receptors.

____ 9. Receptor up-regulation occurs in response to
a. low hormone concentrations.
b. high hormone concentrations.
c. receptor down-regulation.
d. increase in protein membrane complexes.
e. integration.

____ 10. Receptors on the cell surface that convert chemical signals into electrical signals are called
a. G protein-linked receptors.
b. ion channel-linked receptors.
c. enzyme-linked receptors.
d. protein complex receptors.
e. ligand-gated channels.

____ 11. Proteins that regulate the expression of genes are
a. cell surface proteins.
b. cell membrane glycoproteins.
c. transcription factors.
d. intracellular receptors.
e. intranuclear integrators.

____ 12. When gamma-aminobutyric acid (GABA) binds to a ligand-gated chloride ion channel in a neuron
a. the channel opens.
b. adjacent channels close.
c. chloride ions flow into the neuron.
d. neural signaling is inhibited.
e. the neuron's electrical potential changes.

____ 13. Second messengers
a. are ions or small molecules
b. may amplify a neuronal signal.
c. are found mainly in interstitial fluids.
d. are cAMP.
e. are hormones.

____ 14. Calcium ion messengers are not involved in
a. microtubule disassembly.
b. muscle contraction.
c. blood clotting.
d. liver cell activation.
e. initiation of development.

____ 15. Calmodulin
a. combines to only one specific enzyme.
b. produces the binding ezyme calmodulinase.
c. changes the shape of the 4- Ca^{2+} binding complex.
d. is found in eukaryotic cells.
e. is a Ca^{2+} binding protein.

____ 16. Ras proteins
a. are activated when bound to GTP.
b. are small proteins.
c. are a group of G proteins.
d. trigger a series of reactions called the Ras pathway.
e. phosphorylate glucosamine.

____ 17. Integrins
a. are transmembrane proteins.
b. are intracellular signaling transmitters.
c. interface between the cell and the extracellular matrix.
d. are enzymes that catalyze signaling reactions.
e. may be involved in "inside-out" signaling.

____ 18. The components of signal transduction pathways have been identified in
a. protists.
b. fungi.
c. animals.
d. plants.
e. yeasts.

____ 19. Which word pair is not correctly matched?
a. rhodopsin: signal transduction pathway
b. phytochromes: absorption of red light
c. signal transduction: signal conversion
d. cryptochromes: absorption of blue light
e. ligand: enzyme

____ 20. Local regulators of cellular activity include
a. nitrous oxide.
b. histamine.
c. growth factors.
d. helium.
e. dopamine.

____ 21. G proteins
a. bind guanine nucleotides.
b. are interstitial proteins.
c. are members of a large family of transmembrane proteins.
d. associate with the receptor after ligand binding.
e. signal neighboring cells through gap junctions.

____ 22. In the inactive state a G protein consists of ________ subunit(s).
a. 1
b. 2
c. 3
d. 4
e. a GDP and a GTP.

____ 23. cAMP, a signaling molecule, is inactivated by
a. cAMP kinase.
b. diacylglyceol.
c. IP_3.
d. adenylyl cyclase.
e. phosphodiesterase.

VISUAL FOUNDATIONS

Color the parts of this illustration of the mechanism of intracellular reception as indicated.

RED	❑	extracellular signaling molecules
GREEN	❑	signaling molecules moving through the cytosol
YELLOW	❑	intranuclear receptor
BLUE	❑	transcription factor (activated receptor)
ORANGE	❑	DNA
BROWN	❑	m-RNA
PINK	❑	ribosomes
TAN	❑	protein that alters cell activity

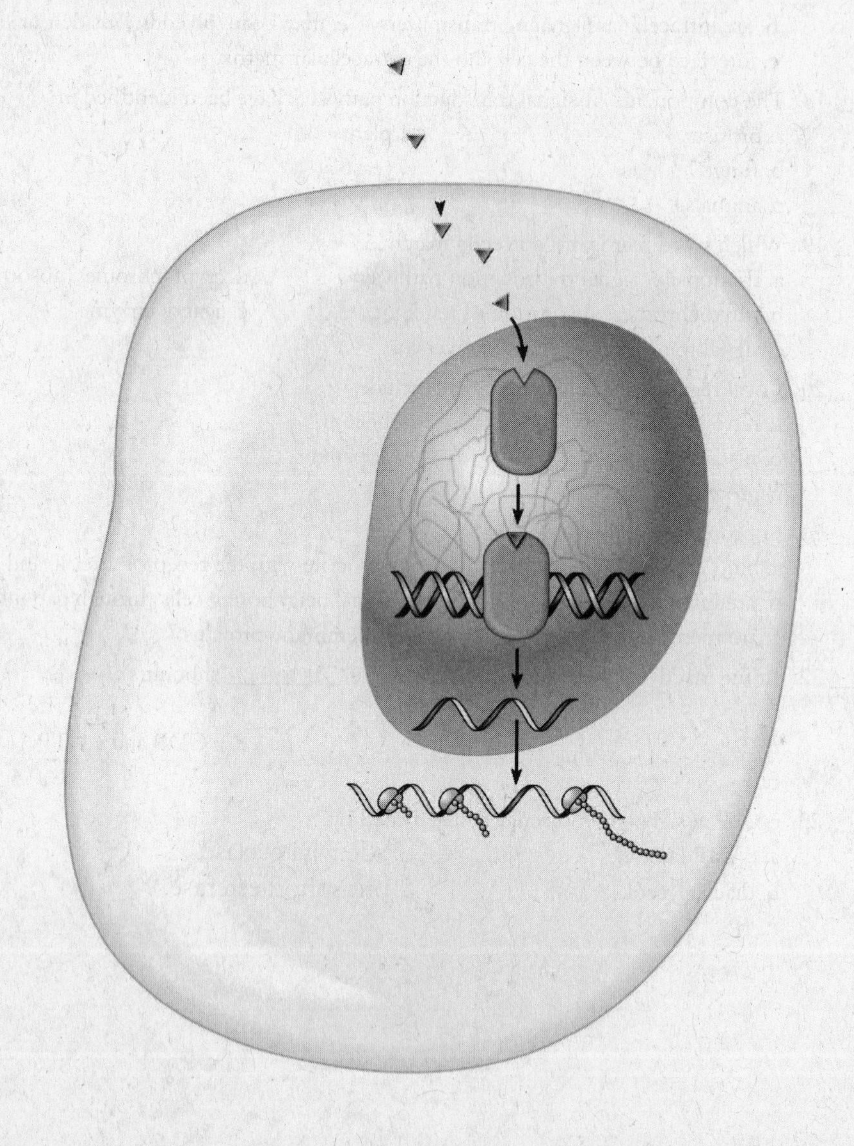

Color the parts of this illustration of types of cell signaling as indicated.

RED	❑	generalized signaling cells
GREEN	❑	generalized target cells
YELLOW	❑	receptors
BLUE	❑	signaling molecules
ORANGE	❑	signaling neuron
BROWN	❑	receptor neuron

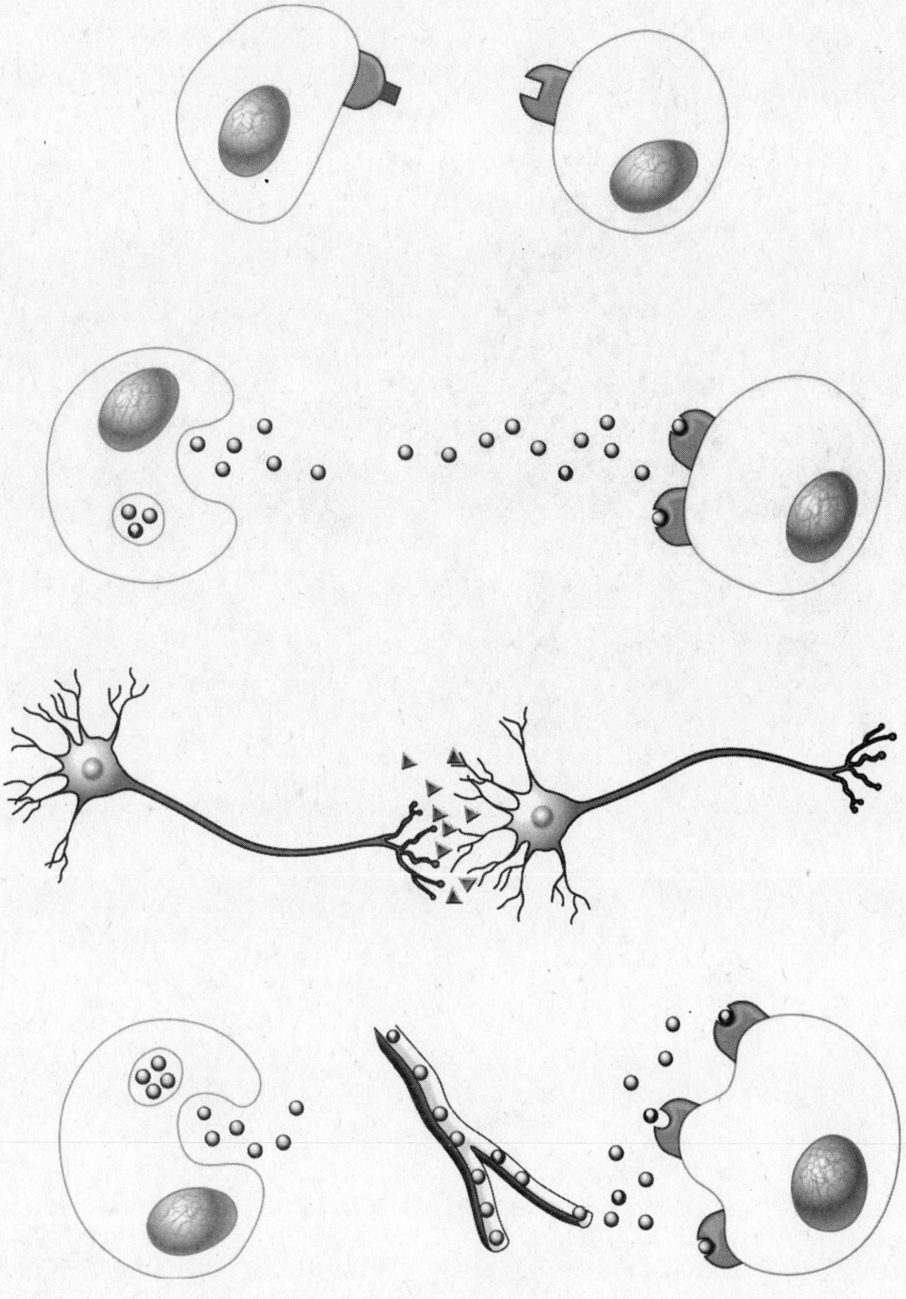

CHAPTER 7

Energy and Metabolism

You should now have an understanding of the chemical materials that function in life's processes and of the structure of a cell. You will now examine one of the basic themes of biology – the flow of energy through organisms. Life depends on a continuous input of energy. The myriad chemical reactions of cells that enable them to function involve energy transformations between potential and kinetic energy. These transformations are governed by the laws of thermodynamics which explain why organisms cannot produce energy but must continuously capture it from somewhere else, and why in every energy transaction, some energy is dissipated as heat. In the metabolism of living things, some chemical reactions occur spontaneously, releasing free energy that is then available to perform work. Other reactions require an input of free energy before they can occur, and sometimes these reactions are coupled together. ATP, the energy currency of the cell, has a structure that easily allows it to perform its vital function. Energy transfers often occur in cells through redox reactions utilizing electron carrier molecules. Chemical reactions in organisms are regulated by enzymes, substances that lower the amount of energy needed to activate reactions. Enzymes have several unique characteristics and are highly regulated. The effectiveness of some drugs is due to their ability to inhibit enzymes that are critical to the normal functioning of certain pathogenic organisms.

REVIEWING CONCEPTS

Fill in the blanks.

INTRODUCTION

1. Photosynthetic organisms and plants capture about ____% of the sun's energy that reaches the earth.
2. Plants convert radiant energy to ______________________ energy.

BIOLOGICAL WORK

3. Energy is the capacity to do (a)__________________ which biologists express as (b)__________________.

Organisms carry out conversions between potential energy and kinetic energy

4. Energy is in one of two forms: (a)__________________ is "stored energy" and (b)__________________ is "energy of motion."

THE LAWS OF THERMODYNAMICS

5. Thermodynamics is the study of (a) __________________and its (b)__________________.

The total energy in the universe does not change

6. The first law of thermodynamics states that energy can be neither (a)__________________ nor __________________, however, it can be (b)__________________________ and changed in form.

The entropy of the universe is increasing

7. The second law of thermodynamics states that in every energy conversion or transfer some energy is dissipated as ________________.

8. The term entropy refers to the ________________________ in the universe.

ENERGY AND METABOLISM

9. Metabolism is the sum of all (a)________________, and (b)________________ that occurs in an organism.

Enthalpy is the total potential energy of a system

10. The energy required to break a chemical bond is referred to as ________________________.

11. The total potential energy of a chemical reaction, or enthalpy, equals the ________________________ of the reactants and products.

Free energy is available to do cell work

12. An increase in ________________ leads to a decrease in the amount of free energy.

Chemical reactions involve changes in free energy

13. The change in free energy during a reaction is equal to the change in (a)________________ minus the product of the absolute temperature multiplied by the change in (b)________________.

Free energy decreases during an exergonic reaction

14. ________________________ are spontaneous and they release energy that can perform work.

15. The mathematical/chemical symbol for change in free energy is ______.

Free energy increases during an endergonic reaction

16. In an endergonic reaction, free energy has a ________________ value.

Diffusion is an exergonic process

17. In a concentration gradient, energy moves from a region of (a)________ concentration to a region of (b)________ concentration.

Free-energy changes depend on the concentrations of reactants and products

18. In a state of ________________________, the rate of the reverse reaction equals the rate of the forward reaction.

19. Cell reactions are virtually never at ________________.

Cells drive endergonic reactions by coupling them to exergonic reactions

20. Exergonic reactions (a)________________[release or require input of?] free energy; endergonic reactions (b)________________[release or require input of?] free energy.

21. In the reaction C → D, where the value of ΔG is negative, the reactant has _______________ (more or less?) free energy than the free energy of the product.

ATP, THE ENERGY CURRENCY OF THE CELL

22. The three main parts of the ATP molecule are (a) _______________, (b)_______________, and (c)_______________.

ATP donates energy through the transfer of a phosphate group

23. Phosphate bonds in ATP are transferred by the process known as _______________.

ATP links exergonic and endergonic reactions

24. Exergonic reactions are generally part of (a)_______________ pathways and endergonic reactions are generally part of (b)_______________ pathways.

The cell maintains a very high ratio of ATP to ADP

25. The cell cannot store _______________ quantities of ATP.

ENERGY TRANSFER IN REDOX REACTIONS

26. The substance that gives up electrons is (a)_______________ and the substance that receives the electrons is (b)_______________.

Most electron carriers transfer hydrogen atoms

27. Redox reactions in cells usually involve the transfer of a _______________ rather than just an electron.
28. The most common acceptor molecule in cellular oxidation reduction reactions is _______________.

ENZYMES

29. _______________ has the highest catalytic rate of any enzyme.

All reactions have a required energy of activation

30. The activation energy of a reaction begins the reaction by using energy to _______________.

An enzyme lowers a reaction's activation energy

31. Enzymes lower the activation energy to _______________ the rate of the reaction.

An enzyme works by forming an enzyme-substrate complex

32. Enzymes lower activation energy by forming an unstable intermediate called the _______________.
33. A substrate binds to an enzyme at the _______________.
34. When the substrate binds to the enzyme molecule, it causes a change, known as _______________.

Enzymes are specific

35. Most enzymes are specific because the shape of the _______________ is closely related to the shape of the substrate.
36. Most enzyme names end in _______________.

Many enzymes require cofactors

37. Some enzymes have two components, a protein referred to as the (a)____________________, and a chemical component called a (b)____________________.
38. An organic nonpolypeptide cofactor is called a ____________________.

Enzymes are most effective at optimal conditions

39. Factors that affect enzyme activity include (a)________________, (b)________________, and (c)________________.

Enzymes are organized into teams in metabolic pathways

40. When enzymes work in teams, the (a)____________________ from one enzyme-substrate reaction becomes the (b)____________________ for the next enzyme-substrate reaction.
41. A series of reactions can be illustrated as A ⟶ B ⟶ C … etc. where each reaction (⟶) is carried out by a specific enzyme. Such a series is referred to as a __________________________.

The cell regulates enzymatic activity

42. If a product in a series of reactions inhibits an earlier reaction in the pathway, the enzymatic activity is being regulated by a process called __________________________.
43. Substances that bind to an enzyme at a specific site other than the active site and alter enzyme activity are called ________________.

Enzymes can be inhibited by certain chemical agents

44. Inhibition is ________________________ when the inhibitor-enzyme bond is weak.
45. __________________________ inhibition occurs when the inhibitor competes with the normal substrate for binding to the active site of the enzyme.
46. __________________________ inhibition occurs when the inhibitor binds to the enzyme at a site other than the active site.

Some drugs are enzyme inhibitors

47. Because sulfa drugs have a chemical structure similar to __________________, they are able to selectively affect bacteria.

BUILDING WORDS

Use combinations of prefixes and suffixes to build words for the definitions that follow.

Prefixes	The Meaning	Suffixes	The Meaning
allo-	other, "another"	-calor(ie)	heat
ana-	up	-ergonic	work, "energy"
cata-	down	-lyst	to loosen
end(o)-	within	-steric	"space"
entrop-	transfromation	-y	state of
ex(o)-	outside, outer, external		
kilo-	thousand		
substrat-	strewn under		

Prefix	Suffix	Definition
________	________	1. Heat energy; the amount of heat required to raise the temperature of 1000 grams (1 kg) of water from 14.5^{O} C to 15.5^{O} C.
________	-bolism	2. In living organisms, the "building up" (synthesis) of more complex substances from simpler ones.
________	-bolism	3. In living organisms, the "breaking down" of more complex substances into simpler ones.
________	________	4. A spontaneous reaction that releases free energy and can therefore perform work.
________	________	5. A reaction that requires an input of free energy from the surroundings.
________	________	6. Refers to a receptor site on some region of an enzyme molecule other than the active site.
________	________	7. A measure of the state of disorder or randomness in the universe.
________	________	8. A substance that increases the speed of a chemical reaction without be consumed in the reaction.
________	-e	9. A substance on which an enzyme acts.

MATCHING

Terms:

a.	Activation	g.	Enthalpy	m.	Kilojoule
b.	Catalyst	h.	Entropy	n.	Open
c.	Closed	i.	Enzyme	o.	Potential energy
d.	Coenzyme	j.	Free energy	p.	Substrate
e.	Dynamic	k.	Heat energy	q.	Thermodynamics
f.	Energy	l.	Kinetic energy		

For each of these definitions, select the correct matching term from the list above.

____ 1. Units used to express an amount of energy.

____ 2. An organic substance that is required for a particular enzymatic reaction to occur.

____ 3. Type of system that can exchange energy with its surroundings.

____ 4. Energy in motion.

____ 5. The study of energy and its transformations.

____ 6. Adjective to describe when a forward and a reverse reaction occur in equilibrium.

____ 7. An organic catalyst that greatly increases the rate of a chemical reaction without being consumed by that reaction.

____ 8. Type of energy that cannot be used to perform work in a living system..

____ 9. Stored energy.

____10. The capacity or ability to do work.

____11. Type of energy required to break existing bonds and begin a chemical reaction.

____12. A carrier molecule that transfers electrons from one substrate to another.

____13. The total potential energy of a system.

MAKING COMPARISONS

Fill in the blanks.

Chemical Reaction	Type of Metabolic Pathway	Energy Required (Endergonic) or Released (Exergonic)
Synthesis of ATP	Anabolic	Endergonic
Hydrolysis	#1	#2
Phosphorylation	#3	#4
ATP → ADP	#5	#6
Oxidation	#7	#8
Reduction	#9	#10
FAD → $FADH_2$	#11	#12

MAKING CHOICES

Place your answer(s) in the space provided. Some questions may have more than one correct answer.

____ 1. Enzymes
a. are lipoproteins.
b. lower required activation energy.
c. speed up biological chemical reactions.
d. become products after complexing with substrates.
e. are regulated by genes.

____ 2. Enzyme activity may be affected by
a. cofactors.
b. temperature.
c. pH.
d. substrate concentration.
e. genes.

____ 3. A kilocalorie (kcal) is
a. a measure of heat energy.
b. an energetic electron.
c. a way to measure energy generally.
d. the temperature of water.
e. an essential nutrient.

Use the following formula to answer question 4:

$ATP + H_2O \rightarrow ADP + P\Delta G = -7.3$ kcal/mole

____ 4. This reaction
a. hydrolyzes ATP.
b. loses free energy.
c. produces adenosine triphosphate.
d. is endergonic.
e. is exergonic.

____ 5. Compliance with the second law of thermodynamics presumes that

a. disorder is increasing.

b. maintaining order in a system requires input of energy.

c. entropy will decrease as order in organisms increases.

d. all energy will eventually be useless to life.

e. heat dissipates in all systems.

____ 6. Kinetic energy is

a. doing work.

b. stored energy.

c. in chemical bonds.

d. energy of motion.

e. energy of position or state.

____ 7. In the formula $\Delta G = \Delta H - T\Delta S$

a. free energy decreases as entropy decreases.

b. entropy and free energy are inversely related.

c. change in enthalpy is greater than change in free energy.

d. temperature increase decreases free energy.

e. increasing enthalpy increases free energy.

____ 8. Chemical energy in molecules is

a. kinetic energy.

b. potential energy.

c. released in an endergonic reaction.

d. stored in chemical bonds.

e. energy in motion.

____ 9. The sum of all chemical activities in an organism is known as

a. anabolism.

b. catabolism.

c. metabolism.

d. entropy.

e. enthalpy.

____10. An exergonic reaction

a. releases energy.

b. can be spontaneous.

c. is a "downhill" reaction.

d. reduces total free energy.

e. produces a negative value for ΔG.

____11. In the reaction $A \rightleftharpoons B$

a. "**A**" is the product molecule.

b. "**B**" is the reactant molecule.

c. double arrows indicate a reversible reaction.

d. free energy is reduced.

e. "**B**" will accumulate

____12. According to the first law of thermodynamics

a. energy conversions are never 100% efficient.

b. during chemical reactions some energy is lost as heat

c. disorganization in the universe is decreasing

d. energy cannot be created or destroyed.

e. total energy in the universe is increasing.

____ 13. The second law of thermodynamics

a. supports why heat is generated by a fire.

b. supports the observation that energy available for doing work is decreasing.

c. states that heat will be released in an endergonic reaction.

d. describes what happens when energy is converted from one form to another.

e. explains why metabolism is the total sum of all chemical activity in an organism.

____ 14. Redox reactions
a. occur because every oxidation reaction must be coupled to a reduction reaction.
b. often occur in a series.
c. consume the energy stored in food molecules so that ATP can be synthesized.
d. may use either NAD or NADP as a hydrogen acceptor.
e. may produce $FADH^2$.

____ 15. The energy barrier in chemical reactions necessary to start a reaction is the energy of
a. catalysis.
b. synthesis.
c. catabolism.
d. hydrolysis.
e. activation.

____ 16. Enzymes are inhibited
a. either competitively or reversibly.
b. allosterically.
c. when an inhibitor competes for the active site on the enzyme.
d. when they bind weakly with an inhibitor.
e. by specific drugs.

VISUAL FOUNDATIONS

Color the parts of the illustration below as indicated.

RED ❑ active sites
GREEN ❑ substrates
YELLOW ❑ enzyme
BLUE ❑ allosteric site
ORANGE ❑ regulator
BROWN ❑ cyclic AMP
TAN ❑ enzyme-substrate complex

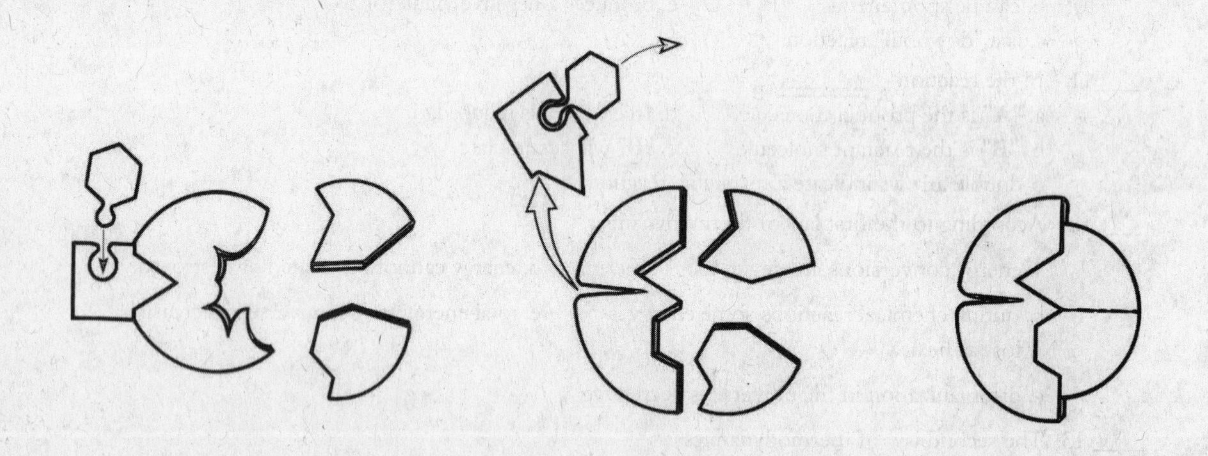

CHAPTER 8

How Cells Make ATP: Energy-Releasing Pathways

This chapter continues your study of the flow of energy through living systems. In cellular respiration, cells break down nutrients one step at a time, releasing energy from chemical bonds and transferring it to ATP where it is available for cellular work. There are three major pathways in which cells extract energy from nutrients: aerobic respiration, anaerobic respiration, and fermentation. Oxidation-reduction reactions are important in all three. Aerobic respiration, the most common pathway, involves a series of reactions in which hydrogen is transferred from a nutrient to oxygen, resulting in the formation of water. In anaerobic respiration, fuel molecules are broken down in the absence of oxygen, and an inorganic compound serves as the final hydrogen (electron) acceptor. Fermentation is a type of anaerobic respiration in which the final electron acceptor is an organic compound, and it is a much less efficient energy releasing mechanism.

REVIEWING CONCEPTS

Fill in the blanks.

INTRODUCTION

1. ______________________________ is the energy releasing aspect of metabolism involving breakdown of complex molecules.
2. ______________________________ is the energy utilizing aspect of metabolism involving the synthesis of complex molecules.
3. Most anabolic reactions are endergonic and require ________________________ or some other energy source to drive them.
4. (a)____________________________ is the process where energy in nutrients is converted to chemical energy stored in (b)____________________________.
5. (a) ________________________________ respiration requires oxygen, while (b)________________________________ respiration and (c)________________________________ do not use oxygen.

REDOX REACTIONS

6. Most cells use the catabolic process called ______________________________to extract free energy from nutrients, such as glucose.

7. In aerobic respiration, a fuel molecule is oxidized, yielding the by-products (a)________________________, and (b)________________________ with the release of the essential (c)________________________ in the form of (d)________________________ that is required for life's ctivities.

THE FOUR STAGES OF AEROBIC RESPIRATION

8. The four stages of aerobic respiration of glucose are (a)________________________, (b)________________________, (c)________________________, and (d)________________________.
9. The electrons removed from glucose during these four stages are transferred from (a)________________ and (b)________________ to a chain of electron acceptors ending in the final transfer to (c)________________ and the formation of (d)________________.
10. Hydrogens are removed to form reduced electron carriers in (a)________________________ and CO_2 is produced during (b)________________________.

In glycolysis, glucose yields two pyruvates

11. Glycolysis does not require (a)________________________ and can proceed under (b)________________, or (c)________________ conditions.
12. Glycolysis reactions take place in the ________________________ of the cell.
13. In the first phase of glycolysis, two ATP molecules are consumed and glucose is split into two ________________________ molecules.
14. In the second phase of glycolysis, each of the molecules resulting from the splitting of glucose is oxidized and transformed into a ________________________ molecule.
15. Glycolysis *nets* ________________ (#?) ATP molecules, made by (b)________________, and (c) ________________ NADH molecules.

Pyruvate is converted to acetyl CoA

16. In eukaryotes, pyruvate is converted to acetyl CoA in the (a)________________ of the cell. In aerobic prokaryotes, the reactions occur in the (b)________________. During these reactions, pyruvate undergoes a process called (c)________________________.
17. (a)________(#?) NADH and (b) ________ CO_2 form during the formation of acetyl CoA from pyruvate.

The citric acid cycle oxidizes acetyl CoA

18. The eight step citric acid cycle completes the oxidation of glucose. For each acetyl group that enters the citric acid cycle, (a)________ (#?) NAD^+ are reduced to NADH, (b)________ (#?) molecules of CO_2 are produced, and (c)__________ (#?) $FADH_2$ is made.
19. ______________(#?) acetyl CoAs are completely degraded with two turns of the citric acid cycle.
20. By the end of the citric acid cycle a total of ________# ATP molecules have been made from a single glucose molecule.

The electron transport chain is coupled to ATP synthesis

21. The hydrogens (electrons) removed during glycolysis, acetyl CoA formation, and the citric acid cycle are first transferred to the primary hydrogen acceptors (a) ______________, and (b) ______________, then transferred down an (c) ______________________________. As the electrons are passed ATP is produced by (d)______________________.
22. The electron transport chain consists of a series of electron acceptors embedded in the (a) ____. (b) ______________________________ is the final acceptor in the chain.
23. The electron transport chain consists of four complexes of electron acceptors. Complex I consists of (a)______________________________, complex II consists of (b)______________________________, complex III consists of (c)______________________________, and complex IV consists of (d) ______________________________.
24. Electron pumping produces a (a) ______________ gradient across the inner mitochondrial membrane. The concentration is high in the (b)____________________ and low in the (c) ____________________.
25. The (a) _________ flow back across the inner membrane through (b)____________________ in an (c) ______________ process.
26. The process of ________________ couples ATP synthesis with electron transport.

Aerobic respiration of one glucose yields a maximum of 36 to 38 ATPs

27. Most ATP is made by (a) ________________ as opposed to a small amount made by (b) ________________________.
28. Mitochondrial shuttle systems harvest the electrons of ________________ produced in the cytosol.

Cells regulate aerobic respiration

29. Aerobic respiration requires a steady input of fuel molecules and ______________________.

30. The rate of respiration is determined by the amount of (a)_______________ and (b) _______________ available.

ENERGY YIELD OF NUTRIENTS OTHER THAN GLUCOSE

31. Human beings and many other animals usually obtain most of their energy by oxidizing ___________________. Amino acids are also used.
32. Amino groups are metabolized by a process called ______________.
33. The (a) __________________, and (b) ______________________ components of a triacylglycerol are used as fuel.
34. Fatty acids are converted into acetyl CoA by the process of ______________________________________. These molecules enter the citric acid cycle.

ANAEROBIC RESPIRATION AND FERMENTATION

35. Anerobic respiration is performed by some prokaryotes that live in anaerobic environments such as (a) ____________________, (b)__________________________, and (c) ______________________.
36. Anerobic respiration is a means of extracting energy that produces reduced (a) ______________ substances and does not require (b)_______________.
37. Fermentation is an anaerobic process in which the final acceptor of electrons from NADH is an (a) ______________________________, therefore regenerating the required (b) __________. The final reduced molecules are often (c) _______________ or (d) _______________ which can be toxic to cells if they are allowed to build up.

Alcohol fermentation and lactate fermentation are inefficient

38. Yeasts are an example of an organism that use anerobic respiration when oxygen is available but they switch to (a)________________________ when deprived of oxygen. As a result of this type of metabolism, these organisms are called (b)_________________________ anerobes.
39. When hydrogens from NADH are transferred to acetaldehyde, _____________________________ is formed.
40. When hydrogens from NADH are transferred to pyruvate, _____________________________ is formed.
41. Fermentation yields a net gain of only (a)_______(#?) ATPs per glucose molecule, compared with about (b)_________(#?) ATPs per glucose molecule in aerobic respiration.

BUILDING WORDS

Use combinations of prefixes and suffixes to build words for the definitions that follow.

Prefixes	**The Meaning**	**Suffixes**	**The Meaning**
aero-	air	-ation	the process of
an-	without, not, lacking	-be (bios)	life
chem(i)-	chemical	-lysis	breaking down, decomposition
de-	indicates removal, separation		
ferment-	ferment	-osmo(sis)	pushing
glyco-	sweet, "sugar"		

Prefix	**Suffix**	**Definition**
________	________	1. An organism that requires air or free oxygen to live.
________	-aerobe	2. An organism that does not require air or free oxygen to live.
________	-hydrogenation	3. A reaction in which hydrogens are removed from the substrate.
________	-carboxylation	4. A reaction in which a carboxyl group is removed from a substrate.
________	-amination	5. A reaction in which an amino group is removed from a substrate.
________	________	6. A sequence of reactions that breaks down a molecule of glucose (a sugar) to two molecules of pyruvate.
________	________	7. In this process, some of the energy of the electrons in the electron transport chain is used to pump protons across the inner mitochondrial membrane.
________	________	8. An anaerobic pathway that does not involve an electron transport chain and breaks down organic compounds into energy, carbon dioxide and alcohol or lactic acid.

MATCHING

Terms:

a.	Aerobic respiration	f.	Electron transport chain	k.	Oxidation
b.	Anaerobic respiration	g.	Ethyl alcohol	l.	Phosphorylation
c.	Chemiosmosis	h.	Facultative anaerobe	m.	Pyruvic acid
d.	Citric acid cycle	i.	Fermentation	n.	Reduction
e.	Cytochromes	j.	Lactic acid		

For each of these definitions, select the correct matching term from the list above.

____ 1. The process by which a proton gradient drives the formation of ATP.

____ 2. Aerobic series of chemical reactions in which acetyl Co-A is completely degraded to carbon dioxide and water with the release of ATP.

____ 3. An organism that can live in either the presence or absence of oxygen.

____ 4. Anaerobic respiration that utilizes organic compounds both as electron donors and acceptors.

____ 5. Oxygen-requiring pathway by which organic molecules are broken down and energy is released that can be used for biological work.

____ 6. Produced along with carbon dioxide during fermentation.

____ 7. The loss of electrons or hydrogen atoms from a substance.

____ 8. The addition of a phosphate group to an organic molecule.

____ 9. A series of chemical reactions during which hydrogens or their electrons are passed along from one receptor molecule to another, with the release of energy.

____10. An end product of glycolysis.

MAKING COMPARISONS

Fill in the blanks.

Reactions in Glycolysis	Catalyzing Enzyme	ATPs Used (–) or Produced (+)
glucose → glucose-6-phosphate	hexokinase	– 1 ATP
glucose-6-phosphate → #1________________	Phospho-glucoisomerase	zero
#1________________ → fructose-1,6-biphosphate	#2	#3
fructose-1,6-biphosphate → #4________________ and glyceraldehydes-3-phosphate (G3P)	#5	zero
G3P → O=C~(P), H—C—OH, H₂C—O—(P)	Glyceraldehydes-3-phosphate dehydrogenase	#6
O=C~(P), H—C—OH, H₂C—O—(P) → #7________________	Phosphoglycero-kinase	#8
#7________________ → #9________________	Phosphoglycero-mutase	zero
#9________________ → phosphoenolpyruvate	Enolase	#10
#11________________ → pyruvate	#12	+2 ATP

MAKING CHOICES

Place your answer(s) in the space provided. Some questions may have more than one correct answer.

____ 1. Complete aerobic metabolism yields

a. 36 to 38 ATPs.
b. 2 ATPs from glycolysis.
c. 32 to 34 ATPs from electron transport and chemiosmosis.
d. 4 ATPs from the citric acid cycle.
e. 2 pyruvates.

____ 2. Anaerobic respiration

a. does not involve an electron transport chain.
b. is performed by certain prokaryotes.
c. uses an inorganic substance as the final hydrogen acceptor.
d. involves chemiosmosis.
e. may use nitrate as the final hydrogen acceptor.

____ 3. A facultative anaerobe

a. is capable of carrying out aerobic respiration.
b. is capable of carrying out alcohol fermentation.
c. is capable of producing ethanol.
d. is capable of producing CO_2.
e. requires oxygen.

____ 4. During substrate-level phosphorylation

a. ATP converts to ADP.
b. ATP forms ADP.
c. some of the ATP from aerobic metabolism is produced.
d. the electron transport chain is directly involved.
e. chemiosmosis is directly involved.

____ 5. Production of acetyl CoA from pyruvate

a. is anabolic.
b. takes place in mitochondria.
c. takes place in cytoplasm.
d. takes place in endoplasmic reticulum.
e. yields CO_2 and NADH.

____ 6. The citric acid cycle begins with

a. 2 H_2O and 4 CO_2.
b. 6 NADH.
c. oxaloacetate.
D. one $FADH_2$.
E. acetyl CoA.

____ 7. Glycolysis

a. generates a net profit of two ATPs.
b. is more efficient than aerobic respiration.
c. takes place in the cristae of mitochondria.
d. produces two pyruvates.
e. reduces glucose to H_2O and CO_2.

____ 8. ATP synthase

a. is a cytochrome.
b. converts ATP to ADP.
c. forms channels across the inner mitochondrial membrane.
d. is a transmembrane protein.
e. couples protons to electrons to form water.

____ 9. Chemiosmosis involves

a. flow of protons down an electrical gradient.
b. flow of protons down a concentration gradient.
c. pumping of protons into the mitochondrial matrix.
d. the production of ATP.
e. the production of ATP.

____10. In the electron transport chain of aerobic respiration

a. oxygen is the final electron acceptor.
b. cytochromes carry electrons.
c. the final electron acceptor has a negative redox potential.
d. glucose is a common carrier molecule.
e. electrons gain energy with each transfer.

____11. Cells may obtain energy from large molecules by means of

a. catabolism.
b. aerobic respiration.
c. anaerobic respiration.
d. fermentation.
e. the Kreb's cycle.

____12. In redox reactions ______________ is/are transferred

a. hydrogens.
b. electrons.
c. energy.
d. water.
e. oxygen.

____13. Place the following in correct order from start to finish.

1. Electron transport and chemiosmosis
2. Formation of acetyl coenzyme A
3. Glycolysis
4. Citric acid cycle
5. Formation of NAD^+ and FAD

a. 1, 2, 3, 4, 5
b. 4, 3, 2, 1, 5
c. 5, 2, 1, 4, 3
d. 3, 2, 4, 1, 5
e. 5, 1, 4, 3, 2

____14. The second phase of glycolysis is sometimes called the

a. dehydrogenation phase.
b. decarboxylation phase.
c. energy capture phase.
d. oxidative phase.
e. the Kreb's phase.

____15. Pyruvate is converted to acetyl CoA as part of the process called

a. oxidative decarboxylation.
b. substrate dephosphorylation.
c. substrate-level phosphorylation.
d. substrate phosphorylation.
e. oxidative dehydrogenation.

____16. The citric acid cycle is also known as the

a. TCA cycle.
b. matrix cycle.
c. carbonyl cycle.
d. mitochondrial cycle.
e. Kreb's cycle.

____17. In an individual cycle, the citric acid cycle produces

a. 1 $FADH_2$.
b. 2 $FADH_2$.
c. 1 NAD^+.
d. 2 NAD^+.
e. 2 NADH.

____18. Cyanide

a. uncouples the electron transport chain.
b. inhibits electron transfer to oxygen.
c. inhibits oxidative phosphorylation.
d. works at the outer mitochondrial membrane.
e. inhibits cytochrome activity.

____19. Deamination of proteins

a. may yield pyruvate as a product.
b. yields CO_2 as a product.
c. yields NH_3 as a product.
d. yields products that enter aerobic respiration.
e. yields products that enter glycolysis.

VISUAL FOUNDATIONS

Enter the appropriate enzyme in the spaces provided.

1. ______________________
2. ______________________
3. ______________________
4. ______________________
5. ______________________
6. ______________________
7. ______________________
8. ______________________

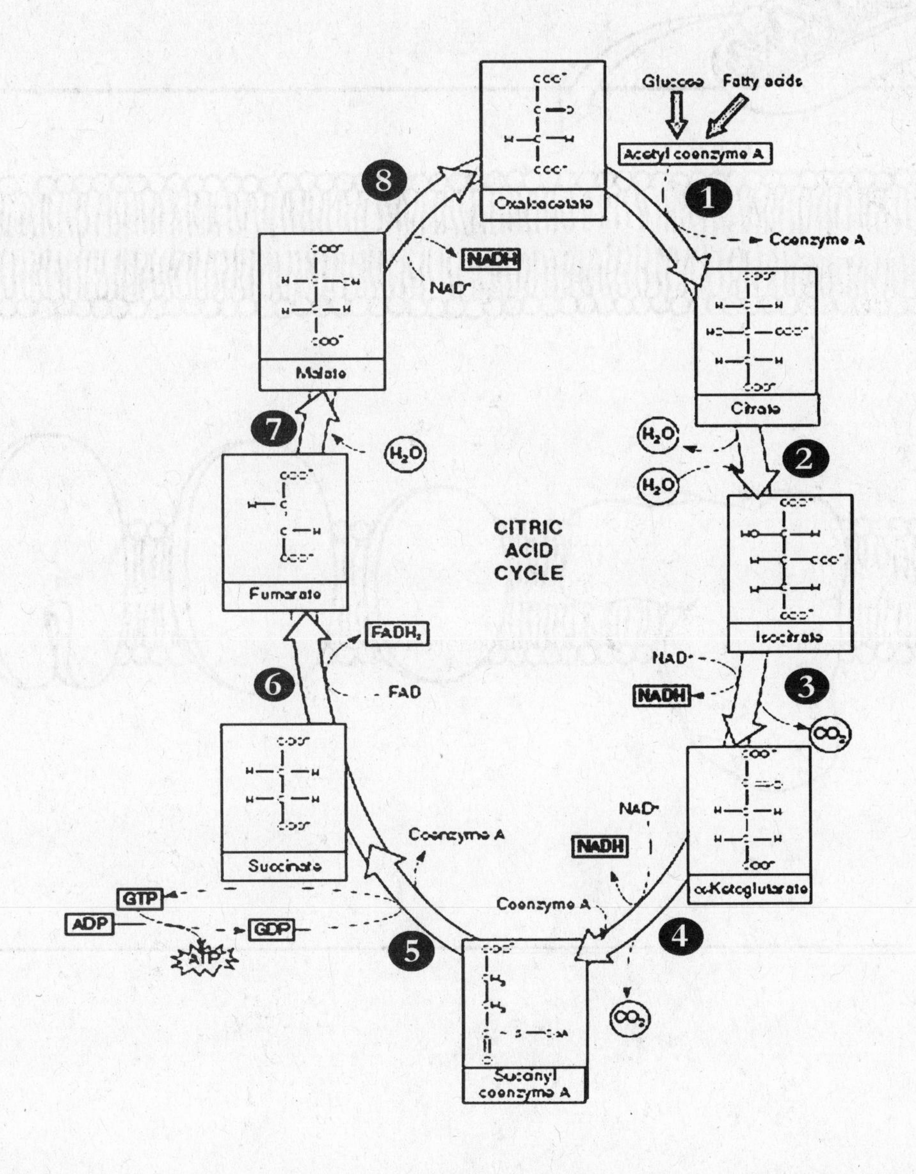

Color the parts of the illustration below as indicated.

RED	❑	electron path
GREEN	❑	complex I
YELLOW	❑	complex II
BLUE	❑	complex III
ORANGE	❑	complex IV
BROWN	❑	inner mitochondrial membrane
TAN	❑	matrix
PINK	❑	intermembrane space
VIOLET	❑	outer mitochondrial membrane

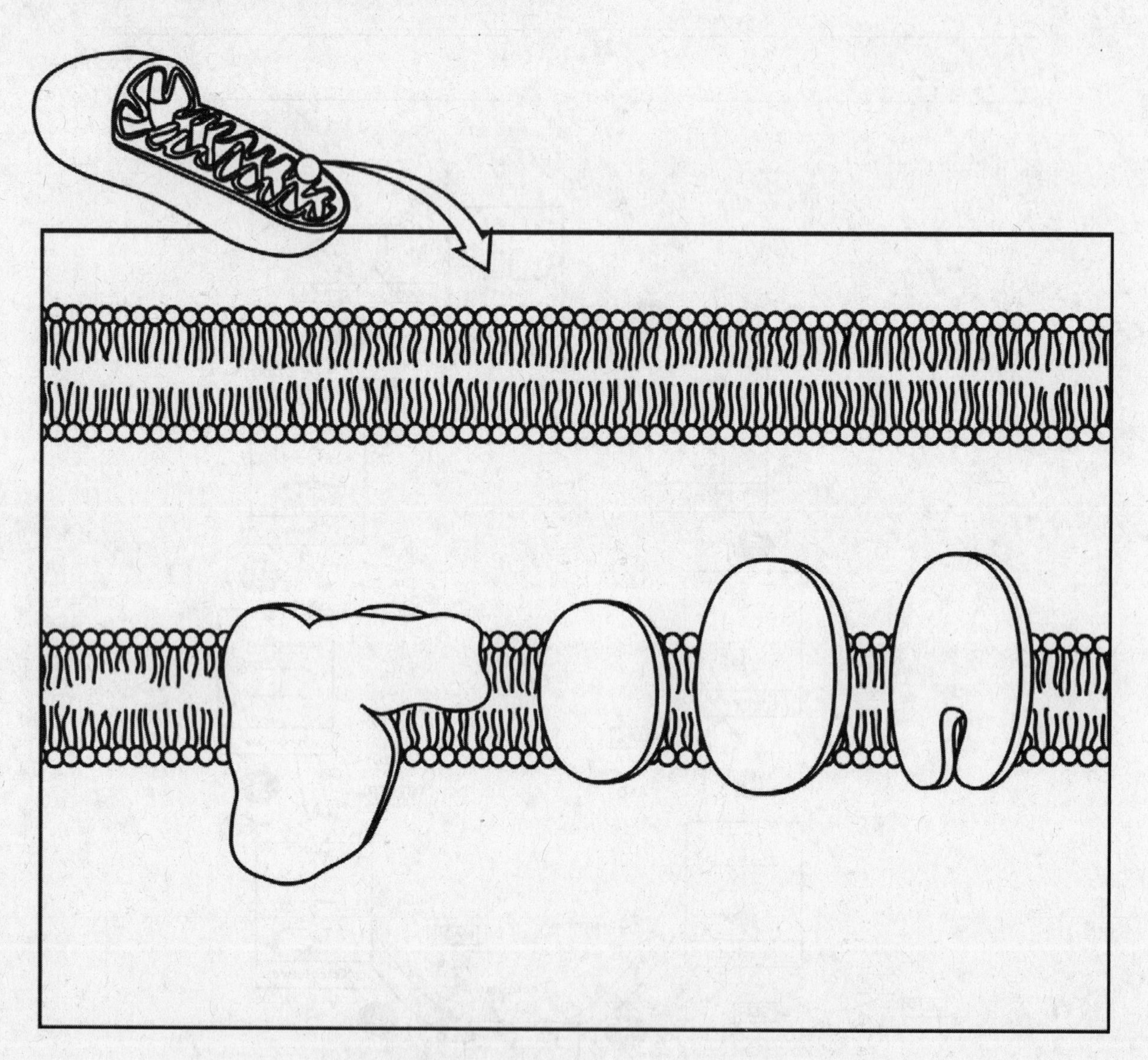

CHAPTER 9

Photosynthesis: Capturing Light Energy

This chapter concludes the overview of energy flow through living systems by examining the sequence of events in which light energy is converted into the stored chemical energy of organic molecules. This energy is what fuels the metabolic reactions discussed in the previous chapter, which sustain all life. Photosynthesis is the first step in the flow of energy through most of the living world and involves capturing solar energy and using it to produce organic compounds from carbon dioxide and water while releasing oxygen as a by-product. Major features of light are discussed as it is key to this process. In eukaryotes, the process occurs in organized structures within chloroplasts. Photosynthesis consists of both the light-dependent and the carbon fixation reactions. There are three different pathways by which carbon dioxide is assimilated into plants. Organisms can be classified metabolically based on their carbon and energy sources. Photosynthesis is a regulated process absolutely essential for the vast majority of life on our planet.

REVIEWING CONCEPTS

Fill in the blanks.

INTRODUCTION

1. Using the basic raw materials (a)_______________, and (b)_____________________, photosynthetic organisms use (c) _____________________ energy to make ATP and other molecules that hold (d)________________________ energy. However, this energy cannot be (e)_________________________ by organisms.

LIGHT AND PHOTOSYNTHESIS

2. A (a)_______________is the distance from one wave peak to the next. Light is composed of packets of energy called (b)_______________. Light with (c)___________________________ has more energy associated with it.
3. When a molecule (a)_____________ a photon an electron becomes energized. (b)_______________________ occurs when energized electrons return to the ground state, emitting their excess energy in the form of visible light.
4. In photosynthesis, energized electrons leave atoms and pass to an (a)_____________________________ molecule. In the process, these molecules become (b)____________________.

CHLOROPLASTS

5. Most chloroplasts are located in the (a)_______________ cells of leaves. The (b)_______________ is the fluid-filled region within the inner membrane of chloroplasts that contains most of the enzymes for photosynthesis.
6. (a)_______________ are flat, disk-shaped membranes in chloroplasts that are arranged in stacks called (b)_________. (c)_______________ is located in these membranes.

Chlorophyll is found in the thylakoid membrane

7. Chlorophyll absorbs light mainly in the (a)__________, and (b)___________ portions of the visible spectrum.
8. Of the several types of chlorophyll in plants, (a)_______________ is the bright green form that initiates the (b)_______________ reactions, and the yellowish-green form, (c)_______________, is an accessory pigment. Other yellow and orange accessory pigments in plant cells are (d)_______________.

Chlorophyll is the main photosynthetic pigment

9. An (a)_______________ is a graph that illustrates the relative absorption of different wavelengths of light by a given pigment. It is obtained with an instrument called a (b)_______________.
10. An (a)_______________ of photosynthesis is a measurement of the relative effectiveness of different wavelengths of light in affecting photosynthesis. This spectrum does not parallel the absorption section of chlorophyll. The difference is accounted for by action of (b)_______________ that transfer energy absorbed from the green wavelengths to chlorophyll.

OVERVIEW OF PHOTOSYNTHESIS

11. Photosynthesis involves the pigment (a)_______________capturing (b)_______________ and converting (c)_______________into (d)_______________and releasing (e)_______________as a by-product.

ATP and NADPH are the products of the light-dependent reactions: An overview

12. Light-dependent reactions convert light energy to (a)___________ energy. These reactions occur in the (b)_______________.
13. Oxygen is released when _______________ is split during the electron transfer process.
14. The products of the light-dependent reactions are (a)_______________, and (b)_______________ both of which are needed in energy-requiring (c)_______________ reactions.

Carbohydrates are produced during the carbon fixation reactions: An overview

15. Carbon fixation reactions transfer the energy from (a)__________, and (b)________ to the bonds in (c)______________ molecules and occur in the (d)__________.
16. Carbon fixation "fixes" carbon atoms from (a)______________ to existing skeletons of (b)______________ molecules.

THE LIGHT-DEPENDENT REACTIONS

17. These reactions use light energy to phosphorylate (a)______ to form (b)______ and reduce (c)__________ forming (d)__________..

Photosystems I and II each consist of a reaction center and multiple antenna complexes

18. The light-dependent reactions of photosynthesis begin when ______________ and/or accessory pigments absorb light.
19. Chlorophylls *a* and *b* and accessory pigment molecules are organized with pigment-binding proteins in the thylakoids membrane into units called ______________.
20. Each antenna complex absorbs light energy and transfers it to the ________________, which consists of chlorophyll molecules and proteins.
21. Light energy is converted to chemical energy in the reaction centers by a series of ________________________.
22. The reaction center of photosystem I is made up of a pair of chlorophyll *a* molecules with an absorption peak of about 700 nm and is referred to as (a)_________. The reaction center of photosystem II is made up of a pair of chlorophyll *a* molecules with an absorption peak of about 680 nm and is referred to as (b)________.
23. When a pigment molecule absorbs light energy, that energy is passed from one pigment molecule to another in a process known as (a)______________ until it reaches the (b)______________.

Noncyclic electron transport produces ATP and NADPH

24. In photosystem I, the energized electron is passed along an electron transport chain from one electron acceptor to another, until it is passed to (a)__________________, an iron-containing protein, which transfers the electron to (b)__________.
25. Like photosystem I, photosystem II is activated when a pigment molecule in an __________________ absorbs a photon of light energy.
26. __________________ is a process that not only yields electrons, but is also the source of almost all the oxygen in the Earth's atmosphere.
27. As the electrons from photosystem II are transferred down their transfer chain, (a)__________is lost, but some is used to pump

(b)__________ across the (c)__________ membrane into the lumen creating a gradient used to generate (d)________.

Cyclic electron transport produces ATP but no NADPH

28. Only (a)________________ is involved in cyclic electron transport. It is cyclic because electrons that originate from (b)__________ at the reaction center return to (c)__________.
29. The enzyme (a)________________ uses the energy in the (b)________________ to make ATP.
30. It is generally believed that ancient bacteria used ________________ to produce ATP from light energy.

ATP synthesis occurs by chemiosmosis

31. Each member of the electron transport chain linking photosystem II to photosystem I can exist in a/an (a)__________ (lower-energy level) and a/an (b)__________(higher-energy level).
32. Coupling ATP production to transfer of electrons energized by photons is called __________.
33. The chemiosmotic model explains the coupling of ATP synthesis and ________________.
34. Protons accumulate in the (a)__________and flow back across the membrane through (b)__________.

THE CARBON FIXATION REACTIONS

35. During carbon fixation reactions, the energy of (a)________ and (b)________ generated in the light-dependent reactions, is used to form organic molecules from (c)__________.

Most plants use the Calvin cycle to fix carbon

36. This process occurs in the __________ of the chloroplast.
37. The Calvin cycle begins with the combination of one CO_2 molecule and one five-carbon sugar, (a)________________, to form a six-carbon molecule and is catalyzed by the enzyme (b)________. The generated molecule instantly splits into two three-carbon molecules called (c)________________.
38. To produce one six-carbon carbohydrate, the light-independent reactions utilize six molecules of (a)________________, hydrogen obtained from (b)__________, and energy from (c)__________.
39. Most of the G3P generated during the Calvin cycle is used to regenerate ________________.

Photorespiration reduces photosynthetic efficiency

40. Photorespiration occurs mainly during (a)__________ days when plant stomata are closed to (b)________________. When this happens, photosynthesis rapidly uses up the remaining (c)_______.

41. Photorespiration reduces photosynthetic efficiency because CO_2 cannot enter the system, causing the enzyme (a)______________________________ to bind RuBP to (b)_______________ instead of CO_2.

The initial carbon fixation step differs in C_4 plants and in CAM plants

42. The C_4 pathway efficiently fixes (a)__________ at low concentrations into the four-carbon molecule (b)______________________
43. Another name for the C_3 pathway is the (a)______________________. The C3 pathway occurs after the (b)______________________.
44. The C_4 pathway occurs in the (a)__________________ cells of the leaf but the C_3 cycle occurs in the (b)______________________cells.
45. The C_4 pathway uses the enzyme (a)__________________________ to catalyze the reaction between CO_2 and (b)______________________.
46. CAM plants open (a) _____________at night and are well adapted to (b)_________ environments.

METABOLIC DIVERSITY

47. (a)______________________ are organisms that cannot make their own food, unlike (b)______________________ that are self nourishing.
48. Nonsulfur bacteria able to use light energy but unable to carryout carbon fixation are called (a)____________________________. Some other bacteria which obtain their energy from the oxidation of reduced inorganic molecules such as hydrogen sulfide are called (b)______________________.

PHOTOSYNTHESIS IN PLANTS AND IN THE ENVIRONMENT

49. Photoautotrophs are the ultimate source of almost all (a)___________________ and they slow climate change by removing (b)_________from the atmosphere.
50. Molecular oxygen is constantly replenished by the ______________________ that releases the oxygen that all aerobic organisms require for respiration.

BUILDING WORDS

Use combinations of prefixes and suffixes to build words for the definitions that follow.

Prefixes	The Meaning	Suffixes	The Meaning
auto-	self, same	-lysis	breaking down, decomposition
carot-	carrot		
hetero-	different, other	-plast	formed, molded, "body"
meso-	middle	-phyll	leaf
photo-	light	-troph	nutrition, growth, "eat"
chloro-	green		

Prefix	Suffix	Definition
____________	-phyll	1. Tissue in the middle of a leaf specialized for photosynthesis.
____________	-synthesis	2. The conversion of solar (light) energy into stored chemical energy by plants, blue-green algae, and certain bacteria.
____________	____________	3. The breakdown (splitting) of water under the influence of light energy trapped by chlorophyll.
____________	-phosphorylation	4. Phosphorylation that uses light as a source of energy.
____________	____________	5. A membranous organelle containing green photosynthetic pigments.
____________	____________	6. A green photosynthetic pigment.
____________	____________	7. An organism that fixes carbon, producing the organic compounds it needs.
____________	____________	8. An organism that is dependent upon the organic molecules produced by other organisms as the building blocks from which it synthesizes the carbon compounds it needs.
____________	-respiration	9. Degradation of Calvin cycle intermediates to CO_2 and water.
____________	-enoid	10. A yellow-orange pigment.

MATCHING

Terms:

a. Action spectrum
b. Absorption spectrum
c. C_3 pathway
d. C_4 pathway
e. Calvin cycle
f. CAM
g. Carotenoid
h. Chemoautotroph
i. Chemoheterotroph
j. G3P
k. Granum
l. Photoheterotroph
m. Photon
n. Photosystem
o. Reaction center
p. Stoma
q. Stroma
r. Thylakoid
s. Wavelength

For each of these definitions, select the correct matching term from the list above.

____ 1. The fluid region of the chloroplast surrounding the thylakoids.

____ 2. Interconnected system of flattened sac-like membranous structures inside the chloroplast where light energy is converted into chemical energy.

____ 3. A stack of thylakoids within a chloroplast.

____ 4. A yellow to orange plant pigment.

____ 5. The usual pathway for fixing carbon dioxide in the synthesis reactions of photosynthesis.

____ 6. A cyclic series of reactions occurring during the light-independent phase of photosynthesis.

____ 7. A highly organized cluster of photosynthetic pigments and electron/hydrogen carriers embedded in the thylakoid membranes of chloroplasts.

____ 8. An autotrophic organism that obtains energy from inorganic compounds.

____ 9. A particle or packet of electromagnetic radiation.

____10. A metabolic pathway that fixes carbon in desert plants.

____11. Intermediate used to make carbohydrates.

____12. Opening in plants for gas and water exchange.

____13. Distance from one wave peak to the next.

____14. A graph of the relative effectiveness of different wavelengths of light.

____15. Chlorophyll molecules and proteins that participate directly in photosynthesis.

MAKING COMPARISONS

Fill in the blanks.

Category	Photosynthesis	Respiration
Eukaryotic cellular site	Chloroplasts	Cytosol and mitochondria
Type of metabolic pathway (anabolic or catabolic)	#1	#2
#3	#4	Cristae of mitochondria
Source of hydrogen (electrons)	#5	#6
#7	NADP+	#8

MAKING CHOICES

Place your answer(s) in the space provided. Some questions may have more than one correct answer.

Use the following formula to answer questions 1-3:

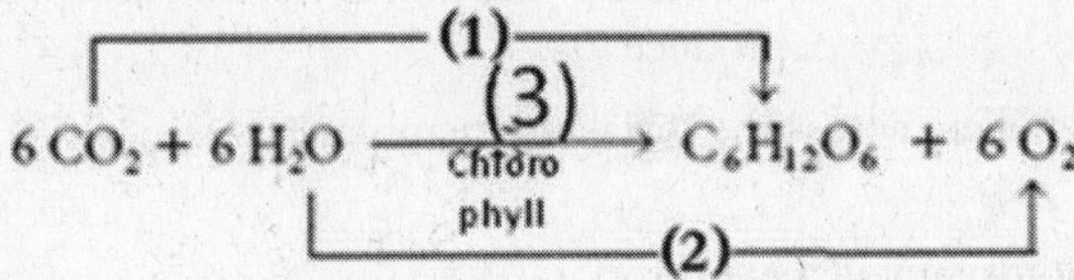

____ 1. What kind of reaction is represented by number one (1)?
a. reduction
b. oxidation
c. neutral
d. substitution
e. replacement

____ 2. What kind of reaction is represented by number two (2)?
a. reduction
b. oxidation
c. neutral
d. substitution
e. replacement

____ 3. What missing component is represented by number three (3)?
a. carotene
b. sunlight
c. wavelengths in the range of 422-492 nm or 647-760 nm
d. palisade layer of a leaf
e. chlorophyll

____ 4. Some of the requirements for the light-dependent reactions of photosynthesis include
a. water.
b. photons of light.
c. carbohydrates.
d. NADP.
e. oxygen.

____ 5. Chloroplasts in eukaryotic cells
a. possess ATP synthase complexes.
b. have double membranes.
c. do not exist.
d. contain chlorophyll.
e. carry out photosynthesis.

____ 6. The carbon fixation reactions of photosynthesis
a. generate PGA.
b. generate oxygen.
c. take place in the stroma.
d. require ATP.
e. require NADPH.

____ 7. Thylakoids
a. comprise grana.
b. are continuous with the plasma membrane.
c. possess the basic fluid mosaic membrane structure.
d. are located in the stroma.
e. carry out the carbon fixation reactions.

____ 8. Chlorophyll
a. contains magnesium in a porphyrin ring.
b. dissolves in water.
c. is green because it absorbs green portions of the light spectrum.
d. is the only pigment found in most plants.
e. is found mostly in the stroma.

____ 9. The leaf tissue in which most of photosynthesis occurs is the
a. periderm.
b. epidermis.
c. mesophyll.
d. stomata layer.
e. grana.

____10. C_4 fixation
a. replaces C_3 fixation.
b. produces PGA.
c. produces oxaloacetate.
d. takes place in bundle sheath cells.
e. supplements C_3 fixation.

____11. The reactions of photosystem II
a. include splitting water with light photons.
b. produce H_2O.
c. assist in producing an electrochemical gradient across the thylakoid membrane.
d. use $NADP^+$ as a final electron acceptor.
e. produce O_2.

____12. Most producers are
a. chemosynthetic heterotrophs.
b. chemosynthetic autotrophs.
c. photoautotrophs.
d. photosynthetic heterotrophs.
e. photosynthetic consumers.

____13. In plants that have chloroplasts, chlorophyll is found mainly in the
a. thylakoid lumen.
b. thylakoid membranes.
c. stomata.
d. stroma.
e. intermembrane spaces.

____14. In photosynthesis, electrons in atoms
a. are excited.
b. produce fluorescence.
c. are accepted by a reducing agent when they escape.
d. are pushed to higher energy levels by photons of light.
e. release absorbed energy as another wavelength of light.

____15. Reactions occurring during electron flow in respiration and photosynthesis are
a. both endergonic.
b. both exergonic.
c. exergonic and endergonic respectively.
d. endergonic and exergonic respectively.
e. neither endergonic nor exergonic.

____16. A/an________spectrum measures wavelengths of light absorbed by a pigment.
a. activity
b. absorption
c. action
d. fluorescence
e. photosynthesis

_____17. In the ground state, an atom

a. is unable to absorb energy.
b. has absorbed the maximum amount of energy.
c. has filled its outer orbital of electrons.
d. has electrons that are fluorescing as they loss energy.
e. has all of its electrons at their lowest energy level.

_____18. The pigment that initiates the light-dependent reactions of photosynthesis is

a. chlorophyll a.
b. chlorophyll b.
c. chlorophyll c.
d. carotenoid.
e. is yellow-green in color.

_____19. Carotenoids

a. expand the spectrum of light for photosynthesis.
b. can transfer energy to chlorophyll.
c. have a methyl group.
d. are antioxidants.
e. absorb different wavelengths than chlorophyll.

_____20. A chlorophyll molecule

a. contains a magnesium atom.
b. is nonpolar.
c. has a porphyrin ring.
d. contains an iron atom.
e. contains carbon and nitrogen atoms.

_____21. Carbon fixation reactions

a. have enzymes that are more active in light.
b. produce carbohydrates.
c.. produce ADP.
d. have no direct requirement for light.
e. require darkness.

_____22. Light energy is passed from one pigment molecule to another in a process known as

a. electron transfer.
b. excitation.
c. absorption.
d. photon transfer.
e. resonance.

_____23. Place the following events of noncyclic electron transport in order from start to finish.

1. Ferrodoxin-NADPH$^+$ reductase catalyzes electron transfer
2. Absorbed energy reaches reaction center
3. Absorption of a photon of light
4. Excitation of P700
5. Electron transfer to a primary electron acceptor

a. 3, 5, 4, 2, 1
b. 4, 3, 5, 2, 1
c. 4, 3, 2, 5, 1
d. 3, 4, 2, 1, 5
e. 3, 2, 4, 5, 1

_____24. Cyclic electron transfer

a. produces ATP.
b. does not produce NADPH.
c. involves only P700.
d. utilizes ATP and NADPH.
e. produces ATP and NADPH.

_____25. ATP synthase

a. is present in the thylakoid lumen.
b. is a transmembrane protein.
c. release ATP into the stroma.
d. uses the proton gradient energy to synthesize ATP.
e. is present in mitochondria.

_____26. Chemiosmosis

a. is a basic mechanism of energy coupling in cells.
b. occurs in aerobic respiration.
c. connects the electron transport chain and ATP formation.
d. builds up protons in the stroma.
e. occurs in cyclic and noncyclic electron transport.

VISUAL FOUNDATIONS

Color the parts of the illustration below as indicated. Also label the entry site of CO_2 and the structure that transports water.

RED	❑	stoma	GREEN	❑	chloroplasts
ORANGE	❑	palisade mesophyll	TAN	❑	upper epidermis
YELLOW	❑	spongy mesophyll			
BLUE	❑	vein			

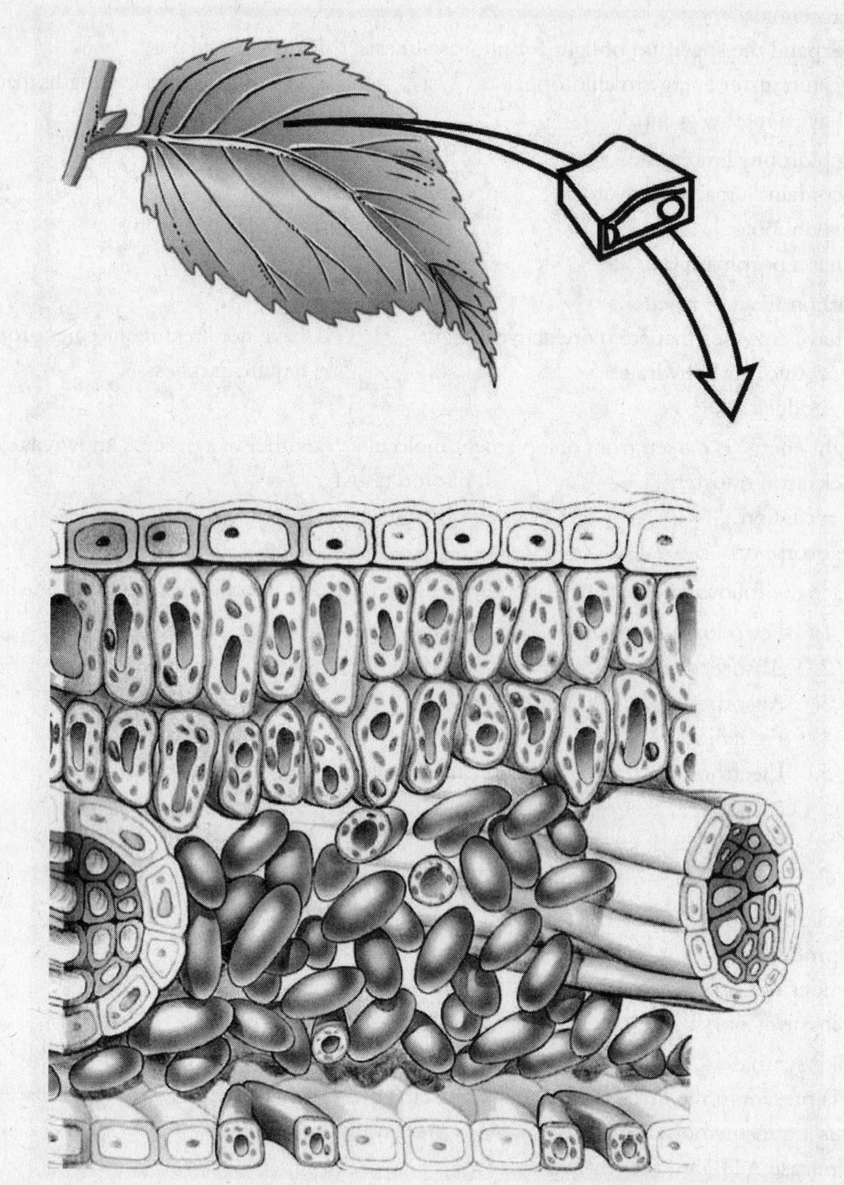

CHAPTER 10

❑

Chromosomes, Mitosis, and Meiosis

This chapter begins an examination of one of the major themes of biology – the transmission of information, specifically, the transmission of information from one generation of cells or organisms to the next. The DNA is organized into informational units called genes that control the activities of a cell. In prokaryotes, the information is contained in a single circle of DNA. In eukaryotes, it is carried in the chromosomes, made up of DNA and protein, contained within the cell nucleus. Each species is unique due to the information specified by its genes. Cell division is used for growth, repair, or reproduction. Genes are passed from parent cell to daughter cell by mitosis, a part of the cell cycle, ensuring that each new nucleus receives the same number and types of chromosomes as were present in the original nucleus. Prokaryotes use a different division process known as binary fission. The cell cycle is a highly regulated process. There are two basic types of reproduction — asexual and sexual. In asexual reproduction, a single parent cell usually splits, buds, or fragments into two or more individuals. In sexual reproduction, sex cells, or gametes, are produced by meiosis decreasing the number of chromosomes in the resulting cells by half and introducing genetic variability. When two gametes fuse, the resulting cell contains the same number of chromosomes as the parent cells. In living things, different types of sexual life cycles exist.

REVIEWING CONCEPTS

Fill in the blanks.

INTRODUCTION

1. Cell division allows for (a)______________________,
   (b)__________________________, and
   (c)__________________________.

EUKARYOTIC CHROMOSOMES

2. DNA and associated proteins form a complex called __________________________, which makes up chromosomes.

DNA is organized into informational units called genes

3. Humans have about ____________ genes that code for proteins.

DNA is packaged in a highly organized way in chromosomes

4. Prokaryotes usually contain one (a) ___________DNA molecule, but eukaryotes have multiple (b) ______________that must be (c) __________ to fit in the nucleus.
5. A human nucleus is about the size of a ____.

6. Positively charged histones associate with (a)______________ charged DNA forming structures called (b)________________________.
7. Nucleosomes prevent DNA from ________________________.
8. The packed nucleosome state occurs when histone (a)________ associates with (b)______________packing nucleosomes together to form a compacted (c)________________ fiber.

Chromosome number and informational content differ among species

9. (a)______________number is not what makes a species unique, instead it is the (b)________they carry.
10. Most human body cells have exactly (a)__________(#?) chromosomes. Humans share this chromosome number with the (b)_______ tree.
11. Most animal and plant species have between (a)_____ and (b)_____ chromosomes per somatic cell.

THE CELL CYCLE AND MITOSIS

12. The period from the beginning of one cell division to the beginning of the next cell division is the ____________________.

Chromosomes duplicate during interphase

13. Cells spend most of their time in ______________.
14. Interphase is divided into the G_1 phase, which stands for (a)______________ phase, the S phase, or (b)________________ phase, and the G_2 phase, or (c)________________ phase.
15. The M phase involves (a)______________ and (b)__________ as well as (c)________________.
16. Mitosis is a (a)____________division while cytokinesis is a (b)____________ division.

During prophase, duplicated chromosomes become visible with the microscope

17. Prophase is the __________stage of mitosis.
18. Chromosomes are duplicated during the (a)____________ phase of the cell cycle. Sister chromatids contain (b)________DNA sequences. Each chromatid includes a constricted region called the (c)________.
19. Sister chromatids are linked by a protein complex called ________.
20. Attached to each centromere is a (a)______________, a multiprotein complex to which (b)________________ can bind.
21. Some of the microtubules radiating from each pole elongate toward the chromosome, forming the (a)____________________, which separates the chromosomes during anaphase. These

microtubules extend from a (b)________________________,
which in animal cells contains a pair of (c)________________.

Prometaphase begins when the nuclear envelope breaks down

22. During prometaphase, the (a)________________ fragments so that the spindle microtubules come into contact with the (b)________________.
23. In prometaphase, sister chromatids of each duplicated chromosome become attached at their (a)________________ to (b)________________ at opposite poles of the cell. The chromosomes then begin to move toward the (c)________________.

Duplicated chromosomes line up on the midplane during metaphase

24. During metaphase, all the cell's chromosomes align at the cell's midplane, or ________________.
25. (a)________________ microtubules attach to the chromosomes while (b)________________ microtubules overlap with each other.

During anaphase, chromosomes move toward the poles

26. Once the (a)________________ chromatids separate, each is again called a (b)________________.
27. Anaphase ends when ________________________________.
28. Microtubules lack (a)________________ or (b)________________ properties.

During telophase, two separate nuclei form

29. Telophase is characterized by the (a)________________________________ and a return to (b)________________________ conditions.

Cytokinesis forms two separate daughter cells

30. Cytokinesis is the division of (a)________________ and is the last step in the (b)______ phase of the cell cycle.
31. In animal cells, actin and myosin filaments form a (a)________________ that contracts to create the (b)________________. In plant cells, cytokinesis occurs by forming a (c)________________, a partition constructed in the equatorial region of the spindle and growing laterally toward the cell wall.

Mitosis produces two cells genetically identical to the parent cell

32. The regularity of the process of cell division ensures that each daughter nucleus receives exactly the same number and kinds of ________________________ the parent cell had.

Lacking nuclei, prokaryotes divide by binary fission

33. The asexual reproduction process in which one cell divides into two daughter cells is named ________________________.
34. In prokaryotes, DNA replication begins at the ________________________________ site on the chromosome.

REGULATION OF THE CELL CYCLE

35. ______________________can temporarily block certain key cell cycle events.
36. Protein kinases are enzymes that regulate by (a)__________________________ other proteins. The kinases that control the cell cycle, called (b)__________, are only active when bound to regulatory proteins called (c)________________.
37. One major cyclin-Cdk complex, (a)__________ promotes the events of mitosis and also activates another enzyme complex called (b)_____________ toward the end of metaphase.
38. (a)____________________ are a group of plant hormones that promote mitosis both in normal growth and in wound healing. In animal cells, hormones such as (b)________________ stimulate growth and mitosis. In some animal cells, (c)____________ stimulate mitosis.

SEXUAL REPRODUCTION AND MEIOSIS

39. Offspring inherit traits that are virtually identical to those of their single parent when reproduction is (a)_______________ and can be called (b)__________.
40. In (a)__________ reproduction, offspring receive genetic information from two parents. Haploid (n) gametes from the parents fuse to form a single (b)___________ (2n) cell called the (c)__________.
41. Homologous chromosomes are members of a pair of chromosomes that are similar in (a)____________, (b)____________, and (c)______________________________________.

Meiosis produces haploid cells with unique gene combinations

42. There are ______major differences between meiosis and mitosis.
43. Meiosis typically consists of two nuclear and cytoplasmic divisions referred to as (a)_______________, and (b)__________________.

Prophase I includes synapsis and crossing-over

44. During prophase I, homologous chromosomes come to lie lengthwise side by side in a process known as ______________.
45. Each homologous pair of chromosomes has a (a)_______________ homologue and a (b)_______________ homologue.
46. Homologous chromosomes exchange genetic material in a process called (a)______________ during the first meiotic (b)________________, providing more (c)_____________________ among gametes and offspring.

During meiosis I, homologous chromosomes separate

47. The haploid condition is established as the members of each pair of homologous chromosomes separate during the first meiotic ____________________.

Chromotids separate in meiosis II

48. In metaphase I, chromatids are arranged in groups of (a)________________, and in metaphase II, chromatids are in groups of (b)________________.

Mitosis and meiosis lead to contrasting outcomes

49. (a)____________________ results in two daughter cells identical to the original cell. (b)____________________ results in (c)________ genetically different,(d)________ daughter cells.

SEXUAL LIFE CYCLES

50. Sex cells (sperm and eggs or ova) are known as (a)______________, therefore, the formation of sex cells is referred to as (b)________________________.
51. The formation of sperm is called ________________________.
52. The formation of eggs or ova is called ____________________.
53. Plants and some algae and fungi have life cycles with (a)______________of generations. A multicellular diploid stage is the (b)______________ generation while a multicellular haploid stage is the (c)______________generation.

BUILDING WORDS

Use combinations of prefixes and suffixes to build words for the definitions that follow.

Prefixes	The Meaning	Suffixes	The Meaning
centro-	center	-gen(esis)	production of
chromo-	color	-mere	part
dipl-	double, in pairs	-on(e)	a particle
gameto-	sex cells, eggs and sperm	-phase	stage
hapl-	single	-phyte	plant
hist-	tissue	-ploid	multiple of
inter-	between	-sis	the process of
mito-	thread	-some	body
poly-	many	-tic	pertaining to
oo-	egg	-tin	chemical substance
som-	body		
spermato-	seed, "sperm"		
sporo-	spore		
meta-	middle		
pro-	first		
ana-	opposite		

Prefix	Suffix	Definition
________	________	1. A dark staining body within the cell nucleus containing genetic information.
________	-phase	2. The stage in the life cycle of a cell that occurs between successive cell divisions.
________	-oid	3. An adjective pertaining to a single set of chromosomes.
________	-oid	4. An adjective pertaining to a double set of chromosomes.
________	________	5. The constricted part or region of a chromosome (often near the center) to which a spindle fiber is attached.
________	________	6. The process by which gametes (sex cells) are produced.
________	________	7. The process whereby sperm are produced.
________	________	8. The process whereby eggs are produced.
________	________	9. The stage in the life cycle of a plant that produces gametes by mitosis.
________	________	10. The stage in the life cycle of a plant that produces spores by meiosis.
________	________	11. The phase in mitosis during which the chromosomes line up along the equatorial plate.
________	________	12. The phase in mitosis where sister chromatids separate.
________	________	13. The phase in mitosis where chromosomes condense.
________	________	14. Body cells of eukaryotes that divide by mitosis.
________	________	15. Thread-like network of DNA, RNA, and nucleoproteins that is present during interphase.
________	________	16. Small nuclear protein that binds DNA and plays a role in chromosome packaging.
________	________	17. Nuclear division that produces two nuclei with the same chromosomes as the parental nucleus.
________	________	10. Having three or more sets of chromosomes.

MATCHING

Terms:

a. Aster
b. Anaphase
c. Cdk
d. Chromatin
e. Condensin
f. Crossing-over
g. Cytokinesis
h. Haploid
i. Interkinesis
j. Kinetochore
k. Meiosis
l. Cyclin
m. Mitosis
n. Nucleosome
o. Polyploid
p. Prophase
q. S phase
r. Synapsis
s. Telophase
t. Z-ring

For each of these definitions, select the correct matching term from the list above.

____ 1. The last stage of mitosis and meiosis.
____ 2. The phase in interphase during which DNA and other chromosomal components are synthesized.
____ 3. Portion of the chromosome centromere to which the mitotic spindle fibers attach.
____ 4. Process whereby genetic material is exchanged between homologous chromatids during meiosis.
____ 5. An enzyme that regulates the cell cycle by phosphorylation.

____ 6. DNA-protein fibers that condense to form chromosomes during prophase.
____ 7. Process during which a diploid cell undergoes two successive nuclear divisions resulting in four haploid cells.
____ 8. Having only a single set of chromosomes per nucleus.
____ 9. Division of the cell nucleus resulting in two daughter cells with the identical number of chromosomes as the parental cell.
____10. Stage of cell division in which the cytoplasm divides into two daughter cells.
____11. Regulatory proteins whose levels fluctuate during the cell cycle.
____12. A combination of DNA and histones.
____13. Binds to DNA and wraps it into coiled loops compacted into chromosomes.
____14. A cluster of microtubules that radiate out from the pericentriole material during late prophase.
____15. Controls cytokinesis in prokaryotes.
____16. The process occurring in prophase I when homologous chromosomes come to lie lengthwise side by side.

MAKING COMPARISONS

Fill in the blanks.

Event	Mitosis	Meiosis
Chromosome compaction (condensation)	Prophase	Prophase I, prophase II
Cytokinesis	#1	#2
Homologous chromosomes move to opposite poles	#3	#4
#5	#6	Prophase I
Cytokinesis occurs	#7	#8
Chromatids separate	#9	#10
#11	Metaphase	#12
Duplication of DNA	#13	#14

MAKING CHOICES

Place your answer(s) in the space provided. Some questions may have more than one correct answer.

____ 1. Typical somatic cells include
a. skin cells.
b. 2n cells.
c. sperm.
d. ova.
e. cells resulting from oogenesis.

____ 2. In mitosis, cells with 16 chromosomes produce daughter cells with
a. 32 chromosomes.
b. 8 chromosomes.
c. 16 chromosomes.
d. 8 pairs of chromosomes.
e. 4 pairs of chromosomes.

____ 3. In meiosis, cells with 16 chromosomes produce daughter cells with
a. 32 chromosomes.
b. 8 chromosomes.
c. 16 chromosomes.
d. 8 pairs of chromosomes.
e. 4 pairs of chromosomes.

____ 4. Gametogenesis typically involves (n=single chromosomes [haploid]; 2n=paired chromosomes [diploid])
a. 2n to 2n.
b. 2n to n.
c. reduction division.
d. mitosis.
e. meiosis.

____ 5. Eukaryotic chromosomes
a. contain DNA and protein.
b. possess asters.
c. are distinctly visible in a light microscope in interphase.
d. are uncoiled in interphase.
e. align in the equator during prophase.

____ 6. Eukaryotic DNA packaging involves
a. nucleosomes.
b. histones.
c. scaffolding proteins.
d. 30nm fiber.
e. polar microtubules.

Use this list to answer questions 6-15 about meiosis:

a. interphase
b. prophase I
c. metaphase I
d. anaphase I
e. telophase I
f. prophase II
g. metaphase II
h. anaphase II
i. telophase II

____ 7. Nuclear membranes break down.
____ 8. Homologous chromosomes synapse.
____ 9. Chromatids separate.
____ 10. Members of tetrad separate.
____ 11. Crossing over takes place.
____ 12. Pairs of chromosomes align at equatorial plane.
____ 13. Reduction of chromosome number from 2n to n.
____ 14. Tetrads form.
____ 15. Diploid to haploid.
____ 16. Single chromosomes align at equatorial plane.

Use this list to answer questions 16-25 about mitosis:

a. interphase
b. prophase
c. metaphase
d. anaphase
e. telophase
f. T phase
g. G_1 phase
h. G_2 phase
i. S phase
j. M phase

____ 17. The time between the synthesis phase and prophase.
____ 18. Chromosomes begin to condense by coiling.
____ 19. Condensed chromosomes uncoil.
____ 20. DNA replicates.
____ 21. Active synthesis and growth.
____ 22. Chromosomes are lined up in a central plane.
____ 23. All stages of mitosis collectively and cytokinesis.
____ 24. Nuclear envelope breaks down.
____ 25. The time between mitosis and start of the synthesis phase.

____26. Chromatids divide.

____27. Which nonhistone proteins help to maintain chromosome structure?

a. nucleosomes
b. condensin
c. scaffolding proteins
d. chromatin
e. elastin

____28. Both animal and plant cells have

a. a microtubule-organizing center.
b. a centriole.
c. asters.
d. a mitotic spindle.
e. two centrioles.

____29. Cohesion

a. is a ring-shaped protein complex.
b. help ensure accurate chromosome separation.
c. extend along the length of the sister chromatid arms.
d. are concentrated at the centromere
e. are included in the pericentriolar material.

____ 30. A chiasma

a. can occur in prophase I.
b. aligns tetrads during meiosis .
c. hold homologous chromatids together during meiosis II.
d. are associated with cohesins.
e. originates at a crossing-over site.

VISUAL FOUNDATIONS

Color the parts of the illustration below as indicated. Also label the phase in which DNA replicates and the phase in which cells that are not dividing are arrested.

RED ☐ M phase

YELLOW ☐ G1

BLUE ☐ S

ORANGE ☐ G2

VIOLET ☐ Interphase

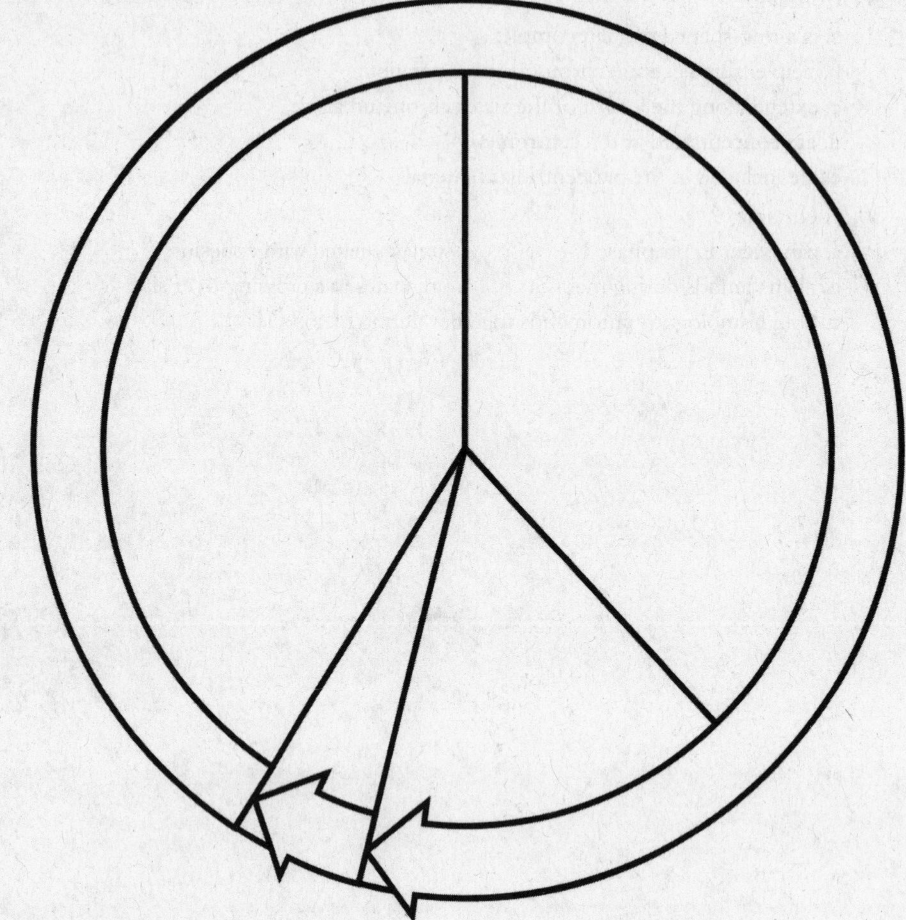

CHAPTER 11

The Basic Principles of Heredity

Now that you have a basic understanding of chromosomal behavior during sexual reproduction, we will examine the process of heredity, in which genetic information is transferred from parent to offspring. Genes that occupy corresponding loci on homologous chromosomes, called alleles, govern variations of the same characteristic. In diploid organisms, these genes exist in pairs. Scientists often study inheritance patterns by performing crosses. A monohybrid cross is a cross between two individuals carrying different alleles for a single gene locus. Similarly, a dihybrid cross is a cross between two individuals carrying different alleles at each of two gene loci. The results of monohybrid and dihybrid crosses illustrate the two basic principles of genetics known as segregation and independent assortment. The laws of probability are used to predict the results of a cross between two individuals. Special chromosomes called sex chromosomes determine the sex of most animal species and can affect inheritance of certain traits. The relationship between a single pair of alleles at a gene locus and the characteristic it controls may be simple, or it may be complex. Even the environment may influence the expression of genes.

REVIEWING CONCEPTS

Fill in the blanks.

INTRODUCTION

1. In the mid-19th century, the study of inheritance as a modern branch of science began with the work of ____________________.

MENDEL'S PRINCIPLES OF INHERITANCE

2. The advantage of using peas in Mendell's experiments was that (a)____________________ could be controlled by removing the (b)______________ which are the male parts of the flower that produce pollen.
3. (a)________________describes the physical appearance of an organism, while (b)________________refers to its genetic makeup.
4. A true-breeding line produces offspring expressing ______________________________
5. Offspring from the P generation cross are heterozygous; they are called the F_1 or (a)________________ generation. Offspring from the F_1 cross are called the (b)________________generation.
6. A (a)______________ gene may mask the expression of a (b)______________ gene.

7. Mendel's hereditary factors, now called (a)________, are a sequence of (b)________ that contains the information to make an (c)__________________.

Alleles separate before gametes are formed: the principle of segregation

8. Alleles are alternative forms of __________.
9. The principle of segregation states that before sexual reproduction occurs, the two alleles carried by an individual parent must (a)______________ during (b)__________.

Alleles occupy corresponding loci on homologous chromosomes

10. The position of a gene on a chromosome is called its __________.
11. Alleles are genes that govern (a) ________________ of the same character and occupy (b)__________________ loci on (c)__________________ chromosomes.

A monohybrid cross involves individuals with different alleles of a given locus

12. An individual is said to be (a)____________________ for a particular feature when the two alleles it carries for that feature are different. They are said to be (b)__________________if the two alleles are the same.
13. A __________________ predicts the ratios of genotypes and phenotypes of the offspring of a cross.
14. The phenotype of an individual does not always reveal its (a)______________. An individual expressing the dominant phenotype could be (b)_______________ or __________________.
15. A test cross is a cross between an individual of unknown genetic composition and a ______________________ individual.

A dihybrid cross involves individuals that have different alleles at two loci

16. A dihybrid cross is a cross between individuals that differ with respect to their alleles at ______________.

Alleles on nonhomologous chromosomes are randomly distributed into gametes: the principle of independent assortment

17. Independent assortment of alleles can result in __________________.
18. The events of ________________ are the basis for independent assortment.

Recognition of Mendel's work came during the early 20th century

19. The chromosome theory of inheritance can be explained by assuming that genes are ______________________ in specific locations along the chromosomes.

USING PROBABILITY TO PREDICT MENDELIAN INHERITANCE

20. Probability can range from (a)________(impossible) to (b)__________(certain).

21. The ____________________ predicts the combined probabilities of independent events.
22. The probability of two independent events occurring together is the ______________ of the probabilities of each occurring separately.
23. The ______________ predicts the combined probabilities of mutually exclusive events.

The rules of probability can be applied to a variety of calculations

24. If events are truly independent, (a)____________________ have no influence on the probability of the occurrence of (b)____________________.

INHERITANCE AND CHROMOSOMES

25. The chromosome theory of inheritance helps explain certain ____________________ to Mendelian inheritance.

Linked genes do not assort independently

26. Genes on the same chromosome are said to be (a)____________ and do not assort independently. Therefore, they are (b)____________ together.
27. Linked genes are recombined when chromatids exchange genetic material, a process known as ____________________ that occurs during meiotic prophase.

Calculating the frequency of crossing-over reveals the linear order of linked genes on a chromosome

28. A chromosome can be genetically mapped by determining the frequency of (a)________________ among genes, which can be used to generate a distance known as a (b)________________.

Sex is generally determined by sex chromosomes

29. The sex or gender of many animals is determined by the X and Y sex chromosomes. The other chromosomes in a given organism's genome are called ____________________.
30. The (a)____________gene causes the formation of the testes and (b)________________production.
31. When a Y-bearing sperm fertilizes an ovum, the result is a(n) (a)____________________, and fertilization by an X-bearing sperm produces a(n) (b)________________.
32. X-linked recessive traits are more common in (a)________ because they have only one (b)____ chromosome.
33. A male is neither (a)________________ or (b)________________ for X-linked alleles. Males are actually (c)________________ with respect to X-linked alleles.
34. The effect of X-linked genes is made equivalent in males and females by dose compensation, which is accomplished by a (a)____________________ X-chromosome in the male or (b)________________________ of one X-chromosome in the female.

35. A dense, metabolically inactive X chromosome at the edge of the nucleus in female mammalian cells is known as the ________________.
36. X chromosome inactivation in animals such as cats results in the appearance of patches of different colors in the fur. This phenomenon is called ________________________________.

EXTENSIONS OF MENDELIAN GENETICS

Dominance is not always complete

37. In genetic crosses involving (a)________________________________, the genotypic and phenotypic ratios are identical and the heterozygote has an (b)________________________phenotype.
38. The AB blood type is an example of ________________because two phenotypes are being expressed by the heterozygote.

Multiple alleles for a locus may exist in a population

39. Multiple alleles are (a)______________(#?) different alleles that can occupy the same (b)_________. However, in a single diploid individual there is a maximum of (c)___different alleles at a particular locus.

A single gene may affect multiple aspects of the phenotype

40. ________________________ refers to the many different effects that can often result from a given gene.

Alleles of different loci may interact to produce a phenotype

41. ________________________ is when one allele of a gene pair determines whether alleles of other gene pairs are expressed.

In polygenic inheritance, the offspring exhibit a continuous variation in phenotypes

42. It is called ____________________________ when multiple independent pairs of genes have similar and additive effects on a phenotype.

Genes interact with the environment to shape phenotype

43. The range of phenotypic possibilities that can develop from a single genotype under different environmental conditions is known as the ________________________________.

BUILDING WORDS

Use combinations of prefixes and suffixes to build words for the definitions that follow.

Prefixes	The Meaning
di-	two, twice, double
hemi-	half
hetero-	different, other
homo-	same
mono-	alone, single, one
poly-	much, many
geno-	genetics
pheno-	to show

Prefix	**Suffix**	**Definition**
____________	-gene	1. Two or more pair of genes that affect the same trait in an additive fashion.
____________	-zygous	2. Having the same (identical) members of a gene pair.
____________	-zygous	3. Having dissimilar (different) members of a gene pair.
____________	-hybrid	4. Pertaining to the mating of individuals differing in two specific pairs of genes.
____________	-hybrid	5. Pertaining to the mating of individuals differing in one pair of genes.
____________	-zygous	6. Having only half (one) of a given pair of genes.
____________	-type	7. The genetic makeup of an individual.
____________	-type	8. Physical display of an individual's genes.

MATCHING

Terms:

a.	Allele	h.	Epistasis	o.	Product rule
b.	Anther	i.	Hybrid	p.	Recessive allele
c.	Barr body	j.	Inbreeding	q.	Stapel
d.	Carpel	k.	Linkage	r.	Sum rule
e.	Character	l.	Locus	s.	True-breeding
f.	Codominance	m.	Overdominance	t.	Trait
g.	Dominant allele	n.	Pleiotropy		

For each of these definitions, select the correct matching term from the list above.

____ 1. Condition in which certain alleles at one locus can alter the expression of alleles at a different locus.

____ 2. An alternative form of a gene.

____ 3. Condition in which a single gene produces two or more phenotypic effects.

____ 4. A condensed and inactivated X-chromosome appearing as a distinctive dense spot in the nucleus of certain cells of female mammals.

____ 5. The allele that is not expressed in the heterozygous state.

____ 6. The place on a chromosome at which the gene for a given trait occurs.

____ 7. Condition in which both alleles of a locus are expressed in a heterozygote.

____ 8. The allele that is always expressed when it is present.

____ 9. The offspring of genetically dissimilar parents.

____ 10. The female part of a flower.

____ 11. A heritable difference observed in organisms.

____ 12. Rule used to determine the probability of having twins in two successive pregnancies.

MAKING COMPARISONS

Fill in the blanks. T = tall; t = short (complete dominance); Y = yellow; y = green (complete dominance)

	T Y	T y	t Y	t y
T Y	tall plant with yellow seeds (TT YY)	tall plant with yellow seeds (TT Yy)	tall plant with yellow seeds (Tt YY)	tall plant with yellow seeds (Tt Yy)
T y	#1	#2	#3	#4
#5	tall plant with yellow seeds (Tt YY)	#6	#7	#8
#9	#10	#11	short plant with yellow seeds (tt Yy)	#12

MAKING CHOICES

Place your answer(s) in the space provided. Some questions may have more than one correct answer.

R and r are genes for flower color. Homozygous dominant and heterozygous genotypes both have red flowers; the homozygous recessive genotype has white flowers. T and t are genes that control plant height. Homozygous dominant plants are tall, heterozygous plants are medium height, and homozygous recessive plants are short. Genes for flower color and height are on different chromosomes. Use these data and the following list to answer questions 1-9. Construct Punnett Squares as needed.

a. RRTT	f. rrtt	k. 1:2:1:2:4:2:1:2:1	p. pink, medium
b. RrTt	g. 1:1	l. red, tall	q. pink, short
c. Rrtt	h. 1:2:1	m. red, medium	r. white, tall
d. rrTT	i. 9:3:3:1	n. red, short	s. white, medium
e. rrTt	j. 1:2:1:1:2:1	o. pink, tall	t. white, short

Plants with the genotypes RRtt and rrTT are mated. Their offspring (F_1) are then mated to produce another generation (F_2). Questions 1-5 pertain to the F_2 generation.

____ 1. Which of the genotypes listed above are found among offspring?

____ 2. What would the phenotypic ratio among offspring be if both flower color and height genes had exhibited incomplete dominance?

____ 3. What are the genotypic and phenotypic ratios among offspring?

____ 4. What are the genotypes found among the offspring that are not in the list?

____ 5. What are all of the phenotypes among offspring?

Questions 6-9 pertain to the following cross: Rrtt x rrTT.

____ 6. What is the phenotypic ratio among offspring?

____ 7. What are the genotypes and phenotypes of the parents?

____ 8. What is the genotypic ratio among offspring?

____ 9. What are the genotypes and phenotypes of the offspring?

____10. Which of the following is/are not consistent with Mendel's principle of dominance?
a. All F_1 offspring express the dominant trait.
b. All F_2 offspring express the dominant trait.
c. Both organisms in the P generation are homozygous.
d. Both P generation parents are true breeding.
e. Only one P generation parent is true breeding.

____11. Phenotypic expression
a. may involves only one pair of genes.
b. may involve many pairs of alleles.
c. is partly a function of environmental influences.
d. refers to the genes responsible for the trait.
e. refers to the appearance of a trait.

____12. Pleiotropy means that a gene
a. exhibits incomplete dominance.
b. masks expression of other genes.
c. has the same effect as another pair of genes.
d. has multiple effects.
e. is sex-influenced.

____13. Which of the following would be false if hairy toes happened to be a recessive X-linked trait?
a. All men would have hairy toes.
b. No women would have hairy toes.
c. Parents with hairy toes could have a child without hairy toes.
d. More women than men would have hairy toes.
e. More men than women would have hairy toes.

____14. Linked genes are
a. inseparable.
b. generally on the same chromosome.
c. in separate homologous chromosomes.
d. in different chromatids.
e. always separated during crossing over.

____15. Given the following information about crossing over, determine the relative positions of three loci (X, Y, Z) on one chromosome: X and Y = 8%, Y and Z = 5%, X and Z = 3 map units.
a. X, Y, Z
b. X, Z, Y
c. Z, Y, X
d. Y, Z, X
e. Z, Y, X

____16. Mendel's "factors"
a. are haploid in gametes.
b. interact in gametes.
c. segregate into separate gametes.
d. are genes.
e. affect flower color, but not seed color.

____17. Epistasis means that alleles
a. are sex-linked.
b. have multiple effects.
c. have the same effect as another pair of genes.
d. mask the expression of another pair of genes.
e. exhibit incomplete dominance.

____18. Which of the following is/are false?
a. XX is usually female.
b. XY is usually male.
c. birds and butterflies do not have sex chromosomes.
d. hermaphrodites are XX or XY.
e. XXY is usually male.

____19. Which of the following applies if two pure-breeding P generation plants are used to ultimately produce an F_2 generation with flower colors in a ratio of 1 red: 2 pink: 1 white? (R is a gene for red; r = white)
a. All F_1 plants were Rr.
b. F_1 plants were red and white.
c. At least one P generation parent was pink.
d. F_2 genotypic ratio is 3:1.
e. F_2 genotypic ratio is 1:2:1.

____20. Polygenic inheritance means that a pair of genes
a. exhibits incomplete dominance.
b. affects expression of other genes.
c. has a similar and additive effect as another independent pair of genes.
d. has multiple effects.
e. is sex-influenced.

____21. Locus
a. may refer to a segment of DNA.
b. is the singular form of loci.
c. may refer to an alleles location on a chromosome.
d. may refer to a gene's location on a chromosome.
e. may be used instead of the term allele.

VISUAL FOUNDATIONS

Use the cells in Row 1 as a reference. In Rows 2 and 3, draw in color the following structures: centrioles (yellow), spindle (black), properly assorted chromosomes in their appropriate colors (red and blue), and centrosomes (green).

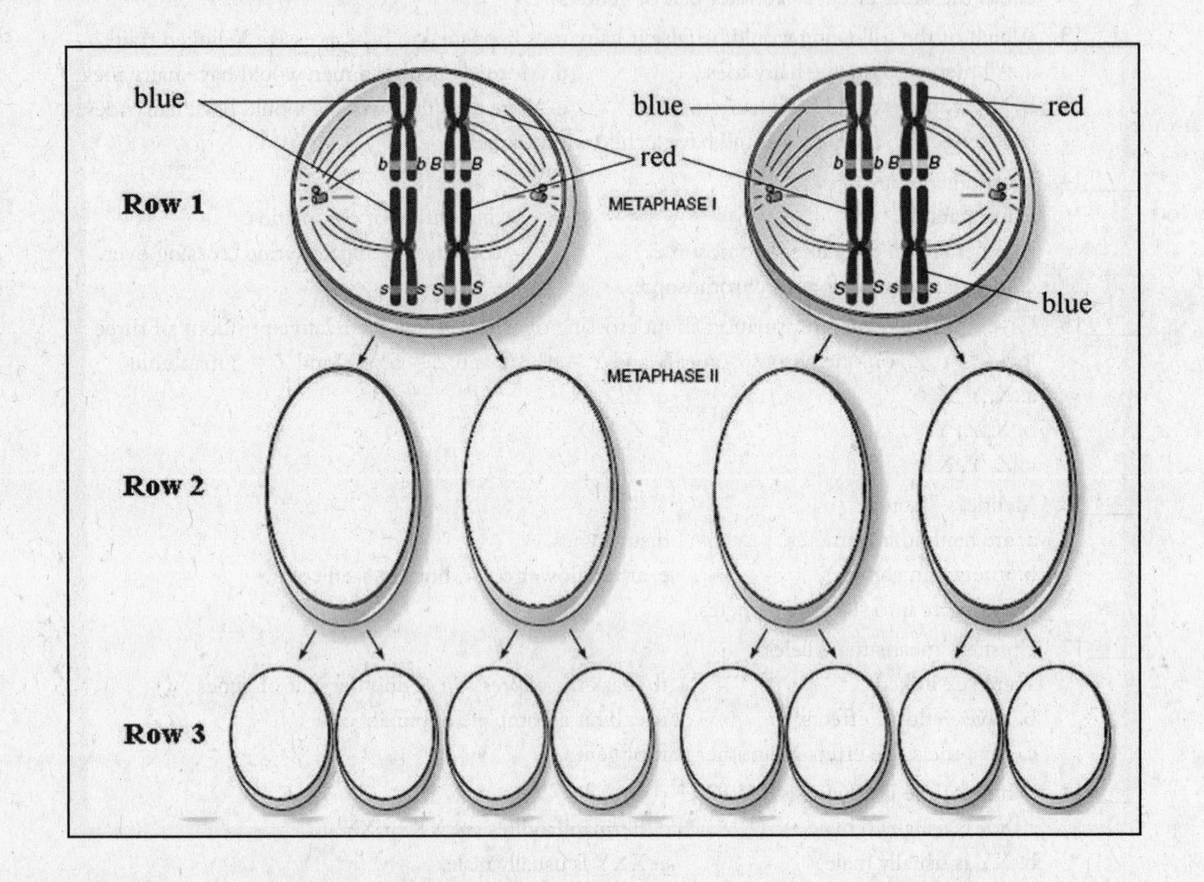

CHAPTER 12

DNA: The Carrier of Genetic Information

In the previous chapter, you learned how genes are transmitted from generation to generation. This chapter addresses the chemical nature of genes and how they function. Genes are made of deoxyribonucleic acid (DNA). Each DNA molecule consists of two polynucleotide chains, held together by hydrogen bonds, arranged in a coiled double helix. The sequence of bases in DNA provides for the storage of genetic information and the uniqueness of individuals. Many proteins and enzymes are involved in the replication of DNA. When DNA replicates, the bonds between the two polynucleotide chains break, and the chains unwind and separate. Each half of the helix is then used as a template to synthesize a new strand of complementary nucleotides. The result is two DNA double helices, each identical to the original and consisting of one original strand from the parent molecule and one newly synthesized complementary strand, which is why the process is semiconservative. Since DNA is the hereditary material, repair mechanisms exist to try and correct any errors in the sequence. Replication in eukaryotic chromosomes is complicated by their linear nature.

REVIEWING CONCEPTS

Fill in the blanks.

INTRODUCTION

1. DNA is a nucleic acid which is the molecular basis of ____________________.

EVIDENCE OF DNA AS THE HEREDITARY MATERIAL

2. Scientists initially thought that a (a)__________was the hereditary material because of the many unique properties each has.
3. Scientists felt this way because they knew DNA was made of only _____nucleotides making them rather uninteresting to study.

DNA is the transforming factor in bacteria

4. A type of permanent genetic change in which the properties of one genetic type are conferred on another genetic type is known as ____________________.

DNA is the genetic material in certain viruses

5. ____________________________ can reproduce by injecting only their DNA into cells, indicating that DNA is the genetic material.

THE STRUCTURE OF DNA

Nucleotides can be covalently linked in any order to form long polymers

6. Nucleotides consist of (a)______________________________,
(b)______________________, and
(c)__________________________.
7. The bases in nucleotides include the two purines,
(a)___________________, and (b) __________________, and
the two pyrimidines (C)_________________, and
(d)______________.
8. Nucleotides are linked by a (a)______________bond called a
(b)______________________.
9. A polynucleotide chain is directional having a 5'end with a
(a)______________ group and a 3' end with a (b)_____________
group.
10. Chargaff's rules about DNA structure state that equal numbers of
(a)_______________ bind (b)______________ and
(c)_______________ bind to (d)______________

DNA is made of two polynucleotide chains intertwined to form a double helix

11. ______________________________ studies by Franklin and
Wilkins showed that DNA has a helical structure with nucleotide
bases stacked like rungs of a ladder.
12. Watson and Crick devised a DNA model based for the most part on
existing data. Their model suggested that DNA was formed from two
(a)____
arranged in a coiled double helix with the
(b)___________________________forming the outside and the
(c)_____associating as pairs along a central axis.
13. Since the two strands of the helix run in (a)__________directions,
they are (b)___________to each other. This places the 5'
(c)_______________ end of one strand across from the 3'
(d)_______________ end of the other

In double-stranded DNA, hydrogen bonds form between A and T and between G and C

14. (a)_______ hydrogen bonds form between adenine and thymine, and
(b)_______ hydrogen bonds form between guanine and cytosine.
15. The sequence of nucleotides in one chain dictates the sequence in the
other because of ________________________________.

DNA REPLICATION

16. Replication of DNA is considered semiconservative because each
"old" strand serves as a (a)__________________for the formation of
a new strand. Therefore, this type of replication is called
(b)______________________.

Meselson and Stahl verified the mechanism of semiconservative replication

17. Using density gradient centrifugation, scientists can separate large molecules like DNA on the basis of differences in their ____________.

Semiconservative replication explains the perpetuation of mutations

18. A change in the sequence of bases in a strand of DNA is known as a ________________.

DNA replication requires protein "machinery"

19. Replication begins at an (a)____________________ where the DNA is unwound by a (b)____________. (c)______________________ stabilize the unwound strands to prevent the helix from reforming before the strands are replicated.
20. To prevent (a)______________ occurring during replication, the enzyme (b)____________ generates breaks in the DNA strands and then rejoins them.
21. ______________________ catalyze the linking together of nucleotide subunits.
22. Replication requires the synthesis of an (a)__________________ by (b) __________________. which is eventually displaced by (c) __________________ that now adds (d)__________________ to the 3' end of the primer.
23. DNA synthesis proceeds in a (a)____' → (b)____' direction. One strand is copied continuously, the other adds (c)____________ fragments discontinuously, which are linked together by (d)________________.
24. Replication is considered (a)________________because every origin of replication generates (b)___(#) replication fork(s).
25. Bacterial DNA is circular resulting in the formation of (a)______(#) fork(s).

Enzymes proofread and repair errors in DNA

26. ____________________can proofread and correct errors in base pairing.
27. Defects in the ____________________enzyme are linked to the development of a certain type of colon cancer.
28. The three enzymes involved in nucleotide excision repair are (a)____________________, (b)____________________, and (c)________.

Telomeres cap eukaryotic chromosome ends

29. Telomeres are (a)______________________DNA sequences generated by (b)__________.
30. Telomere shortening during DNA replication may contribute to (a)________________ and (b)______________.
31. Programmed cell death is known as ____________________.

BUILDING WORDS

Use combinations of prefixes and suffixes to build words for the definitions that follow.

Prefixes	The Meaning	Suffixes	The Meaning
anti-	against, opposite of	-ation	the process of
bacter(io)-	small rod	-ent	having the quality of
muta-	change	-phage	to eat
semi-	half	-tion	the process of
tel(o)-	end		
transform-	through shape		
vir-	poisonous slime		

Prefix	Suffix	Definition
____________	____________	1. Possing the ability to cause a disease.
____________	-parallel	2. The arrangement of the two polynucleotide chains in a DNA molecule, viz., "running" in opposite directions to each other.
____________	-conservative	3. The manner in which DNA replicates; half of the original DNA strand is conserved in each new double helix.
____________	-mere	4. The end of a eukaryotic chromosome.
____________	____________	5. The incorporation of DNA from one cell strain into another.
____________	____________	6. A virus that infects a bacterium.
____________	____________	7. An inheritable change in the DNA sequence.

MATCHING

Terms:

a.	Chromatin	g.	Double helix	m.	Origin of replication
b.	Deoxyribose	h.	Histone	n.	Template
c.	DNA helicase	i.	Mutation	o.	Topoisomerase
d.	DNA ligase	j.	Nucleotide	p.	Transformation
e.	DNA polymerase	k.	Purine	q.	Virulent
f.	DNA primase	jl	Pyrimidine		

For each of these definitions, select the correct matching term from the list above.

____ 1. The ability to cause disease.

____ 2. The enzyme in DNA replication responsible for base pairing.

____ 3. The region of the DNA where the molecule "unzips" for replication to begin.

____ 4. The bases adenine and guanine.

____ 5. Random heritable changes in DNA.

____ 6. A molecule composed of one or more phosphate groups, a 5-carbon sugar, and a nitrogenous base.

____ 7. A pentose sugar lacking a hydroxyl group on carbon-2.

____ 8. The shape of the DNA molecule.

____ 9. Okazaki fragments would not be connected if this enzyme were missing.

____ 10. This enzyme cannot work without an RNA primer.

____ 11. A pattern or guide during replication.

____ 12. These molecules are helix-destabilizing enzymes that bind to DNA at the origin of replication and break hydrogen bonds.

MAKING COMPARISONS

Fill in the blanks.

Researcher(s)	Contribution to the Body of Knowledge About DNA
Watson and Crick	Developed a model that demonstrated how DNA can both carry information and serve as its own template for duplication
Frederick Griffith	#1
#2	Only the DNA of a bacteriophage is necessary for the reproduction of new viruses
Chargaff	#3
Franklin and Wilkins	#4
#5	Chemically identified the transforming principle described by Griffith as DNA

MAKING CHOICES

Place your answer(s) in the space provided. Some questions may have more than one correct answer.

____ 1. Okazaki fragments are
- a. joined by DNA ligase.
- b. comprised of at least 3000 nucleotides.
- c. linked by 3' hydroxyl and 5' phosphate.
- d. assembled on ribosomes.
- e. joined to a growing DNA strand by a phosphodiester linkage.

____ 2. The two new strands in replicating DNA "grow" as follows:
- a. Leading continuously, lagging as fragments.
- b. Leading and lagging continuously.
- c. Leading toward replication fork, lagging away from fork.
- d. Both strands "grow" toward replication fork.
- e. Both strands "grow" away from replication fork.

____ 3. The molecules that "proofread" base pairing in replicated DNA, and correct any errors found are
- a. 3',5' phosphatases.
- b. purine and pyrimidine polymerases.
- c. DNA ligases.
- d. DNA polymerases.
- e. DNA primases.

____ 4. The scientists who demonstrated that DNA replication is semiconservative were
- a. Watson and Crick.
- b. Meselson and Stahl.
- c. Hershey and Chase.
- d. Avery, MacLeod, and McCarty.
- e. Franklin and Wilkins.

____ 5. In double-stranded DNA, adenine is joined to thymine and cytosine is joined to guanine by
- a. ionic bonds
- b. phosphates and sugars.
- c. covalent bonds.
- d. hydrogen bonds.
- e. purines and pyrimidins.

____ 6. In most bacteria, such as *E. coli*, DNA is
- a. a single strand.
- b. circular.
- c. a single double-stranded molecule.
- d. absent.
- e. associated with structural proteins.

____7. Which of the following base pairs is/are incorrect?
a. A – T
b. C – A
c. G – C
d. T – A
e. T – C

____8. The molecules responsible for preventing supercoiling & knot formation in replicating DNA are
a. histones.
b. chromatins.
c. scaffolding proteins.
d. topoisomerases.
e. protosomes.

____9. A mutation is
a. a change in an enzyme.
b. a change in a gene.
c. a change in DNA.
d. an alteration of mRNA.
e. sometimes inheritable.

____10. The 3' end of one DNA fragment is linked to the 5' end of another fragment by means of
a. DNA ligase.
b. helicase enzymes.
c. a DNA polymerase.
d. hydrogen bonds.
e. phosphodiester linkage.

____11. The direction for synthesis of DNA is
a. 5' → 3'.
b. 3' → 5'.
c. 5' → 5'.
d. 3' → 3'.
e. variable.

____12. The genetic code is carried in the
a. DNA backbone.
b. sequence of bases.
c. arrangement of 5', 3' phosphodiester bonds.
d. Okazaki fragments.
e. histones.

____13. The double helix structure of DNA was suggested as a result of X-ray diffraction data collected by
a. Hershey and Chase.
b. Griffith.
c. Avery, MacLeod, and McCarty.
d. Watson and Crick.
e. Franklin and Wilkens.

____14. In one molecule of DNA one would expect the composition of the two strands to be
a. both either old or new.
b. both all new.
c. both partly new fragments and partly old parental fragments.
d. one old, one new.
e. unpredictable.

____15. Chargaff's rules state or infer that
a. [A] = [T].
b. [G] = [C].
c. ratio of purines to pyrimidines = 1.
d. ratio of T to A = 1.
e. ratio of G to C = 1.

____16. Deoxyribose and phosphate are joined in the DNA backbone by
a. one of four bases.
b. purines.
c. pyrimidines.
d. phosphodiester linkages.
e. the 1' carbon of the sugar.

____17. The investigators credited with elucidating the structure of DNA are
a. Hershey and Chase.
b. Messelson and Stahl.
c. Avery, MacLeod, and McCarty.
d. Watson and Crick.
e. Franklin and Wilkens.

____18. DNA polymerase requires
- a. a 3' hydroxyl group.
- b. RNA primer.
- c. nucleotides.
- d. Okazaki fragments.
- e. DNA repair.

____19. Which selection has terms that are associated with one another?
- a. virulent/avirulent
- b. molecules/density gradient centrifugation
- c. Meselson-Stahl/semiconservative DNA replication
- d. phages/transformation
- e. topoisomerases/breaks in DNA

____20. Mismatch repair
- a. involves DNA polymerase enzymes.
- b. precedes excision repair.
- c. corrects errors uncorrected by DNA polymerase.
- d. helps in prevention of colon cancer.
- e. occurs at the replication fork.

VISUAL FOUNDATIONS

Color the parts of the illustration below as indicated.

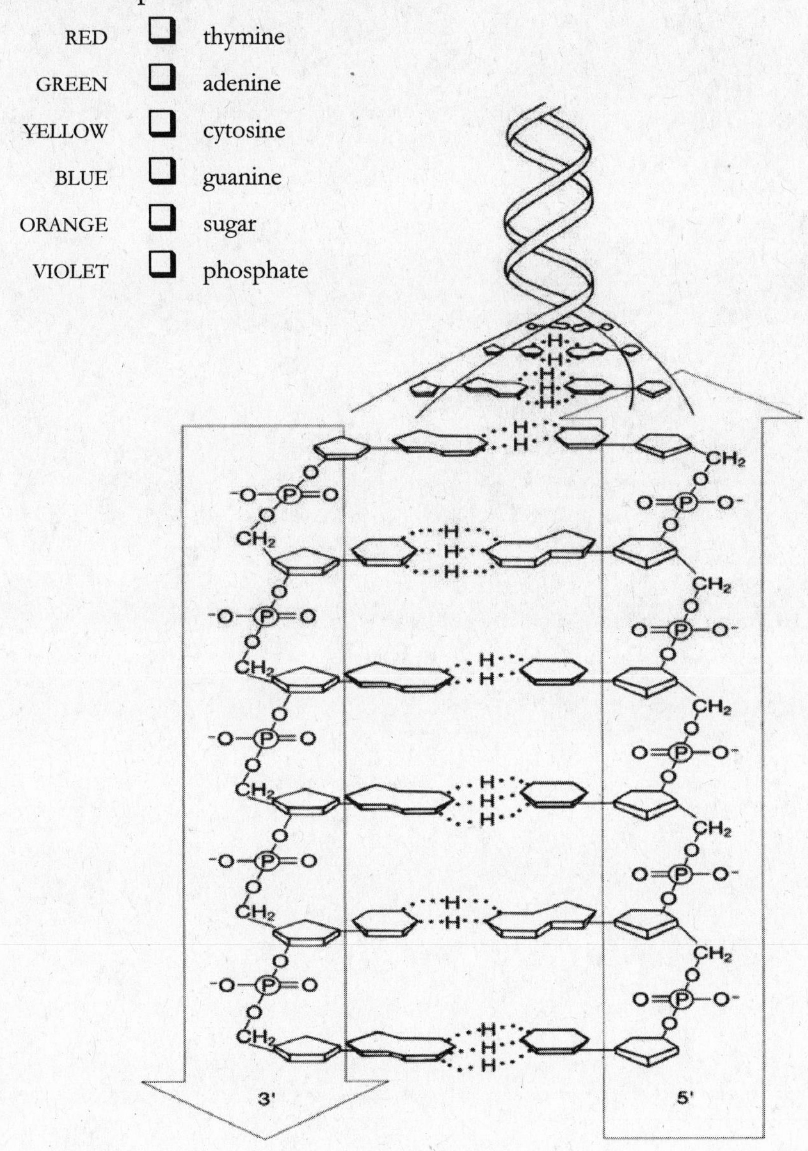

CHAPTER 13

Gene Expression

This chapter examines how cells convert the information in genes into proteins, thereby determining the phenotype of the organism. The manufacture of proteins by cells involves two major steps: the transcription of base sequences in DNA into the base sequences in RNA, and the translation of the base sequences in RNA into amino acid sequences in polypeptides. Of 64 possible codons within the genetic code, 61 code for amino acids and three serve as signals that specify the end of the coding sequence for a polypeptide chain. The genetic code is nearly universal, suggesting that it was established early in evolutionary history. Both transcription and translation involve three steps; initiation, elongation, and termination. Variations exist in gene expression between prokaryotes and eukaryotes. A gene is a DNA nucleotide sequence that carries the information to produce a specific RNA or polypeptide product. A mutation is a change in a DNA nucleotide sequence and many types and causes exist.

REVIEWING CONCEPTS

Fill in the blanks.

INTRODUCTION

1. Gene expression involves a series of steps in which the information in the sequence of ____________________ specifies the makeup of the cell's proteins.

DISCOVERY OF THE GENE-PROTEIN RELATIONSHIP

2. The first clear identification of an enzyme being a protein was the purification of the enzyme____________.

Beadle and Tatum proposed the one-gene, one-enzyme hypothesis

3. Neurospora is an ideal experimental organism because it grows primarily as a haploid organism, allowing researchers to immediately identify a (a)______________________________ because there is no (b)______________________________.

INFORMATION FLOW FROM DNA TO PROTEIN: AN OVERVIEW

4. RNA is different from DNA in three major ways. It is (a)____________________, contains the sugar (b)______________________________ and the pyridine (c)______________.

DNA is transcribed to form RNA

5. DNA is transcribed to form three main kinds of RNA. These are (a)______________, (b)______________, and (c)__________________.

6. Each (a)_______ bonds with only one kind of amino acid which it then transports to a (b)_____________________.

RNA is translated to form a polypeptide

7. Each _____________ in mRNA consists of three-bases that specify one amino acid in a polypeptide chain.
8. The (a)_____________in a (b)__________is complementary to a codon in the (c)___________.

Biologists cracked the genetic code in the 1960s

9. The genetic code is read one triplet at a time from a fixed starting point that establishes the _____________________.
10. There are (a)___________(#?) possible codons. Of these, (b)_________ code for amino acids, and (c)_________(#?) function to terminate protein synthesis.

The genetic code is virtually universal

11. The genetic code being almost universal across species suggests ____.

The genetic code is redundant

12. Of the three nucleotides that make up a triplet code, it appears that the ________ seem to contain specific information. for a specific amino acid.
13. (a)_________________, and (b)_________________ are the only amino acids that are specified by single codons.
14. The wobble hypothesis explains_____________________ ____.

TRANSCRIPTION

15. mRNA synthesis is catalyzed by DNA-dependent ________________.
16. RNA polymerase I catalyzes the synthesis of (a) __________. RNA polymerase II catalyzes the synthesis of (b) __________. RNA polymerase III catalyzes the synthesis of (c) __________.
17. RNA is synthesized in the (a) ________direction while the DNA template is read in the (b) __________direction.
18. Any point closer to the 3' end of transcribed DNA (and therefore toward the 5' end of mRNA) is said to be (a)_______________ relative to a given reference point. Conversely, areas toward the 5' end of DNA and the 3' end of mRNA are (b)_______________.

The synthesis of mRNA includes initiation, elongation, and termination

19. Transcription is initiated at sites containing specific DNA sequences called the ______________ regions.
20. During (a)_________________RNA polymerase uses complementary base pairing to add nucleotides to the (b)_______of the RNA molecule.

21. When RNA polymerase recognizes a (a)_______________sequence in the DNA transcription ends and RNA polymerase separates from the (b)___________________ and newly synthesized (c)________.

Messenger RNA contains base sequences that do not directly code for protein

22. Well before the protein-coding sequences at the 5' end of mRNA, a segment called the (a)_______________________ contains recognition signals for ribosome binding and is followed by the (b)______________.
23. Stop codons are followed by noncoding 3' _______________________.

Eukaryotic mRNA is modified after transcription and before translation

24. The original transcript in eukaryotes is known as _______________________.
25. A cap is added to the (a)__________and a (b)___________to the (c)__________of the pre-mRNA.
26. RNA splicing is mediated by the (a)___________________and removes noncoding sequences called (b)____________and links together the coding (c)_______________.

TRANSLATION

27. The covalent bonds that join amino acids together are known as _______________________.

An amino acid is attached to tRNA before incorporation into a polypeptide

28. The specific enzymes that catalyze the formation of covalent bonds between amino acids and their respective tRNAs are the ____.
29. Aminoacyl-tRNAs bind to the (a)_______________ sequence to align the (b)_______________ in correct order during polypeptide synthesis.

The components of the translational machinery come together at the ribosomes

30. The ribosome has (a)_____binding site(s) for mRNA and (b)_________for tRNA. The (c)_____site contains the tRNA holding the growing polypeptide. The (d)____site is where the tRNA enters the ribosome and the (e)______site is where they leave.

Translation begins with the formation of an initiation complex

31. Initiation begins when (a)____________factors bind to the (b)__________ribosomal subunit.
32. The codon ____________ initiates the formation of the protein-synthesizing complex.
33. The initiator tRNA coding for (a)_______________then binds, followed by the (b)_________ribosomal subunit and the release of remaining initiation factors, which completes the intiation complex.

During elongation, amino acids are added to the growing polypeptide chain

34. Elongation requires (a)_______________factors and energy in the form of (b)________.
35. Peptide bond formation requires the enzyme (a)_______________ which is not a protein but is a (b)_______________..
36. The ribosome moves down the mRNA one codon by a process called (a)_______________in the (b)_______________direction.

One of the stop codons signals the termination of translation

37. Newly formed polypeptides are released and ribosome subunits dissociate when a _______________________recognizes a stop codon.
38. The new polypeptide is assisted in folding by _______________________.

TRANSCRIPTION AND TRANSLATION ARE COUPLED IN BACTERIA

39. A _______________________ consists of an mRNA molecule that is bound to clusters of ribosomes.

MUTATIONS

40. A mutation is a change in the _______________________________ in DNA.
41. Most mutations are either (a)_______________ or (b)_______________ but some may be (c)_______________.

Base substitution mutations result from the replacement of one base pair by another

42. (a)_______________________________ mutations involve a change in only one pair of nucleotides. These can lead to a (b)_______________mutation with no effect.
43. The substitution of one amino acid for another, is called a (a)_______________ mutation and a (b)_______________ mutation involves the conversion of an amino acid-specifying codon to a termination codon.

Frameshift mutations result from the insertion or deletion of base pairs

44. In frameshift mutations, one of two nucleotide pairs are _______________________________from the molecule, altering the reading frame.

Some mutations involve mobile genetic elements

45. Mobile genetic elements can disrupt the functions of some genes because they are (a)_______________________ and "jump" into the middle of a gene which may result in the (b)_______________________ or (c)_______________ of gene activity.
46. Many mobile genetic elements are (a)_______________________. Many of these molecules encode the enzyme (b)_______________________

47. Retrotransporsons increase an organism's ability to ___________.

Mutations have various causes

48. ___________ are regions of DNA that are particularly susceptible to mutations.
49. Agents that cause mutations are called (a)_________________ and many are (b)___________which cause cancer.

VARIATIONS IN GENE EXPRESSION

Many eukaryotic genes produce "noncoding" RNAs with catalytic,regulatory, or other cellular functions

50. ___________ molecules bind to specific proteins to form a small nuclear ribonucleoprotein complex (snRNP).
51. Groups of snRNPs form a (a)_________________ which catalyzes (b)_________________.
52. (a)_________, in combination with other proteins, directs the ribosome-mRNA –polypeptide complex to the (b)_____________.
53. Pre-rRNA are processed by ___________.
54. RNA interference regulates (a)_________________and involves both (b)_____________ and (c) ___________.
55. ___________ are single-stranded RNA molecules that inhibit translation of mRNAs.

The definition of a gene has evolved

56. A gene carries the information needed to produce (a)_______________ or a (b)_________________.

The usual direction of information flow has exceptions

57. Retroviruses are viruses that require _________________ which synthesizes DNA from RNA.

BUILDING WORDS

Use combinations of prefixes and suffixes to build words for the definitions that follow.

Prefixes	The Meaning	Suffixes	The Meaning
anti-	against, opposite of	-gen	production of
cod(e)-	writing tablet	-on	a particle
intr(a)-	within	-some	body
muta-	change	-tion	the process of
poly-	much, many		
transpose	reverse the order of		

Prefix	Suffix	Definition
____________	-codon	1. A sequence of three nucleotides in tRNA that is complementary to ("opposite of"), and combines with, the three-nucleotide codon on mRNA.
ribo-	____________	2. An organelle (microbody) composed of RNA and proteins that functions in protein synthesis.
____________	____________	3. A submicroscopic complex consisting of many ribosomes attached to an mRNA molecule during translation.
muta-	____________	4. A substance capable of producing mutations.
carcino-	____________	5. A substance capable of producing cancer.
____________	____________	6. A segment of DNA that can move from one location into another along the DNA.
____________	____________	7. The fundamental unit of the genetic code.
____________	____________	8. A segment of DNA that does not code for a protein.
____________	____________	9. An inheritable change in DNA sequence.

MATCHING

Terms:

a. Adenine
b. Anticodon
c. Codon
d. Elongation
e. Exon
f. Intron
g. Initiation
h. mRNA
i. Promoter
j. Release factor
k. Retrovirus
l. rRNA
m. Transcription
n. tRNA
o. Translocation
p. Uracil

For each of these definitions, select the correct matching term from the list above.

____ 1. Site on DNA to which RNA polymerase attaches to begin transcription.
____ 2. The making of mRNA from a DNA template.
____ 3. An RNA virus that produces a DNA intermediate in its host cell.
____ 4. A pyrimidine found in RNA.
____ 5. A triplet of mRNA bases that specifies an amino acid or a signal to terminate the polypeptide.
____ 6. The stage of translation during which amino acid chains are built.
____ 7. The RNA responsible for base pairing during protein synthesis.
____ 8. In eukaryotes, the coding region of DNA.
____ 9. RNA that has been transcribed from DNA that specifies the amino acid sequence of a protein.
____ 10. In eukaryotes, the region removed during splicing.
____ 11. Recognizes stop codons.
____ 12. The shifting of a ribosome on a mRNA.
____ 13. The base to which uracil binds.
____ 14. The region of tRNA that pair with a complementary area on mRNA.

MAKING COMPARISONS

Fill in the blanks.

mRNA Codon	DNA Base Sequence	tRNA Anticodon	Amino Acid Specified
5' — AAA — 3'	3'—TTT—5'	3'—UUU—5'	Lysine (Lys)
#1	#2	3'— UGG —5'	Threonine (Thr)
5' — GCG — 3'	#3	#4	Alanine (Ala)
5' — UGU — 3'	#5	3' — ACA — 5'	Cysteine (Cys)
#6	3' — ATT — 5', 3' — ATC — 5', 3' — ACT — 5'	#7	None (stop codon)
5'—AUG—3'	#8	#9	Methionine (Met) (start codon)

MAKING CHOICES

Place your answer(s) in the space provided. Some questions may have more than one correct answer.

____ 1. The following part of a spliceosome particle is involved in intron removal and regulation of transcription:
a. SRP RNA
b. snoRNA
c. siRNA
d. snRNA
e. miRNA

____ 2. The ____________ selectively suppresses gene expression.
a. SRP RNA
b. snoRNA
c. siRNA
d. snRNA
e. miRNA

____ 3. Exons are
a. intervening sequences.
b. special integrated coding sequences.
c. noncoding sequences within a gene.
d. organized codes for specific amino acids.
e. HIV resistant sequences.

____ 4. Energy for transfer of a peptide chain from the A site to the P site is provided by
a. ATP.
b. enzymes.
c. GTP.
d. ADP.
e. guanosine triphosphate.

____ 5. Transcription involves
a. mRNA synthesis.
b. decoding of codons.
c. copying DNA information.
d. copying codes into codons.
e. peptide bonding.

____ 6. Ribosomes attach to
a. 5' end of mRNA.
b. 3' end of mRNA.
c. an mRNA recognition sequence.
d. tRNA.
e. RNA polymerase.

____ 7. Translation involves
a. hnRNA.
b. decoding of codons.
c. adapter molecules.
d. copying codes into codons.
e. copying codons into codes.

____ 8. Which of the following is/are termination codons?
a. UAG
b. UUA
c. UAA
d. GUA
e. UGA

____9. A recognition sequence for ribosome binding to mRNA is/are
a. aminoacyl-tRNA.
b. coding sequence.
c. initiator.
d. upstream from the coding sequence.
e. leader sequence.

____10.An "adapter" molecule containing an anticodon and a region to which an amino acid is bonded is a/an
a. aminoacyl-tRNA.
b. coding sequence.
c. initiator.
d. peptidyl transferase.
e. leader sequence.

____11. Beadle and Tatum concluded that
a. one gene affects one polypeptide.
b. one gene affects one enzyme.
c. a mutation directly affects an enzyme.
d. mutations have no effect on enzymes.
e. proteins mutate when exposed to radiations.

____12. *Neurospora* was a good choice for the Beadle and Tatum studies because
a. it is diploid.
b. its asexual spores enable genetic analysis.
c. it can be cultivated on minimal medium.
d. it had no known mutant forms.
e. the control strain could not use arginine.

____13. Post-transcriptional modification and processing of mRNA takes place in
a. cytoplasm.
b. the nucleus.
c. prokaryotes only.
d. eukaryotes only.
e. both prokaryotes and eukaryotes.

____14. Elongation of protein synthesis involves
a. the code AUG.
b. initiatior tRNA.
c. loading initiatior tRNA on the large ribosomal subunit.
d. adding amino acids to a growing polypeptide chain.
e. peptidyl transferase.

____15. RNA
a. is used by cells to indirectly read information in DNA.
b. can have double stranded regions.
c. contains ribose with a 2" hydroxyl group.
d. is a polymer.
e. is translated from DNA.

____16. 5'-A-T-G-C--3'
a. In this sequence C is upstream of G.
b. RNA synthesis would begin at the 5' end.
c. Transcribes to the sequence T-A-C-G.
d. G is downstream of T.
e. This is a base sequence of DNA.

____17. A stop codon for protein synthesis is.
- a. UUU.
- b. UGA.
- c. UGG.
- d. UUA.
- e. UUG.

____18. Eukaryotic RNA is posttransciptionally modified.
- a. immediately after leaving the nucleus.
- b. leading to increased stability.
- c. by polyadenylation on the 5' end.
- d. by enzymes that add a 5' cap.
- e. in the same way as bacterial RNA.

VISUAL FOUNDATIONS

Color the parts of the illustration below as indicated. Also label the 5' end of mRNA, the start codon, the leader sequence and circle the initiation complex.

RED	❑	formylated methionine tRNA
GREEN	❑	small ribosome subunit
YELLOW	❑	large ribosome subunit
PINK	❑	initiation factors
BLUE	❑	E site
ORANGE	❑	P site
VIOLET	❑	A site

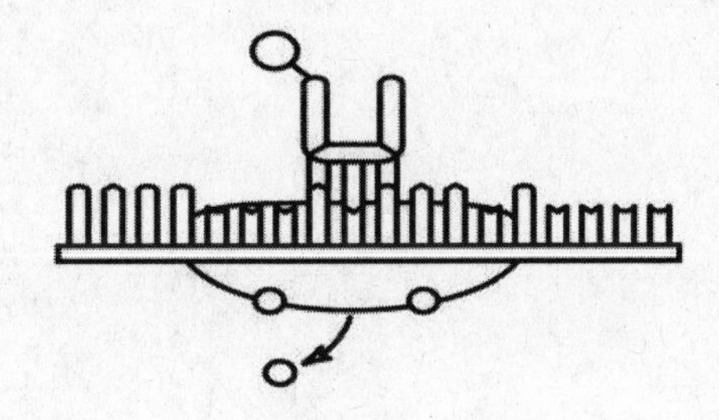

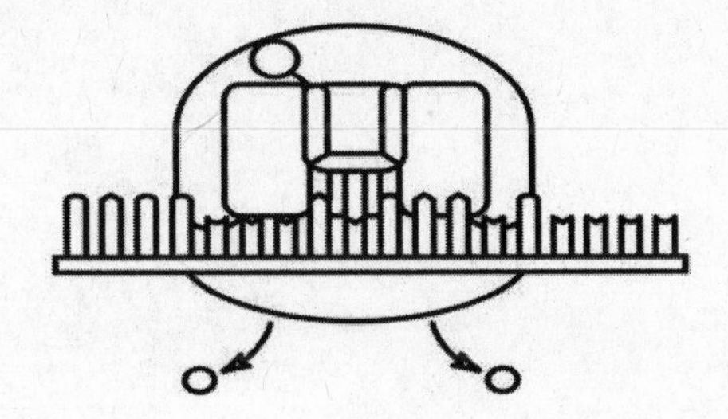

CHAPTER 14

Gene Regulation

This chapter looks at the mechanisms that control gene expression, since all of the genes of a multicellular organism are not expressed all at the same time. Each type of cell in a multicellular organism has a characteristic shape, carries out very specific activities, and makes a distinct set of proteins because only part of the genetic information in a given cell is ever expressed at any given time. The control mechanisms that allow this require information in the form of signals originating either within the cell, other cells, or the environment and that interact with DNA, RNA, or protein. The mechanisms include controlling the amount of mRNA available, the rate of translation of mRNA, and the activity of the protein product. In prokaryotes, most gene regulation involves transcriptional level control of genes, commonly using operons, whose products are involved in resource utilization. In eukaryotes, gene regulation occurs at many levels, including chromatin structure, transcriptional, posttranscriptional, translational, and posttranslational.

REVIEWING CONCEPTS

Fill in the blanks.

INTRODUCTION

1. Mechanisms that regulate gene expression include:
   (a)______________________________,
   (b)______________________________, and
   (c)______________________________.
2. The rate of transcription and the rate of mRNA degradation both control the amount of ________________.

GENE REGULATION IN BACTERIA AND EUKARYOTES: AN OVERVIEW

3. ______________________________ is the most efficient mechanism of gene regulation in bacteria.
4. Multicellular organisms focus on ______________________________ of genes.

GENE REGULATION IN BACTERIA

5. ______________________________ are genes that encode essential proteins that are in constant use.
6. Cell metabolic activity is controlled by regulating
   (a)______________________ and (b)________________________.

Operons in bacteria facilitate the coordinated control of functionally related genes

7. A gene complex consisting of a group of structural genes with related functions and the DNA sequences responsible for controlling them is known as a/an ______________________________. .
8. A group of functionally related genes may be controlled by one ______________________________ that is located upstream from the coding sequences.
9. The ____________ is a sequence of bases that switches mRNA synthesis "on" or "off."
10. A/an ____________________ inactivates a repressor in order to turn "on" a gene or operon.
11. Repressible genes and operons are usually turned (b)________.
12. Repressible operons are turned "off" when a repressor is activated by binding to a __________________.
13. Negative control involves the binding of (a)______________while activators are important for (b)______________control.
14. Constitutive genes are transcribed at different rates based on the strength of their ______________.

Some posttranscriptional regulation occurs in bacteria

15. Protein levels can be controlled by transcriptional level control. Other regulatory mechanisms that occur *after* transcription are referred to as ______________________________.
16. ______________________________ is when a product blocks its own production by binding to an enzyme that is required to generate that product.

GENE REGULATION IN EUKARYOTIC CELLS

17. Eukaryotic genes are typically not arranged in ______________.

Eukaryotic transcription is controlled at many sites and by many regulatory molecules

18. Highly coiled and compacted chromatin containing inactive genes is called (a)__________________________, whereas loosely coiled chromatin containing genes capable of transcription is referred to as (b)__________________________. Changing chromatin type is often done by modifying (c)____________.
19. DNA (a)________________is a way to (b)______________ genes. In mammals it accounts for (c)________________, which is an example of (d)____________.
20. Genes whose products are essential for cell function can be arranged as multiple copies in (a)____. Other genes are replicated in only certain cells by (b)__________________________.

21. RNA polymerase in multicellular eukaryotes binds to a portion of the promoter known as the ______________.
22. (a)______________are DNA sequences that increase the rate of transcription, while (b)________________decrease it.
23. DNA binding proteins that regulate transcription are called ______________________.
24. Many regulatory proteins have more than one ______________, a region with its own tertiary structure and function.

Long non-coding RNAs (IncRNAs) regulate transcription over long distances within the genome

25. _________ is an IncRNA responsible for inactivation of X chromosomes in females.

The mRNAs of eukaryotes are subject to many types of posttranscriptional control

26. In eukaryotes, ____________________is a useful control point.
27. Through ______________________________, the cells in each tissue produce their own version of mRNA corresponding to the particular gene.
28. The stability of _________can also affect how much protein is produced.
29. Expression of mRNA is blocked by small RNAs called __________.

Posttranslational chemical modifications may alter the activity of eukaryotic proteins

30. The addition or removal of phosphate groups is an example of ____________________, a mechanism used by eukaryotic cells to regulate protein activity.
31. Enzymes that modify chemicals by adding phosphate groups are called _______________.
32. Enzymes that modify chemicals by removing phosphate groups are called _______________.
27. Protein degradation is controlled by adding (a)______________to proteins targeting them for destruction by the (b)_______________.

BUILDING WORDS

Use combinations of prefixes and suffixes to build words for the definitions that follow.

Prefixes	The Meaning	Suffixes	The Meaning
co-	with, together, in association	-dimer	a part of two
eu-	good, well, "true"	-on	particle
hetero-	different, other	-repressor	state of keeping back
hist-	tissue		
homo-	same		
intra-	within		
oper-	work		

Prefix	Suffix	Definition
____________	-chromatin	1. Chromatin that appears loosely coiled. Because it is the only chromatin capable of transcription, it is considered to be the "good" or "true" chromatin.
____________	-chromatin	2. The inactive chromatin that appears highly coiled and compacted, and is generally not capable of transcription. (Chromatin that is "different from" or "other than" the euchromatin.)
____________	____________	3. A substance that, together with a repressor, represses protein synthesis in a specific gene.
____________	____________	4. A dimer in which the two component polypeptides are different.
____________	____________	5. A dimer in which the two component polypeptides are the same.
____________	____________	6. A segment of DNA controlled by one operator and one repressor.
____________	____________	7. A small nuclear protein that binds to DNA and plays a role in the formation of chromatin.
____________	____________	8. A segment of DNA that does not code for a protein.

MATCHING

Terms:

a.	Allosteric binding site	f.	Gene amplification	k.	Repressible system
b.	Chaperones	g.	Inducible system	l.	Repressor
c.	Constitutive gene	h.	Negative control	m.	Transcriptional control
d.	Corepressor	i.	Operator	n.	Translational control
e.	Enhancer	j.	Operon	o.	Silencer

For each of these definitions, select the correct matching term from the list above.

____ 1. One of the control regions of an operon.

____ 2. A site located on an enzyme that enables a substance other than the normal substrate to bind to the molecule, and to change the shape of the molecule and the activity of the enzyme.

____ 3. A regulatory protein that represses expression of a specific gene.

____ 4. A regulatory mechanism that controls the rate at which a particular mRNA molecule is translated.

____ 5. Process by which multiple copies of a gene are produced by selective replication, thus allowing for increased synthesis of the gene product.

____ 6. Genes that are constantly transcribed.

____ 7. In prokaryotes, a group of structural genes that is coordinately controlled and transcribed as a single message, plus their adjacent regulatory elements.

____ 8. Positive regulatory elements that can be located long distances away from the actual coding regions of a gene.

____ 9. Control in which the presence of a substrate induces the synthesis of an enzyme.

____ 10. Negative regulatory elements that can be located long distances away from the actual coding regions of a gene.

____ 11. Binds to an allosteric site on a repressor.

____ 12. Molecules that help newly synthesized molecules fold into their proper shape.

MAKING COMPARISONS

Fill in the blanks.

Regulatory Event	Name
Protein dimers held together by leucine	Leucine zipper proteins
#1	Temporal regulation
Assist newly synthesized proteins to fold properly	#2
#3	Genomic imprinting
Degradation of a protein	#4
#5	Feedback inhibition
Help form an active transcription initiation complex thereby increasing the rate of RNA synthesis	#6
#7	Translational controls

MAKING CHOICES

Place your answer(s) in the space provided. Some questions may have more than one correct answer.

____ 1. Genes that code for proteins that are constantly needed for survival of an organism are called
a. promoters.
b. constitutive genes.
c. operons.
d. repressor genes.
e. always "on".

____ 2. A repressor usually controls an inducible gene by
a. keeping it "turned off".
b. suppressing mRNA production.
c. reducing product resulting from an active inducer.
d. inducing an antagonist.
e. blocking DNA replication at specific sites.

____ 3. The basic way(s) that cells control their metabolic activities is/are by
a. regulating enzyme activity.
b. mutations.
c. developing different genes for different purposes.
d. controlling the number of enzyme molecules.
e. transduction.

____ 4. Feedback inhibition is an example of
a. transcriptional control.
b. pretranscriptional control.
c. a control mechanism affecting events after translation.
d. an inducible system.
e. a repressible system.

____ 5. The lactose repressor
a. can convert to an operator.
b. is several bases upstream from the operator.
c. is downstream from RNA polymerase coding sequences.
d. becomes an activator when lactose is present.
e. is continuously "on".

____ 6. The genes and genetic information in different cells of multicellular organisms are
a. all slightly different.
b. distinctly different.
c. identical in the same tissues, different in different tissues.
d. identical.
e. almost all identical.

____ 7. Replication of specific genes in cells that need more of them is an example of
a. magnification.
b. amplification.
c. induction.
d. a positive control system.
e. a regulon.

____ 8. Enhancers and silencers are similar because both
a. increase transcription rate.
b. are DNA sequences.
c. are downstream from promoter.
d. are located far from the promoter.
e. decrease transcription rate.

____ 9. The tryptophan operon in *E. coli* is
a. usually on.
b. on in the absence of corepressor.
c. repressed only when tryptophan binds to repressor.
d. an inducible system.
e. a repressible system.

____10. The proteasome
a. degrades proteins.
b. degrades mRNA.
c. seem to determine the strength of a repressor.
d. recognizes ubiquitin.
e. is a posttranslational control.

____11. The lactose operon in *E. coli*
a. contains three structural genes.
b. is turned on by a repressor.
c. is a repressor.
d. is transcribed as part of a single RNA molecule.
e. is an inducible system.

____12. "Zinc fingers" seem to be involved in
a. unwinding DNA.
b. activating transcription.
c. insertion of a regulator domain into grooves of DNA.
d. binding a regulatory protein to DNA.
e. blocking posttranscriptional events.

____13. A tightly coiled portion of DNA that contains active genes is
a. called heterochromatin.
b. called euchromatin.
c. found in most eukaryotes and few prokaryotes.
d. found only in prokaryotes.
e. found only in eukaryotes.

____14. Translational controls
a. are posttranscriptional controls.
b. use feedback mechanisms to control enzymes.
c. act as "on and off' switches.
d. regulate DNA translation.
e. occur in bacteria.

____15. Cells can change heterochromatin into euchromatin
a. by chemically modifying histones.
b. through acetylation.
c. by adding proteins to histones.
d. through methylation.
e. by adding sugar groups to histones.

____16. Epigenetic inheritance
a. can give rise to new phenotypic traits.
b. will change a gene's sequence.
c. have been linked to the regulation of suppressor genes.
d. have been linked to development of diseases.
e. includes genomic imprinting.

VISUAL FOUNDATIONS

Color the parts of the illustration below as indicated. Also label the lactose operon, ribosomes and identify the enzymes produced in the presence of the inducer.

RED	❑	operator region
GREEN	❑	promoter region
YELLOW	❑	repressor gene
BLUE	❑	mRNA
ORANGE	❑	inducer
BROWN	❑	inactivated repressor protein
VIOLET	❑	active repressor protein
TAN	❑	structural gene
PINK	❑	RNA polymerase

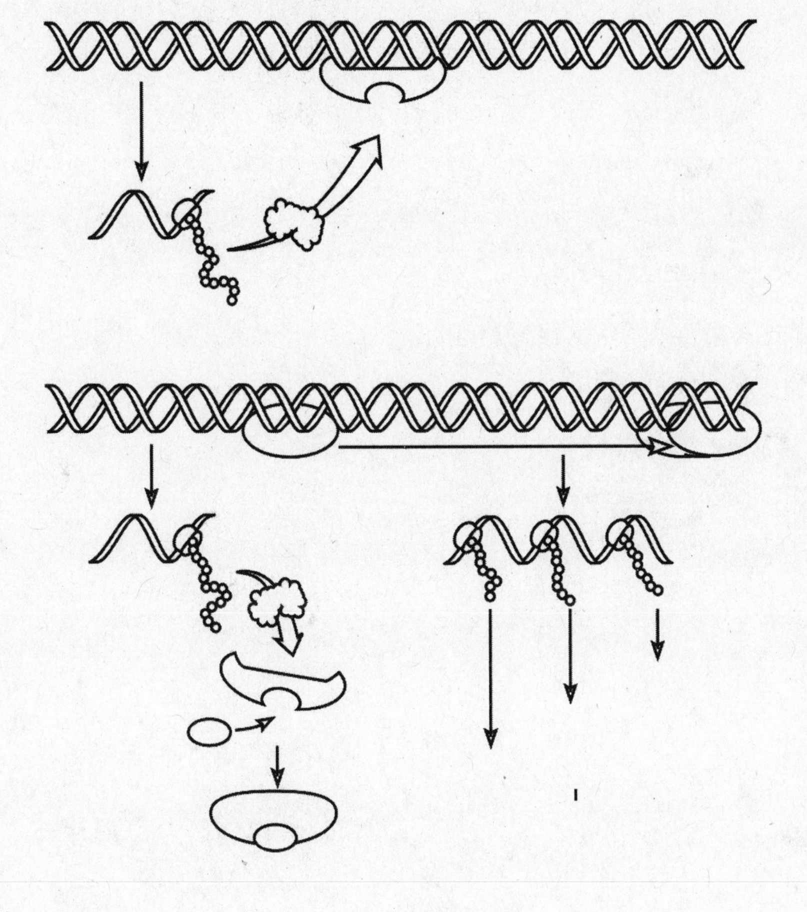

CHAPTER 15

DNA Technology and Genomics

Recombinant DNA technology involves splicing together DNA from different organisms. DNA cloning specifically allows a particular DNA sequence to be amplified to provide millions of identical copies that can be isolated in pure form, and then its functions can be studied and possibly engineered for a particular application. DNA can also be amplified in vitro using the polymerase chain reaction. DNA can be analyzed in many different ways allowing us to better understand the complex structure and control of eukaryotic genes and the roles of genes in development. The newer area of genomics also contributes to this understanding. These various techniques have led to genetic engineering — the modification of the DNA of an organism to produce new genes with new characteristics. Increasingly, DNA technology has important applications in many fields of study and business, including evolution, medicine, pharmacology, and food and forensic sciences. DNA technology has also raised important safety concerns that scientists are continually assessing.

REVIEWING CONCEPTS

Fill in the blanks.

INTRODUCTION

1. The molecular modification of an organism's DNA to produce new genes with new traits is called ________________.
2. ________________ is the commercial or industrial use of cells or organisms.

DNA CLONING

3. ________________________ cut DNA molecules at specific places, producing manageable segments that can be incorporated into vector molecules.
4. Recombinant DNA vectors are usually constructed from ________________________.
5. Researchers introduce plasmids into bacterial cells in a process called ________________.

Restriction enzymes are "molecular scissors"

6. A ________________ sequence reads the same as its complement when both are read in the 5' to 3' direction.
7. Restriction enzymes cut in a staggered fashion creating (a) __________ which can be joined to complementary sequences by (b) ___________.

Recombinant DNA forms when DNA is spliced into a vector

8. Three important features of a vector are
(a)________________________________,
(b)________________________________, and
(c)________________________________.

DNA can be cloned inside cells

9. A ________________________ is a collection of thousands of DNA fragments that represent the entire DNA in the genome.
10. A ______________ is a homologous segment to the DNA of interest and are often radioactively labeled. They are used to identify a particular fragment in a large population of clones.
11. Probes can ____________ to the target DNA.

A cDNA library is complementary to mRNA and does not contain introns

12. The enzyme ____________________ is used to synthesize single-stranded complementary DNA (cDNA) on an mRNA template.

The polymerase chain reaction amplifies DNA in vitro

13. The polymerase chain reaction (PCR) technique for amplifying DNA is advantageous because it side steps the cumbersome, time-consuming process of cloning DNA. PCR is an *in vitro* process that alternately uses (a) ____________________________ to replicate DNA with the use of (b) ______________ to dissociate the replicated DNA strands for further replications.
14. The PCR technique must use a heat stable DNA polymerase named _______ polymerase after the organism from which it is obtained.

TOOLS FOR STUDYING DNA

Gel electrophoresis is used for separating macromolecules

15. In gel electrophoresis, nucleic acids migrate through the gel toward the (a) ______________________ of the electric field because they are negatively charged due to their (b) ________________________. They will be separated by (c) ______.

DNA, RNA, and protein blots detect differences in related molecules separated by gel electrophoresis

16. The method of detecting DNA fragments by separating them by gel electrophoresis and then transferring them to a nitrocellulose or nylon membrane is called a (a) ____________________. When RNA is transferred it is a (b) __________________ and when proteins or polypeptides are transferred it is a (c) ________________.

Automated DNA sequencing methods have been developed

17. Automated DNA sequencing relies on the ____________________________________.

Gene databases are powerful research tools

18. The genomes of over (a) ________ (#) organisms have been sequenced. Geneticists use (b) ________________ to compare to compare newly discovered DNA sequences.

Reverse transcription of mRNA to cDNA is used to measure gene expression in a number of ways

19. For RT-PCR analysis, total cellular (a) ________________ are first copied to (b) ________ molecules using the enzyme (c)________________________.
20. Researchers isolate (a) ________ which they then use to prepare (b)________ that will be tagged with a dye. The dye tagged molecules are loaded onto the array. An analysis of the tagged molecules bound to the array give researchers an indication of gene activity.

GENOMICS

Collaborative genome-wide association studies have radically changed our view of the human genome

21. The ENCODE project showed that up to (a) ____% of the genome contains elements linked to biochemical functions. Most of these encode (b) ________________________ that appear to be involved with the regulation of expression of protein-coding genes.
22. The ENCODE study also showed that many disease-linked mutations were associated with ________________________ ________________________ of DNA.

Comparative genomic databases are tools for uncovering gene functions

23. ________________________ involves sequencing and mapping studies on other organisms to study the human genome.
24. Rats and mice are sufficiently different from humans so that any conserved DNA sequences found in both the rodents and humans would most likely indicate that the sequences are ________________________.
25. Comparing DNA sequences of humans and chimpanzees has helped biologists gain a better understanding of human evolution including which genes govern (a) ________________ and (b)________________________ capabilities.

RNA interference is used to study gene functions

26. Short RNA molecules called (a) ________can be used to determine the function of a gene because these molecules are capable of (b)________________ the target gene.
27. RNAi silencing is used in ________________________ studies to determine gene functions.
28. Gene function can also be revealed in a procedure called ________________________, in which the researcher chooses and "knocks out" a single gene in an organism.

29. Scientists place knockout genes into mouse embryonic stem cells with the intention that the knockout allele will be exchanged for the normal allele on the mouse chromosome. This exchange occurs in a process called ______________________________.

APPLICATIONS OF DNA TECHNOLOGIES

DNA technology has revolutionized medicine

30. ______________________________, the use of specific DNA to treat a genetic disease by correcting the genetic problem, is another application of DNA technology.
31. Genetic engineering allows the production of (a) __________to treat diseases and disorders and (b) __________to help prevent infectious diseases.

DNA fingerprinting has numerous applications

32. STRs are highly ______________________________ because they vary in length from one individual to another, making them useful in identifying individuals.

Transgenic organisms have foreign DNA incorporated into their cells

33. Transgenic animals are generally produced by injecting the DNA of a gene into the nucleus of a (a) ______________________________ or (b)____________________ cells
34. Genetically modified stem cells are injected into isolated ____________________ which are then implanted into the uterus of a foster mother where embryo development will occur.
35. Transgenic animals secreting foreign proteins in their milk is called ________________.
36. Scientists have been using transgenic plants to improve plant survival and agricultural yields. They are using the ________ plasmid as a vector to introduce genes into plants.
37. Genes that have been introduced into plants using a plasmid are passed onto the next generation either through the plants (a)_______________ during (b) _______________ reproductions or (c)__________________ in plant cuttings.
38. Transgenic plants that are resistant to insect pests, viral diseases, drought, heat, cold, herbicides, and salty or acidic soil are known as ______________________________.

DNA TECHNOLOGY HAS RAISED SAFETY CONCERNS

39. The science of ______________________________ uses statistical methods to quantify risks so they can be compared and contrasted.

BUILDING WORDS

Use combinations of prefixes and suffixes to build words for the definitions that follow.

Prefixes	The Meaning
bio-	life
bacterio-	bacteria
gen-	origin
kilo-	thousand
plasm-	formed
retro-	backward
trans-	across

Suffixes	The Meaning
-id	tending to
-ome	mass
-phage	eat, devour

Prefix	Suffix	Definition
________	-genic	1. Pertains to organisms that have had their genome altered by recombinant DNA technology. (Genes are taken from one organism and transferred "across" into another organism.)
________	-virus	2. An RNA virus that makes DNA copies of itself by reverse transcription.
________	________	3. A virus that infects and destroys bacteria.
________	-base	4. A unit of measurement for DNA.
________	-technology	5. The use of cells or organisms commercially.
________	________	6. Small circular DNA molecule found in bacteria.
________	________	7. The total DNA of a cell.

MATCHING

Terms:

a.	Bioinformatics	f.	Palindrome	k.	Reverse transcriptase
b.	Colony	g.	Plasmid	l.	Transformation
c.	DNA ligase	h.	PCR	m.	Vector
d.	Genetic probe	i.	Recombinant DNA	n.	Pharmacogenetics
e.	Hybridization	j.	Restriction enzyme	o.	Metagenomics

For each of these definitions, select the correct matching term from the list above.

____ 1. Enzyme produced by retroviruses to enable the transcription of DNA from the viral RNA in the host cell.

____ 2. Small, circular DNA molecules that carry genes separate from the main bacterial chromosome.

____ 3. A cluster of genetically-identical cells.

____ 4. Bacterial enzymes used to cut DNA molecules at specific base sequences.

____ 5. Method used to produce large amounts of DNA from tiny amounts.

____ 6. An agent, such as a plasmid or virus that transfers genetic information.

____ 7. Any DNA molecule made by combining genes from different organisms.

____ 8. A radioactively-labeled segment of RNA or single-stranded DNA that is complementary to a target gene.

____ 9. Study of microbial community gene expression.

____ 10. An enzyme used to put a DNA fragment into a BAC.

____ 11. Comparing nucleotide sequences.

____ 12. When the base sequence of one DNA strand reads the same as its complement when both strands are read in the 5' to 3' direction.

____ 13. The process in which plasmids are introduced into bacterial cells.

____ 14. The paring of the target DNA and the complementary base pair probe.

MAKING COMPARISONS

Fill in the blanks.

Technique	Definition/Description
genetic engineering	Modifying the DNA of an organism to produce new genes with new traits
#1	Microbial proteins used to cut DNA in specific locations
probe	#2
#3	Method for replicating and isolating many copies of a specific recombinant DNA molecule.
PCR	#4
#5	Recombinant DNA vectors derived from bacterial DNA
genomic DNA library	#6

MAKING CHOICES

Place your answer(s) in the space provided. Some questions may have more than one correct answer.

____ 1. Genes in eukaryotic organisms that do not code for proteins are
a. introns.
b. exons.
c. short tandem repeats (STRs).
d. highly polymorphic.
e. frequently used to diagnose genetic diseases.

____ 2. Copies of DNA made from isolated mRNA by using reverse transcriptase are called
a. genomic DNA.
b. recombinant DNA.
c. PCRs.
d. complementary DNA (cDNA).
e. probes.

____ 3. Analysis of an individual's DNA based on his or her STRs is known as
a. proteomics.
b. DNA blotting analysis.
c. DNA fingerprinting.
d. a cDNA library.
e. a genomic DNA library.

____ 4. Detection of protein fragments transferred to a nitrocellulose or nylon membrane after separating them by gel electrophoresis is known as a(n)
a. Northern blot.
b. Southern blot.
c. Eastern blot.
d. Western blot.
e. fingerprinting blot.

____ 5. The vector(s) commonly used to incorporate a DNA fragment into a carrier is/are
a. prophages.
b. *E. coli.*
c. plasmids.
d. translocation microphages.
e. bacteriophages.

____ 6. "Sticky ends" of a DNA molecule are generated by
a. plasmozymes.
b. restriction enzymes.
c. DNA polymerase.
d. DNA ligase.
e. vector enzymes.

____ 7. Restriction enzymes are normally used by bacteria to
a. defend against viruses.
b. remove extraneous introns.
c. denature antibiotics.
d. cut the large circular chromosome.
e. insert operators into DNA.

____ 8. Palindromic sequences are base sequences in one DNA strand that
a. produce RNA polymerase.
b. form RNA complementary to DNA.
c. can be cut by restriction enzymes.
d. cut specific fragments from DNA.
e. run parallel to its complement strand.

____ 9. cDNA libraries are compiled
a. from introns.
b. DNA copies of mRNA.
c. from multiple copies of DNA fragments.
d. from introns and exons.
e. with the use of reverse transcriptase.

____10. A DNA base sequence that is palindromic to 3'-GCATATGC-5' would read
a. 3'-AGTATACG-5'.
b. 5'-GCATATGC-3'.
c. 3'-GCATATACG-5'.
d. 5'-CGUAUACGC-3'.
e. 3'-CGUAUACG-5'.

____11. The Ti used to insert genes into plant cells is
a. an RNA fragment.
b. not effective in dicots.
c. useful to increase yields of grain foods.
d. a plasmid.
e. a vector.

____12.Which terms are not correctly paired?
a. Northern blot: nucleic acid probe.
b. Western blot: antibody probes.
c. Southern blot: protein separation.
d. Southern blot: RFLP.
e. Northern blot: RNA separation.

____13. DNA fragments for recombination are produced by cutting DNA
a. and introducing a cleaved plasmid.
b. with restriction enzymes.
c. at specific base sequences.
d. with bacteriophages.
e. with DNA ligase.

____14. Transgenic organisms are
a. dead animals.
b. vectors for retroviruses.
c. used to produce recombinant proteins.
d. animals with foreign genes.
e. plants with foreign genes.

_____15. A radioactive strand of DNA that is used to locate a cloned DNA fragment is
a. a probe.
b. the genome.
c. complementary to the targeted gene.
d. a restriction enzyme.
e. complementary to the ligase used.

_____16. RNA viruses that make DNA copies of themselves by reverse transcription are
a. plasmids.
b. retroviruses.
c. double stranded.
d. used as genetic probes.
e. bacteriophages.

_____17. One limitation of PCR is that it
a. requires thermally stable proteins for polymerization.
b. .does not work with bacterial DNA.
c. uses cloned DNA as the primer.
d. amplifies all DNA in the sample.
e. requires a large sample size.

_____18. Microarrays
a. use DNA hybridization techniques.
b. can be used to develop specific drugs.
c. may use fluorescent labeled RNA to detect the probe.
d. are used to study gene expression.
e. are used to identify disease causing genes.

_____19. DNA fingerprinting
a. uses PCR amplification.
b. uses Southern blot hybridization .
c. looks for short tandem repeats in DNA sequences.
d. uses restriction enzymes.
e. uses Western blot hybridization.

VISUAL FOUNDATIONS

Color the parts of the illustration below as indicated. Also label DNA, pre-mRNA, mature mRNA, transcription, RNA processing, and double-stranded cDNA.

- RED ❑ exon
- GREEN ❑ intron
- YELLOW ❑ reverse transcriptase
- BLUE ❑ cDNA copy of mRNA
- VIOLET ❑ mRNA
- PINK ❑ DNA polymerase

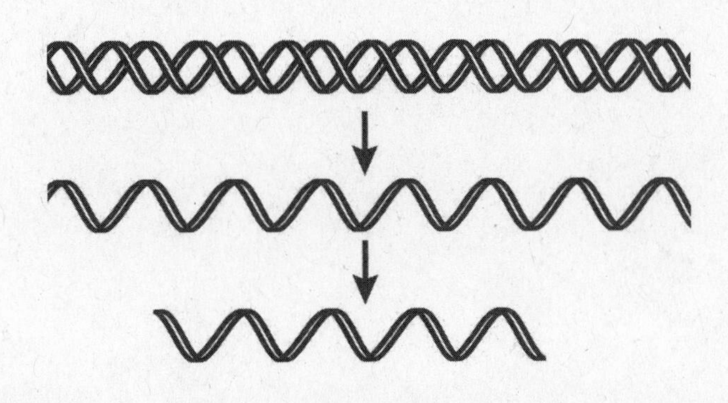

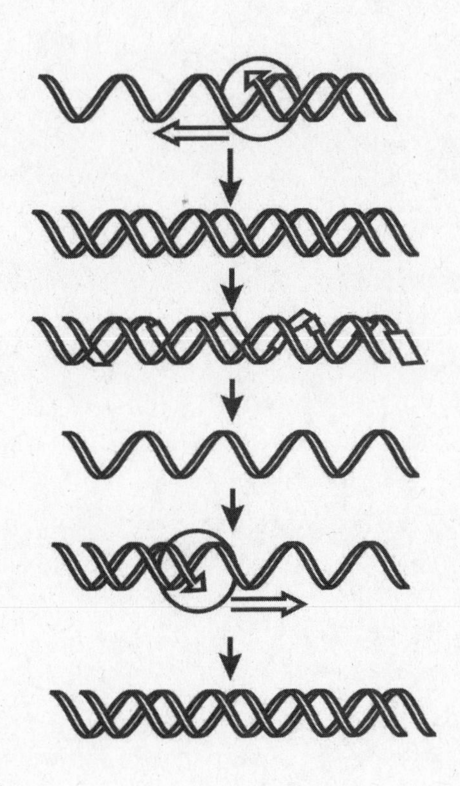

CHAPTER 16

Human Genetics and the Human Genome

The principles of genetics apply to all organisms, including humans. Human geneticists, however, cannot make specific crosses of pure human strains as geneticists do with other organisms. Instead, they must rely on studies of other organisms and population studies of large extended families. Also, knowledge of human genetics has been greatly advanced in recent years by the mapping and sequencing of the human genome and by the study of genetic diseases in humans, which can arise from abnormalities in chromosome number or structure or by single gene mutations. Gene therapy is a promising new treatment for genetic disease, but there are concerns and technical issues with the approach. New tools have been developed for genetic testing, screening, counseling, and therapy. Increasingly, work in the field of human genetics raises many ethical issues.

REVIEWING CONCEPTS

Fill in the blanks.

INTRODUCTION

1. Human genetics studies ______________________ in humans.

STUDYING HUMAN GENETICS

Human chromosomes are studied by karyotyping

2. A ________________ is an individual's chromosome composition.
3. Geneticists often use (a)_________ blood cells for genetic studies because they can be (b)____________________.
4. (a)_________is another methods to analyze a karyotype using (b)______________________to "paint" the chromosomes.

Family pedigrees help identify certain inherited conditions

5. A ________________________________ shows the transmission of genetic traits within a family over several generations and helps determine the exact interrelationships of the DNA molecules from related individuals.

Human gene databases allow geneticists to map the locations of genes on chromosomes

6. Pedigrees can provide information about inheritance of a gene, but it cannot provide information on the (a)____________ of that gene or how that form of the gene (b)_________________.
7. Single nucleotide polymorphisms (SNPs) are (a)______________________ differences in DNA sequences. SNPs can be used to help locate the (b)__________ on DNA where the gene of interest occurs.

8. In humans only about ____% of the genome specifies amino acid sequences of polypeptides.
9. Genome studies have found that almost 90% of SNP variants associated with disease-linked regions are found outside of ________________ genes.
10. ____________________compare genomes of a healthy individual with that of a diseased individual to look for gene variation.
11. About ____________ (#?) DNA segments longer than 200 base pairs are known to be 100% identical between the mouse and human genomes.
12. Researchers have used ____________________________ to produce strains of mice that are homozygous or heterozygous to genetic diseases.

ABNORMALITIES IN CHROMOSOME NUMBER AND STRUCTURE

13. ________________, the presence of multiple sets of chromosomes, is usually lethal in humans, but quite common in plants.
14. The abnormal presence or absence of one chromosome in a set is called (a) ________________. For the affected chromosome, the normal condition, called disomy, becomes a (b) __________ condition with the addition of an extra chromosome, and a (c) ____________ condition when one member of a pair is missing. These abnormalities generally occur due to (d) ____________.

Down syndrome is usually caused by trisomy 21

15. Most people with Down syndrome have (a)_______(#) chromosomes because of autosomal (b)__________ that occurs as a result of (c)______________ during meiosis.
16. Down syndrome incidence increases with (a) ________________ but no explanation for this has yet been supported. However, most Down syndrome babies are born to mothers (b) ____________.

Most sex chromosome aneuploidies are less severe than autosomal aneuploidies

17. Persons with ______________________________ are tall, sterile males, except for an extra X chromosome.
18. Persons with __________________________ are sterile females without Barr bodies.
19. The _____ karyotype produces tall, fertile males.

Abnormalities in chromosome structure cause certain disorders

20. The breakage and rejoining of chromosome parts result in structural changes within or between chromosomes. In a/an (a)____________________ error a segment of the chromosome is repeated one or more times. When the orientation of the chromosome segment is reversed a/an (b)______________ is said

to have occurred. The loss of part of a chromosome occurs as the result of a/an (c)______________________. In the fourth type of structural change, (d) __________________, a chromosome fragment breaks off and attaches to a nonhomologous chromosome.

21. Individuals suffering from translocation Down syndrome have one chromosome (a) _____ (#), one combined (b) __________ chromosome, and two normal copies of chromosome (c) _____ (#).
22. In cri du chat syndrome, part of the short arm of chromosome _____ (#) is deleted.
23. ________________ are weak points where part of a chromatid appears to be attached to the rest of the chromosome by a thin thread of DNA.
24. ______________________________ is the most common cause of inherited intellectual disability.

Genomic imprinting may determine whether inheritance is from the male or female parent

25. Genomic imprinting occurs when the expression of a gene in a given tissue or developmental stage is based on its ________________.
26. (a) ____________________________ refers to changes in how a gene is expressed without any change in the coding of the DNA bases. It may cause (b) ________________________________.

GENETIC DISEASES CAUSED BY SINGLE-GENE MUTATIONS

27. PKU and akaptonuria are examples of disorders involving enzyme defects, collectively known as ________________________________.

Many genetic diseases are inherited as autosomal recessive traits

28. Phenylketonuria results from an ___________ deficiency.
29. Sickle cell anemia results from a defect in ______________.
30. Cystic fibrosis results from defective ____________________.
31. In cystic fibrosis, a mutant protein causes the production of a very heavy __________ that eventually causes severe tissue damage.
32. Tay-Sachs disease is an (a)__________________ disease that affects the (b)________________. It results from abnormal (c)__________ metabolism in the (d)__________.

Some genetic diseases are inherited as autosomal dominant traits

33. Huntington's disease is caused by a rare (a)________________________________ that affects (b)__________________________________.
34. The gene that is responsible for Huntington's Disease is located on the short arm of chromosome _____(#).

Some genetic diseases are inherited as X-linked recessive traits

35. Hemophilia A is caused by the absence of a blood-clotting protein called ________________________.
36. The X chromosome carries a large number of genes that affect (a)______________________________, so (b)__________suffer from mental impairment more frequently.
37. The human X chromosome contains (a)______% of the human genome but (b)____% of the genes in which defects are known to cause some form of intellectual disability.

GENE THERAPY

38. Gene therapy is a strategy that aims to replace a (a)__________________ allele with a (b)__________________ allele to treat some serious genetic diseases.

Performing gene therapy on humans always has inherent risks

39. The main safety concerns with gene therapy is the toxicity of the ____________.

GENETIC TESTING AND COUNSELING

Prenatal diagnosis detects chromosome abnormalities and gene defects

40. (a)______________________ may be used to diagnose prenatal genetic diseases. It analyzes cells in amniotic fluid withdrawn from the uterus of a pregnant woman. The fluid will contain (b)______________________________.
41. One technique designed to detect prenatal genetic defects is ______________________________________, which requires removing and inspecting fetal cells involved in forming the placenta.
42. ______________________________________is available to test for a variety of inherited genetic conditions in individuals using IVF.

Genetic screening searches for genotypes or karyotypes

44. Newborn genetic screening is used primarily as the first step in ____________________________________.
45. In adults, screening is used to find ____________of recessive genetic disorders.

Genetic counselors educate people about genetic diseases

46. Genetic counselors use family histories of each partner to determine the __________________ that any given offspring will inherit a particular condition.

HUMAN GENETICS, SOCIETY, AND ETHICS

47. (a)_________individuals and (b)________ethnic groups carry (c)______________ mutant alleles.

48. Genetic disease rates are low because the chances of reproducing with some carrying the same harmful allele are very low, unless it is a ______________________ between genetically related individuals, which increases the risk.

Genetic discrimination provokes heated debate

49. (a)____________ discrimination is discrimination against an individual or family members because of differences from the (b)"____________" genome in that individual.
50. The GINA prohibits employers and insurance companies from (a)______________based on information from (b)______________.

Many ethical issues related to human genetics must be addressed

51. As human genetics becomes more important issues of ________________________________must be addressed.

BUILDING WORDS

Use combinations of prefixes and suffixes to build words for the definitions that follow.

Prefixes	The Meaning
iso-	equal, "same"
kary(o)	nut, nucleus
poly-	much, many
trans-	across, beyond, through
tri-	three
aneu-	without

Suffixes	The Meaning
-ploid(y)	multiple sets of
-typ(e)	form

Prefix	Suffix	Definition
__________	__________	1. A display of an organism's chromosomes according to number, size and shape.
__________	__________	2. The presence of multiples of complete chromosome sets.
__________	-somy	3. Condition in which a chromosome is present in triplicate instead of duplicate (i.e., the normal pair).
__________	-location	4. An abnormality in which a part of a chromosome breaks off and attaches to another chromosome.
__________	__________	5. The absence or presence of an extra single chromosome.

MATCHING

Terms:

a. Amniocentesis
b. Consanguineous mating
b. Disomic
d. Down Syndrome
e. Fragile X syndrome
f. Gene therapy
g. Huntington's Disease
h. Karyotype
i. Klinefelter Syndrome
j. Monosomic
k. Nondisjunction
l. Nuclear sexing
m. SNP
n. Turner syndrome

For each of these definitions, select the correct matching term from the list above.

____ 1. A technique used to detect birth defects by sampling the fluid surrounding the fetus.

____ 2. Abnormal separation of homologous chromosomes or sister chromatids caused by their failure to disjoin properly during cell division.

____ 3. Genetic disease that causes mental and physical deterioration, uncontrollable muscle spasms, personality changes, and ultimately death.

____ 4. Syndrome in which afflicted individuals have only 47 chromosomes.

____ 5. Techniques that involve introducing normal copies of a gene into some of the cells of the body of a person afflicted with a genetic disorder.

____ 6. Birth defect caused by an extra copy of human chromosome 21.

____ 7. General term for the abnormal situation in which one member of a pair of chromosomes is missing.

____ 8. Matings of close relatives.

____ 9. A single-letter difference in DNA sequences.

____10. Syndrome in which an individual is XO instead of XX or XY.

____ 8. This syndrome results in malformed dendrites.

MAKING COMPARISONS

Fill in the blanks.

Genetic Disease	Abnormality	Method of Transmission	Clinical Description
Cystic fibrosis	Membrane proteins that transport chloride ion malfunction	Autosomal recessive trait	Characterized by abnormal secretions in respiratory and digestive systems
Sickle cell anemia	#1	Autosomal recessive trait	#2
#3	Abnormal sex chromosome expression	#4	Small testes, sterile, abnormal breast development
Phenylketonuria (PKU)	#5	Autosomal recessive trait	Retardation caused by toxic phenylketones that damage the developing nervous system
#6	Lack of blood clotting Factor VIII	#7	Severe bleeding from even slight wounds
Tay-Sachs	#8	Autosomal recessive trait	#9
#10	A mutated nucleotide triplet (CAG) that is repeated many times	#11	Severe mental and physical deterioration, uncontrollable muscle spasms, personality changes, and ultimately death
Down syndrome	#12	Autosomal trisomy 21	#13

MAKING CHOICES

Place your answer(s) in the space provided. Some questions may have more than one correct answer.

____ 1. The inactive X-chromosome seen as a region of darkly stained chromatin is called a
a. translocated anomalie.
b. Barr body.
c. artifact.
d. chromatic body.
e. Klinefelter chromosome.

____ 2. An individual with Turner syndrome would likely have a sex chromosome composition of
a. XXX.
b. XXY.
c. XO.
d. YO.
e. XYY.

____ 3. An individual with an XXY karyotype is
a. male.
b. female
c. fertile.
d. sterile.
e. likely to be tall.

____ 4. IVF is a popular abbreviation for
a. intravenous filtration.
b. *in vitro* fertilization.
c. a branch of stem cell research.
d. a technique for fertilization in a laboratory dish.
e. insufficient placental development leading to miscarriage.

____ 5. A person with XYY karyotype will
a. have an enlarged head.
b. be female.
c. have 47 chromosomes.
d. be sterile.
e. have mild intellectual disability.

____ 6. Amniocentesis involves
a. examination of maternal blood.
b. insertion of a needle into the uterus.
c. examination of fetal cells.
d. examination of karyotypes.
e. appraisal of paternal abnormalities.

____ 7. Translocations may involve
a. loss of genes.
b. acquisition of extra genes.
c. duplication of genes.
d. loss of a chromosome.
e. acquisition of an extra chromosome.

____ 8. A trisomic individual
a. is missing one of a pair of chromosomes.
b. may result from nondisjunction.
c. has at least three X chromosomes.
d. is a male.
e. has an extra chromosome.

____ 9. A person with Down syndrome will likely
a. have a father with Down syndrome.
b. have trisomy 21.
c. be mentally retarded.
d. be born of a mother in her teens.
e. have 47 chromosomes.

____10. The most accurate statement about the frequency of abnormal genes is that they are found in
a. certain ethnic groups.
b. certain cultural groups.
c. individuals over 45
d. teenagers.
e. everyone.

____11. An ideal organism for genetic studies would be one that
a. is polygenic.
b. is heterozygous.
c. has a short generation time.
d. produces one offspring per mating.
e. is subjected to controlled breeding.

____12. The study of human genetics relies mainly on
a. pedigrees.
b. Punnett squares.
c. fetal karyotypes.
d. distribution of a trait in a population.
e. nondisjunction.

____13. The drug colchicine is used
a. to tag DNA prior to florescent hybridization.
b. arrests cells in anaphase.
c. to reveal the banding pattern on chromosomes.
d. in karyotyping.
e. arrests cells in telophase.

____14. A pedigree can reveal
a. who in a family carries a specific trait.
b. abnormalities in chromosome number.
c. .the likelihood of producing male offspring.
d. can reveal genetic information about deceased relatives.
e. X-linked carriers.

____ 15. Which term is incorrectly paired with its description?
a. GWAS: procedure comparing persons with a disease with those without the disease
b. FISH: procedure used to distinguish chromosomes
c. Dosage compensation: observed in mammalian cells compensating for an extra Y chromosome
d. Inversion: a chromosome with reversed orientation
e. cridu chat syndrome: a condition associated with chromosome 5 abnormalities

VISUAL FOUNDATIONS

Questions 1-3 show pedigrees of the type that are used to determine patterns of inheritance. Indicate carriers by placing a dot in the appropriate squares (⊡) and circles (⊙).

Use the following list to answer questions 1-3:

a. x-linked recessive
b. x-linked dominant
c. autosomal dominant
d. autosomal recessive
e. cannot determine the pattern of inheritance

1. What is the pattern of inheritance for the following pedigree?

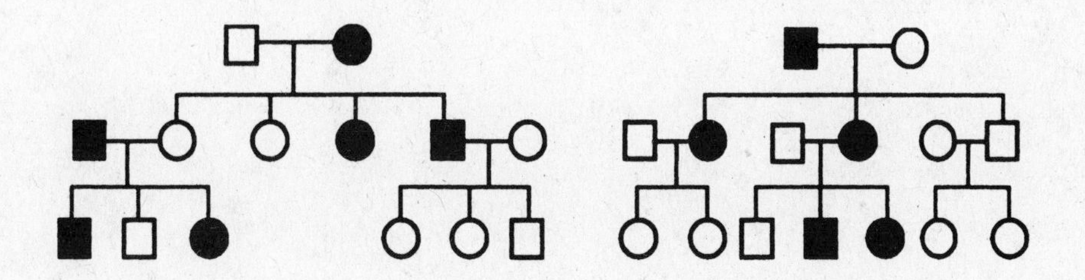

2. What is the pattern of inheritance for the following pedigree?

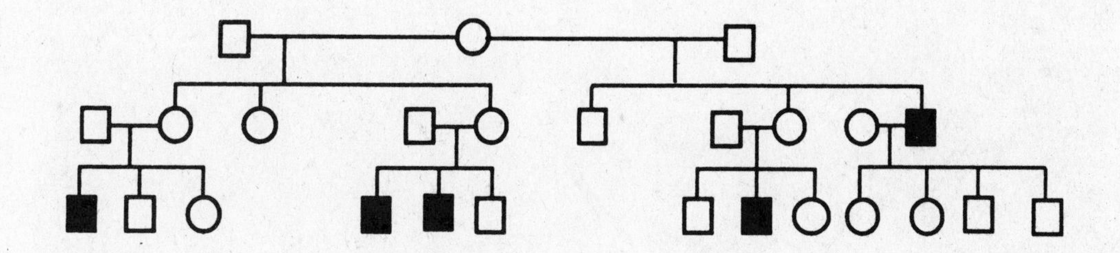

3. What is the pattern of inheritance for the following pedigree?

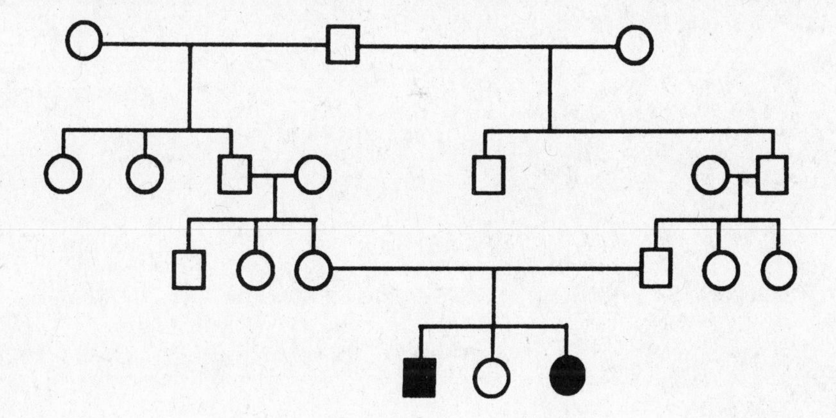

CHAPTER 17

❑

Developmental Genetics

Development encompasses all the changes that occur in the life of an individual organism. Of particular interest is the process by which cells specialize and organize into a complex organism. Groups of cells become gradually committed to specific patterns of gene activity and progressively organized into recognizable structures. Cellular specialization is not due to the loss of genes during development, but rather to differential gene activity. Stem cells are important in this differentiation process. A variety of organisms and methodologies are used today to identify genes that control development and to determine how those genes work. Many genes important in development are quite similar in a wide range of organisms, including humans. Cancer involves altered gene expression of specific genes which leads to uncontrolled cell division.

REVIEWING CONCEPTS

Fill in the blanks.

INTRODUCTION

1. Developmental genetics is the study of the genes involved in (a)______________________________ and the (b)____.
2. The technique to localize proteins in which a fluorescent dye is joined to an antibody that binds to a specific protein is called ________________________.

CELL DIFFERENTIATION AND NUCLEAR EQUIVALENCE

3. Development involves cell specialization, a developmental process involving a gradual commitment by each cell to a specific pattern of gene activity. First, as a result of cell (a)__________________, the fate of embryonic cells becomes limited. This is followed by an event leading to the final step in the process of cell specialization which is called cell (b)______________________.
4. The distinctive organizational pattern, or form, that characterizes an organism is acquired through a process known as (a)__________________________. It proceeds through a series of steps known as (b)__________________________, by which groups of cells organize into identifiable structures.
5. A body of data supports the assertion that, in at least some organisms, almost all nuclei of differentiated cells are identical to each other and to the nucleus of the single cell from which they are

descended, which has given rise to the concept of
_______________________________.

Most cell differences are due to differential gene expression

6. Much of the regulation that is important in development occurs at the ___________________________ level.

A totipotent nucleus contains all the instructions for development

7. Totipotent cells are fully ______________________, yet, since they can still develop, they apparently still contain all of the original genetic equipment of their common single ancestral cell.
8. Genetically identical organisms, cells, or DNA molecules are called _________.

The first cloned mammal was a sheep

9. The scientists who cloned Dolly were ultimately successful because they recognized their fused cells need synchronous _____________.
10. The main focus of cloning research is the production of ____, in which foreign genes have been incorporated.

Stem cells divide and give rise to differentiated cells

11. Embryonic and adult stem cells are known as ___________________________ stem cells because they can give rise to many, but not all, of the types of cells in an organism.
12. Embryonic stem cells are more versatile than adult stem cells. However they are not totipotent because they cannot form cells of the _______________.
13. Induced pluripotent stem cells have been produced by adding _______________________________to cells such as fibroblasts and skin cells.
14. (a)________________________cloning produces a newborn human while (b)__________________________cloning only produces human ES or iPS cells.

THE GENETIC CONTROL OF DEVELOPMENT

A variety of model organisms provide insights into basic biological processes

15. A model organism is a species chosen for biological studies because it has____
____
____
__.
16. A powerful research tool is to isolate _____________of model organisms.

Many genes that control development have been identified in the fruit fly

17. The *Drosophila* life cycle includes (a)__________, (b)_____________, and (c)______________, stages and the (d)____________ stage generated by (e)______________________.

18. The initial stages of development in *Drosophila* are under the control of ________________ genes.
19. Once maternal genes have initiated pattern formation in the embryonic fruit fly, additional genes begin to act on development. Among these are the (a)________________ that begin to organize body segments and the (b)________________________ that specify the developmental plan for each segment.
20. A ________________________ is a chemical agent that affects the differentiation of cell and the development of form.
21. The (a)________________ is a short DNA sequence that is consistently found in association with known developmental genes and codes for a protein functional region called a/an (b)________________.
22. __________ genes are clusters of homeobox-containing genes that specify the anterior-posterior axis during development.

Caenorhabditis elegans has a relatively rigid developmental pattern

23. These roundworms are easy to cross and study because individuals are males or (a)________________________. This sexual makes it easy to perform (b)________________________.
24. The fates of cells in adult roundworms are predetermined by a few ____ that form in the early embryo.
25. Development in *Caenorhabditis elegans* is rigidly fixed in a predetermined pattern; that is, specific cells throughout the early embryo are already irrevocably "programmed" to develop in a particular way. This pattern of development is said to be ________________.
26. Sometimes, cell differentiation is influenced by neighboring cells in a process called ________________.
27. (a)________________, genetically programmed cell death, depends on proteolytic enzymes called (b)______________.

The mouse is a model for mammalian development

28. Early mice embryos can be fused to produce a ______________, a term for offspring that exhibit a diversity of characteristics derived from two or more kinds of genetically dissimilar cells from different zygotes.
29. A self-regulating embryo exhibits a form of development referred to as ____________________. Such an embryo can still develop normally when it has extra cells or missing cells.
30. The _______receptor seems to play a role in mammalian aging, based on the knockout mouse.

Arabidopsis is a model for studying plant development, including transcription factors

31. In addition to the ABC genes, another class of genes designated ______________________ interacts with the B and C genes to specify the development of petals, stamens, and carpels.

CANCER AND CELL DEVELOPMENT

32. A mass of cancer cells is called a (a) ______________ and when those cells spread it is referred to as (b)______________.
33. Cancer-causing genes are known as ______________.
34. A gene that interacts with growth-inhibiting factors to block cell division is a ______________ gene.

Oncogenes are usually altered components of cell signaling pathways that control growth and differentiation

35. Oncogenes arise from changes in the expression of ______________.
36. (a)________ is a G-protein that serves as an "on/off" switch for a number of different signaling pathways. This membrane bound protein is activated by (b)______________.

In many familial cancers, tumor suppressor genes must be inactivated before cells progress to cancer

37. The development of cancer is usually a slow, multistep process in which cells acquire mutations or epigenetic changes that activate (a)______________ and also mutation that inactivate (b)______________________ that normally interact with (c)______________________ to block cell division in abnormally growing cells.

BUILDING WORDS

Use combinations of prefixes and suffixes to build words for the definitions that follow.

Prefixes	The Meaning	Suffixes	The Meaning
morpho-	form	-gen(esis)	production of
onco-	mass, tumor		
toti-	all		
pluri-	many		
neo-	new		

Prefix	Suffix	Definition
__________	__________	1. The development and differentiation (production) of the form and structures of the body.
__________	-gene	2. A cancer-causing gene.
__________	-potent	3. The ability of some nuclei from differentiated plant and animal cells to provide information for the development of an entire organism; such cells contain

		all of the genetic material that would be present in the nucleus of a zygote.
____________	-potent	4. The ability to give rise to many but not all cells in an organism
____________	-plasm	5. A mass of tissue
____________	____________	6. A chemical agent that affects the differentiation of cells and the development of form.

MATCHING

Terms:

a. Apoptosis
b. Determination
c. Differentiation
d. Homeobox
e. Homeotic gene
f. Induction
g. Oncogene
h. Pattern formation
i. Protein kinase
j. Somatic cell
k. Stem cell
l. Totipotent
m. Transgenic
n. Tumor
m. Tumor suppressor gene
n. Zygomatic gene

For each of these definitions, select the correct matching term from the list above.

____ 1. An organism that has incorporated foreign DNA into its genome.
____ 2. Mass of tissue that grows in an uncontrolled manner.
____ 3. Development toward a more mature state; a process changing a young, relatively unspecialized cell to a more specialized cell.
____ 4. The ability of a cell (or nucleus) to provide the information necessary for the development of an entire organism.
____ 5. A phenomenon in which differentiation of a cell is influenced by interactions with particular neighboring cells.
____ 6. A gene that controls the formation of specific structures during development.
____ 7. Specific DNA sequence found in many genes that are involved in controlling the development of the body plan.
____ 8. The progressive limitation of a cell line's potential fate during development.
____ 9. Any of a number of genes that play an essential role in blocking cell division.
____10. A body cell not involved in reproduction.
____11. Programmed cell death.
____12. Enzyme that activates a membrane-bound Ras protein.
____13. The organization of cells into three-dimensional structures.

MAKING COMPARISONS

Fill in the blanks.

Germ Layer	Organ, Organ System or Tissue	Specialized Cell
Mesoderm	Skeletal muscles	Striated muscle cell
#1	#2	Skin cell

Germ Layer	Organ, Organ System or Tissue	Specialized Cell
#3	Nervous system	#4
None	#5	Egg and sperm (gametes)
Endoderm	#6	Tracheal cell
#7	Urinary bladder	Epithelial cell
#8	Blood	Red blood cell

MAKING CHOICES

Place your answer(s) in the space provided. Some questions may have more than one correct answer.

____ 1. Gap genes in *Drosophila*
a. determine segment polarity.
b. act on all embryonic segments.
c. are the first segmentation genes to act.
d. are complementary to maternal genes.
e. are a type of segmentation gene.

____ 2. The concept of nuclear equivalence holds that
a. adult cells are identical to one another.
b. embryonic cells are identical to one another.
c. adult somatic nuclei are identical to the nucleus of the fertilized egg.
d. adult cells are identical to egg cells.
e. all adult somatic nuclei are identical.

____ 3. If destruction of a single cell early in development results in the absence of a structure in the adult, the embryo of that organism
a. is mosaic.
b. is a fruit fly.
c. contains predetermined cells.
d. contained founder cells.
e. is controlled to some extent by homeotic genes.

____ 4. Embryos that act as a self-regulating whole and can accommodate missing parts
a. are transgenic.
b. are mosaics.
c. have regulative development patterns.
d. have founder cells that give rise to adult parts.
e. contain many genomic rearrangements.

____ 5. Morphogens are
a. chemical agents.
b. derived from a homeobox.
c. determined by embryonic segmentation genes.
d. thought to exist in *Drosophila* eggs.
e. unraveled portions of active DNA.

____ 6. A group of cells grown in liquid medium from a single root cell is
a. formed by a predetermined pattern.
b. a mosaic.
c. totipotent.
d. predetermined.
e. an embryoid body.

____ 7. An embryo in which the fate of cells is determined early in development is said to be
a. induced.
b. mosaic.
c. homeotic.
d. transgenic.
e. differentiated.

____ 8. If differentiated cells can be induced to act like embryonic cells, the cells are
a. formed by a predetermined pattern.
b. mosaics.
c. totipotent.
d. predetermined.
e. embryoids.

____ 9. Pattern formation is a series of steps leading to
a. determination.
b. differentiation.
c. predetermination.
d. morphogenesis.
e. the final step in cell specialization.

____10. Differences in the molecular composition of different cells in a multicellular organism are due to
a. loss of genes during development.
b. transformation.
c. regulation of the activities of different genes.
d. nuclear equivalency.
e. differential gene activity.

____11. The process by which developing cells become committed to a specialization is
a. determination.
b. differentiation.
c. transgenesis.
d. morphogenesis.
e. the first step in cell specialization.

____12. The process by which the differentiation of an embryonic cell is influenced by neighboring embryonic cells is
a. transformation.
b. induction.
c. mosaic development.
d. an example of programmed cell death.
e. determinate morphogenesis.

____13. Apoptosis
a. involves caspases.
b. occurs in all animals.
c. helps form the human hand.
d. plays a role in development.
e. is best studied in the mouse.

____14. Human therapeutic cloning involves
a. creating a newborn human.
b. supplying replacement tissues.
c. duplicating iPS cells.
d. world-wide opposition.
e. duplicating ES cells.

____ 15. Imaginal discs
a. occur following differentiation in flies.
b. are under genetic control.
c. are linked to a homeobox.
d. form specific structures in fruit flies.
e. are regulated by hox genes.

____ 16. Segmentation genes
a. include gap genes.
b. regulate the production of embryonic RNA.
c. respond to morphogens.
d. include pair-rule genes.
e. include segment polarity genes.

CHAPTER 18

❑

Introduction to Darwinian Evolution

The concept of evolution is based upon past and current evidence that organisms living today evolved from earlier organisms, and it links all fields of the life sciences into a unified body of knowledge. Evolution is firmly based on the genetic principles you learned in the previous section of the book. Fieldwork conducted by Charles Darwin was instrumental in establishing that natural selection is the actual mechanism of evolution. This mechanism works because each species produces more offspring than will survive to maturity, genetic variation exists among the offspring, organisms compete for resources, and individuals with the most favorable characteristics are most likely to survive and reproduce. As a result, favorable characteristics are passed to succeeding generations. Over time, changes occur in the gene pools of geographically separated populations that may lead to the evolution of new species. A subsequent modification of Darwin's theory, the synthetic theory of evolution, explains Darwin's observation of variation among offspring in terms of mutation and recombination. An enormous body of scientific observations and experiments supports the concept of evolution. The testing of evolutionary hypotheses has demonstrated that not only is evolution a reality, but it occurs daily and hourly all around us.

REVIEWING CONCEPTS

Fill in the blanks.

INTRODUCTION

1. New species evolve from earlier species in a process called_______________.
2. The concept of evolution is the cornerstone of biology, because it ____ _______________________.

WHAT IS EVOLUTION?

3. Evolution is the _______________________________________.
4. The genetic changes that bring about evolution do not occur in (a)_______________________________________, but rather in (b)_______________________.
5. Evolution occurs over ______________________.
6. A group of organisms that are capable of interbreeding are classified as a ___________.
7. The use of bacteria to digest oil from an oil tanker spill is an example of ___________.

PRE-DARWINIAN IDEAS ABOUT EVOLUTION

8. Aristotle was one of the first individuals to group and arrange organisms. To do this he used what he called a ______________________________.
9. ______________________________ was probably the first to recognize that fossils are the remains of extinct organisms.
10. __________________________ was the first to propose that organisms change over time as a result of some natural phenomenon.
11. __________________________ proposed the theory that traits *acquired* during an organism's lifetime could be passed along to their offspring.

DARWIN AND EVOLUTION

12. Charles Darwin served as the naturalist on the ship (a)______________________________. Darwin's theory of evolution used data he collected on the similarities and differences among organisms he observed in the (b)______________________________________, a chain of islands west of mainland Ecuador.
13. When horse breeders mate female and male horses to produce the strongest and fastest racing horse they are using ______________________________ to produce these traits.
14. ______________________________ made the important mathematical observation that populations increase in size geometrically until checked by factors in the environment.
15. An evolutionary modification that improves the chances of survival of oak trees in North America would be considered to be an ______________________________.
16. Both Darwin and ____ arrived at the conclusion that evolution occurred by natural selection.
17. In 1859, Darwin published his monumental book, ______________________________.

Darwin proposed that evolution occurs by natural selection

18. Darwin's mechanism of natural selection consists of four premises based on observations about the natural world. In summary, these are:
(a) ____
(b)____
(c) ____
(d) ____.

The modern synthesis combines Darwin's scientific theory of evolution with genetics

19. The modern synthesis theory incorporates the ideas and observations of (a) ____ and
(b) ____.

20. The synthetic theory of evolution explains variation in terms of ____________________, which are inheritable and therefore potential explanations for *how* traits are passed from one generation to another.

Biologists study the effect of chance on evolution

21. Although it cannot be proven, it appears that (a) ______________________________ is a more important agent of evolutionary change than (b) ____________________.
22. Evolution of major taxonomic groups is called (a)______________________________ and evolution of populations is called (b)____________________.

EVIDENCE FOR EVOLUTION

The fossil record provides strong evidence for evolution

23. The fossil record demonstrates that life ______________________________.
24. The formation of fossils is most likely to occur under conditions that favor rapid (a) ____________________ of the dead body and preventing body (b) ____________.
25. Of the many places that organisms live, the body of those organisms living in ____________________ are least likely to become fossilized.
26. The earliest fossils of *H. sapiens* with anatomically modern features appeared in the fossil record approximately ________________ years ago.
27. ____________________ are the remains of organisms that existed over a relatively short geological time period and that died and were preserved as fossils in large numbers. They are used to identify specific sedimentary layers and to arrange strata in chronological order.
28. Fossils can be dated by determining the ratio (a)____________________ in the environment and possibly the organism. Each one has its own rate of decay and its (b) ____________ does not change regardless of the environmental conditions.

The distribution of plants and animals supports evolution

29. The study of the past and present geographic distribution of plants and animals is called ____________________.
30. In 1915 Alfred Wegener proposed that a single land mass called Pangaea had broken apart in a process known as ______________________________.

Comparative anatomy of related species demonstrates similarities in their structures

31. (a)____________________ features are those derived from the same structure in a common ancestor and the condition is known as (b)________________.

32. There are times when organisms living in similar environments but not in the same geographical region evolve similar adaptations. The independent evolution of similar structures is referred to as ________________________________.
33. Non-homologous features that have similar function but evolved independently are called ________________.
34. The occasional presence of "remnant organs" in organisms is to be expected as ancestral species evolve and adapt to different modes of life. Such organs or parts of organs, called ________________, are degenerate and have no apparent function.

Molecular comparisons among organisms provide evidence for evolution

35. The molecular evidence for evolution incudes the universal (a)________________ and the conserved (b)________________________________ and of (c)________________________.
36. Codons code for a particular ________________ in a polypeptide chain.
37. Determining the order of nucleotide bases in DNA is known as ________________________.

Developmental biology helps unravel evolutionary patterns

38. The evolution of new features depends on modification in ________________________________ that already exist.

Evolutionary hypotheses are tested experimentally

39. Rapid evolution is actually occurring on a time scale of ____________ rather than centuries.

BUILDING WORDS

Use combinations of prefixes and suffixes to build words for the definitions that follow.

Prefixes	The Meaning	Suffixes	The Meaning
bio-	life	-ent	that which
converg-	come together		
homo-	same		

Prefix	Suffix	Definition
________	-logous	1. Pertains to having a similarity of form due to having the same evolutionary origin.
________	-geography	2. The study of the geographical distribution of living things.
________	-plastic	3. Pertains to features with the same or similar functions in distantly related organisms.
________	________	4. Evolution of similar structures in distantly related organisms.

MATCHING

Terms

a. Adaptation
b. Artificial selection
c. Continental drift
d. Convergent evolution
e. Evolution
f. Fossil
g. Gene pool
h. Half-life
i. Homoplastic
j. Index fossil
k. Natural selection
l. Range
m. Radioisotope
n. Synthetic theory of evolution
o. Vestigial

For each of these definitions, select the correct matching term from the list above.

____ 1. Used by seed companies to produce plants that are especially resistant to drought.
____ 2. The independent evolution of structural or functional similarity in two or more organisms of widely-different, unrelated ancestry.
____ 3. Parts or traces of an ancient organism preserved in rock.
____ 4. Carbon-14 and uranium-235.present in different rock formations.
____ 5. Is based upon a combination of Darwin's theory and Mendelian genetics.
____ 6. The remains of a once functional pelvic bone.
____ 7. The remains of an animal that can be found only in rock that is 1 million years old.
____ 8. The genetic change in a population of organisms that occurs over time.
____ 9. Similar in function or appearance, but not in origin or development.
____10. The F2 generation is more reproductively fit than the F1 generation and the F1 generation is more fit than the parental generation.
____11. Inherited variations that are favorable for survival are preserved in a population while unfavorable ones are eliminated.
____ 12. A measurement of radioactive decay.

MAKING COMPARISONS

Fill in the blanks.

Evidence for Evolution	How the Evidence Supports Evolution
Fossils	Direct evidence in the form of remains of ancient organisms
#1	Indicate evolutionary ties between organisms that have basic structural similarities, even though the structures may be used in different ways
#2	Indicate that organisms with different ancestries can adapt in similar ways to similar environmental demands
Vestigial organs	#3
Biogeography of plants and animals	#4
All organisms use a genetic code that is virtually identical	#5
#6	The best-adapted individuals produce the most offspring, whereas individual that are less well adapted die prematurely or produce fewer or inferior offspring
#7	Whales descended from land dwelling ancestors

Evidence for Evolution	How the Evidence Supports Evolution
#8	Some mammals found in Australia have developed similar features to mammals found in North America
DNA sequencing	#9

MAKING CHOICES

Place your answer(s) in the space provided. Some questions may have more than one correct answer.

____ 1. Which of the following statements is/are consistent with Darwin's theory of evolution?
a. Organisms evolve by a process of natural selection.
b. Evolution progresses toward a more perfect state.
c. Accumulation of adaptations in a population can lead to new species.
d. Better adapted individuals are more likely to survive.
e. New species result from a need to survive.

____ 2. The fossil record
a. supports contemporary theories of evolution.
b. includes evidence of an animal's life, such as how they walked.
c. demonstrates that life evolved through time.
d. is consistent with Darwin's theories.
e. includes evidence of both land and water dwelling organisms

____ 3. The earliest fossils of *Homo sapiens* with modern anatomical features appear around
a. 1,000 years ago.
b. 10,000 years ago.
c. 200,000 years ago.
d. 5,000 BC.
e. recorded human history.

____ 4. The first comprehensive theory of evolution, which proposed that organisms change as a result of natural phenomenon and not divine intervention, was proposed by
a. Aristotle.
b. Lamarck.
c. Darwin.
d. Wallace.
e. da Vinci.

____5. The wings of a butterfly and a bat are
a. homoplastic but not homologous.
b. homologous but not homoplastic.
c. neither homoplastic nor homologous.
d. both homoplastic and homologous.
e. probably derived from a common ancestor.

____ 6. The humerus in a bird and human are
a. homoplastic but not homologous.
b. homologous but not homoplastic.
c. neither homoplastic nor homologous.
d. both homoplastic and homologous.
e. probably derived from a common ancestor.

____ 7. The person(s) who suggested that populations increase geometrically and that their numbers must eventually be checked was
a. Malthus.
b. Lamarck.
c. Darwin.
d. Wallace.
e. Lyell.

____8. The person who supplied evidence that the earth was old enough to provide adequate time for evolution and the appearance of new species to have occurred is
a. Aristotle. d. Wallace.
b. Lamarck. e. Lyell.
c. Darwin.

____9. The two people who proposed that the mechanism for the evolution of organisms is by natural selection were
a. Malthus. d. Wallace.
b. Lamarck. e. Lyell.
c. Darwin.

____10. Structures of different organisms that have a similar form due to a common origin are
a. homoplastic or analogous. d. usually vestigial.
b. artifacts. e. found in fossils but never in contemporary organisms.
c. homologous.

____11. If Darwin had known and employed current biochemical techniques, it most likely would have resulted in
a. a different theory of evolution. d. confirmation of his theories.
b. no theory of evolution. e. an argument with Malthus.
c. his conviction of immutable species.

____12. Using microorganisms to clean up waste sites is known as
a. biological control. d. hazardous waste removal.
b. reclamation. e. bioincineration.
c. bioremediation.

____13. The genetic code
a. specifies a triplet of nucleotides in DNA. d. has been used to discredit the fossil record.
b. is virtually universal. e. has provided evidence used to build cladograms.
c. provides convincing evidence that all organisms arose from a common ancestor.

____14. A species
a. is a group of interbreeding organisms. d. may have multiple ancestors.
b. is a group of organisms with similar structures. e. will exhibit common behaviors.
c. is the basic group of a population.

____15. The person who proposed that all continents at one time were connected into a single land mass was
a. Verde. d. Wegener.
b. Lyell. e. Berger.
c. Aristotle.

____16. If the half-life of the radioactive atom biology 103 is 5 years, the approximate percentage of the original amount of the isotope that will remain after thirty years is
a. 6.0. d. 5.0.
b. 33.3. e. 1.56.
c. 6.25.

____17. Darwin's thoughts about evolution were influenced by
a. the breeding of domestic animals. d. the many varieties of plants of the same species.
b. the work of geologists. e. the work of Mendel.
c. a brief stop in the Malay Archipelago.

____18. Homology in plants has been studied using their

a. leaves.
b. roots
c. bark
d. stems.
e. seeds.

VISUAL FOUNDATIONS

Color the parts of the illustration below as indicated.

RED ☐ metacarpels
GREEN ☐ ulna
YELLOW ☐ carpels
BLUE ☐ radius
BROWN ☐ phalanges
VIOLET ☐ humerus

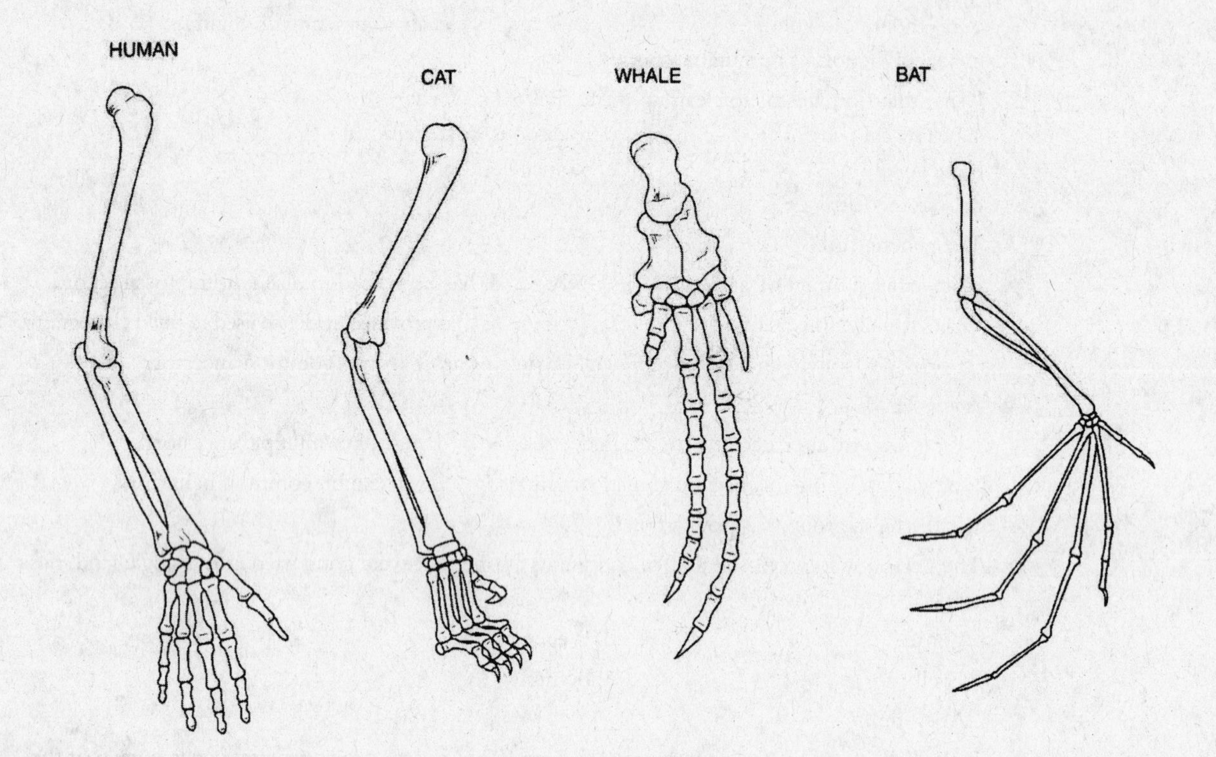

CHAPTER 19

Evolutionary Change in Populations

In the previous chapter you learned that natural selection results from differential survival and reproduction of individuals, that evolutionary change is inherited from one generation to the next, that populations not individuals evolve as a result of these processes, and that the principles of inheritance you learned in Chapter 11 underlie Darwinian evolution. This chapter examines the genetic variability that occurs within a population and the evolutionary forces that act on it. The factors responsible for evolutionary change are nonrandom mating, mutation, genetic drift, gene flow, and natural selection. Genetic variation in a gene pool is the raw material for evolutionary change. Considerable variability can be observed in the gene pools of most populations.

REVIEWING CONCEPTS

Fill in the blanks.

INTRODUCTION

1. Population genetics is the study of (a) ______________________________ within the population and the (b) __________________ forces that act on it.

GENOTYPE, PHENOTYPE, AND ALLELE FREQUENCIES

2. The gene pool of a population will include all ________________________ ____in that population.
3. (a)______________________________ is the proportion of a particular genotype in a population, (b)_____________________________ is the proportion of a particular phenotype in a population, and (c)________________ ______________________ is the proportion of a specific allele in a population..

THE HARDY-WEINBERG PRINCIPLE

4. In order to say that evolution is occurring within a population (a)____ _____.
5. For any given locus the frequency of an allele can range from (a) _____ to (b)_____.

6. When the distribution of genotypes in a population conforms to the binomial equation (a)______________________________, the population is (b)______________________________ and is not evolving.

Genetic equilibrium occurs if certain conditions are met

7. The proportion of alleles in successive generations does not change in a population when certain conditions are met. Among them, briefly stated, are the following five:
(a)____,
(b)____,
(c)____,
(d)____.
(e)____.

Human MN blood groups are a valuable illustration of the Hardy-Weinberg principle

8. The alleles for the MN blood group are of special interest to geneticists because they are ________________________.
9. Medical evidence suggests that the MN characteristic is not subject to (a)______________________________ and it does not produce a (b)______________________________ that might affect random mating.

MICROEVOLUTION

10. Generation-to-generation changes in genotype frequencies within a population describe a process called ________________________.

Nonrandom mating changes genotype frequencies

11. Inbreeding refers to the mating ____.
12. The average number of survivors among offspring from a given genotype compared to other genotypes is a measure of ______________________________.
13. Selection of mates based on a phenotype is known as ________________________.
14. Assortative mating changes allele frequencies only (a)______________________________ while inbreeding affects genotype frequencies (b) ____.

Mutation increases variation within a population

15. A ______________________________ is an unpredictable change in DNA.
16. Mutations in ______________________________ are not inheritable.
17. When a polypeptide is sufficiently altered to change its function, the mutation is ____.

In genetic drift random events change allele frequencies

18. Production of random evolutionary changes in small breeding populations is known as ________________.
19. Genetic drift tends to (a)____________ genetic variations within a population, although it tends to (b) ____________ genetic differences among different populations.
20. There are times when a population may suddenly and randomly decrease in size. Such a dramatic change in total population number can lead to a (a)______________ which will have an impact on (b)________________ as the population increases in size again.
21. The founder effect occurs when ______________________________ ______________________________.

Gene flow generally increases variation within a population

22. When individuals of breeding age migrate from one country to another there is a corresponding movement of alleles. This movement of alleles referred to as ______________ has significant evolutionary consequences.

Natural selection changes allele frequencies in a way that increases adaptation

23. Natural selection results in the differential reproduction of individuals with different (a)______________, or (b)______________ in response to (c)______________________________________.
24. Selection against phenotypic extremes, thereby favoring intermediate phenotypes, is called ______________________________.
25. Selection that favors one particular phenotype over another is known as ______________________________.
26. Selection that favors phenotypic extremes is called ____________________.

GENETIC VARIATION IN POPULATIONS

Genetic polymorphism can be studied in several ways

27. Much of genetic polymorphism is not evident, because it doesn't produce distinct ______________________________.
28. Sequencing the nucleotides in DNA from individuals in a population provides a direct estimate of (a)______________________________ because it is sensitive enough to detect (b)______________________________________.

Balanced polymorphism exists for long periods

29. ____________________ occurs when two or more alleles persist in a population over many generations as a result of natural selection.
30. ______________________________ occurs when a genotype such as Aa has a higher degree of fitness than either AA or aa.

31. Selection that acts to decrease the frequency of the more common phenotypes and increase the frequency of the less common phenotypes is called ______________________________.

Neutral variation may give no selective advantage or disadvantage

32. Variation that does not alter the ability of an individual to survive and reproduce, and is therefore not adaptive, is called ______________________________.

Populations in different geographical areas often exhibit genetic adaptations to local environments

33. Along with population variation, genetic differences often exist among different populations within the same species, a phenomenon known as ______________________________.

BUILDING WORDS

Use combinations of prefixes and suffixes to build words for the definitions that follow.

Prefixes	The Meaning	Suffixes	The Meaning
gen(o)-	origin	-morphism	the process of form
micro-	small	-typ(e)	form
poly-	many		
pheno-	visible		

Prefix	Suffix	Definition
________	-evolution	1. Changes in allele frequencies over successive generations; involves small changes within a population.
________	________	2. An organism's visible form or characteristics.
________	________	3. The genetic makeup of an individual as determined by their alleles.
________	________	4. The type of genetic variation that occurs among individuals in a population.

MATCHING

Terms:

a. Assortative mating
b. Balanced polymorphism
c. Directional selection
d. Disruptive selection
e. Founder effect
f. Gene flow
g. Gene pool
h. Genetic drift
i. Genetic polymorphism
j. Hardy-Weinberg principle
k. Heterozygote advantage
l. Natural selection
m. Population
n. Stabilizing selection

For each of these definitions, select the correct matching term from the list above.

____ 1. Natural selection that favors intermediate variants and acts against extreme phenotypes.

____ 2. A phenomenon in which organisms that are heterozygotes are more likely to demonstrate reproductive success than individuals that are homozygotes.

____ 3. A phenomenon that results in changes in allele frequency in a population from one generation to the next.

____ 4. A phenomenon that occurs as a result of individuals from Canada migrating and interbreeding with individuals from Finland.

____ 5. Natural selection that favors one phenotypic extreme shifting the phenotypic mean..

____ 6. A phenomenon in which natural selection results in multiple alleles being maintained in a population over several generations.

____ 7. Genetic drift that results from a small number of individuals colonizing a new area.

____ 8. A phenomenon in which individuals who have black hair and are six feet tall interbreed with other individuals having the same phenotype.

____ 9. The principle that in a randomly-mating large population the relative frequencies of allelic genes do not change from generation to generation.

____10. A group of organisms of the same species that live in the same geographical area at the same time.

____ 11. A phenomenon in which organisms in a population with certain characteristics are more likely to show reproductive success than other organisms in the same population.

MAKING COMPARISONS

Fill in the blanks.

Category	Contributes to Evolutionary Change (yes or no)
Natural selection	#1
Gene flow	#2
Random mating	#3
Genetic drift	#4
Mutation	#5
Small population size	#6
Geographic stability	#7
Genetic polymorphism	#8
Neutral variation	#9

MAKING CHOICES

Place your answer(s) in the space provided. Some questions may have more than one correct answer.

____ 1. The proportion of alleles in a population does not change if there is/are
a. only 10% of the alleles mutating in each generation.
b. random mating.
c. a large number of individuals in the population.
d. mating with individuals in a similar population.
e. natural selection of advantageous alleles.

____ 2. What will occur over time when gradual environmental changes favor intermediate phenotypes in a normal distribution curve?
a. stabilizing selection
b. no selection
c. a shift in allele frequencies
d. disruptive selection
e. directional selection

____ 3. Changes in allele frequencies within a population occur during
a. genetic drift.
b. macroevolution.
c. microevolution.
d. p^2 shifts.
e. q^2 shifts.

____ 4. What will occur over time when an extreme environmental change favors several phenotypes at the expense of the mean phenotype?
a. stabilizing selection
b. no selection
c. a shift in allele frequencies
d. disruptive selection
e. directional selection

____ 5. Positive reproductive success of individuals who are Dd compared to those who are DD best describes
a. positive assortative mating.
b. natural selection.
c. disruptive selection.
d. genetic polymorphism.
e. a heterozygote advantage.

____ 6. In populations, all new alleles occur as a result of
a. gene flow.
b. natural selection.
c. mutations.
d. genetic drift.
e. the heterozygote advantage.

____ 7. Random sexual reproduction among members of a large population that occurs in the absence of any selective pressure on the gene pool ultimately leads to
a. a Hardy-Weinberg value of 1.
b. changes in gene frequencies.
c. generations of unchanged allele frequencies.
d. non assortative mating.
e. a change in the frequency of 2pq, but not p^2 or q^2.

____ 8. What will be the ultimate result if 75% of a beekeeper's bees freeze to death in a severe snow storm?
a. a bottleneck
b. frequency-dependent selection
c. genetic drift
d. stabilizing selection
e. disruptive selection

____ 9. The gradual change in the size and shape of a species of berry bush that occurs as you go from the top of the Grand Canyon into the valley is an example of
a. a cline
b. a neutral variation
c. frequency-dependent selection
d. genetic drift
e. disruptive selection

____ 10. Genetic equilibrium occurs as a result of
- a. migration
- b. nonrandom mating
- c. a large population size
- d. natural selection
- e. no net mutations occurring

Use the following information, the Hardy-Weinberg equation, and the list below to answer questions 11-16. The phenotype coded for by the genotype "bb" is found among 900 individuals in a population of 10,000 randomly mating individuals.

a. 0.01	f. 0.12	k. 0.39	p. 0.60	u. 4.0
b. 0.02	g. 0.18	l. 0.40	q. 0.65	v. 9.0
c. 0.04	h. 0.22	m. 0.42	r. 0.70	w. 35.0
d. 0.08	i. 0.30	n. 0.49	s. 0.81	x. 49.0
e. 0.09	j. 0.33	o. 0.50	t. 2.0	y. 91.0

____ 11. Frequency of homozygously recessive individuals.

____ 12. Frequency of homozygously dominant individuals.

____ 13. Frequency of q.

____ 14. Percent of individuals containing one or more of the dominant alleles.

____ 15. Frequency of p.

____ 16. Frequency of heterozygous individuals.

VISUAL FOUNDATIONS

Color the parts of the illustration below as indicated. Also label directional selection, disruptive selection, and stabilizing selection.

- RED ❑ axis indicating number of individuals
- GREEN ❑ axis indicating phenotype variation
- YELLOW ❑ normal distribution
- BLUE ❑ less favored phenotypes
- ORANGE ❑ favored phenotypes

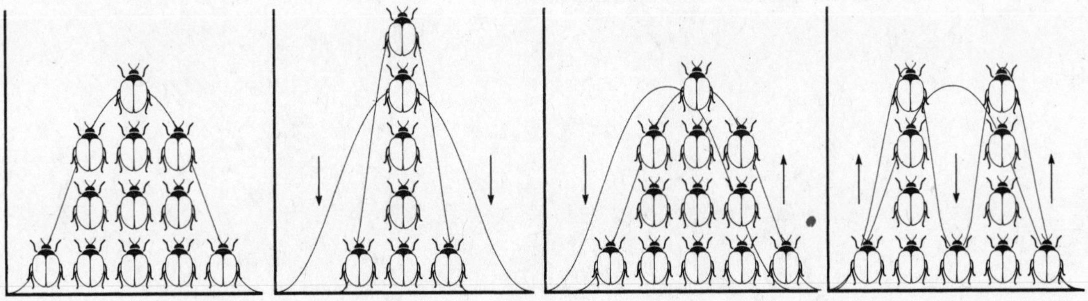

CHAPTER 20

❑

Speciation and Macroevolution

In the previous chapter, you learned how populations evolve. This chapter examines the evolution of species and macroevolution – the dramatic evolutionary changes that occur over long time spans. A biological species is generally viewed as a group of sexually reproducing organisms that share a common gene pool and that are reproductively isolated from other organisms. Most species have two or more mechanisms that serve to isolate and preserve the integrity of their gene pool. If two different species should attempt to mate, prezygotic mechanisms prevent fertilization from taking place. Other, postzygotic, mechanisms, ensure reproductive failure should fertilization occur. Speciation, or the development of a new species, most commonly occurs when a group of reproducing individuals of a population becomes geographically separated from the rest of the species and subsequently evolves. A new species can also evolve, however, within the same geographical region as its parent species. There are two theories related to the pace of evolution. One is gradualism, which says that there is a slow, steady change in species over time, and the other is the punctuated equilibrium, which says there are long periods of little evolutionary change followed by short periods of rapid speciation. Macroevolution includes the origin of large-scale phenotypic changes and evolutionary novelties (e.g., wings with feathers), the evolution of many species from a single species, and the extinction of species.

REVIEWING CONCEPTS

Fill in the blanks.

INTRODUCTION

1. Although there is an estimated (a) ______________ species living today, more than (b) ______% of all species that ever existed are extinct.
2. Life today is the product of ______________billion years of evolution.

WHAT IS A SPECIES?

The biological species concept is based on reproductive isolation

3. In 1942, biologist Ernst Mayr proposed the biological species concept which states that a species is a
____
____.

The phylogenetic species concept defines species based on such evidence as molecular sequencing

4. Biologists may use the phylogenetic species concept to describe a population if the population has undergone
____
____.

5. Many biologists believe that the phylogenetic species concept cannot be applied if the ______________________________ of a taxonomic group has not been carefully studied.

REPRODUCTIVE ISOLATION

6. Reproductive isolating mechanisms preserve the genetic integrity of individual species because
____
____.

Prezygotic barriers interfere with fertilization

7. Prezygotic barriers prevent the ______________________________ ______________________________ from occurring.
8. Prezygotic isolating mechanisms prevent the formation of an __________ zygote.
9. Prezygotic isolating mechanisms include (a) ____________________, which occurs because two groups reproduce at different times; (b)____________________, which occurs when individuals display different, incompatible courtship patterns; (c)____________________, which occurs when anatomical differences thwart successful mating; and (d)____________________, which occurs when chemical differences between gametes prevents interspecific fertilization.

Postzygotic barriers prevent gene flow when fertilization occurs

10. The union of gametes of two closely related species produces an ______________________________.
11. The genes from parents belonging to different species do not interact properly in regulating (a)____________________ and so embryos are frequently (b)____________________. In this case, reproductive isolation occurs as a result of (c)____________________.
12. ____________________ occurs when the gametes produced by interspecific hybrids are abnormal and nonfunctional.
13. On occasion an interspecific hybrid is formed and the F1 generation is able to reproduce. However, the continued reproductive success of the hybrid is prevented as a result of ____________________.

Biologists are discovering genes responsible for reproductive isolating mechanisms

14. ____________ is a sperm protein in abalone that attaches to a specific receptor protein located on the egg envelope and then produces a hole in the envelope that permits the sperm to penetrate the egg.

SPECIATION

15. Two new species are formed from a single, original species when two (a) ____________ of that species become (b) ____________ from one another and the (c)____________ diverge.

Long physical isolation and different selective pressures result in allopatric speciation

16. Allopatric speciation occurs when one population becomes ______________ from the rest of the species and subsequently evolves.
17. ______________________ is a random change in allele frequency resulting from the effects of chance on survival and reproduction in small breeding populations.
18. Only in natural selection is the change in ____________ adaptive.
19. Allopatric speciation is the (a) ______________________ method of speciation and accounts for (b)__________________________ evolution of new animal species.
20. Speciation is more likely to occur in ________________ populations.

Two populations diverge in the same physical location by sympatric speciation

21. Sympatric speciation usually occurs *within* a geographical region in at least two ways: (a) __________________ and (b)_____________________.
22. Of all the organisms, sympatric speciation is most common in (a)______________ as a result of (b) _________________.
23. Possession of more than two sets of chromosomes is called (a)____________ which is common in plants.
24. (a) ____________________ describes the condition when offspring arising from the mating of a single species have multiple sets of chromosomes and (b) _______________ describes the condition when offspring arising from the mating of different species have multiple sets of chromosomes,.
25. Allopolyploidy occurs in conjunction with (a) ________________ and produces (b) _________________ individuals.
26. Many examples of sympatric speciation in animals involve (a)_____________________ and rely on genetic mechanisms other than (b)__________________________

The study of hybrid zones has made important contributions to what is known about speciation

27. The __________________ is an area of overlap in which populations, subspecies, or species come into contact and can interbreed.
28. (a)_________________ describes an increase in reproductive isolation that can occur over time in a hybrid zone because the hybrids are (b)______________ than the parental populations.
29. _________________ describes the hybrids in a hybrid zone when the hybrids are as fit as the parental populations.
30. ____________ is said to occur when the hybrids in a hybrid zone have greater reproductive fitness than either of the two parental species.
31. ______________ are areas of transition between two different environments.

THE RATE OF EVOLUTIONARY CHANGE

32. The theory of ________________________________ supports the concept that evolution proceeds in rapid bursts of changes which are followed by long periods of inactivity or stasis.
33. The theory of ________________________________ supports the concept that populations slowly and steadily diverge from one another by the accumulation of adaptive characteristics within a population.

Evolutionary novelties originate through modifications of pre-existing structures

34. Evolutionary novelties are variations of pre-existing structures called ____________________________.
35. Evolutionary "novelties" originate from mutations that alter developmental pathways. For example, (a) ________________ occurs when developing body parts grow at different rates, and (b)__________ results from differences in the *timing* of development.

Adaptive radiation is the diversification of an ancestral species into many species

36. (a)__________________________ are new ecological roles made possible by an adaptive advancement. When an organism with a newly acquired evolutionary advancement assumes a new ecological role made possible by its advancement(s), its diversification is known as (b)______________________.
37. Adaptive radiation appears to be more common during periods of ____.

Extinction is an important aspect of evolution

38. Continuous, ongoing, and relatively low frequency extinction is called ________________________.
39. ____________________ extinction is a relatively rapid and widespread loss of numerous species.

Is microevolution related to speciation and macroevolution?

40. Biologists hypothesize that (a)________________________ processes explain (b)______________________________ patterns.

BUILDING WORDS

Use combinations of prefixes and suffixes to build words for the definitions that follow.

Prefixes	The Meaning	Suffixes	The Meaning
allo-	other, different	-metric	unit of length
macro-	large, long, great, excessive	-morph(ic)	form
paedo-	child	-patri(c)	native land
poly-	much, many	-ploidy	number of chromosome sets in a genome
speci-	kind		
sym-	together		

Prefix	Suffix	Definition
______	______	1. Retaining juvenile characteristics as an adult
______	______	2. The presence of multiples of complete chromosome sets.
______	______	3. Originating in or occupying different (other) geographical areas.
______	______	4. Having multiple sets of chromosomes from two different species
______	-evolution	5. Large-scale evolutionary change.
______	______	6. Varied rates of growth for different parts of the body during development (some parts grow at rates different from other parts)
______	-es	7. A group of organisms that freely interbreed and produce fertile offspring.

MATCHING

Terms:

a.	Adaptive radiation	h.	Gradualism	o.	Phylogenetic species
b.	Allopatric speciation	i.	Habitat isolation	p.	Punctuated equilibrium
c.	Behavioral isolation	j.	Hybrid breakdown	q.	Reinforcement
d.	Biological species	k.	Hybrid inviability	r.	Reproductive isolation
e.	Extinction	l.	Hybrid sterility	s.	Speciation
f.	Fusion	m.	Hybridization	t.	Sympatric speciation
g.	Gametic isolation	n.	Mechanical isolation	u.	Temporal isolation

For each of these definitions, select the correct matching term from the list above.

____ 1. A postzygotic isolating mechanism in which the F1 hybrids reproduce but the F2 generation cannot.

____ 2. The rapid appearance of new species in the fossil record.

____ 3. The evolution of a new species within the same geographical region as the parent species.

____ 4. The evolution of many related bird species from a single ancestral bird species in a relatively short period of time.

____ 5. A prezygotic isolating mechanism in which sexual reproduction between two individuals cannot occur because of differences in compatibility of the reproductive organs.

____ 6. A model of evolution in which the hybrids have less reproductive success than either parental group.

____ 7. Sexual reproduction between individuals from closely related species.

____ 8. Small chickadee birds nesting in a tree do not mate with small chickadee birds nesting on the ground.

____ 9. A group of individuals determined by the number of appendages they have.

____10. The end of a lineage, occurring when the last individual of a species dies.

____11. The zygote produced as the result of a seal mating with a walrus dies before it can fully develop.

MAKING COMPARISONS

Fill in the blanks.

Reproductive Isolating Mechanism	How It Works
Behavioral isolation	Similar species have distinctive courtship behaviors
#1	Prevents the offspring of hybrids that are able to reproduce successfully from reproducing past one or a few generations
Gametic isolation	#2
#3	Interspecific hybrid survives to adulthood but is unable to reproduce successfully
Mechanical isolation	#4
#5	Interspecific hybrid dies at early stage of embryonic development
#6	Similar species reproduce at different times
Habitat isolation	#7
#8	Barrier to reproduction that prevents the formation of a zygote after two different species have mated

MAKING CHOICES

Place your answer(s) in the space provided. Some questions may have more than one correct answer.

____ 1. When large-scale phenotypic changes in populations justify placing them in taxonomic groups at the species level or higher it is known as
a. macroevolution.
b. paedomorphosis.
c. polymorphic speciation.
d. macromorphism.
e. adaptive radiation.

____ 2. The process by which an ancestral species evolves into many new species is known as
a. macroevolution.
b. microevolution.
c. polymorphic speciation.
d. macromorphism.
e. adaptive radiation.

____ 3. A species typically has
a. a common gene pool.
b. an isolated gene pool.
c. the capacity to reproduce with other similar species.
d. members with different morphological characteristics.
e. become reproductively isolated.

____ 4. The fact that dogs come in many sizes, shapes, and colors supports the contention that
a. dogs have one gene pool.
b. dogs can produce hybrids.
c. physical appearance alone is not enough to define a species.
d. all dogs can reproduce together.
e. all dogs belong to one species.

_____ 5. A population that evolves within the same geographical region as its parent species is an example of
a. cladogenesis.
b. character displacement.
c. punctuated equilibrium.
d. sympatric speciation.
e. allopatric speciation.

_____ 6. If two closely related species produce a fertile interspecific hybrid, the hybrid is the result of
a. anagenesis.
b. phyletic evolution.
c. cladogenesis.
d. diversifying evolution.
e. allopolyploidy.

_____ 7. If two distinct populations of organisms occasionally interbreed in the wild, they are considered to be
a. separate species.
b. one species.
c. related species.
d. one gene pool.
e. evolving.

_____ 8. If flower-loving female botany majors were required to smell a flower before male botany majors would mate with them, and if male zoology majors would only mate with females that did not sniff flowers, zoology majors and botany majors would likely become
a. separate species.
b. one species.
c. reproductively isolated.
d. one gene pool.
e. behaviorally isolated.

_____ 9. Evolution of organisms occurs by means of
a. changes in individual organisms.
b. uniformitarianism.
c. changes in gene frequencies in the gene pool.
d. extinctions.
e. changes in populations.

_____10. A population that evolves as the result of its geographic separation from the rest of the species is an example of
a. cladogenesis.
b. character displacement.
c. adaptive radiation.
d. sympatric speciation.
e. allopatric speciation.

_____11. If college-educated persons mated only at night and noncollege-educated persons mated only at the lunch hour and these habits remained unchanged, these two populations would be considered
a. separate species.
b. one species.
c. reproductively isolated.
d. gametically isolated.
e. temporally isolated.

_____12. Hybrid breakdown affects the
a. P_1 generation.
b. F_1 generation.
c. F_2 generation.
d. F_3 generation.
e. course of evolution.

_____13. The death of interspecific embryos during development is an example of
a. hybrid sterility.
b. a postzygotic barrier.
c. anagenesis.
d. hybrid inviability.
e. allopatric dissociation.

_____ 14. Containing multiple sets of chromosomes from two or more species would make an organism a/an
a. hybrid.
b. autopolyploid.
c. polyploidy.
d. allopolyploid.
e. a species formed by hybridization.

_____ 15. Ectones are
a. transition areas between two different environments.
b. stable hybrid zones.
c. areas of parental reproductive success.
d. endemic to North America.
e. an example of polyploidy in action.

_____ 16. Reproductive isolating mechanisms
a. prevent gene flow.
b. will prevent fertilization.
c. more commonly occur prior to mating.
d. prevent two species from mating.
e. preserve genetic integrity of a species.

_____ 17. Reproductive isolating mechanisms
a. have a genetic basis.
b. involve speciation genes.
c. are not known to occur in invertebrate animals.
d. include differences in egg surface receptors.
e. prevent hybrid formation.

VISUAL FOUNDATIONS

Color the parts of the illustration below as indicated. Also label gradualism and punctuated equilibrium.

RED ❑ original species
GREEN ❑ last species to evolve
YELLOW ❑ time axis
BLUE ❑ first species to evolve
ORANGE ❑ structural change axis

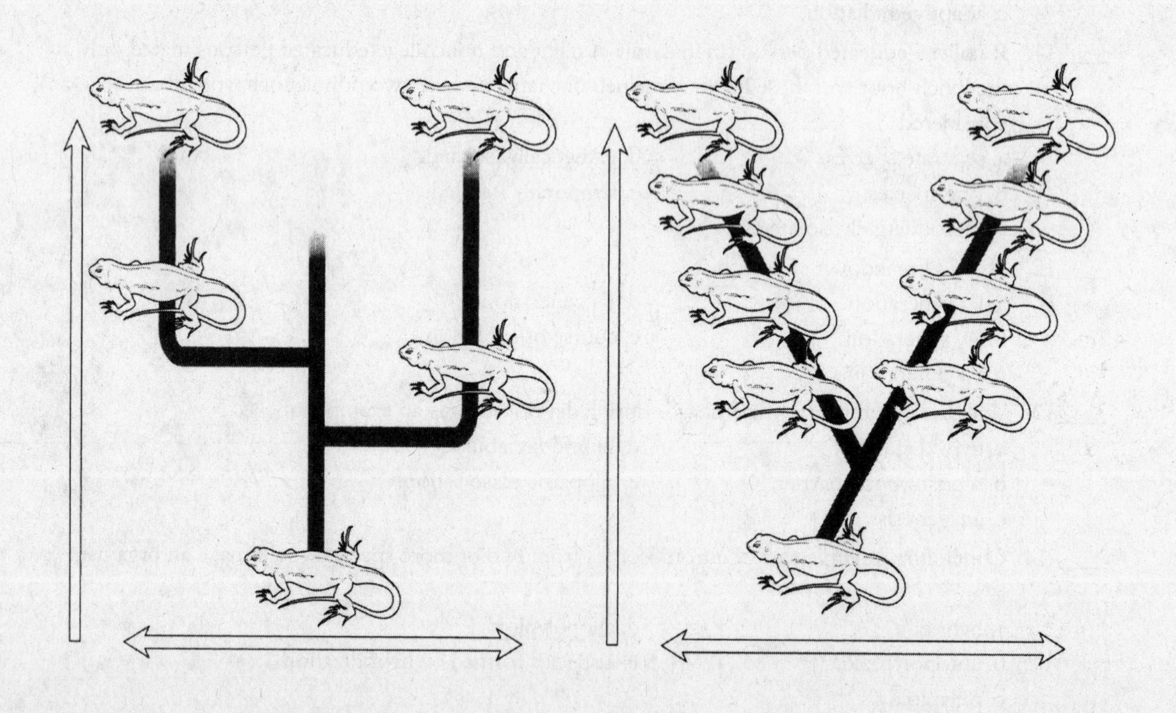

CHAPTER 21

❑

The Origin and Evolutionary History of Life

The previous three chapters dealt with the evolution of organisms, but what about the evolution of life in the first place? This chapter addresses how life began and traces its long evolutionary history. Life on earth developed from nonliving matter. Exactly how this process, called chemical evolution, occurred is not certain. Current models suggest that small organic molecules formed spontaneously, accumulated, and became organized into complicated assemblages. Cells then evolved from these macromolecular assemblages. This process occurred in the presence of little or no oxygen and in the presence of energy, the chemical building blocks of organic molecules, and sufficient time for the molecules to accumulate and react. The first cells to evolve were anaerobic and prokaryotic. They are believed to have obtained their energy from organic compounds in the environment. Later, cells evolved that could obtain energy from sunlight. Photosynthesis produced enough oxygen to change the atmosphere significantly and thereby alter the evolution of early life. Organisms evolved that had the ability to use oxygen in cell respiration. Eukaryotes are believed to have evolved from prokaryotes. Mitochondria, chloroplasts, and certain other organelles may have originated from symbiotic relationships between two prokaryotic organisms.

REVIEWING CONCEPTS

Fill in the blanks.

INTRODUCTION

1. ______________________ is the process by which life developed from nonliving matter.
2. The first multicellular eukaryotes to evolve were ______________________.
3. Some jawed fish gave rise to ______________.
4. The first vertebrates with limbs and capable of moving on land were (a)______________________ which then gave rise to (b)______________.

CHEMICAL EVOLUTION ON EARLY EARTH

5. Earth is estimated to be about ________________billion years old.
6. The four basic requirements for the chemical evolution of life are:
   (a)____,
   (b)____,
   (c)____,
   (d)____.

Organic molecules formed on primitive Earth

7. Two hypotheses have been proposed to explain how organic molecules may have originated. The (a) ______________________________ hypothesis proposes that these molecules formed near the surface of the earth and the (b) ______________________________ hypothesis proposes that the organic precursors formed near cracks on the ocean floor.
8. The idea that organic molecules could form from nonliving components on primitive Earth originated in the early 20th century with the Russian biochemist (a)____________________ and the Scottish biologist (b)____________________.
9. In the 1950s, (a)______________ and (b)______________ constructed conditions in the laboratory that were thought to prevail on primitive Earth. Their experiment investigated an atmosphere of water and the three gases (c)________________, (d)________________, (e)________________________. These and other experiments have produced a variety of organic molecules.

THE FIRST CELLS

10. ____________________ are assemblages of organic polymers that form spontaneously, are organized, and to some extent resemble living cells.
11. ____________________ are formed by adding water to abiotically produced peptides.

The origin of a simple metabolism within a membrane boundary may have occurred early in the evolution of cells

12. The ______________________________ hypothesis proposes that life began as a self-sustaining, organized system consisting of chemical reactions between molecules enclosed within a boundary separating them from the external environment.
13. One of the most significant parts of the origin of living cells from pre-cells was the evolution of ______________________________.

Molecular reproduction was a crucial step in the origin of cells

14. (a)__________ may have been the first polynucleotide to carry "hereditary" information. Interestingly, some forms of this molecule, called (b)________________, can catalyze biological reactions and are hypothesized to have catalyzed synthesis of the first (c)________.
15. Double stranded DNA is hypothesized to have evolved from double stranded (a)__________. DNA then evolved into the information transfer system because of the stability of (b)______________.

Biological evolution began with the first cells

16. (a)____________________________, or ancient remains of microscopic life, suggest that cells may have been thriving as long as (b)__________ billion years ago.
17. ____________________ are rocklike columns of fossilized prokaryotic cells and provide fossil evidence of the earliest cells.

The first cells were probably heterotrophic

18. The first cells were most likely (aerobic or anaerobic?) (a)___________________ and (prokaryotic or eukaryotic?) (b)____________________.
19. Organisms that use photosynthesis to produce their own raw materials such as glucose are called _____________________________.
20. The first photosynthetic organisms to obtain hydrogen electrons by splitting water were the _____________________________.
21. The evolution of photosynthesis ultimately changed early life and afforded vastly new opportunities for diversity because it generated ________________.

Aerobes appeared after oxygen increased in the atmosphere

22. ____________________________________ are organisms that can only survive in the absence of oxygen.
23. The evolution of aerobic respiration stabilized (a)___________________ and (b)___________________________ levels in the biosphere.

Eukaryotic cells descended from prokaryotic cells

24. Eukaryotes appeared in the fossil record about ______billion years ago.
25. The ______________________________ hypothesis proposes that organelles such as mitochondria and chloroplasts may have evolved from a symbiotic relationship between two prokaryotes.

THE HISTORY OF LIFE

26. ____________________________________are used to date the same layer of rock that were deposited around the world.
27. The largest division of geologic time is called an ______________.
28. Geologic eras are divided into (a)____________________ which are further divided into (b) ____________________.
29. Life originated on Earth during the _____________________ eon.

Rocks from the Ediacaran period contain fossils of cells and simple animals

30. The _____________________________________eon came after the Archean eon and in this eon the rocks are less altered by heat and pressure.
31. The oldest known fossils of multicellular animals are called ___________ fossils.

A diversity of organisms evolved during the Paleozoic era

32. The Paleozoic era began about (a)_______million years ago with a profound burst of evolution during the (b)__________________ period..
33. Later, in the (a)______________________ period, the jawless, bony-armored fish called (b)____________________________ appeared.
34. Jawed fishes first appeared in the __________________________ period. This period became known as the age of the fishes.
35. During the (a)_____________________ period the great swamp forests arose and today their remains are present in the form of (b)___________.

36. Reptiles diversified during the ______________ period and came to dominate the carnivorous and herbivorous terrestrial lifestyles.

Dinosaurs and other reptiles dominated the Mesozoic era

37. The Mesozoic era began about _____ million years ago.
38. The Mesozoic era consisted of the three periods: (a)______________ when reptiles underwent adaptive radiation and mammals appeared, (b)______________ when large dinosaurs and the first toothed birds appeared, and (c)______________ during which large dinosaurs and toothed birds became extinct.

The Cenozoic era is the Age of Mammals

39. The Cenozoic era can be divided into three periods: (a)______________ (b) ______________ (c) ______________
40. An explosive radiation of primitive mammals occurred during the ______________ epoch.
41. An explosive radiation of birds occurred during the ______________epoch.
42. The first apes appeared in Africa during the ______________epoch.
43. Human ancestors appeared in Africa during the late (a)______________ and early (b)______________ epochs.

BUILDING WORDS

Use combinations of prefixes and suffixes to build words for the definitions that follow.

Prefixes	The Meaning	Suffixes	The Meaning
aero-	air	-be (bios)	life
an-	without, not, lacking	-bio(nt)	life
auto-	self, same	-troph	nutrition, growth, "eat"
Pre-	before, prior to		
hetero-	different, other		
proto-	first, earliest form of		

Prefix	Suffix	Definition
__________	__________	1. An organism that produces its own food.
__________	__________	2. Spontaneous assemblages of organic polymers that may have been involved in the evolution of the earliest forms of life.
__________	__________	3. An organism that is dependent upon other organisms for food, energy, and oxygen.
__________	-cambrian	4. The geologic time period prior to the Cambrian.
__________	__________	5. An organism that requires air or free oxygen to live.
__________	-aerobe	6. An organism that does not require air or free oxygen to live.

MATCHING

Terms:

a.	Autotroph	g.	Era	m.	Paleozoic era
b.	Cenozoic era	h.	Heterotroph	n.	Period
c.	Cyanobacteria	i.	Iron-sulfur world hypothesis	o.	Prebiotic soup hypothesis
d.	Chemical evolution	j.	Mesozoic era	p.	Ribozyme
e.	Endosymbiont	k.	Microfossil	q.	Stromatolite
f.	Epoch	l.	Microsphere		

For each of these definitions, select the correct matching term from the list above.

____ 1. A geological time period that is a subdivision of an era, and is divided into epochs.

____ 2. Term applied to divisions of geological time that are divided into periods.

____ 3. The "Age of Mammals."

____ 4. A rocklike column composed of many minute layers of prokaryotic cells; a type of fossil evidence.

____ 5. A type of protobiont formed by adding water to abiotically-formed polypeptides.

____ 6. A major interval of geological time; a subdivision of a period.

____ 7. An organism that lives symbiotically inside a host cell.

____ 8. The remains of a microscopic organism.

____ 9. A molecule of RNA that has catalytic ability.

____10. Organism that consume preformed molecules.

____11. Hypothesis that proposes that organic precursors formed on the ocean floor.

____12. Era that began with the appearance of bacteria and cyanobacteria and ended with the diversification of conifers.

MAKING COMPARISONS

Fill in the blanks.

Biological Event	Period	Era
First fishes appear	#1	#2
The first mammals and first dinosaurs	#3	#4
Dinosaurs peak and become extinct	#5	#6
Age of Marine Invertebrates	#7	#8
Age of *Homo sapiens*	#9	#10
Amphibians and wingless insects appear	#11	#12
Apes appear	#13	#14
Reptiles appear	#15	#16

MAKING CHOICES

Place your answer(s) in the space provided. Some questions may have more than one correct answer.

____ 1. The earth's surface is protected from the mutagenic effects of ultraviolet radiation by a layer of
a. CO_2.
b. N_2.
c. H_2.
d. O_3.
e. H_2O.

____ 2. It is thought that the mitochondra in eukaryotic cells evolved from
a. anaerobic bacteria.
b. aerobic bacteria.
c. endosymbionts.
d. primitive stem cells.
e. intracytoplasmic protozoans.

____ 3. The Earth is thought to be about how many years old?
a. 10-20 billion
b. 4,000
c. 3.5 billion
d. 4.6 billion
e. 10-20 million

____ 4. About how many years passed between the Earth's formation and the appearance of the earliest life?
a. 10-20 million
b. 100-200 million
c. 800-900 million
d. five billion
e. 10 billion

____ 5. Earth's early atmosphere became oxygenated through the photosynthetic activity of
a.. purple sulfur bacteria.
b. cyanobacteria.
c. water-splitting autotrophs.
d. green sulfur bacteria.
e. hydrogen-sulfide bacteria.

____ 6. Aerobes are generally better competitors than anaerobes because aerobic respiration
a. is more efficient.
b. splits water.
c. is confined to symbiotic mitochondria.
d. extracts more energy from a molecule.
e. has had a longer period of evolution.

____ 7. The early Earth's atmosphere included
a. CO_2.
b. N_2.
c. H_2.
d. CO.
e. H_2O vapor.

____ 8. The basic requirements for chemical evolution include
a. DNA.
b. oxygen.
c. energy.
d. water.
e. time.

____ 9. Experimenters that simulated conditions thought to have existed on early Earth found that these conditions could produce
a. DNA.
b. RNA.
c. nucleotide bases.
d. proteins.
e. amino acids.

____10. Strong support for the endosymbiotic theory of eukaryotic cell origins is derived from the fact that mitochondria and chloroplasts have
a. unique DNA.
b. two membranes.
c. ribosomes.
d. tRNA.
e. endosymbionts.

____11. The consensus among scientists is that the first cells were
a. aerobic autotrophic prokaryotes.
b. anaerobic autotrophic prokaryotes.
c. aerobic heterotrophic prokaryotes.
d. anaerobic heterotrophic prokaryotes.
e. anaerobic heterotrophic eukaryotes.

____12. Which of the following era/period combinations is/are mismatched?
a. Cenozoic/Paleogene
b. Paleozoic/Miocene
c. Mesozoic/Jurassic
d. Cenozoic/Cambrian
e. Paleozoic/Devonian

____13. The Paleozoic era
a. began about 542 mya.
b. lasted about 291 million years.
c. consists of six periods.
d. includes the Cambrian period.
e. gave rise to many new species.

____14. Life evolving from nonliving matter describes
a. protobionts.
b. how the first cells evolved.
c. microspheres.
d. the appearance of microfossils.
e. chemical evolution.

____ 15. Stromatolites
a. are found in the U.S. and Canada.
b. may be 3 billion years old.
c. are evidence of eukaryotes.
d. are fossils.
e. are evidence of early prokaryotes.

____16. The first photosynthetic organisms appeared about
a. 2.bya.
b. 3 bya.
c. 4 bya.
d. 2.5 bya.
e. 3.5 bya.

____ 17. Experiments investigating the hypothesis that molecules could form in a reducing atmosphere like that present on the primitive Earth were conducted by
a. Haldane.
b. Urey.
e. Clay.
d. Miller.
e. Oparin.

____ 18. The oldest eon is
a. Ordovician.
b. Phanerozoic.
c. Archaean.
d. Proterozoic.
e. Cenozoic.

____19. The prebiotic soup hypothesis
a. proved the existence of protobionts.
b. was supported by experiments conducted by Haldane.
c. proposes life's organic molecules evolved near thermal vents.
d. proposes life's organic molecules evolved in cracks along the ocean floor
e. propose life's organic molecules evolved near Earth's surface.

____20. Protobionts
a. include microspheres.
b. are molecular aggregates.
c. may show catalytic activity.
d. are early cells.
e. are vesicle-like.

____21. Ancient remains of early life forms can be seen in
- a. the Holocene epoch.
- b. in Pilbara rocks.
- c. microfossils.
- d. stromatolites.
- e. steranes.

VISUAL FOUNDATIONS

Color the parts of the illustration below as indicated.

- RED ❑ DNA
- GREEN ❑ chloroplast
- YELLOW ❑ aerobic bacterium
- BLUE ❑ nuclear envelope
- ORANGE ❑ mitochondrion
- BROWN ❑ endoplasmic reticulum
- TAN ❑ photosynthetic bacterium

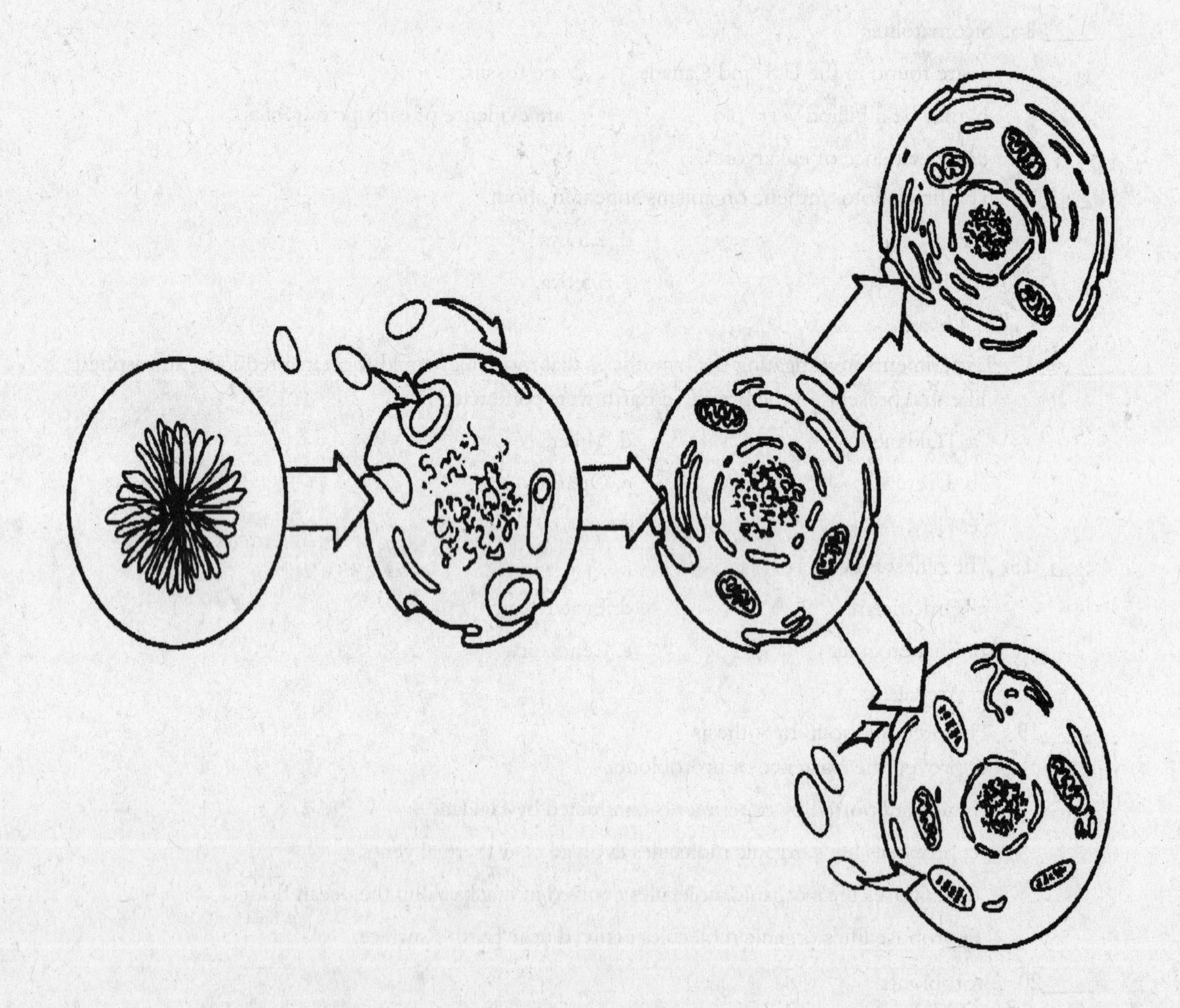

CHAPTER 22

❑

The Evolution of Primates

This is the last of five chapters that address the concept of evolution – the concept that links all fields of the life sciences into a unified body of knowledge. This chapter addresses the evolution of primates, with particular focus on humans and their ancestors. Primates evolved from small, tree-dwelling, shrew-like placental mammals. The first primates to evolve were the lemurs, galagos, and lorises. These were followed by the tarsiers and anthropoids. The early anthropoids branched into two groups, the New World Monkeys and the Old World Monkeys. The latter gave rise to apes that, in turn, gave rise to the hominins (humans and their ancestors). Hominin evolution began in Africa. The first hominin to have enough human features to be placed in the same genus as modern humans is *Homo habilis*. Anatomically modern *Homo sapiens* appeared approximately 195,000 years ago. The evolutionary increase in human brain size made cultural evolution possible.

REVIEWING CONCEPTS

Fill in the blanks.

INTRODUCTION

1. The study of human evolution is called ______________________.
2. Paleontologists hypothesize that the first primates descended from small, shrew-like ______________________.
3. Many traits of the living primate species are a reflection of their ______________________.

PRIMATE ADAPTATIONS

4. Humans and other primates have (a) ______________________ hands and feet with (b) ______________________.
5. The first primates appeared on Earth about __________ million years ago.
6. Primates are able to grasp objects because they have an ______________________.
7. Eyes located on the front of the head provides primates with (a)______________________ which allows them to (b)______________________ ______________________.

PRIMATE CLASSIFICATION

8. The three suborders in the order Primates are
   (a)________________________ which includes lemurs, lorises, and galagos,
   (b)________________________ which includes tarsiers, and
   (c)________________________ which includes monkeys, apes, and humans.

Suborder Anthropoidea includes monkeys, apes, and humans

9. Fossil evidence indicates that anthropoids originated in either
   (a) ________________________ or (b) ________________________.
10. A significant difference between anthropoids and other primates is the size (a)________________________. The (b)________________________ in particular is more developed in monkey, apes, and humans.
11. A few monkeys have ________________________tails capable of wrapping around branches and serving as fifth limbs.
12. ________________________ animals are capable of walking on all fours.

Apes are our closest living relatives

13. Apes and humans comprise a group called the ________________.
14. The five living genera of hominoids are
   (a)________________________,
   (b)________________________,
   (c)________________________,
   (d)________________________ and
   (e)________________________.
15. Gibbons can arm-swing from limb to limb. Another name for this activity is ________________.
16. Gorillas use quadrupedal locomotion also known as ________________________.

HOMININ EVOLUTION

17. In apes the large hole in the base of the skull through which the spinal cord connects to the brain is called the ________________________.
18. Apes have prominent ________________________above the eye sockets which are absent in humans.

The earliest hominins may have lived 6 mya to 7 mya

19. Hominid evolution probably began on the continent of ____________.
20. The oldest known homini fossil was discovered in ____________.

Ardipithecus, Australopithecus, and Paranthropus are australopithecines, or "southern man apes"

21. ______________________________ is considered to be close to the "root" of the human family tree.
22. ______________________________ has a mixture of apelike and humanlike features.
23. The hominin species *Australopithecus anamensis* exhibits distinct phenotypic differences between the two sexes, a condition known as ______________________________.
24. "Lucy" was one of the most ancient hominins, a member of the species ______________________________.

Homo habilis* is the oldest member of genus *Homo

25. *Homo habilis* was discovered in the country of ____________ in Africa.
26. *Homo habilis* appeared about _____ million years ago. *H. habilis* was the first hominin to consciously design useful tools.

Homo ergaster may have arisen from H. habilis

27. New discoveries indicate that the early fossils of *H. erectus* really represent two different species. (a) ______________________________ an earlier African species and (b) ______________________________ a later eastern Asian offshoot.

Homo erectus* probably evolved from *H. ergaster

28. Fossils of *Homo erectus* were originally found in ____________________.
29. *Homo erectus* is about _____ million years old. It had a larger brain than *H. habilis* and was the first hominin to have fewer differences between the sexes. *H. erectus* made more advanced tools and used fire.

Archaic *Humans date from about 1.2 mya to* 200,000 years ago

30. Archaic humans are descendants of ______________________________.
31. The oldest archaic human fossils, *H. antecessor,* were discovered in ____________.

Neanderthals appeared approximately 250,000 years ago

32. Neanderthal fossils were first discovered in ______________________________.
33. Neanderthals are considered to be an evolutionary ____________.

Scientists have reached a near consensus on the origin of modern *Homo sapiens*

34. The "______________________________" model is the main explanation for the origin of modern humans living around the world.

CULTURAL CHANGE

35. At the DNA level, humans are approximately (a) ____% identical to gorillas and (b) _____% identical to chimpanzees.
36. Human culture is generally divided into three stages:
(a)____
__________, (b)______________________________,
(c)______________________________.

Development of agriculture resulted in a more dependable food supply

37. Evidence has shown that humans had begun to cultivate crops approximately ____ thousand years ago.

Human culture has had a profound impact on the biosphere

38. The Industrial Revolution began in the ________________.
39. Advances in agriculture encouraged ____________.

BUILDING WORDS

Use combinations of prefixes and suffixes to build words for the definitions that follow.

Prefixes	The Meaning
arbor-	tree-like
anthrop(o)-	human
bi-	twice, two
di-	two
homin-	man
pro-	"before"
quadr(u)-	four
supra-	above

Suffixes	The Meaning
-eal	pertaining to
-morphic	pertaining to shape
-oid	resembling, like
-ped(al)	foot

Prefix	Suffix	Definition
________	________	1. Member of the suborder Anthropoidea; animals that "resemble" humans.
________	________	2. Pertains to an animal that walks on four feet.
________	________	3. Pertains to an animal that walks on two feet.
________	________	4. Pertains to animals that live in trees.
________	-orbital	5. Situated above the eye socket.
________	________	6. Member of the suborder Anthropoidea; animals that "resemble" monkeys, apes, and humans.
________	________	7. Organisms having two distinct forms such as male and female humans.

MATCHING

Terms:

a.	Arboreal	e.	Hominin	i.	Neandertals
b.	*Australopithecus*	f.	*Homo erectus*	j.	Orrorin
c.	Brachiate	g.	*Homo habilis*	k.	Paleoanthropologist
d.	Foramen magnum	h.	*Homo sapiens*	l.	Prehensile

For each of these definitions, select the correct matching term from the list above.

____ 1. Any of a group of extinct and living humans.
____ 2. The first hominin in the line leading up to *H. sapiens.*
____ 3. Tree-dwelling.
____ 4. One of the earliest know hominins.
____ 5. Adapted for grasping by wrapping around an object.
____ 6. A scientist that studies human evolution.
____ 7 The first hominid to have enough human features to be placed in the same genus as modern humans.
____8. To swing arm to arm, from one branch to another.
____9. The opening in the base of the vertebrate skull through which the spinal cord passes.
____10. This species could survive in cold areas, obtained food by hunting, and lived in shelters.

MAKING COMPARISONS

Fill in the blanks.

Hominid	Time of Origin	Characteristic(s)
Orrorin	About 6 million years ago (mya)	One of the earliest known hominins
#1	About 3.6 mya	Small brain, walked upright, large canine teeth, jutting jaw, no evidence of tool use (famous 3.2 mya example was nicknamed "Lucy")
#2	250,000 years ago	Disappeared mysteriously approximately 28,000 years ago
#3	About 4.8 mya	Close to the "root" of the human family tree
#4	3 mya	Walked erect, humanlike hands and teeth, likely omnivorous, likely derived from *A. afarensis*
Homo habilis	#5	#6
Sahelanthropus sp.	#7	The earliest known hominid, close to the last common ancestor of hominids and chimpanzees

Hominid	Time of Origin	Characteristic(s)
Homo erectus	#8	#9
Homo antecessor	#10	Oldest archaic human fossils found in Europe, practiced cannibalism
#11	4.2 mya	Apelike and human like features, marked phenotypic differences show sexual dimorphism

MAKING CHOICES

Place your answer(s) in the space provided. Some questions may have more than one correct answer.

____ 1. A suborder of primates is
a. Hominoids.
b. Hominins.
c. Prosimii.
d. Tarsiiformes.
e. Anthropoidea.

____ 2. The group(s) to which humans and their ancestors belong is/are
a. Hominoids.
b. Hominins.
c. Prosimii.
d. Tarsiiformes.
e. Anthropoidea.

____ 3. The group(s) to which apes belong is/are
a. Hominoids.
b. Hominins.
c. Prosimii.
d. Tarsiiformes.
e. Anthropoidea.

____ 4. Both apes and humans
a. are hominoids.
b. are hominids.
c. can brachiate.
d. lack tails.
e. have opposable thumbs.

____ 5. The first hominin that migrated to Europe and Asia was
a. *H. habilis.*
b. *H. erectus.*
c. *H. sapiens.*
d. australopithecines.
e. "Lucy" and her descendants.

____ 6. Archaic humans appeared about how many years ago?
a. 20,000-40,000
b. 100,000-150,000
c. 200,000-400,000
d. 800,000-900,000
e. one million

____ 7. The Peking man and Java man are classified as
a. *H. habilis.*
b. *H. erectus.*
c. *H. sapiens.*
d. apes.
e. hominids.

____ 8. The "Neandertal man" appeared about ________________ years ago.
a. 800,000
b. 250,000
c. 30,000
d. 1 million
e. 2 million

____ 9. The earliest hominin species to be placed in the genus *Homo* is
a. *habilis.*
b. *erectus.*
c. *sapiens.*
d. australopithecines.
e. "Lucy" and her descendants.

____10. The earliest hominins
a. had short canines.
b. are in the genus *Homo.*
c. are in the species *sapiens.*
d. appeared about 6-7 mya.
e. evolved in Africa.

____11. Based on molecular similarities and other characteristics, it is thought that the nearest living relative of humans is the
a. gorilla.
b. monkey.
c. gibbon.
d. chimpanzee.
e. ape.

____ 12.Which of the following support(s) the "out-of-Africa hypothesis?"
a. modern *H. sapiens* evolved from *Homo neanderthalensis*
b. newly evolved *H. sapiens* migrated to Europe
c. results from recent molecular analyses
d. racial differences
e. paternal mitochondrial DNA

____13. The immediate ancestor to the genus *Homo* is
a. Prosimii.
b. Tarsiiformes.
c. Australopithecines.
d. primitive apes.
e. Therapsids.

____14. The earliest Hominins may have lived as long as
a. 3-4 mya.
b. 4-5 mya.
c. 5-6 mya.
d. 6-7 mya.
e. 7-8 mya.

____ 15. Old world monkeys
a. have a prehensile tail.
b. have a fully opposable thumb.
c. are found in tropical environments.
d. have nostrils that are close together.
e. are social animals.

____ 16. In a cladogram demonstrating primate evolution, the first outgroup is the
a. tarsiers.
b. chimpanzees.
c. New World monkeys.
d. gibbons.
e. lemurs.

____ 17. Placental mammals
a. had ground dwelling ancestors.
b. are the largest group of mammals.
c. gave rise to the first primates.
d. are primates.
e. are the most successful group of mammals.

____ 18. Primates
a. reach sexual maturity relatively late in life.
b. have stereoscopic vision.
c. have young that need a longer period of protection.
d. include lemurs and lorises.
e. have a relatively small brain for their size.

____ 19. In a cladogram of arthropods, the common hominoid ancestor arose
a. after gorillas.
b. after New World monkeys.
c. before chimpanzees.
d. before Old World monkeys.
e. before gibbons.

____ 20. Gibbons
a. brachiate.
b. knuckle walk.
c. lack a tail.
d. are adapted for life on the ground.
e. are adapted for quadrupedal walking.

____ 21. Major divisions in human culture include
a. a learned component.
b. an inherited component.
c. gathering societies.
d. hunting societies.
e. an industrial revolution.

VISUAL FOUNDATIONS

Color the parts of the illustration below as indicated.

VIOLET	❑	cerebrum
GREEN	❑	jaw (mandible bone)
BLUE	❑	supraorbital ridge
YELLOW	❑	canine teeth
ORANGE	❑	incisor teeth

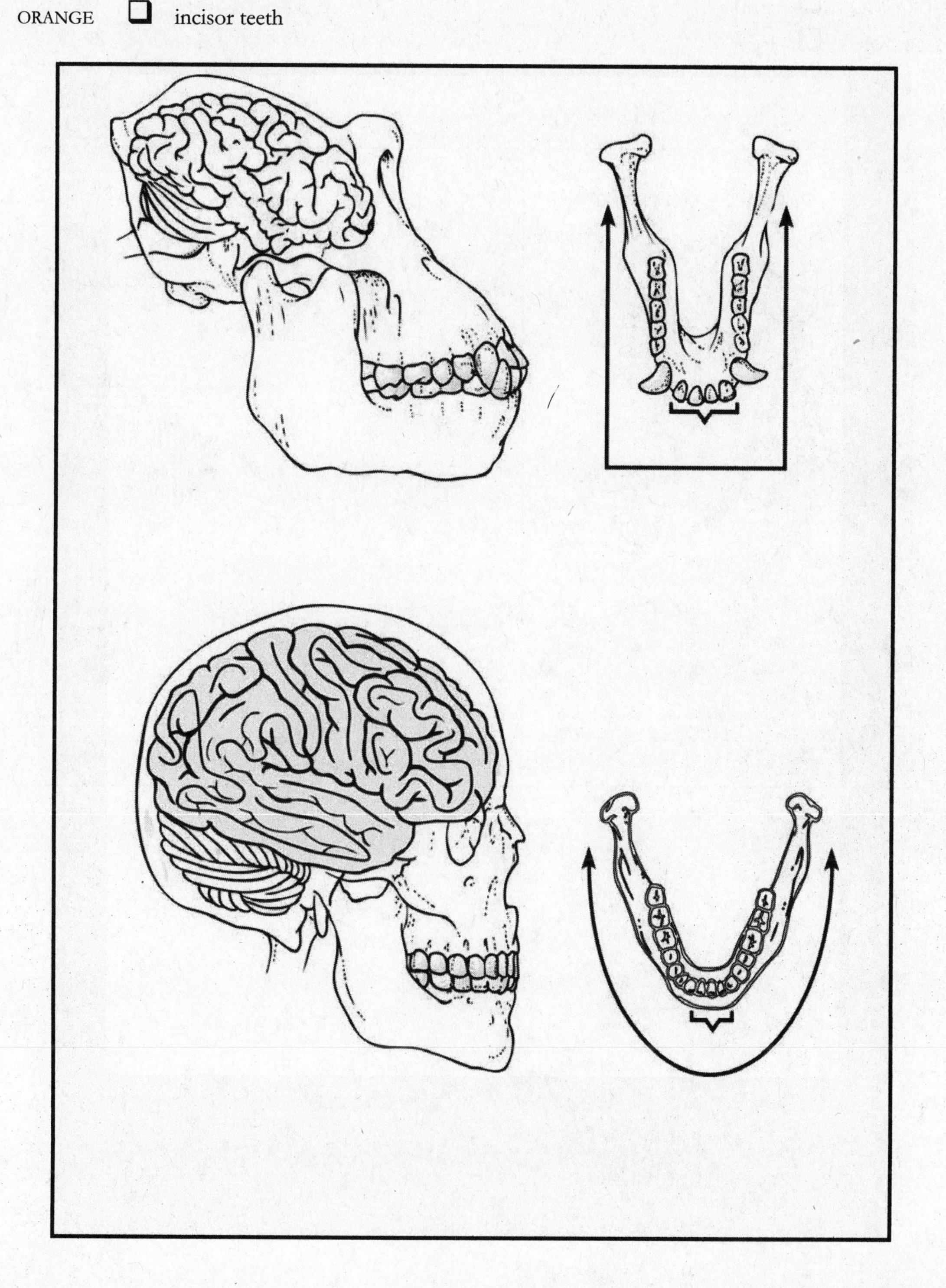

Color the parts of the illustration below as indicated. Label the anatomical difference between human and gorilla on each skeletal element.

RED	❑	pelvis
GREEN	❑	skull
YELLOW	❑	first toe
BLUE	❑	spine
ORANGE	❑	arm
BROWN	❑	leg

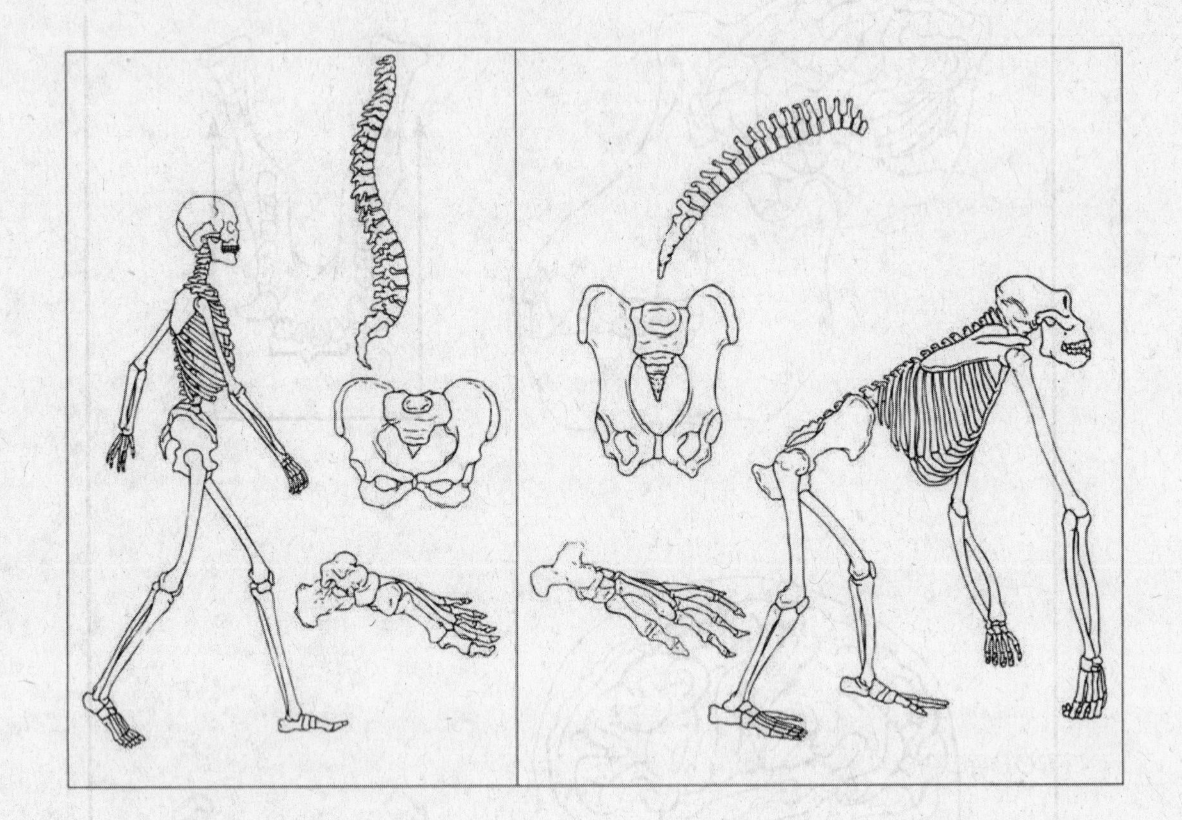

CHAPTER 23

❑

Understanding Diversity: Systematics

Evolution, the cumulative changes that occur in the gene pool of a population over time, has been the central focus of diversity. This section looks at The Diversity of Life which occurs as a result of differences in the gene pool and gene expression between species. This chapter provides an overview of some of the methods and approaches scientists can use to classify organisms. Biologists rely on a combination of molecular data and structural comparisons to infer relationships between organisms and reconstruct evolutionary history. Systematics is the scientific study of the diversity of organisms and their evolutionary relationships. Taxonomy, the science of naming, describing, and classifying organisms, is an important part of systematics. Since the mid-18th century biologists have classified organisms using the binomial system developed by Carolus Linnaeus. This system uses a two-part name to describe an individual species. The first name, a noun, is the genus and the second, an adjective, is the specific epithet. To accurately identify a species both names must be used together and never alone. Classification is hierarchical beginning with a single organism, the species, and ending with the largest, most inclusive group, the domain. The intermediate groups are made up of related individuals. For example a group of genera constitutes a family and a group of families constitutes an order. In establishing a "tree of life" systematists rely heavily on evolutionary change and relationships and they look for evidence that suggests groups of organisms are related to a common ancestor. It is important to remember that taxonomists continue to make changes and adjustments to phylogenetic relationships as new information becomes available. Most recently, scientists began to use cladograms, diagrams of hypothetical evolutionary relationships based upon shared derived characters, to classify organisms.

REVIEWING CONCEPTS

Fill in the blanks.

INTRODUCTION

1. To date, biologists estimate they have identified less than (a) _____% of the terrestrial species and (b) _____% of the marine species.
2. The variety of living organisms and ecosystems is referred to as ____.
3. More than _______% of the prescriptions dispensed in the United States derive from living organisms.
4. ________________ is the study of diversity of organisms and their evolutionary relationships.

CLASSIFYING ORGANISMS

5. ____________ is the science of naming, describing, and classifying organisms.

6. Living species are "related" via the characteristics they share and the traits they share are ____________________________.

Organisms are named using a binomial system

7. Carolus Linnaeus designed a classification system, the ____________________________ of nomenclature, in which each species or organism is identified using a unique two-part name.
8. For each species the two-part name consists first of the (a)__________, which is capitalized, followed by the (b)____________________ which is not capitalized.

Each taxonomic level is more general than the one below it

9. The classification system consists of several levels of organization. A group of closely related species is placed in the same genus and a group of closely related genera are grouped together in the same (a)____________ which in turn are grouped and placed in orders. Orders would be grouped in the next level to form a (b)__________, which would be grouped to form a (c)____________.
10. The broadest, most inclusive group is called a ______________.
11. In the classification system, each grouping, or level, such as the species, genus, order, is called a ____________.

DETERMINING THE MAJOR BRANCHES IN THE TREE OF LIFE SYSTEMATICS IS AN EVOLVING SCIENCE

12. Before the development of the microscope, organisms were only placed into either kingdom: (a)______________ or kingdom (b)____________.
13. In the late 1800s a third kingdom, ____________, was proposed that was to include bacteria and microorganisms.
14. ____________ are non-photosynthetic organisms that absorb nutrients and are placed in a kingdom of the same name.
15. Kingdom ____________________ includes organisms that lack a distinct, membrane bound nucleus.

The three domains form the three main branches of the tree of life

16. The three domains include two domains under kingdom Prokaryotae (a)____________________ and (b)____________________ and a third domain which includes the eukaryotes (c)____________________.

Some biologists are moving away from Linnaean categories

17. New biological techniques including ________________________ are providing scientists with a better method of determining how organisms are related. This has resulted in a transition from the more traditional methods of determining taxonomic groups.

Phylogenetic trees show hypothesized evolutionary relationships

18. A phylogenetic tree that uses clades to show evolutionary relationships is called a (a)______________. A branch point on these diagrams is called a (b)____________. The root at the base of

the these diagrams represents the
(c)________________________ of the entire grouping.

Systematists continue to consider other hypotheses

19. Genes are transmitted from parents to offspring in a process called
______________.
20. Gene swapping between organisms in one taxon and related organisms in another taxon is called
________________________________.

RECONSTRUCTING EVOLUTIONAY HISTORY

21. Modern classification is based on the reconstruction of
________________________, or phylogeny, as it is called.
22. Phylogenies are testable (a)______________ supported by
(b)________________________.
23. A ________________ describes a group of individuals of the same species living in the same area.

Homologous structures are important in determining evolutionary relationships

24. ________________________ describes the independent evolution of similar structures in distantly related organisms.
25. A characteristic that superficially appears homologous but is actually independently acquired by convergent evolution or reversal is described as exhibiting ________________.

Shared derived characters provide clues about phylogeny

26. Features that appeared a long time ago and remain in the descendants of that organism are called ________________.
27. Features that occurred in a recent common ancestor and remain in the descendants of that organism are called
________________________.

Systematists base taxonomic decisions on recent shared ancestry

28. Organisms are typically classified on the basis of a
____________ of traits rather than on any single trait.

Molecular homologies help clarify phylogeny

29. An organism's unique DNA and RNA sequences can be used as a
__________ to identify that organism.
30. The science of ________________ focuses on molecular structure to identify evolutionary relationships.
31. Macromolecules that have a similar subunit structure and that are functionally similar in two different groups are considered to be
________________.
32. Many systematists use similarities in (a)__________ structure and in (b) .__________ sequences to determine phylogenies.

Taxa are grouped based on their evolutionary relationships

33. When the organisms in a given taxon include the common ancestral species and all of its descendants, the group is said to be ______________.
34. When the organisms in a given taxon do not include all of the descendants of a common ancestral species, the group is said to be ______________.
35. When the organisms in a given taxon do not share a common ancestor, the group is said to be ______________.

CONSTRUCTING PHYLOGENETIC TREES

36. Three main schools of systematics are (a)______________, also known as numerical taxonomy, (b)______________ and (c)______________, also known as cladistics.

Outgroup analysis is used in constructing and interpreting cladograms

37. An (a) ______________ is a taxon that is considered to have diverged earlier than the taxa under investigation which is called the (b)______________.

A cladogram is constructed by considering shared derived characters

38. To be a valid monophyletic clade, all members must share at least one ______________
39. Membership in a clade cannot be established by shared ______________.

Each branch point represents a major evolutionary step

40. The farther a node is located from the base of a cladogram, the more ______________.
41. Relationships between taxa are determined only by tracing along the branches to the ______________.
42. A clodogram does not establish direct ______________ relationships among taxa.

Systematists use the principles of parsimony and maximum likelihood to make decisions

43. The principle of ______________ is based on the experience that the simplest explanation is probably the correct one.
44. ______________ is a statistical method of systematics that depends on probability.

APPLYING PHYLOGENETIC INFORMATION

45. The phylogenetic relationship among organisms can provide useful information to many disciplines including (a)______________, (b)______________, (c)______________.

BUILDING WORDS

Use combinations of prefixes and suffixes to build words for the definitions that follow.

Prefixes	The Meaning
mono-	alone, single, one
para-	alongside, beside, near
poly-	much, many
sub-	under, below

PrefixSuffix		Definition
____________	-phylum	1. A distinctive subunit of a phylum.
____________	-phyletic	2. Refers to a taxon in which the group includes some but not all the descendants of a common ancestor.
____________	-phyletic	3. Refers to a taxon consisting of several evolutionary lines and not including a common ancestor.

MATCHING

Terms:

a. Ancestral character
b. Clade
c. Classification
d. Derived character
e. Genus
f. Horizontal gene transfer
g. Kingdom
h. Order
i. Parsimony
j. Phenetics
k. Phylum
l. Species
m. Specific epithet
n. Systematics
o. Taxon
p. Taxonomy
q. Vertical gene transfer

For each of these definitions, select the correct matching term from the list above.

____ 1. The arranging of organisms into groups using similarities and evolutionary relationships among lineages.

____ 2. The science of naming, describing, and classifying organisms.

____ 3. The noun part of the binomial system used to describe organisms.

____ 4. A taxon that comprises related classes.

____ 5. A formal grouping of organisms such as a class or a family.

____ 6. A monophyletic group of organisms sharing a common ancestor.

____ 7. The systematic study of organisms based on similarities of many characters.

____ 8. The transfer of genes between different species.

____ 9. A recently evolved characteristic found in a clade.

____10. Using the simplest explanation of the available data to classify organisms.

MAKING COMPARISONS

Fill in the blanks.

Domain	Kingdom	Characteristics
Bacteria	Bacteria	Unicellular prokaryotes, generally with cell walls composed of peptidoglycan
#1	#2	Unicellular prokaryotes lacking peptidoglycan in cell walls
#3	#4	Heterotrophic eukaryotes with cell walls
#5	#6	Heterotrophic eukaryotes without cell walls
#7	#8	Autotrophic eukaryotes with cell walls

MAKING CHOICES

Place your answer(s) in the space provided. Some questions may have more than one correct answer.

____ 1. The study of the diversity of organisms and their evolutionary relationships is
a. taxonomy.
b. nomenclature.
c. systematics.
d. determinism.
e. classification.

____ 2. ________ is the classification of organisms based on the number of characters they share.
a. Taxonomy
b. Phylogeny
c. Systematics
d. Phenetics
e. Evolutionary systematics

____ 3. Which of the following groups includes the largest number of organisms?
a. class
b. phylum
c. order
d. phylum
e. family

____ 4. The systematist would be most interested in
a. lumping taxa.
b. splitting taxa.
c. evolutionary relationships.
d. ontogeny.
e. ancestries.

____ 5. Which organisms are placed in domain Eukarya?
a. bacteria
b. plants
c. fungi
d. archea
e. protists

____ 6. In what way(s) are fungi different from plants?
a. Fungi are not photosynthetic.
b. Fungi have multicellular reproductive organs.
c. Fungi have a different mode of reproduction.
d. Fungi have cellulose in their cell walls.
e. Fungi have a different body structure.

____ 7. Shared ancestral characters
a. can be used to determine when two groups diverged.
b. can be used to distinguish various classes of vertebrates.
c. are evidence of convergent evolution.
d. will be homoplastic.
e. are used in establishing barcodes.

____ 8. Which of the following is/are used extensively by the molecular biologist to determine relationships?
a. carbohydrates.
b. nucleic acids.
c. lipids.
d. proteins.
e. hormones.

____ 9. The system of binomial nomenclature was developed by
a. Charles Darwin.
b. Carolus Linnaeus.
c. Paul Hebert.
d. Carl Woese.
e. Watson and Crick.

____10. Which statement is incorrect?
a. A cladogram tells us how recently two groups shared a common ancestor.
b. Systematists use statistical probability, maximum likelihood, in establishing taxonomic relationships.
c. A specific epithet can be used in more than one binomial classification.
d. Protists are no longer placed in a single kingdom.
e. A taxon is a formal grouping of organisms that cannot be separated into subgroups.

____ 11. The archosaurs include
a. crodiles.
b. flying reptiles.
c. dinosaurs.
d. .birds.
e. .diapsid animals.

VISUAL FOUNDATIONS

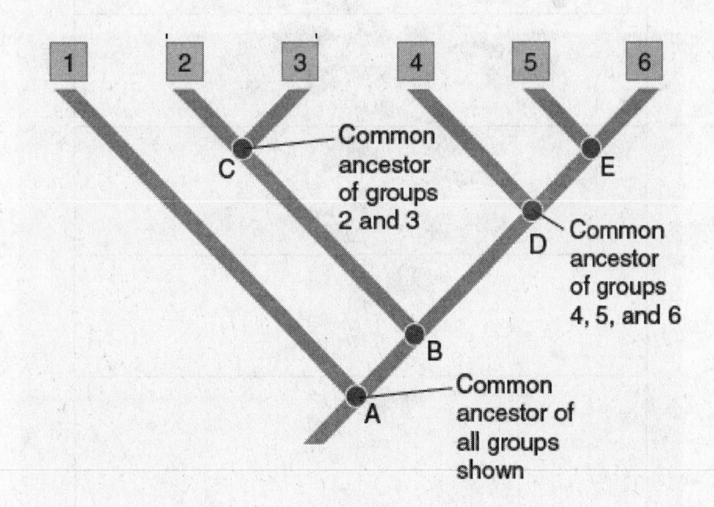

Use the above figure to answer the following questions.

1. B would be the common ancestor for which monophyletic taxon?

2. D would be the common ancestor for which monophyletic taxon?

3. Why does placing groups 5 and 6 into a single taxon make it a monophyletic taxon?

4. Why does placing groups 1 and 2 into a single taxon make it a paraphyletic taxon?

5. Why does placing groups 2 and 4 into a single taxon make it a polyphyletic taxon?

Color the parts of the illustration below as indicated. Inside each large box, write the specific name of each category used in classifying the depicted organisms, and also indicate the specific epithet of the organism used to illustrate this hierarchical organization in the species box.

- RED ❑ species
- TAN ❑ domain
- GREEN ❑ phylum
- YELLOW ❑ class
- BLUE ❑ kingdom
- ORANGE ❑ order
- VIOLET ❑ family
- PINK ❑ genus

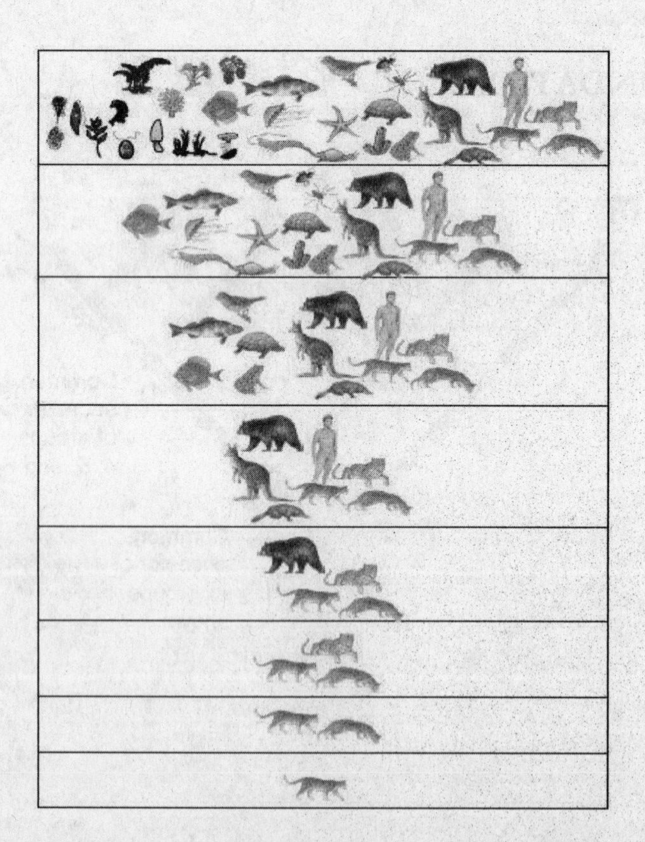

CHAPTER 24

Viruses and Subviral Agents

This chapter examines the diversity and characteristics of viruses and the still smaller subviral agents, satellites, viroids and prions. Viruses and subviral agents impact living organisms. Viruses themselves have some characteristics of living things, for example they are able to replicate, but they cannot survive and exist independently in the way cells can. Viruses consist of a core of DNA or RNA surrounded by a protein coat. Some are surrounded by an outer envelope too. Whereas some viruses may or may not kill their hosts, others invariably do. Some integrate their DNA into the host DNA, conferring new properties on the host. Viruses cause serious diseases in plants and animals. Satellites are subviral agents that depend upon co-infection of a host cell with a helper virus to facilitate their own replication. Viroids are the smallest known pathogens and they consist of RNA with no protein coat. Viroids are generally found within the host cell nucleus and appear to interfere with gene replication. Prions appear to consist of only protein. When present, prions can cause cells to malfunction.

REVIEWING CONCEPTS

Fill in the blanks.

INTRODUCTION

1. Pathogens are infective agents such as viruses or bacteria that cause ______________________.
2. Viruses adapt to their environment through (a) ______________________. Viruses can transfer (b) ______________________ into the genetic material of their host. They can also transfer (c)______________________ from one organism to another.

THE STATAUS AND STRUCTURE OF VIRUSES

3. The core of a virus is (a)______________________, and it is dependent upon (b)______________________________.
4. Viruses are considered as nonliving particles because they (a)____ and they (b)______________________________ nor can they (c)________________ on their own.

Viruses are very small

5. Viruses range in size from (a)______ to (b)________ nm.

A virus consists of nucleic acid surrounded by a protein coat

6. The core of a virus is surrounded by a protein coat called a ________.

7. A typical viral core is either (a)_______ or (b)_________ but (c)______________ hybrids have recently been discovered.

The capsid is a protective coat

8. The capsid coat consist of protein subunits called (a)______________ which determine the (b)____________ of the virus.
9. The three common shapes of viruses are (a)________________, (b)______________________ or (c)________________________.

Some viruses are surrounded by an envelope

10. The viral envelope surrounds the ____________________.
11. Typically, the virus obtains its envelope from ________________________________ as it leaves the host.
12. The viral envelope contains (a)______________, and (b)______________ from the host cell's plasma membrane and (c)______________ produced by the virus.

CLASSIFICATION OF VIRUSES

13. Viruses can be classified based on the types of organisms they infect or on their ____________________.

VIRAL REPLICATION

14. Viruses replicate inside host cells by taking over the (a)______________________ and (b)______________________ mechanisms of the host cell.

Bacteriophages infect bacteria

15. The most common structure of a bacteriophage consists of a long nucleic acid molecule coiled within a ______________________.
16. Most viruses have a __________ which may be contractile and may function to penetrate the host cell.
17. Phages have been used to treat ______________. This function has been replaced in many countries where antibiotics are available.

Viruses replicate inside host cells

18. The two types of reproductive cycles observed among viruses are (a)____________________ and (b)____________________.
19. Viruses that have only a lytic cycle are said to be ______________.
20. The five steps that are typical of lytic viral reproduction are (a)______________________, (b)__________________________, (c)______________, (d)______________________ (e)______________.

21. Bacteria protect themselves from viral infections by producing (a)_______________________________ that (b)____________ the phage DNA which prevents the phage DNA from replicating. The bacterial cell protects its own DNA by (c)_______________________ after replication
22. Viruses that do not always destroy the host cell are referred to as (a)________________________ viruses. These viruses alternate between a (b)__________________ cycle and a (c)____________________ cycle.
23. Viral DNA that is integrated into host DNA is called a (a)____________, and bacterial cells carrying integrated viral DNA are called (b)________________________.
24. ____________________________________ occurs when bacteria exhibit new properties as a result of integration of temperate viral DNA.

VIRAL DISEASES

25. Viruses that cause a disease are called _____________________________.

Viruses cause serious plant diseases

26. Most plant viruses have (a) ______________________ but do not have (b) ________________________________.
27. The genome of most plant viruses consists of ____ .
28. Viruses can only penetrate plant cells if ____

Viruses cause serious diseases in animals

29. The intentional use of microorganisms or toxins to cause death or disease in humans or in the organisms upon which humans depend is called ____.
30. A virus' survival depends upon the virus ____
31. RNA viruses called __________________________ transcribe their RNA genome into a DNA strand that is then used as a template to produce more viral RNA. HIV is one such virus.
32. The DNA polymerase used by retroviruses in transcription is called ______________________________________.
33. Viruses are not killed by ______________________.

EVOLUTION OF VIRUSES

34. According to the ________________________ hypothesis, viruses may have originated as mobile genetic elements such as transposoms or plasmids.

35. According to the ________________ hypothesis, viruses evolved from small independent cells that were parasites in larger cells.
36. According to the ________________ hypothesis, viruses appeared early in the history of life, even before the three domains diverged.
37. Particles that consist of multiple circles of dsDNA encased in capsid proteins and an envelope are called ________________.

SUBVIRAL AGENTS

Satellites depend on helper viruses

38. Satellites are ________________ agents that depend on co-infection of a host cell with a helper virus.

Viroids are short, single strands of naked RNA

39. Viroids have no (a)________________ and no associated proteins to assist in (b)________________.
40. Viroids are hardy and resist heat and ultraviolet radiation because of the condensed folding of their ______.
41. Viroids are generally found within the host cell (a) ____________ and appear to interfere with (b)________________.

Prions are protein particles

42. Prions are found in the brains of patients with fatal degenerative brain diseases called ________________.
43. Like viruses, prions exhibit ________________.

BUILDING WORDS

Use combinations of prefixes and suffixes to build words for the definitions that follow.

Prefixes	The Meaning	Suffixes	The Meaning
bacterio-	bacteria	-ate	characterized by having
caps-	box	-gen	production of
oblig-	obliged	-id	tending to
patho-	suffering, disease, feeling	-oid	resembling
vir-	poisonous slime	-phage	eat, devour
		-retro	backward

Prefix	Suffix	Definition
________	________	1. A virus that infects and destroys bacteria.
________	________	2. A short, circular, single strand of naked RNA.
________	________	3. A protein coat surrounding the nuclear core of a virus.
________	________	4. An intracellular parasite that can only survive using the resources of a host cell.
________	-virus	5. Viruses that transcribe the RNA genome into a DNA intermediate.
________	________	6. Any disease-producing organism.

MATCHING

Terms:

a. Attachment
b. Coevolution hypothesis
c. Capsid
d. Satillite
e. Retrovirus
f. Lysogenic conversion
g. Tamiflu
h. Regressive hypothesis
i. Envelope
j. Temperate
k. Virulent
l. Penetration
m. Viroid
n. Prion

For each of these definitions, select the correct matching term from the list above.

____ 1. States that viruses evolved from small independent cells that were parasites in larger cells.
____ 2. Found in the host cell nucleus and appear to interfere with replication.
____ 3. Viruses that cause disease and often death.
____ 4. Bacterial cells containing temperate viruses may exhibit new properties.
____ 5. Subviral agent dependent upon co-infection of a host cell with a helper virus.
____ 6. The protein covering of a virus.
____ 7. Outer membranous covering surrounding a capsid.
____ 8. An antiviral drug.
____ 9. Viruses that do not always destroy their host.
____10. Phage DNA enters the bacterial cell.

MAKING COMPARISONS

Fill in the blanks.

Characteristic	Viruses	Prions	Viroids
May contain RNA	yes	no	#1
May be covered by an envelope			
Can arise spontaneously			
Pathogenic			
Infect plants			

MAKING CHOICES

Place your answer(s) in the space provided. Some questions may have more than one correct answer.

____ 1. Viruses
a. are acellular.
b. do not carry on metabolic activities.
c. can only reproduce when occupying a cell.
d. do not produce rRNA.
e. are not included in any of the three domains.

____ 2. A virulent virus
a. has a lytic cycle.
d. degrades its host cell's nucleic acids.

b. can reproduce outside of cells. e. invade but do not destroy their host cell.
c. causes disease.

____ 3. Viruses

a. can be cultured in the laboratory using bacteria.

b. gave rise to prions.

c. respond to treatment with antibiotics.

d. are called prophages when their genome is integrated into the host DNA.

e. may have an envelope made of the host-cell plasma membrane.

____ 4. When a host bacterium exhibits new properties because of a prophage, the phenomenon is called

a. lysogenic conversion. d. viroid induction.
b. assembly. e. reverse transcription.
c. transduction.

____ 5. Viruses are usually grouped (classified)

a. from species to orders. d. into families with the suffix *viridae.*
b. based on their host range. e. in the traditional Linnean system.
c. into cellular and acellular groups.

____ 6. The characteristic(s) of life that viruses do not exhibit is/are

a. presence of nucleic acids. d. a cellular structure.
b. independent movement. e. independent metabolism.
c. reproduction.

____7. A small, circular piece of DNA that is separate from the main chromosome is

a. called a plasmid. d. found in archaebacteria and cyanobacteria, but not eubacteria.
b. a mesosome. e. found in gram-negative, but not gram-positive bacteria.
c. responsible for photosynthesis.

____8. Prions

a. are protein particles. d. have either a DNA or RNA core.

b. can arise spontaneously as the result of a mutation. e. can cause cells to malfunction.

c. are misfolded PrP proteins.

____9. The coat surrounding the nucleic acid core of a virus is

a. a capsid. d. a viral envelope.
b. made of protein. e. composed of capsomeres.
c. a virion.

____10. An important difference between virulent and temperate viruses is that only the temperate virus

a. destroys the host cell. d. contains DNA.
b. lyses the host cell. e. infects eukaryotic cells.
c. does not always lyse host cells in the lysogenic cycle.

____11. The virus that causes AIDS and some types of cancer is a/an

a. paramyxovirus. d. retrovirus.
b. herpesvirus. e. RNA virus.
c. adenovirus.

VISUAL FOUNDATIONS

Color the parts of the illustration below as indicated. Also label helical shape, polyhedral shape, and combination (helical and polyhedral) shape.

- RED ❑ DNA
- GREEN ❑ RNA
- VIOLET ❑ capsid
- ORANGE ❑ fibers
- BROWN ❑ tail

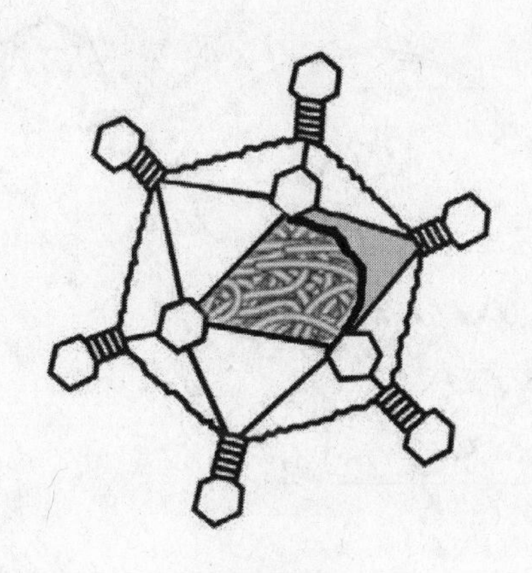

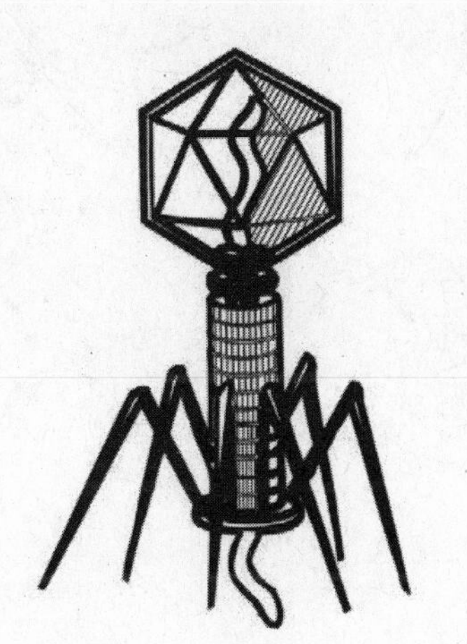

Color the parts of the illustration below as indicated. Also label these components inside of the nucleus: mRNA viral proteins, ribosomes, nucleus, glycoproteins.

RED ❑ envelope
GREEN ❑ capsid (everyplace it is present)
VIOLET ❑ mRNA
ORANGE ❑ endoplasmic reticulum
BROWN ❑ viral proteins
BLUE ❑ viral nucleic acid

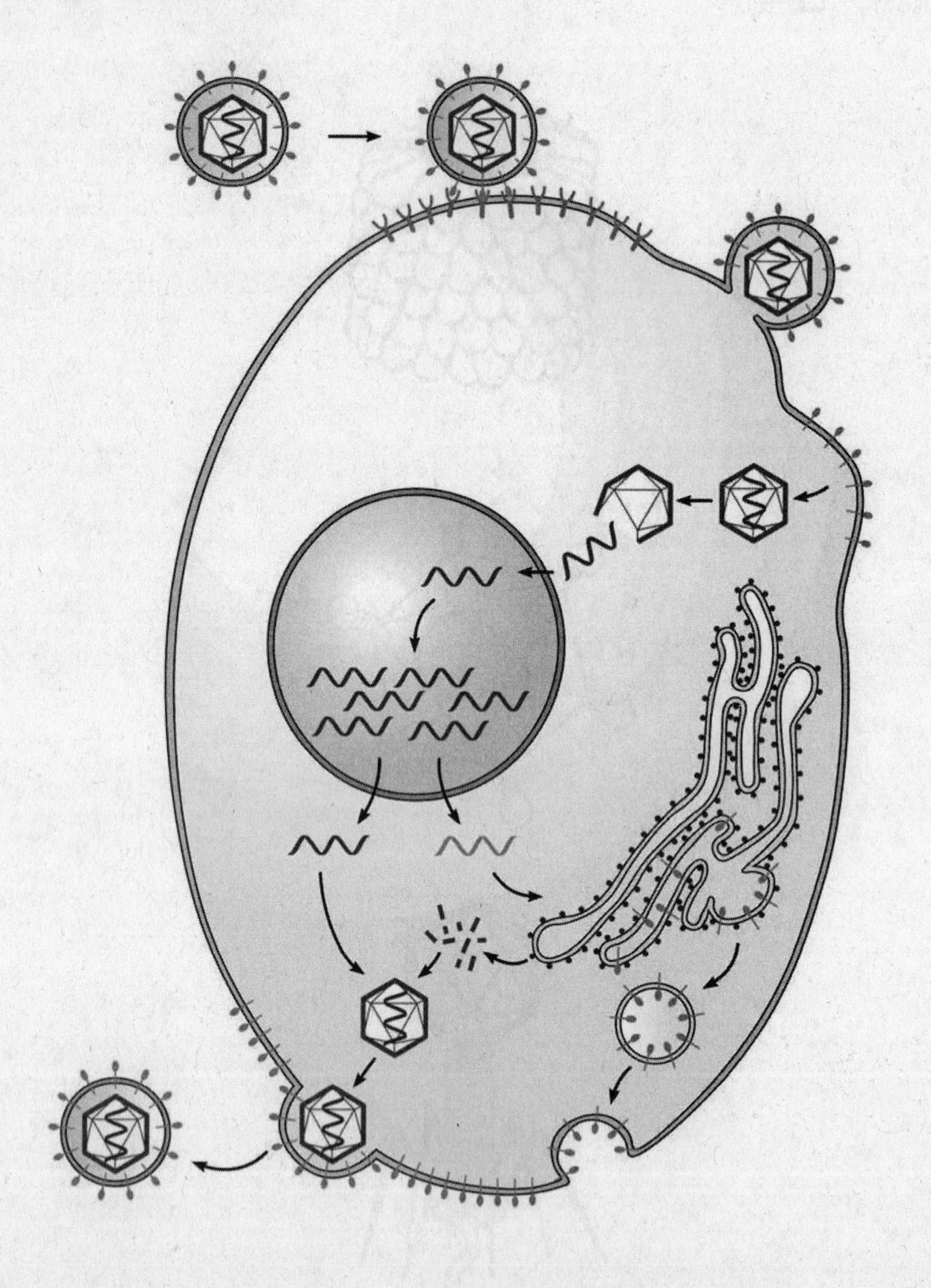

CHAPTER 25

❑

Bacteria and Archaea

This chapter examines two of the three domains, Bacteria and Archaea, often referred to as prokaryotes. These organisms are mostly unicellular, but some form colonies or filaments. They have ribosomes, but lack membrane-bounded organelles. In most the genetic material is a circular DNA molecule, but in some bacteria it is linear. Some have flagella. Whereas most prokaryotes get their nourishment either from dead organic matter or symbiotically from other organisms, some manufacture their own organic molecules. Although most prokaryotes reproduce asexually, some exchange genetic material. The former are thought to be the original prokaryotes from which all cellular life descended. Although most prokaryotes are harmless, a few are notorious for the diseases they cause. Prokaryotes have important medical and industrial uses.

REVIEWING CONCEPTS

Fill in the blanks.

INTRODUCTION

1. Pathogens are microorganisms such as bacteria, fungi, and protozoa that cause ______________.
2. Prokaryotes and fungi are nature's chief ______________.

THE STRUCTURE OF BACTERIA AND ARCHEA

3. Organisms in domains Archaea and Bacteria are ______________

Prokaryotes have several common shapes

4. Spherical prokaryotes are called (a)______________. They may occur in pairs called (b)______________, in chains called (c)______________, or in clumps or bunches called (d)______________.
5. Rod-shaped prokaryotes are known as ______________.
6. A rigid spiral-shaped prokaryote is known as a (a)______________, and a flexible spiral-shaped prokaryote is called a (b)______________.
7. A spirillum shaped like a comma is called a ______________.

Prokaryotic cells do not have membrane-enclosed organelles

8. In prokaryotes, there is a nuclear area that contains DNA and resembles the nucleus in eukaryotes; it is referred to as the ______________.

A cell wall protects most prokaryotes

9. The cell wall surrounds the (a)______________________. The rigid cell wall (b)______________ the cell, maintains the cell's (c)__________, and keeps the cell from (d)______________ under hypotonic stress.
10. The Eubacterial cell wall includes ______________, a polymer that consists of two unusual types of sugars linked with short polypeptides.
11. Bacteria that absorb and retain crystal violet stain are referred to as (a)______________ bacteria. Bacteria that do not retain the stain when rinsed with alcohol are called (b)____________________ bacteria.
12. Penicillin works most effectively against (a)__________________ bacteria because it interferes with (b)____________ synthesis.

Some bacteria produce capsules or slime layers

13. In some bacteria a capsule or slime layer surrounds the __________.
14. In free-living species of bacteria, the outer covering may provide the cell with protection against phagocytosis by ________________.
15. In disease-causing bacteria, a capsule or slime layer may protect against phagocytosis by ______________________________.

Some prokaryotes have fimbriae or pili

16. Fimbriae are made of (a) ______________ and are shorter than (b) ______________.

Some bacteria survive unfavorable conditions by forming endospores

17. Some bacteria form dormant, durable cells called endospores. However, unlike in fungi and plants, in bacteria the formation of endospores is not a form of ______________.

Many types of prokaryotes are motile

18. Prokaryotes use (a)__________________ consisting of a basal body, a hook, and a single filament as their main method of movement. Movement of bacteria in response to chemicals in the environment is called (b)______________.
19. In addition to swimming, some prokaryotes use their flagella to (a)__________, (b) to ______________, or to (c)______________________________.

PROKARYOTE REPORDUCTION AND EVOLUTION

20. In addition to their genomic DNA, many bacteria have small amounts of genetic information present as one or more ________________, which are circular fragments of DNA.

Rapid reproduction contributes to prokaryote success

21. The three different types of reproduction in prokaryotes are (a)________________________, (b)________________________, and (c)________________________.

Prokaryotes transfer genetic information

22. In prokaryotes when genetic material is transferred from parent to offspring (a)________________ gene transfer occurs. If the organism transfers genetic material to another organism that is not its offspring (b)________________ gene transfer occurs.
23. Horizontal gene transfer occurs in prokaryotes in at least three ways. Prokaryotic cells take up fragments of foreign DNA or RNA released by other prokaryotes in a process called (a)________________________. In (b)________________________, a phage carries bacterial genes from one bacterial cell into another. In the third way, prokaryotes of two different "mating types" may exchange genetic material during (c)________________________.

Evolution proceeds rapidly in bacterial populations

24. Because bacteria reproduce by binary fission, ________________________ are quickly passed on to new generations, and natural selection effects are quickly evident.

NUTRITIONAL AND METABOLIC ADAPTATIONS

25. Organisms that are able to use inorganic compounds as a source of carbon for making organic compounds are called ________________.
26. (a) ________________ obtain energy from chemical compounds and (b) ________________________ obtain energy from sunlight.
27. ________________ obtain their energy from dead organic matter.

Most propkaryotes require oxygen

28. Most bacterial cells are (a)______________, or require oxygen for cellular respiration. Some other bacteria, however, are (b)________________________ that use oxygen if it's available, or (c)________________________ that carry out anaerobic respiration.

Some prokaryotes fix and metabolize nitrogen

29. Organisms require nitrogen for the synthesis of (a) ______________ and (b) ________________.
30. Animals obtain nitrogen from organic compounds when they eat ________________________.

THE PHYLOGENY OF THE TWO PROKARYOTE DOMAINS

31. Under a microscope, most prokaryotes appear to be similar in (a)________ and (b) ____________.
32. The two domains of prokaryotes are (a) ________________________ and (b) ________________

Key characteristics distinguish the three domains

33. When compared to bacteria, archaea do not have ________________ in their cell walls.
34. In some ways, archaea are actually more like ______________ than bacteria.

Taxonomy of archaea and bacteria continuously changes

35. Prokaryote taxonomy is now based largely on molecular data, RNA sequencing and more recently on ______________________________.

Many archaea inhabit harsh environments

36. Archaea that require either a very high or a very low temperature for growth are considered to be ________________________.
37. ____________________________ are a group of archaea that inhabit oxygen-free environments such as sewage or swamps.
38. Heterotrophs that require large amounts of Na^{+} for growth are called ____________________________.

IMPACT ON ECOLOGY, TECHNOLOGY, AND COMMERCE

Prokaryotes form intimate relationships with other organisms

39. The three types of symbiotic relationships that can be observed when prokaryotes interact with other organisms are (a)__________________________ when both partners benefit from the relationship, (b) ____________________________ when one partner benefits and the other is neither harmed nor helped, and (c)__________________________ when one benefits and the other is harmed in some way as a result of the relationship.
40. Many bacteria that inhabit watery environments form dense films called (a) __________________ that attach to solid surfaces. (b)________________________________ that forms in the mouth is one example.

Prokaryotes play key ecological roles

41. The most numerous prokaryotes found in soil are _______________.
42. Plant growth is dependent upon (a) ______________ which must be continually added to the soil. (b) ____________________ bacteria that form mutualistic relationships with legumes are involved in this process.
43. During photosynthesis, many prokaryotes such as ________________ fix large amounts of carbon dioxide into organic molecules.

Prokaryotes are important in many commercial processes and in technology

44. Compounds produced by microorganisms that either inhibit the growth of or destroy other microorganisms are called ______________.

45. ____________________________ uses microorganisms to detoxify, the process of removing toxic chemicals, from the environment.

BACTERIA AND DISEASE

46. The community of harmless bacteria that inhabit the surface of the human skin is called a ____________________.

Many scientists have contributed tour understanding of infectious disease

47. The set of guidelines known as ______________________ are used to demonstrate that a specific pathogen causes specific disease symptoms.

Many adaptations contribute to pathogen success

48. To cause a disease, a pathogen must do the following three things:
(a) ____,
(b) ________________, and (c) ________________________________.

Antibiotic resistance is a major public health problem

49. Plasmids that have genes for antibiotic resistance to drugs are called ____________________.

50. A high percentage of infections acquired in hospitals involve bacteria inhabiting ___________________________.

BUILDING WORDS

Use combinations of prefixes and suffixes to build words for the definitions that follow.

Prefixes	The Meaning	Suffixes	The Meaning
an-	without, lacking	-ate	characterized by having
chem(o)-	chemical	-gen	production of
endo-	within	-ive	tending to
eu-	good, well, "true"	-karyo(te)	nucleus
exo-	outside, outer, external	-phil(e)	love
facult	faculty	-taxis	ordered movement
halo-	salt		
oblig-	obliged		
patho-	suffering, disease, feeling		

Prefix	Suffix	Definition
____________	____________	1. Movement toward or away from chemicals in the environment.
____________	-toxin	2. A poison that is secreted by bacterial cells into the external environment.
____________	-toxin	3. A poison that is a component "within" the cell wall of most gram-negative bacteria.

methano-	________	4. A bacterium that produces methane from carbon dioxide and water.
________	-spore	5. A thick-walled spore that forms within a bacterium in response to adverse conditions.
________	________	6. Any disease-producing organism.
________	-aerobe	7. An organism that metabolizes only in the absence of (without) molecular oxygen.
pro-	________	8. An organism that lacks a nuclear membrane.
________	________	9. An organism with a distinct nucleus surrounded by nuclear membranes.
________	________	10. Anerobic organisms that cannot survive in an environment containing oxygen.
________	________	11. Organisms that survive in environments that are high in salt.

MATCHING

Terms:

a. Bacillus
b. Binary fission
c. Capsule
d. Conjugation
e. Flagellum
f. F factor
g. Lysogenic conversion
h. Mycoplasma
i. Nucleoid
j. Peptidoglycan
k. Pilus
l. Plasmid
m. Saprobe
n. Transduction

For each of these definitions, select the correct matching term from the list above.

____ 1. An area of a prokaryote that contains DNA.
____ 2. A rod-shaped bacterium.
____ 3. An eubacterium that lacks cell walls.
____ 4. A structure used for locomotion in some bacteria.
____ 5. A process whereby one cell divides into two similar cells.
____ 6. A layer covering the cell wall of many prokaryotes.
____ 7. A component of eubacterial cell walls.
____ 8. A method of reproduction in single-celled organisms in which two cells link and exchange nuclear material.
____ 9. A type of genetic recombination resulting from transfer of genes from one organism to another by a virus.
____10. A hair-like structure associated with prokaryotes.
____11. A small, circular fragment of DNA
____12. A DNA sequence that is necessary for a bacterium to serve as a donor during conjugation.

MAKING COMPARISONS

Fill in the blanks.

Characteristic	Bacteria	Archaea	Eukarya
Peptidoglycan in cell wall	Present	Absent	#1
70S ribosomes	#2	#3	#4
Nuclear envelope	#5	#6	#7
Simple RNA polymerase	#8	#9	#10
Membrane-bounded organelles	#11	#12	#13

MAKING CHOICES

Place your answer(s) in the space provided. Some questions may have more than one correct answer.

____ 1. Prokaryotes
a. are cellular organisms.
b. may have fimbriae.
c. may transfer genetic information horizontally.
d. are divided into three domains.
e. may be gram-positive organisms.

____ 2. Conjugation
a. is a form of asexual reproduction.
b. occurs in bacteria.
c. is the exchange of plasmids.
d. is an example of horizontal gene transfer.
e. contributes to genetic variation.

____ 3. Autotrophs include
a. the majority of prokaryotes.
b. organisms relying on organic compounds as a carbon source.
c. most bacterial pathogens.
d. decomposers.
e. cyanobacteria.

____ 4. Prokaryote taxonomy relies on
a. genomic sequencing.
b. RNA sequencing
c. the use of bioflims.
d. molecular data.
e. nutritional source.

____ 5. The pathogen responsible for causing stomach ulcers is
a. Bordetella pertussis.
b. Bacillus anthracis.
c. Salmonella typhi.
d. Yersinia pestis.
e. Helicobacter pylori.

____ 6. The disease MRSA is caused by
a. Staphylococcus aureus.
b. plasmid transfer between gram negative bacteria.
c. a combination of methicillin- and vanomycin-resistant bacteria.
d. a drug resistant bacterium.
e. organisms that use horizontal gene transfer.

____ 7. The possible shape(s) for a prokaryote organism is/are
a. bacillus.
b. vibrio.
c. spirochete.
d. coccus.
e. spirillum.

____ 8. The majority of bacteria are
a. autotrophs.
b. heterotrophs.
c. pathogens.
d. saprobes.
e. photosynthesizers.

____ 9. Exchange of genetic material between bacteria sometimes occurs by
a. fusion of gametes.
b. conjugation.
c. incorporation by a bacterium of DNA fragments from another bacterium.
d. transfer of genes in bacteriophages.
e. transduction.

____10. A small, circular piece of DNA that is separate from the main chromosome is
a. called a plasmid.
b. called a mesosome.
c. responsible for photosynthesis.
d. found in archaebacteria and cyanobacteria, but not eubacteria.
e. found in gram-negative, but not gram-positive bacteria.

____11. Bacteria that retain crystal violet stain differ from those that do not retain the stain in having a
a. thick lipoprotein layer.
b. thick lipopolysaccharide layer.
c. thicker peptidoglycan layer.
d. cell wall.
e. mesosome.

____12. Prokaryotic cells are distinguished from eukaryotic cells by
a. absence of a nuclear envelope.
b. absence of mitochondria.
c. absence of DNA in the genetic material.
d. absence of a plasma membrane.
e. bacteriorhodopsin present in cell walls.

____ 13. The antibiotic penicillin
a. works best on gram-positive bacteria.
b. works best on gram-negative bacteria.
c. interferes with lipid synthesis.
d. interferes with peptidoglycan synthesis.
e. works best on bacteria with two protective layers.

____ 14. Many prokaryotes
a. have a capsule.
b. form endospores.
c. have outer layers made of polysaccharide or protein.
d. have pilli.
e. have a slime layer.

____ 15. Which word pair is incorrectly matched?
a. chemoheterotroph: energy source organic compounds
b. chemoautotroph: most bacterial pathogens
c. photoautotrophs: cyanobacteria
d. chemoautotrophs: oxidization of organic chemicals for energy source
e. prokaryotes: nitrification

____ 16. A common sexually transmitted disease in the U.S. is
a. chlamydia.
b. syphilis.
c. Hansen disease.
d. salmonella.
e. gonorrhea.

VISUAL FOUNDATIONS

Color the parts of the illustration below as indicated. Also label cell wall, flagellum, and pili.

RED	❑	plasma membrane
BROWN	❑	outer membrane
YELLOW	❑	peptidoglycan layer
BLUE	❑	DNA
TAN	❑	capsule
GREEN	❑	plasmid

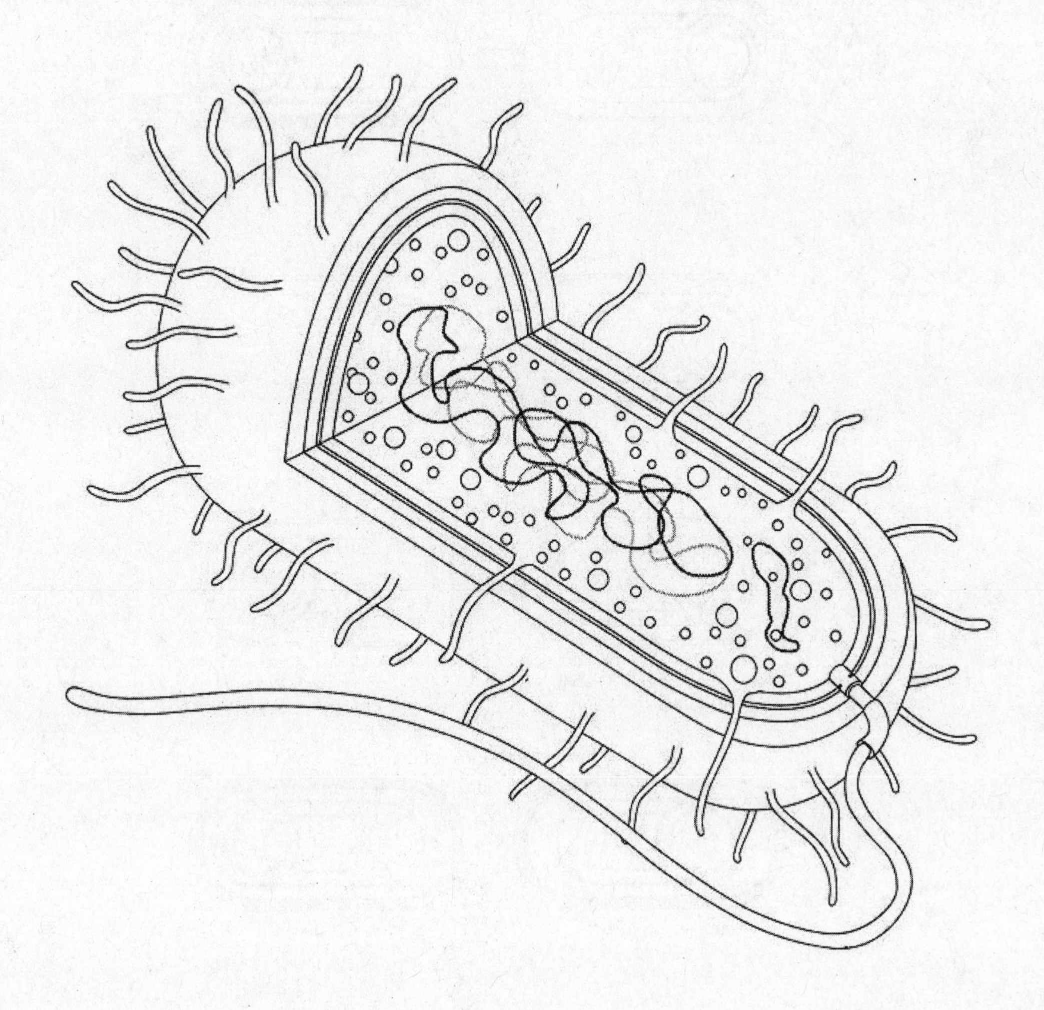

Color the parts of the illustration below as indicated.

RED	❑	bacterial chromosome
GREEN	❑	F plasmid
VIOLET	❑	outline of the F^+ donor cell
ORANGE	❑	outline of the F^- recipient cell
BROWN	❑	the conjugation tube

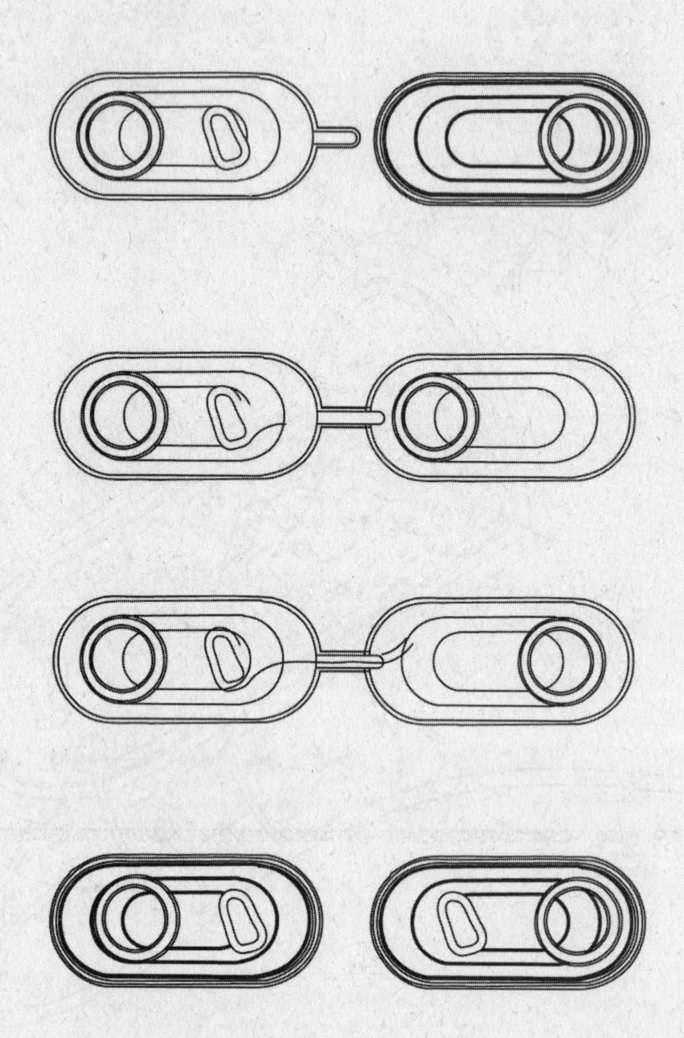

CHAPTER 26

Protists

The protist kingdom consists of a vast assortment of primarily aquatic eukaryotic organisms with diverse body forms, types of reproduction, modes of nutrition, and lifestyles. Biologists currently recognize dozens of protist taxa which are included in domain Eukarya and are part of the five "supergroups" of eukaryotes. Protists range in size from microscopic single cells to meters-long multicellular organisms. They do not have specialized tissues. Protists obtain their nutrients autotrophically or heterotrophically. Whereas some protists are free-living, others form symbiotic associations ranging from mutualism to parasitism. Reproductive strategies vary considerably; most reproduce both sexually and asexually, others only asexually. Whereas some are nonmotile, most move by flagella, pseudopodia, or cilia. Protists are believed to be the first eukaryotes. Evidence suggests that mitochondria and chloroplasts may have originated as prokaryotes that became incorporated into larger cells.

REVIEWING CONCEPTS

Fill in the blanks.

INTRODUCTION

1. Protists are included in domain (a) ________________ which means that protists have a (b) ________________.

DIVERSITY IN THE PROTISTS

2. Most protists have a body plan that is (a) ________________ and some form (b)________________.
3. Some protists are ________________ consisting of a multinucleate mass of cytoplasm.
4. Protists have several means of locomotion including cytoplasmic extensions called (a)________________, some flex (b)________________, glide over the substrate, wave (c)________________ or lash (d)________________.
5. Many protists are (a)________________, but some protists form (b)________________ relationships with other, unrelated organisms

HOW DID EUKARYOTES EVOLVE

6. Eukaryotes may have appeared in the fossil record ________ million years ago.

Mitochondria and chloroplasts probably originated from endosymbionts

7. In the ________________________hypothesis, eukaryotic organelles such as mitochondria and chloroplasts arose

from a symbiotic relationship between larger cells and smaller prokaryotes that were incorporated and lived in them.

8. Protists in the group ______________________ have a nonfunctional chloroplast surrounded by four membranes. This group includes *Plasmodium* which causes malaria.

A consensus in eukaryote classification is beginning to emerge

9. In some cases electron microscopy supports molecular data suggesting that certain protist taxa are (a)____________________ because they evolved from a common ancestor. However, the data show that other protists are actually a (b) ____________________ group because they do not share a common ancestor.

EXCAVATES

10. Excavates are so named because many have a deep, or excavated ____________________.
11. Unlike other protists, evacuates have atypical, greatly modified ____________________.

Diplomonads are small, mostly parasitic flagellates

12. Diplomonads are characterized as having one or two nuclei, no (a)______________________________, no (b)______________________________, and up to eight (c)____________________.
13. ______________________________ is a major cause of water-borne diarrhea throughout the world.

Parabasilids are anaerobic endosymbionts that live in animals

14. Trichonymphs live in the guts of (a) ____________________ and (b) ______________________________.

Euglenoids and trypanosomes include both free-living species and parasites

15. *Euglenoids* and *trypanosomes* have typical eukaryote flagella with a 9 + 2 microbular arrangement. Their flagella are unique because they contain a (a)____________________. Like other excavates these organisms also have (b)____________________.
16. About one-third of all euglenoids are ____________________.
17. Some euglenoids have a flexible outer covering called a ____________________.
18. Autotrophic euglenoids are single-celled, flagellated protists containing the same ____________________ pigments as those found in green algae and higher plants.
19. Trypanosomes have a single ____________________ that contains a deposit of DNA called a kinetoplastid.
20. The genus and species of the parasitic flagellate that causes African sleeping sickness is ______________________________.

CHROMALVEOLATES

21. Unifying features of alveolates include similar ribosomal DNA sequences and alveoli which are ________________________________ located just inside the plasma membrane.

Most dinoflagellates are a part of marine plankton

22. A typical dinoflagellate has two (a) ____________________ and alveoli containing interlocking (b) ____________________ plates.
23. Dinoflagellates that live in the bodies of marine invertebrates are called (a) ____________________. Using photosynthesis they produce (b) ________________________ for the invertebrate hosts. These dinoflagellates are particularly important to the survival of (c) ____________________ found in tropical shallow waters.

Apicomplexans are spore-forming parasites of animals

24. Apicomplexans contain the unpigmented remnant of a ____________________ derived from a red algae.
25. Apicomplexans lack specific structures such as flagella which are necessary for ________________________ and instead flex.
26. Apicomplexans have a specialized (a)____________________________ that is used to attach to its host cell. They also form a (b)____________________________ which protects them as they invade the host cell.
27. At some point in their life cycle, apicomplexans produce ____________________ which are small infective agents that are transmitted to a host.
28. An apicomplexan in the genus ____________________ can enter human RBCs and cause malaria

Ciliates use cilia for locomotion

29. Ciliates are distinct in having one or more small (a)____________________ that function during the sexual process and a larger (b)________________________ that regulates other cell functions.
30. "Cross fertilization" occurs among many ciliates during the sexual reproduction process known as ____________________.

Water molds produce biflagellate reproductive cells

31. The water molds, like the fungi, have a body called the ______________.
32. Water molds reproduce asexually by forming a structure called (a)____________________________ in which biflagellated (b)____________________ are produced and then released into the environment.
33. Water molds reproduce sexually when male and female nuclei fuse to form an __________________.

Diatoms are stramenophiles with shells composed of two parts

34. Diatoms are mostly single-celled, with ________________ impregnated in the cell walls that form the shell
35. Diatoms may have either (a) ______________________ or (b)______________________ symmetry.
36. Diatoms may live on rocks or sediments but they are also part of ______________________________________.
37. When diatoms die, their shells accumulate over time, forming sediment called ______________________ when it becomes exposed on land.

Brown algae are multicellular stramenopiles

38. Brown algae is the largest and most complex of all algae commonly called ____________
39. Brown algae reproduce using biflagellated cells. When reproducing sexually the cells are called (a) ______________ and when reproducing asexually the cells are called (b) ______________.
40. Brown algae spend part of their life as multicellular haploid organisms and part as multicellular diploid organisms. This type of a life cycle is an example of ____.

Most golden algae are unicellular biflagellates

41. Most golden algae are flagellated, ________________ organisms.
42. Golden algae may be covered in tiny scales of either (a)________________ or (b) ____________________________.
43. Golden algae compose much of the ocean's ____________________.

RHIZARIANS

44. Rhizarians often have a hard outer shell which is also called a ______.

Forams extend cytoplasmic projections that form a threadlike, interconnected net

45. Almost all foraminiferans live in a ________________ environment.
46. The dead bodies of foraminiferans are eventually transformed into ____________.
47. Foram fossils are used as ____________________ to help biologists identify rock layers.

Actinopods project slender axopods

48. The cytoplasmic extensions of actinopods are called (a) ____________. These extensions are strengthened by (b) ______________________.

ARCHAEPLASTIDS

49. The group archaeplastids include the (a)________________ and the (b)__________________________ as well as the (c)__________________________ which are in a separate kingdom.

Red algae do not produce motile cells

50. Most multicellular red algae attach to surfaces by means of a structure called the ________________.
51. Some red algae incorporate calcium carbonate into their cell walls and are called ____.
52. Two extracts of commercial value prepared from red algae are (a)__________________________, a polysaccharide which can be used as a food thickener, and (b) __________________________ another polysaccharide which is used as a food stabilizer in soft processed foods such as ice cream and chocolate milk.

Green algae share many similarities with land plants

53. Green algae have pigments, energy reserve products, and cell walls that are chemically identical to those found in ______________________.
54. Multicellular green algae does not have cell differentiated into __________ which is a feature that separates them from green plants.
55. In single celled forms of green algae, asexual reproduction occurs by (a) ____________ followed by (b)____________________________.
56. In multicellular forms of green algae, asexual reproduction occurs by ____________________________.
57. Some forms of green algae produce zoospores which are ____.

UNIKONTS

58. Unikonts are a supergroup of organisms that share the characteristic of having flagellated cells with a singular __________________________ although some extant members have lost this feature.
59. Bikonts are characterized as having __________________________.

Amoebozoa are unikonts with lobose pseudopodia

60. Pseudopodia that are *lobose* will be (a)______________ and (b)______________ as opposed to long and slender.
61. Because they have an extremely flexible outer membrane, amoebas have an ______________________ body form.
62. Amoebas glide along surfaces when their cytoplasm flow into extensions called __________________, meaning "false feet."

63. The parasitic species ______________________________ causes amoebic dysentery.
64. Plasmodial slime molds form intricate stalked reproductive structures called ________________, within which meiosis occurs.
65. When environmental conditions are suitable, spores open and one-celled haploid gametes of two types, one a flagellated form called a (a)________, the other an ameboid (b)________________, emerge and fuse to form a diploid zygote.
66. Slime molds have close affinities to two other organisms (a)________________________ and (b)________________.
67. The single flagellum of the choanoflagellates is surrounded at the base by a delicate collar of ______________.

BUILDING WORDS

Use combinations of prefixes and suffixes to build words for the definitions that follow.

Prefixes	The Meaning	Suffixes	The Meaning
conju-	to join together	-ation	the process of
cyto-	cell	-cle	little
iso-	equal, "same"	-cytosis	cellular process
macro-	large, long, great, excessive	-ile	having the character of
micro-	small	-pod(ium)	foot, footed
pell(i)-	skin	-zoa	animal
phag(o)-	to eat		
proto-	first, earliest form of		
pseudo-	false		
sess-	without a stem		

PrefixSuffix		Definition
________	________	1. Single-celled, animal-like protists including amoebae, ciliates, flagellates, and sporozoans; members of this group are believed to have given rise to the earliest form of animal life.
________	________	2. An outer flexible covering associated with euglenoids.
________	________	3. A temporary protrusion of the cytoplasm of an ameboid cell that the cell uses for feeding and locomotion; a "false foot."
________	-nucleus	4. A small nucleus found in ciliates.
________	-nucleus	5. A large nucleus found in ciliates.
________	________	6. A sexual union in which genetic material is exchanged between cells.
________	________	7. Species that are attached to the substrate as opposed to being free living.
________	________	8. Describes what occurs when one organism surrounds and ingests another for food.

MATCHING

Terms:

a.	Actinopod	f.	Hypha	k.	Zoosporangium
b.	Axopod	g.	Mycelium	l.	Sporozoite
c.	Conjugation	h.	Oospore	m.	Excavates
d.	Foraminiferan	i.	Plankton	n.	Trichonympha
e.	Holdfast	j.	Plasmodium	o.	Giardia

For each of these definitions, select the correct matching term from the list above.

____ 1. One of the long, filamentous cytoplasmic projections that protrude through pores in the skeletons of actinopods.

____ 2. A structure that produces tiny biflagellate zoospores.

____ 3. The vegetative body of fungi; consists of a mass of hyphae.

____ 4. Free-floating, mainly microscopic aquatic organisms found in the upper layers of the water.

____ 5. Marine protozoans that produce chalky, many-chambered shells or tests.

____ 6. A sexual process in which two individuals come together and exchange genetic material.

____ 7. One of the filaments composing the mycelium of a fungus.

____ 8. The basal root-like structure in multicellular algae that anchors the organism to a solid surface.

____ 9. Multinucleate mass of cytoplasm that constitutes the feeding stage in the life cycle of slime molds.

____10. Small infective agents produced by apicomplexans that are transmitted to the next host.

____11. A diverse group of unicellular protists with flagella.

____12. A major cause of water-borne diarrhea around the world.

MAKING CHOICES

Place your answer(s) in the space provided. Some questions may have more than one correct answer.

____ 1. Protists may be
a. unicellular.
b. colonial.
c. simple multicellular organisms.
d. eukayotic.
e. Archaea.

____ 2. Mitochondria and chloroplasts are thought to have been derived from
a. complex viruses.
b. protozoa.
c. endosymbionts.
d. intrasymbiotic eukaryotes.
e. aerobic bacteria.

____ 3. Excavates
a. have ciliate.
b. are colonial organisms.
c. are generally endosymbionts.
d. obtain energy through fermentation.
e. are flagellated.

____ 4. The *Paramecium*
a. is a ciliate.
b. is eukaryotic.
c. is colonial, sometimes multicellular.
d. has a pellicle.
e. has multiple nuclei.

____ 5. The parasite that causes malaria is in the phylum
a. Ciliophora.
b. Apicomplexa.
c. Sarcomastigophora.
d. Dinoflagellata.
e. Euglenophyta.

____ 6. A protozoan whose cells bear a striking resemblance to specialized cells in sponges is the
a. foraminiferan.
b. diatom.
c. choanoflagellate.
d. ameba.
e. radiolarian.

____ 7. One of the reasons that protists are placed in eukaryote super groups rather than kingdom Protista or Plantae is that many of them possess both plantlike and animal-like characteristics, which is particularly well-illustrated in the genus
a. *Paramecium.*
b. *Euglena.*
c. *Plasmodium.*
d. *Ameba.*
e. *Didinium.*

____ 8. Members of the genus *Euglena*
a. are flagellated.
b. do not possess a pellicle.
c. possess chlorophyll.
d. are heterotropic.
e. are autotrophic.

____ 9. A symbiotic relationship in which both "partners" benefit is known as
a. commensalism.
b. mutualism.
c. parasitism.
d. endosymbiosis.
e. fraternism.

____10. A symbiotic relationship in which one "partner" benefits and the other is unaffected is known as
a. commensalism.
b. mutualism.
c. parasitism.
d. endosymbiosis.
e. fraternism.

____11. Protists can be found in the supergroup
a. Chromalveolates.
b. Fungi.
c. Amoebozoans.
d. Rhizarians.
e. Ciliates.

____ 12. Red tide
a. is caused by a toxin produced by dinoflagellates.
b. is caused by an organism in the supergroup excavates.
c. is responsible for large fish kills.
d. occurs in nutrient-rich, warm waters.
e. occurs when apicomplexans infect marine invertebrates.

____13. The diploid stage of a *Plasmodium* life cycle occurs in
a. humans.
b. red blood cells.
c. liver cells.
d. mice.
e. mosquitoes.

____ 14. Because of their similarity to other organisms, water molds were once classified as

a. bread molds.
b. unikonts.
c. fungi.
d. algae.
e. Rhizarians.

____15. Cellular slime molds

a. are unikonts.
b. reproduces by meiosis.
c. reproduce using spores.
d. are classified as amoebozoa.
e. form a slug when conditions are not optimal.

VISUAL FOUNDATIONS

Color the parts of the illustration below as indicated. Also label cilia.

RED ❑ micronucleus
ORANGE ❑ oral groove
BLUE ❑ contractile vacuole
YELLOW ❑ food vacuole
TAN ❑ anal pore
VIOLET ❑ macronucleus

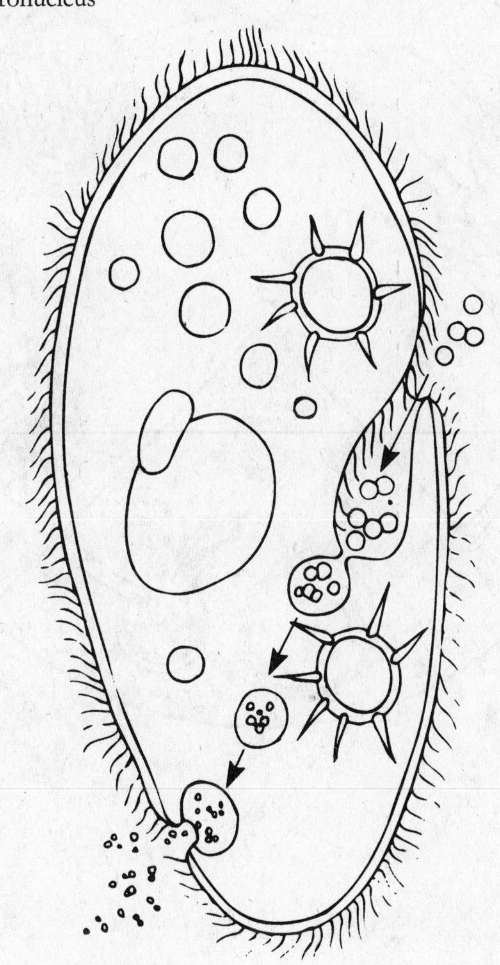

Color the parts of the illustration below as indicated. Label the portions of the illustration depicting sexual reproduction, and the portions depicting asexual reproduction. Also label diploid generation, haploid generation, mitosis, meiosis, and fertilization.

- RED ❑ positive strain
- GREEN ❑ negative strain
- YELLOW ❑ zygote

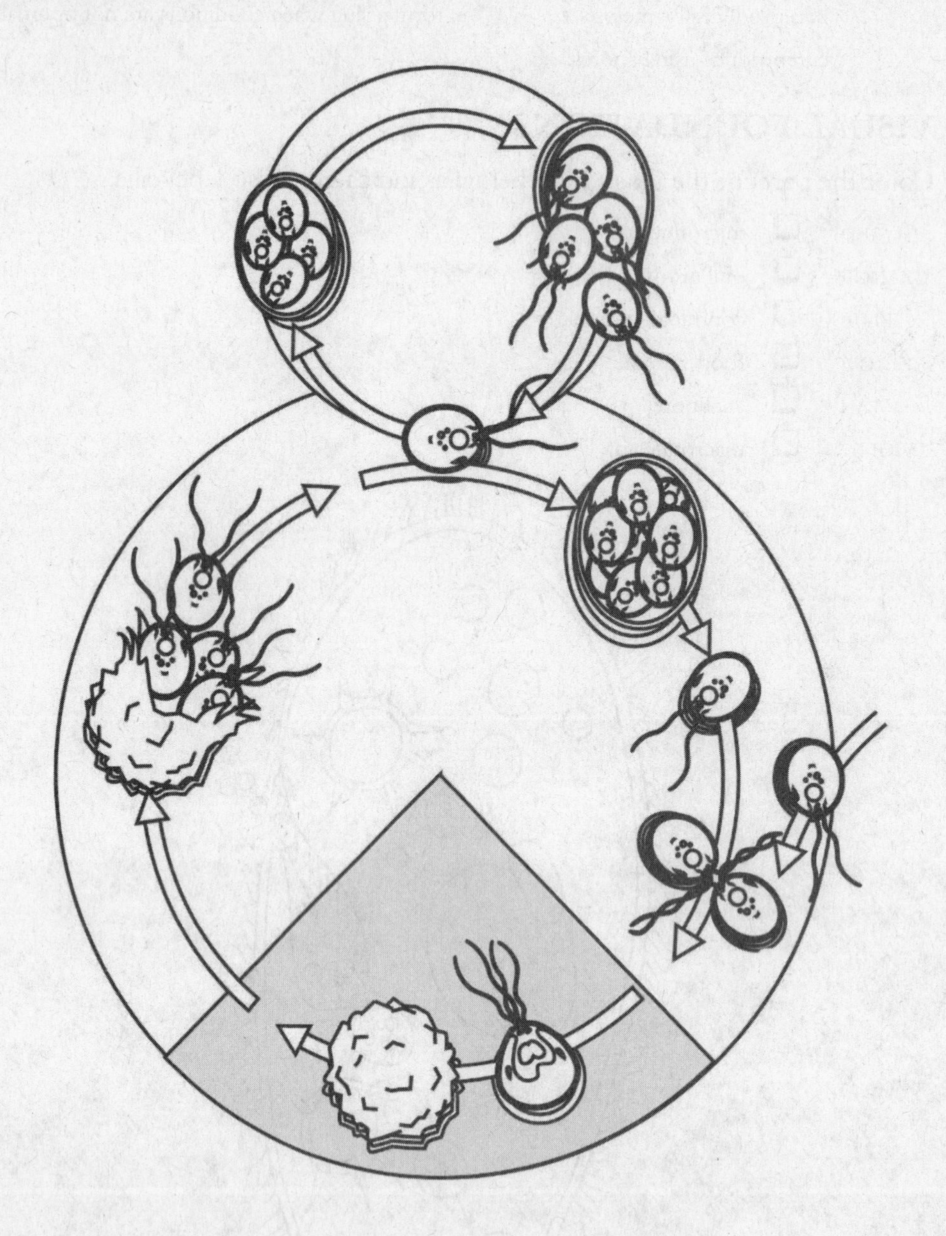

CHAPTER 27

❑

Seedless Plants

Plants, believed to have evolved from green algae, are multicellular and exhibit great diversity in size, habitat, and form. They photosynthesize to obtain their energy. To survive on land, plants evolved 1) a waxy cuticle to protect against water loss, 2) openings in the surface layer that allow for gas exchange, 3) a strengthening polymer in the cell walls that enables them to grow tall, and 4) multicellular sex organs that protect the developing embryo. The plant kingdom consists of four major groups — bryophytes, seedless vascular plants, gymnosperms, and angiosperms. Except for the bryophytes, all plants have a vascular system. A vascular system allows plants to achieve larger sizes because water and nutrients can be transported over great distances to all parts of the plant. The focus of this chapter is on the seedless plants, i.e., the bryophytes and the ferns and their allies. Unlike the bryophytes, the ferns and their allies also have a larger and more dominant diploid stage, which is a trend in the evolution of land plants. Most of the seedless plants produce only one kind of spore as a result of meiosis. A few, however, produce two types of spores, an evolutionary development that led to the evolution of seed plants.

REVIEWING CONCEPTS

Fill in the blanks.

INTRODUCTION

1. Because of the numerous characteristics that they share, plants are thought to have evolved from ancient forms of ____________________.
2. Plants and the group from which they are thought to be derived have the same photosynthetic pigments. They both contain chlorophylls (a)________and ________, and the carotenoids including (b)________________________ (yellow pigments) and (c)________________________ (orange pigments).
3. Green algae and plants store excess carbohydrates as (a) ____________ and have (b) ______________ as a major component of their cell walls.
4. Plants have a character that distinguishes them from green algae. That characteristic is development from multicellular embryos enclosed in ____.

ADAPTATIONS OF PLANTS TO LIFE ON LAND

5. The carbon used by plants to make sugars and other organic molecules is obtained from ______________________ present in the atmosphere.

6. The adaptation of plants to land environments involved a number of important adaptations. Among them was the (a) ________________ that covers aerial parts, protecting the plant against water loss. Openings in this covering also developed. These (b) ___________ facilitate the gas exchange that is necessary for photosynthesis.
7. The sex organs of plants, or _________________, as they are called, are multicellular structures containing gametes.

The plant life cycle alternates between haploid and diploid generations

8. Plants live one part of their lives in a multicellular haploid stage and another part in a multicellular diploid stage. Having two stages to the life cycle is referred to as ______________________________.
9. The haploid part of the life cycle is called the (a)______________________________________ and the diploid part of the life cycle is called the (b) ______________________.
10. In plants, the male gametes form in the (a) _________________, and the female gametes in the (b) _______________.
11. In the plant life cycle, the first stage in the sporophyte generation is the (a) ______________________________, and the haploid (b)___________ are the first stage of the gametophyte generation.

Four major groups of plants exist today

12. Molecular data and structural indicate that land plants probably descended from a group of green algae known as ________________.
13. Of the four major groups of plants only the (a)____ lack a vascular (conducting) system. This group reproduces via (b)_____________. The vascular plants include the (c)___________________________, (d)___________________________, and (e)___________________________.
14. The two vascular tissues present in vascular plants are the (a)_____________________ and the (b)_____________________.
15. ___________ is a polymer in the cell walls of large, vascular plants that strengthens and supports the plant and its conducting tissues.
16. The ___________________________ are vascular plants that reproduce using seeds.

BRYOPHYTES

17. Bryophytes can be separated into three phyla. The phyla are (a)____, (b)____, and (c)____.

Moss gametophytes are differentiated into "leaves" and "stems"

18. Each moss plant has tiny, hair like absorptive structures called ______________.
19. In mosses, fertilization occurs in the (a) ________________________. The transfer of gametes relies upon (b) ____________________.
20. The diploid zygote of mosses develops into a mature ______________.
21. After it germinates, a moss spore grows into a filament of cells called a ____________________
22. Among the most important mosses are those with large water storage vacuoles in their cells. These mosses are in the genus (a) ____________, and are known by their common name, the (b) ____________.

Liverwort gametophytes are either thalloid or leafy

23. The body of a liverwort is a flattened, lobed structure called a ____________.
24. Liverworts reproduce sexually and asexually. In sexual reproduction, the haploid gametangia are called (a)___________________________, and (b)___________________________.
25. During asexual reproduction in liverworts, tiny balls of tissue called____ form on the thallus, and after they are dispersed by raindrops they will develop into a new liverwort.

Hornwort gametophytes are inconspicuous thalloid plants

26. Hornworts belong to the phylum ____________________________ and live in disturbed habitats such as fallow fields and roadsides.
27. A unique feature of hornworts is that the sporophytes continue to grow for the remainder of the gametophyte's life. Botanists refer to this characteristic as ____.

Bryophytes are used for experimental studies

28. Botanists use certain bryophytes as experimental models to study ________________________, i.e., plant responses to varying periods of night and day length.

Recap: details of bryophyte evolution are based on fossils and on structural and molecular evidence

29. Plants are considered to be a __________________________ group.
30. Fossil evidence indicates that ______________________ are probably the first group of plants to arise from the common plant ancestor.

SEEDLESS VASCULAR PLANTS

31. There are two basic types of true leaves: the small (a)_______________ with its single vascular strand and the larger (b)__________________ with multiple vascular strands. Of these two types of leaves, the (c)__________________ represent the majority of leaves we see today.
32. The two main clades of seedless vascular plants are (a)________________ and (b)____________________.

Club mosses are small plants with rhizomes and short, erect branches

33. Club mosses were major contributors to our present-day ____________________.

Ferns are a diverse group of spore-forming vascular plants

34. In ferns, the sporophyte is composed mainly of an underground stem called the (a) ______________, from which extend true roots and leaves called (b) _____________.
35. (a)______________, the spore cases on the leaves of ferns, often occur in clusters called (b)________. The mature fern gametophyte is a tiny, green, often heart-shaped structure called a (c)______________.

Whisk ferns are classified as reduced ferns

36. The main organs of photosynthesis in Psilotum are the (a)____________________ that exhibit (b)____________________ branching (dividing into two equal halves).
37. The small haploid gametophyte of the whisk fern is a non-photosynthetic subterranean plant that apparently obtains nourishment through a symbiotic relationship with a ______________.

Horsetails are an evolutionary line of ferns

38. Horsetails are found on every continent except ________________.
39. The stems of horsetails are impregnated with ________________.
40. The sporangia of horsetails are located in a conelike region at the terminal end of a reproductive branch. This region is called a ____________________.

Some ferns and club mosses are heterosporous

41. Bryophytes, horsetails, whisk ferns, and most ferns and club mosses produce only one type of spore, a condition known as (a)______________. Some ferns and club mosses exhibit (b)__________________, the production of two different types of spores. The spores are either (c)______________________ which develop into male gametophytes or (d)

_________________________ which develop into female gametophytes.

Seedless vascular plants are used for experimental studies

42. The ______________________________ is the area at the tip of the root or shoot where growth occurs.
43. Ferns and other seedless plants have a single large _______________ which is the source of all cells that make up the root or shoot.

Seedless vascular plants arose more than 420 mya

44. Microscopic spores of early vascular plants appear in the fossil record earlier than ________________________________, suggesting that even older megafossils of simple vascular plants may be discovered.

BUILDING WORDS

Use combinations of prefixes and suffixes to build words for the definitions that follow.

Prefixes	The Meaning	Suffixes	The Meaning
arch(e)-	primitive	-angi(o)(um)	vessel, container
bryo-	moss	-gon(o)(ium)	sexual, reproductive
gamet(o)-	sex cells, eggs and sperm	-(o)logy	study of
hetero-	different, other	-phyte	plant
homo-	same	-spor(e)(y)	spore
mega-	large, great	-us	thing
micro-	small		
spor(o)-	spore		
strobil-	a cone		
thall-	a young shoot		
xantho-	yellow		

Prefix	Suffix	Definition
____________	-phyll	1. Yellow plant pigment.
____________	____________	2. Special structure (container) of plants, protists, and fungi in which gametes are formed.
____________	____________	3. The female reproductive organ in primitive land plants.
____________	____________	4. The gamete-producing stage in the life cycle of a plant.
____________	____________	5. The spore-producing stage in the life cycle of a plant.
____________	-phyll	6. A small leaf that contains one vascular strand.
____________	-phyll	7. A large leaf that contains multiple vascular strands.
____________	____________	8. Special structure (container) of certain plants and protists in which spores and sporelike bodies are produced.
____________	____________	9. Production of one type of spore in plants (all spores are the same type).
____________	____________	10. Production of two different types of spores in plants, microspores and megaspores.

__________ __________ 11. Large spore formed in a megasporangium.

__________ __________ 12. Small spore formed in a microsporangium.

__________ __________ 13. Mosses, liverworts, and their relatives.

__________ __________ 14. The study of mosses.

__________ __________ 15. A plant body not developed into true roots, stems, and leaves.

__________ __________ 16. An elongated, cone-like reproductive structure.

MATCHING

Terms:

a. Antheridium
b. Dichotomous
c. Gametangia
d. Lignin
e. Liverwort
f. Megaphylls
g. Phloem
h. Prothallus
i. Protonema
j. Rhizoid
k. Rhizome
l. Sorus
m. Stoma
n. Strobilus
o. Thallus
p. Xylem

For each of these definitions, select the correct matching term from the list above.

____ 1. Vascular tissue that conducts dissolved organic molecules in plants.

____ 2. A strengthening polymer found in the walls of cells that function for support and conduction..

____ 3. Hair-like absorptive structures similar in function to roots that extend from the base of the stem of mosses, liverworts, and fern prothallia.

____ 4. A cluster of sporangia (in the ferns).

____ 5. A type of branching in which the branches or veins always branch into two more or less equal parts.

____ 6. The heart-shaped, haploid gametophyte plant found in ferns, whisk ferns, club mosses, and horsetails.

____ 7. The male gametangium in certain plants.

____ 8. The sex organs of plants.

____ 9. Vascular tissue that conducts water and dissolved minerals in plants.

____10. A horizontal underground stem.

____11. A bryophyte that looks like an internal human organ.

____12. Reduced leaves.

MAKING COMPARISONS

Fill in the blanks.

Plant	Nonvascular or Vascular	Method of Reproduction	Dominant Generation
Hornworts	Nonvascular	Seedless, reproduce by spores	Gametophyte
Angiosperms	#1	Seeds enclosed within a fruit	#2
Ferns	#3	Seedless, reproduce by spores	#4
Club mosses	#5	Seedless, reproduce by spores	#6

Plant	Nonvascular or Vascular	Method of Reproduction	Dominant Generation
Gymnosperms	#7	#8	Sporophyte
Mosses	#9	#10	#11
Horsetails	Vascular	#12	Sporophyte
Whisk ferns	Vascular	#13	#14

MAKING CHOICES

Place your answer(s) in the space provided. Some questions may have more than one correct answer.

____ 1. Spores grow
a. into gametophyte plants.
b. into sporophyte plants.
c. into a haploid plant.
d. to form a plant body by meiosis.
e. to form a plant body by mitosis.

____ 2. Plants all have
a. chlorophyll a.
b. chlorophyll b.
c. xanthophyll.
d. carotene.
e. yellow pigments.

____ 3. A strobilus is
a. on a diploid plant.
b. on a haploid plant.
c. on a vascular plant.
d. found on horsetails.
e. found on plants that do not have true leaves.

____ 4. The gametophyte generation of a plant
a. is diploid.
b. is haploid.
c. produces haploid spores.
d. produces haploid gametes by mitosis.
e. produces haploid gametes by meiosis.

____ 5. When a gamete produced by an archegonium fuses with a gamete produced by a male gametophyte plant, the result is
a. an anomaly.
b. a diploid zygote.
c. the first stage in the sporophyte generation.
d. the first stage in the gametophyte generation.
e. called fertilization.

____ 6. Plant sperm cells form in
a. diploid gametophyte plants.
b. haploid sporophyte plants.
c. haploid gametophyte plants.
d. antheridia.
e. archegonia.

____ 7. The leafy green part of a moss is the
a. sporophyte generation.
b. gametophyte generation.
c. rhizoid.
d. product of buds from a protonema.
e. thallus.

____ 8. The sporophyte generation of a plant
a. is diploid.
b. is haploid.
c. produces haploid gametes.
d. produces haploid spores by mitosis.
e. produces haploid spores by meiosis.

____ 9. Liverworts
a. are bryophytes.
b. are vascular plants.
c. can produce archegonia and antheridia on a haploid gametophyte.
d. contain a medicine that cures liver disease.
e. are in the same class as hornworts.

____10. The spore cases on a fern are
a. usually on the fronds.
b. formed by the haploid generation.
c. called sporangia.
d. often arranged in a sorus.
e. precursors to the fiddlehead.

____11. The leaves of vascular plants that evolved from stem branches
a. are megaphylls.
b. are microphylls.
c. evolutionarily derived from stem tissue.
d. contain one vascular strand.
e. contain more than one vascular strand.

____12. Land plants are thought to have evolved from
a. fungi.
b. green algae.
c. bryophytes.
d. charophytes.
e. mosses.

____13. All plants
a. have a cuticle covering the aerial portion of the plant.
b. develop from multicellular embryos.
c. are a monophyletic group.
d. store starch.
e. evolved from a green algae.

____14. The water conducting portion of the plant vasculature is the
a. phloem.
b. protonema.
c. microphyll.
d. xylem.
e. gemmae.

____15. Red algae, green algae, and land plants are collectively classified as
a. charophytes.
b. archaeplastids.
c. Anthocerophyta.
d. bryophytes.
e. Hepatophyta.

____16. Whisk ferns
a. are found mainly in the tropics and subtropics.
b. are extinct.
c. should be classified as reduced ferns.
d. are in phhlum Pteridophyta.
e. have vascularized stems.

VISUAL FOUNDATIONS

Color the parts of the illustration below as indicated. Also label haploid gametophyte generation, diploid sporophyte generation, fertilization, and meiosis.

RED	❑	zygote	BROWN	❑	capsule
GREEN	❑	archegonium	TAN	❑	gametophyte plants
YELLOW	❑	egg	PINK	❑	sporophyte
BLUE	❑	sperm	VIOLET	❑	spore
ORANGE	❑	antheridium			

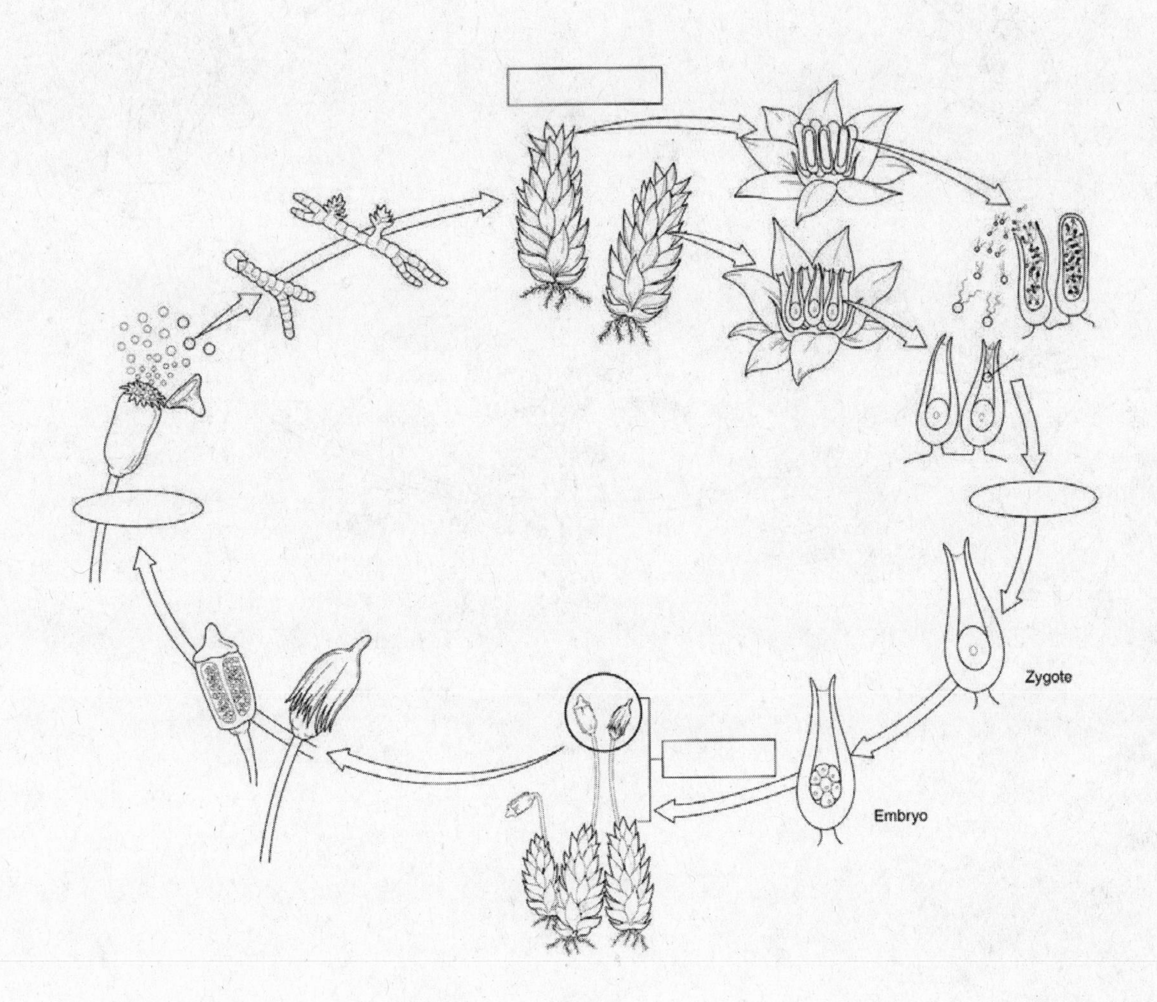

Color the parts of the illustration below as indicated.

RED	❑ zygote	BROWN	❑ sporangium
GREEN	❑ archegonium	TAN	❑ gametophyte
YELLOW	❑ egg	PINK	❑ mature sporophyte
BLUE	❑ sperm	VIOLET	❑ spores

CHAPTER 28

❑

Seed Plants

The most successful plants on earth are the seed plants, also known as the gymnosperms and angiosperms (flowering plants). Whereas gymnosperm seeds are totally exposed or borne on the scales of cones, those of angiosperms are encased within a fruit. All seeds are multicellular, consisting of an embryonic root, stem, and leaves. They are protected by a seed coat that allows them to survive until conditions are favorable for germination, and they contain a food supply that nourishes the seed until it becomes self-sufficient. Seed plants have vascular tissue, and they exhibit alternation of generations in which the gametophyte generation is significantly reduced. They produce both microspores and megaspores. There are four phyla of gymnosperms and one of angiosperms. Seeds and seed plants have been intimately connected with the development of human civilization. Human survival, in fact, is dependent on angiosperms, for they are a major source of human food. Angiosperm products play a major role in world economics. The organ of sexual reproduction in angiosperms is the flower. Flowering plants have several evolutionary advancements that account for their success. The fossil record indicates that seed plants evolved from seedless vascular plants.

REVIEWING CONCEPTS

Fill in the blanks.

INTRODUCTION

1. The primary means of reproduction in plants is by ________________.
2. Seeds are reproductively superior to spores because seeds are (a)________________________________, contain (b)____, and seeds are protected by (c)________________________________.
3. Although most of the seeds eaten by humans are produced by flowering plants, the seeds of the ____________________________, a gymnosperm, may also be eaten.

AN INTRODUCTION TO SEED PLANTS

4. The dominant stage of seed plants is the (a)________________________. The (b)________________________ stage is greatly reduced in size.
5. Seed plants produce (a)____________ which are individual sporangium. After fertilization, the ovule develops into a (b)____________.
6. The two groups of seed plants are the (a)__________________, from the Greek for "naked seed" because the seeds are not surrounded by

an (b)__________________ and the (c)________________, meaning "seed enclosed in a vessel or case."

7. Seed plants all have two types of vascular tissues: (a)_________ for conducting water and minerals and (b)___________ for conducting dissolved sugar.

GYMNOSPERMS

8. The four phyla of plants with "naked seeds" are: (a)______________________, (b)______________________, (c)______________________, and (d)______________________.

Conifers are woody plants that produce seeds in cones

9. Many conifers produce a clear substance called ______________ which helps to protect the plants from attack by insects and fungi.
10. The conifers are the largest group of gymnosperms. Most of them are ______________, which means that the male and female reproductive organs are in different places on the same plant.
11. The anticancer drug ______________ is produced from conifers.

Pines represent a typical conifer life cycle

12. The largest genus (a) ___________ contains the largest number of conifers and pine trees are mature (b) ______________.
13. The leaf-like scales that bear sporangia on male cones of conifers are called (a)______________, at the base of which are two microsporangia containing many "microspore mother cells," also called (b)__________________. These cells develop into the (c)______________________.
14. Female cones are also called (a)______________have (b)______________ on the upper surface of each cone scale, within which are megasporocytes. Each of these cells produces four haploid (c)______________________, three of which disintegrate, while the fourth develops into the egg producing (d)______________.
15. One of the pollen grains that adheres to the sticky surface of the ovule grows to form a ______________________ which digests its way through the megasporangium until it reaches the egg where two sperm cells will be released.
16. A major adaptation of the pine life cycle was the elimination of the need for (a)__________________ and the use of (b) ________ as a means of dispersing the sperm.
17. Reproduction in conifers is totally adapted for life __________.

Cycads have seed cones and compound leaves

18. Cycads are in the phylum ____________________.
19. The cycads reproduce in a manner similar to pines, except cycads are ____________, meaning that male cones and female cones are on *separate* plants.

***Ginkgo biloba* is the only living species in its phylum**

20. *Ginko biloba* is in the phylum ____________________.
21. Unlike conifers, the seeds of *G. biloba* are ____________.

Gnetophytes include three unusual genera

22. Gnetophytes belong to the phylum ______________.
23. The gnetophytes share a number of advances over the rest of the gymnosperms, one of which is the presence of distinct ____________ in their xylem tissues.
24. The gnetophyte genus ______________ contains only one species which is found in southwestern African deserts.

FLOWERING PLANTS

25. Flowering plants are (a) ______________ plants that reproduce (b)____________________ by forming flowers.
26. Flowering plants have efficient water-conducting cells called (a)____________________ in their xylem and efficient sugar-conducting cells called (b)____________________ in their phloem..

Monocots and eudicots are the two largest classes of flowering plants

27. Monocots and eudicots are in the phylum ____________.
28. A few examples of monocots include ______________ ______________________________.
29. A few examples of eudicots include ______________ ______________________________.
30. The monocots have floral parts in multiples of (a)________, and their seeds contain (b)________(#?) cotyledon(s). The nutritive tissue in their mature seeds is the (c)____________.
31. The eudicots have floral parts in multiples of (a)____________, and their seeds contain (b)______(#?) cotyledons. The nutritive tissue, (c)______________, is usually absent having been absorbed the cotyledons during seed development.

Sexual reproduction takes place in flowers

32. The four main organs in flowers are the (a)____________, (b)____________, (c)______________ and (d)____________, each of them arranged in whorls. The "male" organs are the (e)____________ and the "female" organs are the (f)____________. Flowers with both "male" and "female" parts are said to be (g) ____________. A flower that has all four parts is said to be (h) ______________.

33. The (a) ________________ are the lowermost and outermost whorl on the floral shoot and they are green and (b) ____________________ in their appearance.
34. Collectively, the sepals form the flower's (a)____________________. The petals form the (b)________________________.
35. Pollen forms in the (a)____________, a saclike structure on the tip of a stamen, and ovules form within the (b)__________ which are part of the (c)____________.

The life cycle of flowering plants includes double fertilization

36. The megasporocyte in an ovule produces four haploid (a)______________, three of which disintegrate while the remaining one develops into the female gametophyte or (b)________________, as it is also called.
37. Microsporocytes in the anther produce four haploid (a)____________, each of which develops into a (b)________________.
38. Double fertilization is a phenomenon that is unique to flowering plants. It results in the formation of two structures the (a)________________ and the (b)______________________.

Seeds and fruits develop after fertilization

39. As a seed develops from an ovule following fertilization, the ovary wall surrounding it enlarges dramatically and develops into a ________________.

Flowering plants have many adaptations that account for their success

40. Pollen transfer results in (a) ________________________, which mixes the genetic material and promotes (b)____________________ among the offspring.
41. The stems and roots of flowering plants are often modified for (a)____________ and (b)______________________ features that help flowering plants survive in severe environments.

Floral structure provides insights into the evolutionary process

42. Sepals are specialized (a)________________ and petal are modified (b)________________.
43. Many botanists have concluded that stamens and carpels are probably derived from ______________________________________.

THE EVOLUTION OF SEED PLANTS

44. (a)____________________ is an extinct group of plants descended from ancestral seedless vascular plants. This group probably gave rise to conifers and to another group of extinct plants called the (b)______________, which in turn are thought to be the ancestors of cycads and possibly ginkgo.

Our understanding of the evolution of flowering plants has made great progress in recent years

45. Flowering plants evolved from ______________________.
46. Currently, many botanists think that the closest living gymnosperms related to flowering plants are ______________________.

The basal angiosperms comprise three clades

47. 170 species of ____________________________ represent groups thought to be ancestral to all flowering plants.

The core angiosperms comprise magnoliids, monocots, and eudicotes

48. DNA sequence comparisons have shown that the manoliids are neither (a)________________________nor (b)__________________________.

BUILDING WORDS

Use combinations of prefixes and suffixes to build words for the definitions that follow.

Prefixes	The Meaning	Suffix	The Meaning
angio-	vessel, container	-sperm	seed
di-	two, twice, double	-zation	the process of
endo-	within		
fertil-	fertile		
gymn(o)-	naked, bare, exposed		
mon(o)-	alone, single, one		

Prefix	Suffix	Definition
____________	____________	1. A plant that produces seeds that are totally exposed or on the scales of cones.
____________	____________	2. A plant that produces their seeds within a fruit.
____________	-oecious	3. Pertains to plant species in which the male and female reproductive parts are in different locations on one plant.
____________	-oecious	4. Pertains to plant species in which the male and female reproductive parts are on two different plants.
____________	____________	5. The fusion of two gametes producing a zygote.
____________	____________	4. The nutritive tissue that results from double fertilization and is found in a mature seed.

MATCHING

Terms:

a. Anther
b. Calyx
c. Carpel
d. Corolla
e. Cotyledon
f. Integument
g. Ovary
h. Ovule
i. Petals
j. Pistil
k. Pollination
l. Sepals
m. Style
n. Tracheid

For each of these definitions, select the correct matching term from the list above.

____ 1. The part of the stamen in flowers that produces microspores and, ultimately, pollen.
____ 2. An embryonic seed leaf.
____ 3. One of the layers of sporophyte tissue that surrounds the megasporangium.
____ 4. The colored cluster of modified leaves that constitute the next-to-outermost portion of a flower.
____ 5. In seed plants, the transfer of pollen from the male to the female part of the plant.
____ 6. The neck connecting the stigma to the ovary of a carpel.
____ 7. The part of a plant that develops into seed after fertilization.
____ 8. The outermost parts of a flower, usually leaflike in appearance, that protects the flower as a bud.
____ 9. The collective term for the sepals of a flower.
____10. The female reproductive unit of a flower that bears the ovules.
____11. A long, tapering cell with pits through which water flows.

MAKING COMPARISONS

Fill in the blanks.

Plant Features	Monocot	Eudicot
Vascular bundles in stems	Scattered vascular bundles	Vascular bundles arranged in a circle
Growth tissue type	Herbaceous	#1
Nutritive material in mature seeds	#2	#3
Leaf shape	Long, narrow leaves	#4
Seeds	#5	2 cotyledons
Flowers	Floral parts in multiples of 3	#6
Leaf venation	#7	#8

MAKING CHOICES

Place your answer(s) in the space provided. Some questions may have more than one correct answer.

____ 1. The major difference between gymnosperms and angiosperms is that in gymnosperms
a. self fertilization occurs.
b. seeds are exposed ("naked").
c. there is no ovary wall surrounding ovules.
d. flowers generally have 4 or 5 sepals and petals.
e. the mature ovary is a fruit.

____ 2. The petals of a flower are collectively known as the
a. corolla.
b. calyx.
c. gynoecium.
d. androecium.
e. florus perfecti.

____ 3. The sepals of a flower are collectively known as the
a. corolla.
b. calyx.
c. gynoecium.
d. androecium.
e. florus perfecti.

____ 4. The triploid endosperm of angiosperms develops from fusion of
a. two sperm and one polar nucleus.
b. two polar nuclei and one sperm.
c. three polar nuclei.
d. one diploid ovule and one haploid sperm.
e. one haploid egg and one diploid sperm.

____ 5. Which of the following is/are correct about a pine tree?
a. Sporophyte generation is dominant.
b. Male gametophyte produces an antheridium.
c. Gametophyte is dependent on sporophyte for nourishment.
d. Nutritive tissue in seed is gametophyte tissue.
e. Pollen grain is an immature male gametophyte.

____6. A perfect flower has
a. stamens only.
b. carpels only.
c. both stamens and carpels.
d. anthers.
e. ovules.

____ 7. In gymnosperms, the pollen grain develops from
a. microspore cells.
b. spores.
c. the male gametophyte.
d. the gametophyte generation.
e. meiosis of cells in the microsporangium.

____ 8. Dioecious plants with naked seeds, motile sperm, and whose pollen is carried by air or insects are
a. ginkgoes.
b. cycads.
c. extinct.
d. gnetophytes.
e. gymnosperms.

____ 9. A pine tree has
a. sporophylls.
b. megasporangia on the female cones.
c. separate male and female parts on the same tree.
d. two sizes of spores in separate cones.
e. female cones that are larger than male cones.

____10. In angiosperms, of the four haploid megaspores that result from meiosis of the megaspore mother cell, one of them becomes the
a. male gametophyte generation.
b. female gametophyte generation.
c. embryo sac.
d. archegonium.
e. antheridium.

____11. Seeds are reproductively superior to spores because seeds
a. are single cells with stored nutrients.
b. seeds live for an extended period of time.
c. seeds have a reduced rate of metabolism.
d. are protected by a seed coat.
e. a seed is further developed than a spore.

____12. Like bryophytes and seedless plants, seed plants
a. have a life cycle with an alternation of generation.
b. have free living gametophytes.
c. a female gametophyte attached to the sporophyte generation.
d. have a dominate sporophyte generation.
e. are heterosporous.

____13. Humans use conifers
a. for their wood.
b. for the production of chemicals.
c. in some cases as a source of food.
d. for medical purposes.
e. for landscaping.

____14. Many botanists think that female cones are modified
a. megasporangia.
b. sporophylls.
c. sieve tubes.
d. branch systems.
e. cycads

____15. The ancestors of all other flowering plants are thought to be
a. seed ferns.
b. progymnosperms.
c. basal angiosperms.
d. core angiosperms.
e. magnoliids.

VISUAL FOUNDATIONS

Color the parts of the illustration below as indicated. Also label haploid gametophyte generation, diploid sporophyte generation, fertilization, and meiosis.

- RED ☐ zygote
- GREEN ☐ female cone
- YELLOW ☐ megasporangium, megaspore,
- BLUE ☐ pollen grain
- ORANGE ☐ microsporangium, microspore
- BROWN ☐ male cone
- TAN ☐ sporophyte
- PINK ☐ gametophyte
- VIOLET ☐ embryo

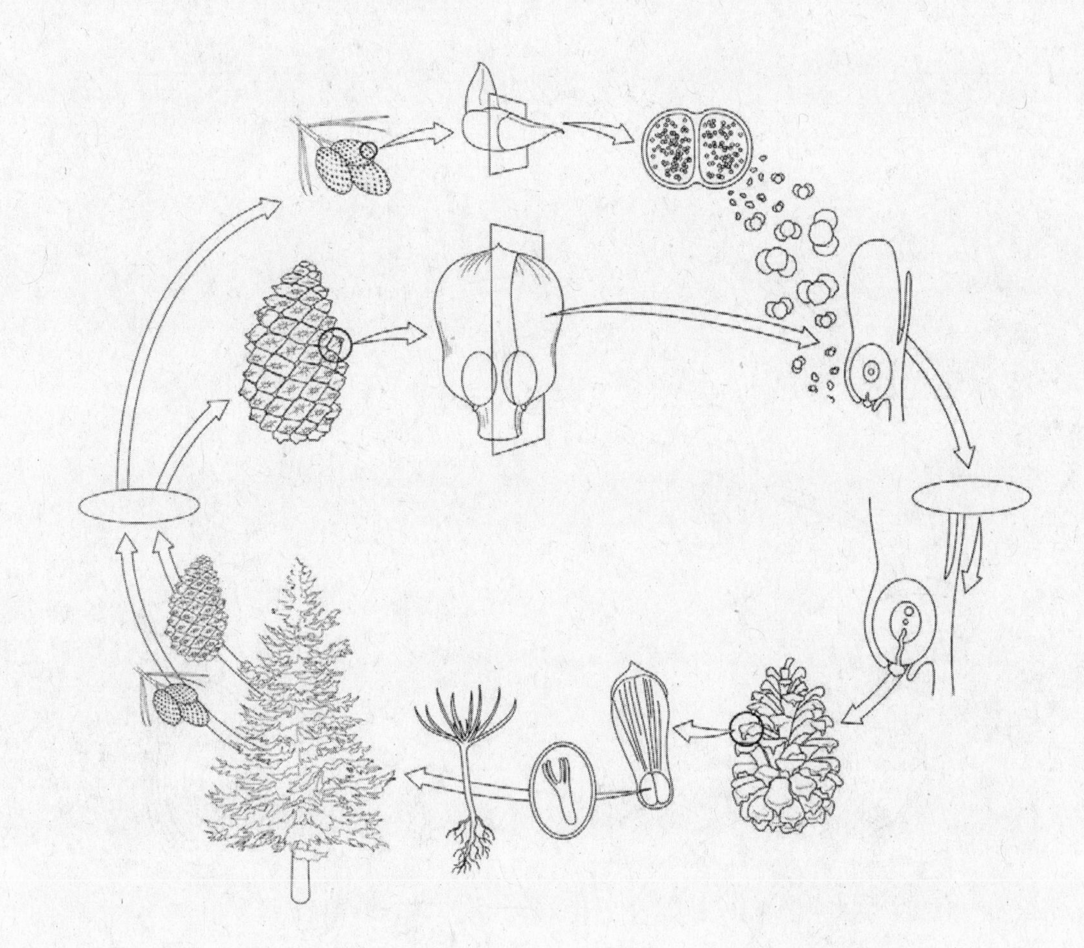

Color the parts of the illustration below as indicated. Also label haploid gametophyte generation, diploid sporophyte generation, double fertilization, and meiosis.

RED	❑ zygote	BROWN	❑ male floral part
GREEN	❑ female floral part	TAN	❑ male and female sporophytes
YELLOW	❑ megasporangium, megaspore	PINK	❑ male and female gametophytes
BLUE	❑ pollen tube	VIOLET	❑ embryo
ORANGE	❑ microsporangium, microspore		

CHAPTER 29

❑

The Fungi

Fungi are eukaryotes with cell walls. They digest their food outside their body by secreting digestive enzymes onto the food source. Fungi then absorb the nutrients into their own body. Fungi are nonmotile and reproduce by means of spores that are formed either sexually or asexually. Some fungi are unicellular (the yeasts) but most are multicellular, having a filamentous body plan. Fungi are classified into five phyla. Most fungi are decomposers, and as such play a critical role in the cycle of life. Others, however, form symbiotic relationships with other organisms. Fungi are of both positive and negative economic importance. Some damage stored goods and building materials. Some provide food for humans. Some function in brewing beer and in baking. Some produce antibiotics and other drugs and chemicals of economic importance, and some cause serious diseases in humans and economically important animals and plants.

REVIEWING CONCEPTS

Fill in the blanks.

INTRODUCTION

1. ____________________ are biologists who study fungi.
2. Like prokaryotes, most fungi are ______________________________ that obtain nutrients from dead organic matter.
3. Many fungi form ______________________________ with other organisms such as terrestrial plants.

CHARACTERISTICS OF FUNGI

4. All fungi are ____________________.
5. Fungal species can survive in a pH range of (a) _____ to _______ Many fungi are (b)_______________(more or less?) sensitive to osmotic pressures than are bacteria.

Fungi absorb food from the environment

6. Fungi share a key characteristic in how they obtain nutrients. They are ___________________________
7. Digestion in fungi occurs ________________ (intracellularly or extracellulary?).
8. Fungi store excessive amounts of nutrients as ____.

Fungi have cell walls that contain chitin

9. At some stage in their life, fungi have a feature present in bacteria, certain protists, and plants. This feature is ______________________.

10. Cell walls of most fungi contain (a) ________, which is highly resistant to (b) ___________ breakdown. This same material is also part of the protective covering of organisms in phylum (c) ____________.

Most fungi consist of a network of filaments

11. The simplest fungi are the ______________.
12. Multicellular fungi have a body consisting of threadlike filaments called ______________.
13. Molds form a tissuelike network called a (a) ______________. Fungi that form these networks are called (b) ________________.
14. The fungi called coenocytes have hyphae that lack ______________.

FUNGAL REPRODUCTION

15. Most fungi reproduce by microscopic structures called ___________.
16. The large part of the common mushroom is called a (a)________________ in which (b) ______________ are produced.

Many fungi reproduce asexually

17. Yeasts reproduce asexually by forming (a) ____________. Many species of multicellular fungi reproduce asexually by producing (b)____________.

Most fungi reproduce sexually

18. Most fungal cells contain a ________________ nucleus.
19. Hyphae that contain one nucleus per cell are called (a)____________. Those hyphae with cells containing two genetically distinct sexually compatible nuclei are called (b)________________.
20. Fungi communicate chemically by secreting signaling molecules called ____________.

FUNGAL DIVERSITY

21. Unlike plants, the walls of fungal cells do not contain ____________.

Fungi are assigned to the opisthokont clade

22. It is hypothesized that the common ancestor of all plants, fungi, and animals was a ____________________________________.
23. Based upon their characteristics, fungi are considered to be more closely related to _________________ (plants or animals?).

Diverse groups of fungi have evolved

24. Based on the characteristics of sexual spores, fruiting bodies, and molecular data, fungi are generally assigned to the five phyla
(a)______________________________,
(b)______________________________,
(c)______________________________,
(d)______________________________, and
(e)______________________________.
25. When individuals within a given taxa share a common ancestor they are said to be ____________________.
26. About 95% of all fungi have been assigned to phylum (a) __________ or phylum (b)____________________. These phyla have hyphae with (c) ______________ and a (d) __________________ stage during their sexual life cycle.
27. Some fungi are classified as deuteromycetes simply because no one has observed them to have a ____________________ in their life cycles.

Chytrids have flagellate spores

28. Most chytrids are unicellular or composed of a few cells that form a simple body called a (a) ____________________, which may have slender extensions called (b) __________________________.
29. Chytrids are the only group of fungi to produce a ________________ cell.
30. Molecular evidence suggests that chytrids were the ____________ fungal group to evolve.

Zygomycetes reproduce sexually by forming zygospores

31. Most zygomydetes are ____________________ that feed on decaying plant or animal matter.
32. The zygomycete species ____________________________ is the well-known black bread mold.
33. Black bread mold is (a) ____________________, meaning that an individual fungal hypha mates only with a hypha of a different mating type. So a (b) ______ strain mates only with a (c)__________strain.

Microsporidia have been a taxonomic mystery

34. Microsporidia are (a) __________________ pathogens. They infect people with (b) ____.
35. Gene sequence data from microsporidia indicate that they originally had (a) __________________________. This is an example of an organism that became (b) ________________________ as it evolved a parasitic life style

Glomeromycetes are symbionts with plant roots

36. The symbiotic relationships between fungi and the roots of plants are called ______________________________.

37. Glomeromycetes and plants exchange nutrients in structures called ___________________.
38. When both partners in a symbiotic relationship benefit it is said to be a __________________ relationship.

Ascomycetes reproduce sexually by forming ascospores

39. Ascomycetes produce sexual spores in sacs called (a) __________, and asexual spores called (b) _______________ at the tips of (c)________________________.
40. Some species of ascomyetes are said to be _______________________ which means that they are self-fertile.
41. As the asci develop, they are surrounded by intertwining hyphae that develop into a fruiting body known as an _____________.
42. Asexual reproduction in yeast occurs primarily by ______________.

Basidiomycetes reproduce sexually by forming basidiospores

43. Basidiomycetes develop an enlarged, club-shaped hyphal cell called a (a) ________________, on the surface of which four spores called (b)_______________________ develop.
44. What we call a mushroom grows from a compact mass of hyphae along the mycelium called a (a) ____________. A mushroom is more formally referred to as a (b) ____________________.
45. Many basidiomycetes produce a common pattern on a lawn or garden called a ______________________.

ECOLOGICAL IMPORTANCE OF FUNGI

46. Most fungi are free-living decomposers that absorb nutrients from organic wastes and dead organisms. In the process, they release (a)________________________ to the atmosphere and return (b)________________ to the soil.

Fungi form symbiotic relationships with some animals

47. Symbiotic relationships have formed between certain fungi and animals; the fungi break down (a) _________________ and (b)___________ that the animals ingest but are incapable of digesting.

Mycorrhizae are symbiotic relationships between fungi and plant roots

48. Glomeromycetes form (a) ___________________ connections inside the roots of plants, but species of acomycetes and basidiomycetes form (b)___________________ that coat the surface of roots.

A lichen consists of two components: a fungus and a photoautotroph

49. The phototroph portion of the lichen is usually a (a) ______________, or a (b) ____________, or (c) ____________. The fungus is sometimes a basidiomycete, but usually an (d) ______________.

50. Lichens typically assume one of three forms: (a) _______________, a low, flat growth; (b) _______________, a flat growth with leaf-like lobes; or (c) _______________, an erect and branching growth.
51. Lichens reproduce asexually by fragmentation, wherein pieces of lichen called _______________ break off and begin growing after landing on a suitable substrate.

ECONOMIC, BIOLOGICAL, AND MEDICAL IMPACT OF FUNGI

52. Cellulose is the most abundant organic compound on earth. The second most abundant is _______________.

Fungi provide beverages and food

53. (a)_______________ are the fungi used to make wine and beer, and to produce baked goods. Wine is produced when the fungus ferments (b) _______________, beer results from fermentation of (c)_______________, and (d)_______________ bubbles cause bread to rise.
54. Roquefort and Camembert cheeses are produced using the genus (a)_______________, and the genus (b)_______________ is used to make soy sauce from soybeans.
55. Some of the most toxic mushrooms, such as the "destroying angel" and "death cap," belong to the genus _______________.
56. The chemical _______________ found in some mushrooms causes hallucinations and intoxication.

Fungi are important to modern biology and medicine

57. Alexander Fleming noticed that bacterial growth is inhibited by the mold (a) _______________, and this discovery eventually led to the development of the most widely used of all antibiotics, (b) _______________.
58. An ascomycete that infects cereal plant flowers produces a structure called an _______________ where seeds would normally form. When livestock or humans eat grain or grain products contaminated with this fungus, they may be poisoned by the extremely toxic substances contained therein.

Fungi are used in bioremediation and to biologically control pests

59. Some fungi can biodegrade pesticides, herbicides, (a)_______________ and (b) _______________.
60. Scientists are investigating the use of micropsoridia to control of (a)_______________, and (b)_______________.

Some fungi cause diseases in humans and other animals

61. Some species of *Aspergillus* produce potent mycotoxins called _______________ that harm the liver and are known carcinogens.

Fungi cause many important plant diseases

62. All plants are susceptible to fungal diseases. Fungi enter plants through stomata, or wounds, or by dissolving a portion of cuticle with the enzyme _______________.

63. Parasitic fungi often extend specialized hyphae called ________________ into host cells in order to obtain nutrients from host cell cytoplasm.

BUILDING WORDS

Use combinations of prefixes and suffixes to build words for the definitions that follow.

Prefixes	The Meaning	Suffixes	The Meaning
basid(io)-	pedestal	-carp	fruiting body
coeno-	common	-cyt(ic)	cell
hetero-	different	-gam(y)	union
homo-	same	-(el)ium	region
mono-	alone, single, one	-karyo(tic)	nucleus
myc-	fungus	-oid	resembling
pher-	to carry	-phore	bearer
plasm-	formed	-troph	nutrition, growth, "eat"
rhiz-	root		
sapro-	rotten		

Prefix	Suffix	Definition
________	________	1. Pertains to an organism made up of a multinucleate, continuous mass of cytoplasm enclosed by one cell wall (all nuclei share a common cell).
________	________	2. Pertains to hyphae that contain only one nucleus per cell.
________	-thallic	3. Pertains to an organism that has two different mating types.
conidio-	________	4. Hyphae in the ascomycetes that bear conidia.
________	-thallic	5. Pertains to an organism that can mate with itself.
________	________	6. All the hyphae of a fungus.
________	________	7. The fusion of two mating type cells forming one larger cell.
________	-omone	8. Signaling molecule used by fungi to communicate chemically.
________	________	9. A hypha used by the fungus as an anchor and to absorb nutrients.
________	________	10. The basic mushroom structure consisting of a base and a cap.

MATCHING

Terms:

a.	Ascocarp	g.	Conidium	l.	Mycorrhizae
b.	Ascospore	h.	Dikaryotic	m.	Mycotoxin
c.	Basidiocarp	i.	Haustorium	n.	Psilocybin
d.	Basidium	j.	Lichen	o.	Soredium
e.	Budding	k.	Mycelium	p.	Zygospore
f.	Chitin				

For each of these definitions, select the correct matching term from the list above.

____ 1. Compound organism composed of a photosynthetic green alga or cyanobacterium and a fungus.

____ 2. The vegetative body of fungi; consists of a mass of hyphae.

____ 3. The sexual spores produced by an ascomycete.

____ 4. The fruiting body of a basidiomycete.

____ 5. Asexual reproductive body produced by lichens.

____ 6. Mutualistic associations of fungi and plant roots that aid in the absorption of materials.

____ 7. Asexual reproduction in which a small part of the parent's body separates from the rest and develops into a new individual.

____ 8. The club-shaped spore-producing organ of certain fungi.

____ 9. Cells having two nuclei.

____10. A component of fungal cell walls; a polymer that consists of subunits of a nitrogen-containing-sugar.

____11. Poisonous compound produced by some fungal species.

MAKING COMPARISONS

Fill in the blanks.

Phylum	Mode of Sexual Reproduction	Mode of Asexual Reproduction	Representative(s) of the Group
#1	Zygospores	Haploid spores	Black bread mold
#2	Ascospores	Conidia	Yeasts, cup fungi, morels, truffles, and blue-green and pink molds
#3	Basidiospores	Uncommon	Mushrooms, puffballs, rusts, and smuts
#4	Unknown	Blastospores	Symbiotic fungi

MAKING CHOICES

Place your answer(s) in the space provided. Some questions may have more than one correct answer.

____ 1. Fungi
a. are heterotrophs.
b. are photoautotrophs.
c. are eukaryotes.
d. possess cell walls.
e. digest food outside their bodies.

____ 2. Fungal hyphae that contain two genetically distinct nuclei within each cell are known as
a. dihaploid.
b. disomic.
c. dikaryotic.
d. $2n$.
e. $n + n$.

____ 3. A lichen can be composed of a/an
a. alga and fungus.
b. photoautotroph and fungus.
c. cyanobacterium and ascomycete.
d. alga and basidiomycete.
e. alga and ascomycete.

____ 4. A common fungal infection of the mucous membranes of the mouth, throat, or vagina is
a. an autoimmune response.
b. St. Anthony's fire.
c. histoplasmosis.
d. ergotism.
e. candidiasis.

____ 5. Fungi can reproduce
a. sexually.
b. asexually.
c. by spore formation.
d. by simple division.
e. by budding.

____ 6. The black fungus growing on a piece of bread
a. is heterothallic.
b. has male and female strains.
c. has coenocytic hyphae.
d. has sporangia on the tips of stolons.
e. is in the phylum Zygomycota.

____ 7. The genus *Penicillium*
a. is an ascomycete.
b. is a sac fungus.
c. produces the flavor in Roquefort cheese.
d. produces the antibiotic penicillin.
e. produces the flavor in Brie cheese.

____ 8. Lichens can be used as indicators of air pollution because they
a. cannot excrete absorbed elements.
b. tolerate sulfur dioxide.
c. can endure large quantities of toxins.
d. do not grow well in polluted areas.
e. overgrow polluted areas.

____ 9. A mass of filamentous hyphae is called a(n)
a. hypha.
b. mycelium.
c. conidium.
d. thallus.
e. ascocarp.

____10. Yeast participates in the brewing of beer by
a. adding vital amino acids.
b. fermenting grain sugars.
c. fermenting fruit sugars.
d. producing ethyl alcohol.
e. converting barley to hops.

____11. If you eat just any mushroom that you find in the wild, there's a chance that you will
a. die.
b. become intoxicated.
c. see colors that aren't really there.
d. not become nauseated or die.
e. ingest the hallucinogenic drug psilocybin.

____12. A fungus infection throughout the body obtained by exposure to bird droppings is likely
a. an autoimmune response.
b. St. Anthony's fire.
c. histoplasmosis.
d. ergotism.
e. candidiasis.

____13. Chytrids (aka chytridiomycetes)
a. are fungi.
b. are funguslike protists.
c. inhabit damp or wet environments.
d. have flagellated spores.
e. are among the latest in their kingdom to evolve.

____ 14. Fungi can be found growing

a. in moist habitats.
b. in tree and plant roots.
c. in the grasses of geothermal hot springs.
d. where organic material is available.
e. in the wood of buildings.

____ 15. The cell walls of fungi contain

a. carbohydrates.
b. lignin
c. cellulose.
d. cutinase.
e. chitin.

____ 16. Coenocytic fungi

a. are elongated.
b. multinucleated.
c. lack septa.
d. round.
e. have a single nucleus.

____ 17. The clade opisthokonts includes

a. plants.
b. fungi.
c. amoebozoa
d. animals.
e. choanoflagellates.

____ 18. Terms associated with sexual reproduction in fungi include

a. plasmogamy.
b. karyogamy.
c. conidiophores.
d. zygote nucleus.
e. conidia.

____ 19. Features some fungi share with some plants include

a. a cell wall.
b. growth to enormous size.
c. asexual reproduction.
d. active growth in dry conditions.
e. alternation of generation.

____ 20. Microsporidia

a. infect eukaryotic cells.
b. were once assigned to protozoa.
c. produce a polar tube.
d. are opportunistic pathogens.
e. have two developmental stages.

____ 21. Species of yeast that are self-fertile are referred to as being

a. heterothallic.
b. polythallic.
c. unithallic.
e. multithallic.
e. homothallic.

____ 22. Terms associated with basidiomycetes include

a. primary mycelium.
b. cytoplasmic streaming.
c. karyogamy.
d. gills.
e. "fairy rings"

____ 23. Mycorrhizae are symbiotic associations between

a. fungi and bacteria.
b. fungi and grasses.
c. fungi and fruits.
d. fungi and plant roots.
e. fungi and grains.

VISUAL FOUNDATIONS

Color the parts of the illustration below as indicated.

- RED ❑ haustoria
- GREEN ❑ hypha
- YELLOW ❑ epidermal cells
- BROWN ❑ spore
- VIOLET ❑ stoma

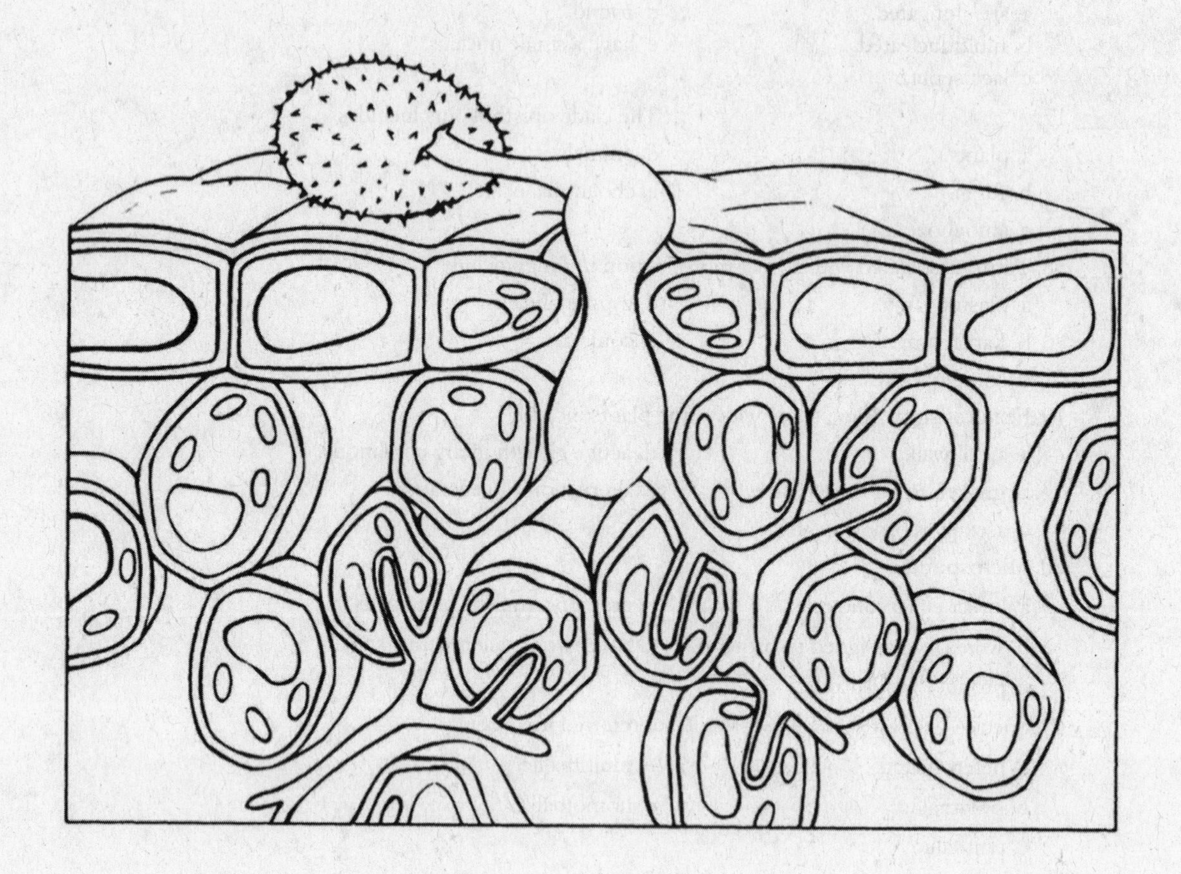

Color the parts of the illustration below as indicated. Also label asexual reproduction, sexual reproduction, and the diploid and haploid generations.

RED ❑ gamete type A
GREEN ❑ gamete type B
YELLOW ❑ haploid zoospore
BLUE ❑ diploid thallus
ORANGE ❑ haploid thallus
BROWN ❑ motile zygote
TAN ❑ resting sporangium
VIOLET ❑ diploid zoospore

CHAPTER 30

❑

An Introduction to Animal Diversity

This is the first of three chapters discussing the animal kingdom. Animals are eukaryotic, multicellular, heterotrophic organisms. They are made up of localized groupings of cells with specialized functions. One grouping, for example, might function in locomotion, another might respond to external stimuli, and another might function in sexual reproduction. Animals inhabit virtually every environment on earth. As discussed in this chapter, animals have adapted to living in ocean, freshwater, and terrestrial habitats. Animals may be classified based upon the characters they posses such as radial or bilaterally symmetry; the absence of a body cavity (the acoelomates); the presence of a body cavity between mesoderm and endoderm (the pseudocoelomates); the presence of a body cavity within mesoderm (the coelomates); those in which the first opening that forms in the embryonic gut becomes either the mouth (the protostomes); or those in which the first opening in the embryonic gut forms the anus (the deuterostomes). This chapter discusses all of these characters.

REVIEWING CONCEPTS

Fill in the blanks.

INTRODUCTION

1. An estimated _______% of all animal species that ever inhabited our planet are extinct.
2. Molecular data indicate that many groups of animals are __________.

ANIMAL CHARACTERS

3. Animals lack cell walls and instead rely on ____________ for structural support.
4. Animals are ____________________, or consumers that depend on producers for their raw materials and energy.
5. Most animals are (a) ____________ organisms that reproduce sexually. The fusion of a sperm and egg cell produces a (b)________.
6. During embryoinic development the zygote undergoes a process called ______________________, during which mitotic divisions occur.

ADAPTATIONS TO OCEAN, FRESHWATER, AND TERRESTRAIL HABITATS

Marine habitats offer many advantages

7. (a) ____________ and (b) ____________ balance are more easily maintained in a salt water environment than a freshwater environment.

8. Marine animals must deal with the water's (a) ________________ and (b) ________________.

Some animals are adapted to freshwater habitats

9. Freshwater environments are ________________ to the tissue fluids of animals living there.
10. Freshwater organisms must remove excess (a) ________________ and at the same time retain (b) ________________.
11. A freshwater environment is much less ________________ environment than the ocean.

Terrestrial living requires major adaptations

12. Biologists hypothesize that the first air-breathing terrestrial animals were ________________.
13. Land environments tend to ________________ animals.
14. Many land animals have evolved (a) ________________ which allows them to reproduce in a dry environment. In addition, some species of land dwelling animals adapted to produce a (b)________________ which also helps avoid the problem of desiccation.

ANIMAL EVOLUTION

15. The common ancestor of animals was a ________________.
16. According to the principle of ________________ complex molecules such as RNA are likely to have evolved only once.

Molecular systematics helps biologists interpret the fossil record

17. The rapid appearance of an amazing variety of body plans in the fossil record is known as the ________________.
18. Molecular data suggest that most animal clades diverged during the ________________ eon.

Biologists develop hypotheses about the evolution of development

19. Mutations in ________ genes could have resulted in rapid changes in body plans.

RECONSTRUCTING ANIMAL PHYLOGENY

20. Biologists use (a) ________________ and differences in (b)________________ to infer relationships between animals.

Animals exhibit two main types of body symmetry

21. (a) ________________ and (b)________________ are two adaptations for locomotion.
22. There are many anatomical terms that are used to describe the locations of body parts. Among them are: (a) ________________ for front and (b) ________________ for rear; (c) ________________ for a back surface and (d) ________________ for the under or "belly" side.

23. Other locations of body parts include (a) ____________ for parts closer to the midline and (b) ____________ for parts farther from the midline, towards the sides; (c) ____________ toward the head end; and (d) ____________ when away from the head, toward the tail.

Animal body plans are linked to the level of tissue development

24. In the early development of all animals except sponges, cells form layers called (a) ____________ that give rise to linings, tissues, and body structures. These layers are the (b) ____________, (c) ____________, and (d) ____________. Animals with all three are said to be (e) ____________.

Most bilateral animals have a body cavity lined with mesoderm

25. Animals can also be grouped according to the presence or absence of a body cavity and, when present, the type of body cavity. For example, the flatworms are called ____________ because they do not have a body cavity.
26. Animals with body cavities are either (a) ____________, possessing a body cavity between the mesoderm and endoderm, or (b) ____________, having a true body cavity within the mesoderm.

Bilateral animals form two main clades based on differences in development

27. Coelomate animals can be grouped by shared developmental characteristics. In mollusks, annelids, and arthropods, all protostomes, the blastopore develops into the (a) ____________, whereas in echinoderms and chordates, both deuterostomes, the blastopore becomes the (b) ____________.
28. Early cell divisions in protostomes and deuterostomes occur in patterns. A pattern of division that is either parallel to or at right angles to the polar axis is known as (a) ____________. In protostomes, early cell divisions are diagonal to the polar axis, a type of division referred to as (b) ____________.

Biologists have identified major animal clades based on structure, development, and molecular data

29. Animals are referred to as ____________.
30. Animals that have germ layers are classified as ____________.
31. The two major clades of protostomes are (a) ____________ and (b) ____________.

Segmentation apparently evolved three times

32. One important innovation of body forms has been ____________, a body plan in which certain structures are repeated producing body compartments.
33. Segmentation is found within the three major clades of ____________.

BUILDING WORDS

Use combinations of prefixes and suffixes to build words for the definitions that follow.

Prefixes	The Meaning
blast-	bud, sprout
cleav-	divide
deutero-	second
ecto-	outer, outside, external
entero-	intestine
gastro-	stomach
meso-	middle
proto-	first, earliest form of
pseudo-	false
schizo-	split

Suffixes	The Meaning
-age	collection of
-coel(om)(y)	cavity
-derm	skin
-stome	mouth
-ula	little

Prefix	Suffix	Definition
________	________	1. A body cavity between the mesoderm and endoderm; not a true coelom.
________	________	2. The outermost germ layer.
________	________	3. The middle layer of the three basic germ layers.
________	________	4. Major division of the animal kingdom in which the mouth forms from the first opening in the embryonic gut (the blastopore).
________	________	5. Major division of the animal kingdom in which the mouth forms from the second opening in the embryonic gut.
________	________	6. The process of coelom formation in which the mesoderm splits into two layers.
________	________	7. The process of coelom formation in which the mesoderm forms as "outpocketings" of the developing intestine.
________	________	8. The hollow ball of cells that forms during early embryonic development.
________	________	9. The tissue lining the gut cavity ("stomach") that functions in digestion in certain phyla.
________	________	10. A series of mitotic cell divisions.

MATCHING

Terms:

a. Acoelomate
b. Autotroph
c. Cephalization
d. Cleavage
e. Coelom
f. Cuticle
g. Dorsal
h. Eumetazoa
i. Gastrulation
j. Heterotroph
k. Invertebrate
l. Metazoa
m. Pseudocoelom
n. Segmentation
o. Sessile
p. Ventral

For each of these definitions, select the correct matching term from the list above.

____ 1. A body cavity not completely lined with mesoderm.

____ 2. Depends upon a producer to provide molecules for energy and raw materials.

____ 3. Without a body cavity (coelom).

____ 4. The main body cavity of most animals.

____ 5. Describes animals with either two or three germ cell layers.

____ 6. The clustering of neural tissues at the anterior (leading) end of an animal.

____ 7. Referring to the belly aspect of an animal's body.

____ 8. A body plan which results in a series of body compartments.

____ 9. Permanently attached to one location.

____ 10. The process that forms and segregates the three germ layers during embryonic development.

MAKING COMPARISONS

Fill in the blanks.

Deuterostome	Protostome	Character
#1	#2	Pattern of cleavage
#3 cleavage	#4 cleavage	Developmental fate of the early cells
#5	#6	Developmental fate of the blastopore
#7	#8	Method of coelom formation

MAKING CHOICES

Place your answer(s) in the space provided. Some questions may have more than one correct answer.

____ 1. Eumetazoans that have bilateral symmetry and are in the clade Lophotrochozoa include

a. echinoderms.
b. arthropods.
c. flatworms.
d. nematodes.
e. annelids.

____ 2. Features of arthropods include

a. radial symmetry.
b. a protostome pattern of development.
c. a coelom.
d. molting.
e. a deuterostome pattern of development.

____ 3. Examples of acoelomate eumetazoans include

a. sponges.
b. insects.
c. flatworms.
d. those groups whose coelom developed from a blastocoel.
e. nematodes.

____ 4. The clade Bilateria includes

a. sponges.
b. cnidarians.
c. ctenophores.
d. arthropods.
e. mollusks.

____ 5. When a group of cells moves inward to form a sac in early embryonic development, and that pore ultimately develops into a mouth, the group of animals is considered to be a
a. protostome.
b. deuterostome.
c. pseudocoelomate.
d. parazoan.
e. schizocoelomate.

____ 6. The one invertebrate phylum with radial symmetry and distinctly different tissues is
a. Porifera.
b. Cnidaria.
c. Platyhelminthes.
d. Nemertea.
e. Echinodermata.

____ 7. The roundworms
a. are pseudocoelomates.
b. have a mastax.
c. are in the phylum Platyhelminthes.
d. share a characteristic with rotifers.
e. are nematodes.

____ 8. Animals with a water vascular system include
a. hemichordates.
b. chordates.
c. sponges.
d. nematodes.
e. echinoderms.

____ 9. The taxon that has epidermal stinging cells is
a. Scyphozoa.
b. Hydrozoa.
c. Porifera.
d. Cnidaria.
e. Ctenophora.

____10. Segmentation occurs in
a. arthropods.
b. nematodes.
c. vertebrates.
d. chordates.
e. earthworms.

____ 11. Animals with bilateral symmetry include
a. ecydyzoans.
b. protostomes.
c. sponges.
d. deuterostomes.
e. lophotrochozoans.

____12. Animls with a muscle layer derived from mesoderm include
a. coelomates.
b. flatworms.
c. vertebrates.
d. nematodes.
e. pseudocoelomates.

____13. A monophyletic grouping
a. can be used to describe the protists.
b. can be used to describe the clade Deuterostomia.
c. can be used to describe the clades Prostomia and Bilateria.
d. includes all the descendents of a common ancestor.
e. is supported using molecular data.

____14. Clade Lophotrochozoa includes
a. the nematodes and arthropods.
b. platyhelminths.
c. protostomes.
d. animals that molt.
e. animals with radial symmetry.

VISUAL FOUNDATIONS

Color the parts of the illustration below as indicated. Label the animals depicted as acoelomate, pseudocoelomate, or coelomate.

- RED ❑ mesoderm
- GREEN ❑ endoderm
- YELLOW ❑ coelom
- BROWN ❑ pseudocoelom
- BLUE ❑ ectoderm
- ORANGE ❑ mesenchyme

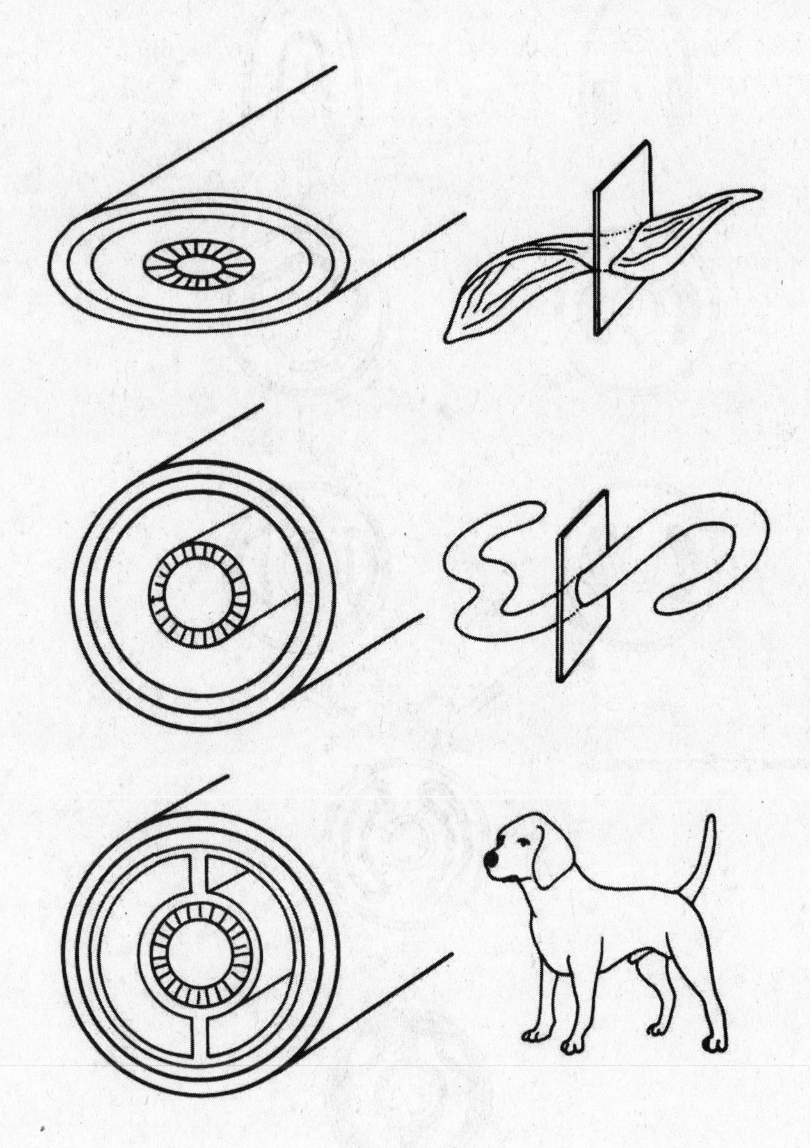

Color the parts of the illustration below as indicated. Label the ectoderm, mesoderm, endoderm, gut, and coelom in the last figure.

BLUE ❑ Ectoderm

RED ❑ Mesoderm

YELLOW ❑ Endoderm

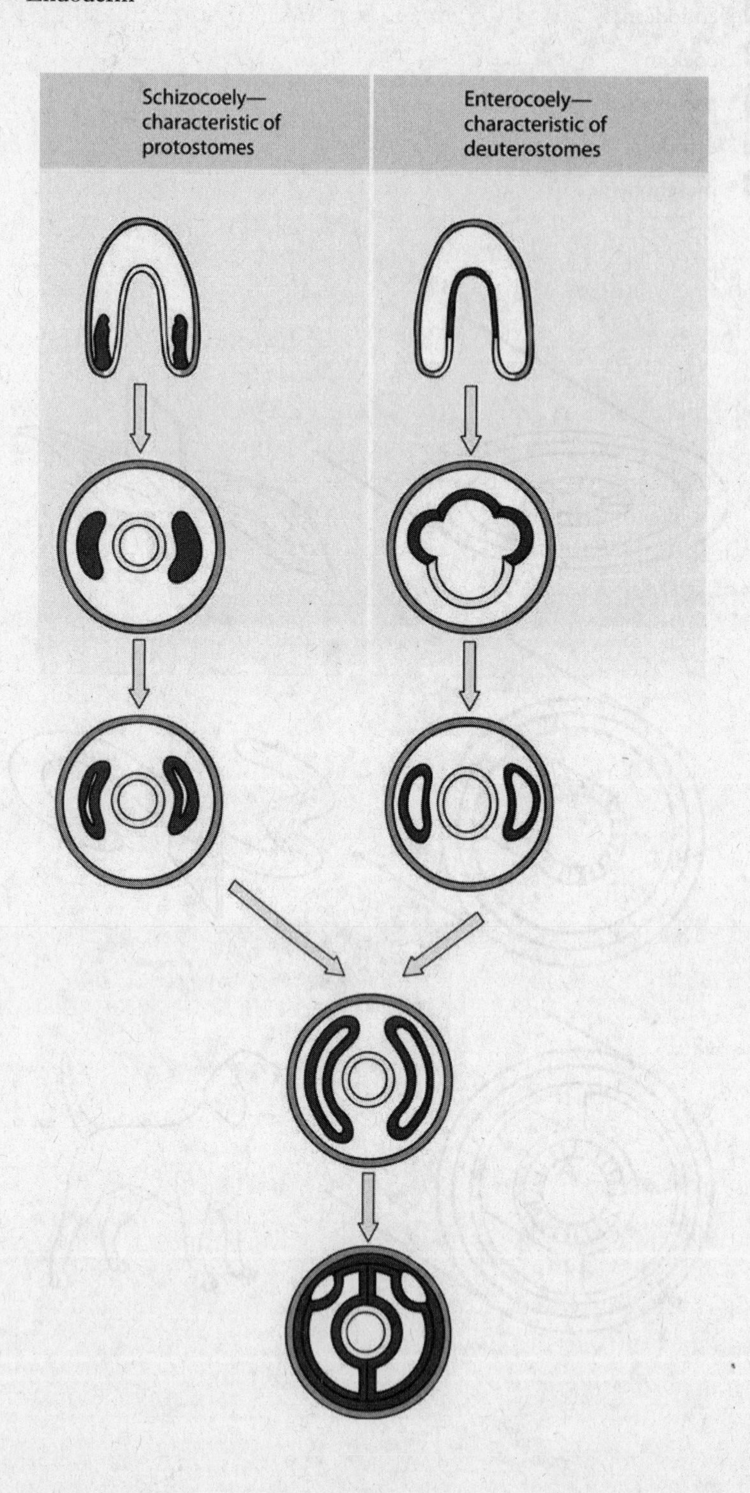

CHAPTER 31

❑

Sponges, Cnidarians, Ctenophores, and Protostomes

This chapter discusses the three phyla that diverged early in the evolutionary history of the animal kingdom: the Parazoa (sponges), the Cnidaria (hydras, jelly fish, sea anemones, and corals), and the Ctenophora (comb jellies), and it concludes with a discussion of the protostomes. The sponges are at the cellular level of organization and lack true tissues. The cnidarians and ctenoporans have tissues and a body that is radially symmetrical. The protostomes are animals with coeloms in which the first opening that forms in the embryonic gut gives rise to the mouth, a complete digestive tract, and most have well-developed circulatory, excretory, and nervous systems. The coelom separates muscles of the body wall from those of the digestive tract, permitting movement of food independent of body movements, and it serves as a space in which organs develop and function. The fluid it contains helps transport food, oxygen, and wastes. Additionally, it can be used as a hydrostatic skeleton, permitting a greater range of movement than in acoelomate animals and conferring shape to the body of soft animals. The coelomate protostomes include the clades Lophotrochozoa (platyhelminthes, nemerteans, mollusks, annelids, and the lophophorate phyla) and Ecdysozoa (nematodes and arthropods).

REVIEWING CONCEPTS

Fill in the blanks.

INTRODUCTION

1. An estimated ________% of all animal species belong to clade Bilateria.
2. Bilateral animals have been so successful because they evolved with three major adaptations that facilitated their ability to (a)________________________, (b)________________________, and (c)________________________.
3. The two major clades of bilateral animals are (a)________________________, and (b)________________________.

SPONGES, CNIDARIANS, AND CTERNOPHORES

4. Sponges have been assigned to the group (a) ________________ because they are (b) ________________ and have a (c)______________ body plan.
5. The cells of sponges do not ________________________.

Sponges have collar cells and other specialized cells

6. Based upon the structure of their collar cells, sponges are thought to have evolved from ________________________.

7. Sponges have three types of canals systems through which water flows to enter the sponge. In the (a) _____________________ system water enters through porocytes and the choanocytes lining the spongocoel trap particles suspended in the water. In the (b)________________ system the choanocytes line channels through which the water flows. In the (c) ________________ system the wall of the sponge is highly folded and the increased system of canals lined with choanocytes provides more surface area for capturing food.
8. In asconoid sponges, water enters the central cavity or (a) __________ through the porocytes and then exits through the (b) __________.
9. Between the outer and inner cell layers of the sponge body is a gelatin-like layer called the (a) ________________ supported by slender skeletal spikes called (b) ____________.
10. The cells of a sponge can react to (a) __________ but sponges do not have (b) ______________ cells.
11. The three main groups of sponges are (a) ______________, (b) ____________________, and (c)________________.

Cnidarians have unique stinging cells

12. Cnidarians have two body shapes (a) __________ and (b)__________ and they have (c)______________ symmetry.
13. The epidermis of cnidarians contains specialized "stinging cells" called ______________.
14. Stinging "thread capsules" in the stinging cells of cnidarians are called ____________.
15. Cnidarians have two definite tissue layers, the outer (a) __________ and the inner (b) ____________________ separated by a gelatinous region called the (c) ______________.
16. The phylum Cnidaria is divided into four classes: Hydras and hydroids are in the class (a) __________, jellyfish are in class (b)__________, sea anemones and corals are in class (c)__________, and the "box jellyfish" are in class (d)__________.
17. Hydra can be found living in (a) ____________________ and during the fall they reproduce (b) ________________.
18. Among the jellyfish, scyphozoa, the (a) ____________ stage is the dominant body form. The largest jellyfish, (b) ______________, can be over 2m in diameter.
19. Cubozoans are food for ____________________.
20. Anthozoan polyps produce eggs and sperm, and the fertilized egg develops into a small, ciliated larva called a __________.
21. Many tropical species of corals have a symbiotic relationship with a photosynthetic algae, ____________________, that lives within the coral cells.

Comb Jellies have adhesive glue cells that trap prey

22. Ctenophores have ______________ symmetry.

23. Ctenophores are similar to cnidarian medusa in that they have a thick jellylike layer, the mesoglea. However they differ from cnidarians because they lack (a) ________________ and they have a digestive system with (b) ________________ and (c)________________.
24. The similarities between ctenophores and cnidarians may be due to (a)________________ evolution as a result living in (b)________________________________.

THE LOPHOTROCHOZOA

25. Lophotrochozoans are one of two clades of (a)________________. They have either a (b)________________ (ciliated ring of tentacles surrounding the mouth) or a (c) ________________ larva.
26. Most lophotrochozoans have (a) ________________ symmetry, (b) ____________ primary germ layers, a (c) ____________ coelom, and a (d)________________ body plan.

Flatworms are bilateral acoelomates

27. Flatworms are in phylum (a) ________________ which is further divided into four classes. These classes are (b)____________, the free-living flatworms; two classes of flukes (c) ________________ and (d) ________________ and (e)____________, the tapeworms.
28. Most flatworms have a simple a simple nervous system in which the brain consists of ________________.
29. In addition to having sensory structures, turbellarians have ________________________ that function in osmoregulation.
30. Planarians have sensory cells located on extensions called (a)____________ that resemble "ears" and they use an extendable (b)________________ to suck liquefied prey into the digestive tube.
31. Flukes have a body plan similar to (a)________________________ but they are parasites and use (b) ________________________ to attach to the host.
32. Most tapeworms attach to the host using suckers and/or hooks on the ________________.
33. In tapeworms, each segment, or ________________, contains both male and female reproductive organs.

Nemerteans are characterized by their proboscis

34. The proboscis organ is a long, hollow, muscular tube that can be rapidly (a)________________________________.
The proboscis is a (b) ________________ character.

Mollusks have a muscular foot, visceral mass, and mantle

35. Mollusks possess a ventral foot for locomotion and a ____________ that covers the visceral mass.
36. Most mollusks have a (a) ____________ located in the mouth region, and an open circulatory system, meaning the blood flows through a

network of open sinuses called the blood cavity or (b)________________. The (c) ____________________ however have a closed circulatory system.

37. Most marine mollusks have larval stages, the first a free-swimming, ciliated form called a (a) ________________ larva. Some, the snails and bivalves go through a second stage a (b) __________ larva with a shell, foot, and mantle.
38. A distinctive feature of chitons is their shell, which is composed of (a)______ (#?) overlapping plates. The chiton head is (b)___________________ and they have not (c) ____________ or (d)______________.
39. Class Gastropoda, the largest and most successful group of mollusks, includes the snails, slugs, and their relatives. It is a class second in size only to the ______________.
40. The gastropod shell (when present) is coiled, and the visceral mass is twisted, a phenomenon known as ____________.
41. Foreign matter lodged between the (a) ___________ and mantle of bivalves may cause the secretion of (b) ________________ by the epithelial cells of the mantel. The result is the formation of a pearl.
42. Class Cephalopoda includes the squids and octopods, which are active predatory animals. The foot is divided into (a) ______________ that surround the mouth, (b) _________ (#?) of them in squids, (c) _______ (#?) of them in octopods.
43. The term Cephalopoda literally means ____________________.
44. Cephalopods have a _____________ which is used to kill and tear prey.

Annelids are segmented worms

45. Phylum Annelida is divided into three classes; (a)____________________, the sandworms and tubeworms, (b)______________________, the earthworms, and (c)______________________, the leeches.
46. Respiration in annelids occurs across the (a) __________ or (b)_________. Excretion occurs in excrectory tubules called (c)__________________.
47. In polychaetes and earthworms, body segments are separated from one another by partitions called (a) ____________. Also, each body segment bears bristle-like structures called (b) ____________ that function to anchor the body during locomotion.
48. Most polychaetes are (a) __________ worms and each body segment bears a pair of paddle-shaped appendages called (b)__________________ used for locomotion and gas exchange.
49. The two parts of the earthworm's "stomach" are the (a) _________, where food is stored, and the (b) _____________, where ingested materials are ground into bits.
50. The respiratory pigment in earthworms is (a) ______________; the excretory system consists of paired (b) ______________

in most segments; and the gas exchange surface is the (c) ________.

The lophophorates are distinguished by a ciliated ring of tentacles

51. There are three lophophorate phyla: (a) ________________, or lampshells; (b) ________________, wormlike sessile animals; and (c) ________________, or microscopic aquatic animals.

Rotifers have a crown of cilia

52. Rotifers have a nervous system which includes a (a) ____________ and (b) ____________. They have protonephridia with (c)____________ that remove excess water from the body and may also excrete metabolic wastes.
53. Rotifers grow through an ________________________.

THE ECDYSOZOA

54. The Ecdysozoa include (a) ________________ and (b)__________. The group names refers to the fact that these organisms (c) __________..

Roundworms are of great ecological importance

55. Members of the phylum Nematoda play key ecological roles as (a)________________, (b)________________, and (c) ________________.
56. Some nematodes are human parasites. The genus ____________ which includes the hookworms, pinworms and trichina worm is an example.

Arthropods are characterized by jointed appendages and an exoskeleton of chitin

57. More than _____% of all known animals are arthropods.
58. Arthropods are characterized as having an armor-like body covering, an (a)________________ composed of (c) __________ and (d)__________.
59. Most insects and many crustaceans have ____________ eyes.
60. Gas exchange in terrestrial arthropods occurs in a system of thin, branching tubes called ______________.
61. Among the earliest arthropods were the ______________ which inhabited seas and are now extinct.
62. The members of the class (a) ______________, or "centipedes," have (b)______ pair(s) of legs per body segment. These animals are active (c)____________ feeding on other animals. They also have (d)__________ claws used to capture and kill prey.
63. The members of the class (a) ______________, or “millipedes,” have (b)_______ pair(s) of legs on most body segments. These animals are (c)____________________ and do not active hunt prey.
64. Animals in the arthropod subphylum, ________________, lack antennae and include the horseshoe crabs and arachnids.

65. The arachnid body consists of a (a) ____________________ and abdomen. They have (b) ________ (#?) pairs of jointed appendages which include (c) ______ (#?) pairs of legs used for walking.
66. Gas exchange in arachnids occurs in tracheae and/or across thin, vascularized plates called ________________.
67. Silk glands in spiders secrete an elastic protein that is spun into web fibers by organs called ________________.
68. The subphylum (a) ______________ includes lobsters, crabs, shrimp, and barnacles. This subphylum uses its (b) ________________ for biting food and its (c) ________________ for manipulating and holding food. Members have (d) ____________ (branched) appendages and (e) ______ (#?) pairs of antennae.
69. ______________ are the only sessile crustaceans.
70. Lobsters, crayfish, crabs, and shrimp are members of the largest crustacean order, the _______________.
71. Among the many highly specialized appendages in decapods are the (a)__________, two pairs of feeding appendages just behind the mandibles; the (b) ____________ that are used to chop food and pass it to the mouth; the (c) ____________ or pinching claws on the fourth thoracic segments; and the pairs of (d) ______________ on the last four thoracic segments.
72. Appendages on the first abdominal segment of decapods are part of the (a)________________ system. (b) ______________ on the next four abdominal segments are paddle-like structures adapted for swimming and holding eggs.
73. In terms of numbers of individuals, number of species, and geographic distribution, animals in the class (a) ______________ and subphylum (b)________________ are considered to be the most successful group of animals on the planet.
74. Adult insects typically have (a) _________ (#?) pairs of legs, (b)________ (#?) pairs of wings, and (c) ______ (#?) pair(s) of antennae. The excretory organs are called (d) ______________.

BUILDING WORDS

Use combinations of prefixes and suffixes to build words for the definitions that follow.

Prefixes	The Meaning	Suffixes	The Meaning
arthro-	joint, jointed	-coel	abdominal cavity, hollow
aur-	ear	-cyte(s)	cell
bi-	twice, two	-icle	little
cephalo-	head	-pod	foot, footed
choan(o)-	funnel	-stat(ic)	to control
exo-	outside, outer, external		
hexa-	six		
hydr(o)-	water		
tri-	three		
uni-	one		

Prefix	Suffix	Definition
____________	-skeleton	1. An external skeleton, such as the shell of arthropods.
____________	-valve	2. A mollusk that has two shells (valves) hinged together.
____________	-ramous	3. Consisting of or divided into two branches.
____________	-ramous	4. Unbranched; consisting of only one branch.
____________	-lobite	5. An extinct marine arthropod characterized by two dorsal grooves that divide the body into three longitudinal parts (or lobes).
____________	-thorax	6. Anterior part of the body in certain arthropods consisting of the fused head and thorax.
____________	____________	7. Having six feet; an insect.
____________	____________	8. An animal with paired, jointed legs.
spongo-	____________	9. The central cavity in the body of a sponge.
____________	____________	10. Cells used by sponges to trap food particles.
____________	____________	11. The type of skeleton used by cnidarians to shorten and elongate.
____________	____________	12. A flap-like extension on the head of a planarian that is used to sense the environment.

MATCHING

Terms:

a. Acoelomate
b. Auricle
c. Cephalization
d. Choanocyte
e. Cnidocyte
f. Coelom
g. Cuticle
h. Dorsal
i. Ecdysis
j. Hermaphroditic
k. Invertebrate
l. Mandible
m. Mesohyl
n. Metamorphosis
o. Nematocyst
p. Pseudocoelom
q. Radula
r. Sessile
s. Spongin
t. Trachea
u. Trochophore
v. Ventral

For each of these definitions, select the correct matching term from the list above.

____ 1. The fibrous part of a sponge skeleton.
____ 2. Possessing sex organs of both the male and the female.
____ 3. A rasplike structure in the mouth region of certain mollusks.
____ 4. A fluid-filled space lined by mesoderm that lies between the digestive tube and the outer body wall.
____ 5. A unique cell having a flagellum surrounded by a collar of microvilli.
____ 6. The clustering of neural tissues at the anterior (leading) end of an animal.
____ 7. The first pair of appendages in arthropods.
____ 8. A stinging structure in cnidarians.
____ 9. Permanently attached to one location.
____10. The shedding and replacement of the arthropod exoskeleton.
____11. A larval form found in mollusks and many polychaetes.
____12. Transition from one developmental stage to another, such as from a larva to an adult.
____13. Internal branching air tubes throughout the body of many terrestrial arthropods and some terrestrial mollusks.

MAKING COMPARISONS

Fill in the blanks.

Phylum	Examples	Body Plan	Key Characteristics
Porifera	Sponges	Asymmetrical; sac-like body with pores, central cavity, and osculum	Choanocytes
Cnidaria	#1	Radial symmetry; diploblastic; gastrovascular cavity with one opening	Tentacles with cnidocytes (stinging cells); polyp and medusa body forms;
Ctenophora	Comb jellies	#2	Eight rows of cilia; tentacles with adhesive glue cells
#3	Flatworms, tape worms, planarians, flukes	#4	No body cavity; some cephalization
#5	Clams, snails, squids	Bilateral symmetry; triploblastic; organ systems; complete digestive tube, coelom	#6
Annelida	#7	Bilateral symmetry; triploblastic; organ systems; complete digestive tube, coelom	#8
#9	Wheel animals	Bilateral symmetry; triploblastic; organ systems; complete digestive tube, coelom	#10
Nematoda	Roundworms: Ascaris, hookworms, trichina worms	Bilateral symmetry; triploblastic; organ systems; complete digestive tube, pseudocoelom	#11
Arthropoda	#12	Bilateral symmetry; triploblastic; organ systems; complete digestive tube, coelom	#13
#14	Starfish, sea urchins, sand dollars	#15	Endoskeleton; water vascular system; tube feet

MAKING CHOICES

Place your answer(s) in the space provided. Some questions may have more than one correct answer.

____ 1. Marine carnivores with a proboscis that can be everted to capture prey are
a. members of Nemertea.
b. are protostomes.
c. are deuterostomes.
d. known as ribbon worms.
e. members of Mollusca.

____ 2. The taxon that has biradial symmetry, tentacles with adhesive glue cells, and eight rows of cilia is
a. Scyphozoa.
b. Hydrozoa.
c. Ctenophora.
d. Cnidaria.
e. Porifera.

____ 3. Examples of acoelomate eumetazoans include
a. sponges.
b. insects.
c. flatworms.
d. sea stars.
e. cnidarians.

____ 4. The phylum/phyla that does/do not have specialized nerve cells is/are
a. Echinodermata.
b. Cnidaria.
c. Platyhelminthes.
d. Nemertea.
e. Porifera.

____ 5. When a group of cells moves inward to form a sac in early embryonic development, and that pore ultimately develops into a mouth, the group of animals is considered to be a
a. protostome.
b. deuterostome.
c. pseudocoelomate.
d. parazoan.
e. schizocoelomate.

____ 6. The one invertebrate phylum with radial symmetry and distinctly different tissues is
a. Porifera.
b. Cnidaria.
c. Platyhelminthes.
d. Nemertea.
e. Echinodermata.

____ 7. The roundworms
a. are pseudocoelomates.
b. have a mastax.
c. are in the phylum Platyhelminthes.
d. include hookworms.
e. are nematodes.

____ 8. Sessile, marine animals with no medusa stage and a partitioned gastrovascular cavity include
a. hydrozoans.
b. scyphozoans.
c. anthozoans.
d. jellyfish.
e. corals and sea anemones.

____ 9. The taxon that has "thread capsules" in epidermal stinging cells is
a. Scyphozoa.
b. Hydrozoa.
c. Porifera.
d. Cnidaria.
e. Ctenophora.

____10. Deuterostomes characteristically have
a. radial cleavage.
b. spiral cleavage.
c. indeterminate cleavage.
d. schizocoely.
e. enterocoely.

____11. The excretory organ(s) in earthworms is/are the
a. parapodia.
b. kidneys.
c. metanephridia.
d. lophophores.
e. crop.

____12. The so-called "wheel animals"
a. are unicellular.
b. are generally aquatic.
c. have no cell mitosis after completing embryonic development.
d. move by spinning head over tail, hence the name "wheel animals".
e. are rotifers.

____ 13. Leeches are
a. in the group Hirudinida.
b. are annelids.
c. .have setae
d. may be active predators.
e. parasites.

____ 14. Earthworms are
a. nematodes.
b. mollusks.
c. chilopods.
d. annelids.
e. cephalopods.

____ 15. The animal that has a shell consisting of eight separate, overlapping dorsal plates is
a. an insect.
b. a chiton.
c. in the same class as squids.
d. in the class Polyplacophora.
e. a mollusk.

____ 16. Earthworms are held together in copulation by mucous secretions from the
a. clitellum.
b. typhlosole.
c. seminal receptacles.
d. epidermis.
e. prostomium.

____ 17. Earthworms are
a. annelids.
b. oligochaets.
c. in the phylum Polychaeta.
d. protostomes.
e. acoelomate animals.

____ 18. Butterflies and black widows
a. are insects.
b. are arthropods.
c. have paired, jointed appendages.
d. have closed circulatory systems.
e. have an abdomen.

____ 19. A marine animal with parapodia and a trochophore larva might very well be
a. a tubeworm.
b. an oligochaet.
c. bilaterally symmetrical.
d. a leech.
e. a polychaet.

____20. An octopus is a
a. vertebrate.
b. mollusk.
c. chilopod.
d. protostome.
e. cephalopod.

____21. Animals with most of their organs in a visceral mass covered by a mantle are
a. annelids.
b. oligochaets.
c. onychophorans.
d. mollusks.
e. cheliceratans.

____22. Clams, scallops, and oysters are
- a. nematodes.
- b. mollusks.
- c. chilopods.
- d. marine.
- e. cephalopods.

____23. Insects are
- a. hexapods.
- b. arthropods.
- c. chelicerates.
- d. in the same subphylum as lobsters and crabs.
- e. in the same subphylum as spiders.

____24. Spiders are
- a. hexapods.
- b. arthropods.
- c. chelicerates.
- d. in the same subphylum as lobsters and crabs.
- e. in the same subphylum as insects.

VISUAL FOUNDATIONS

Color the parts of the illustration below as indicated.

RED ❑ digestive tract

GREEN ❑ shell

YELLOW ❑ foot

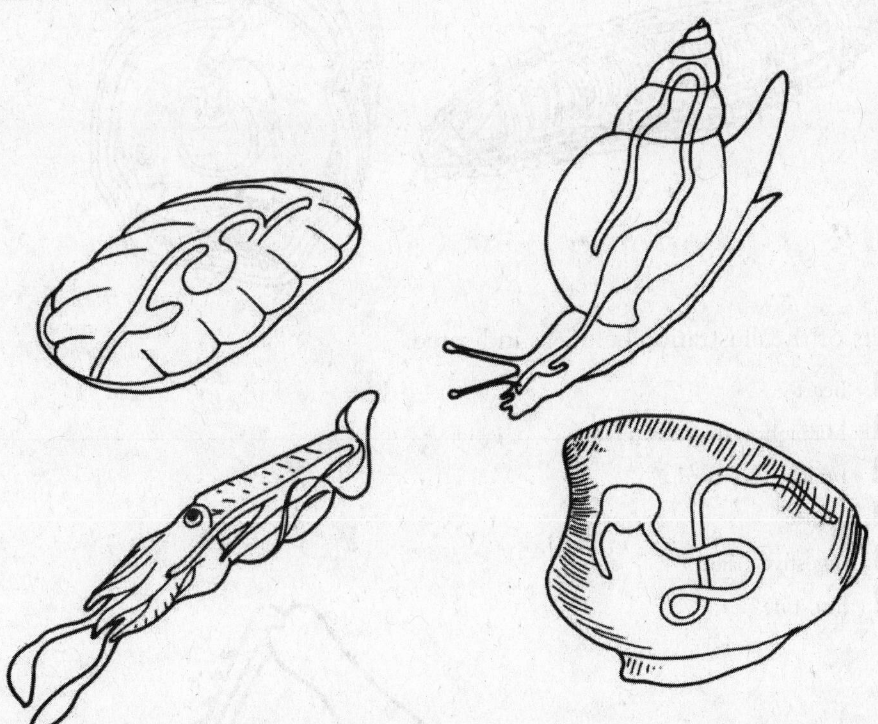

Color the parts of the illustration below as indicated.

RED	❑	dorsal and ventral blood vessels
GREEN	❑	metanephridium
YELLOW	❑	nerve cord
BLUE	❑	cerebral ganglia
ORANGE	❑	pharynx
BROWN	❑	intestine
TAN	❑	coelom

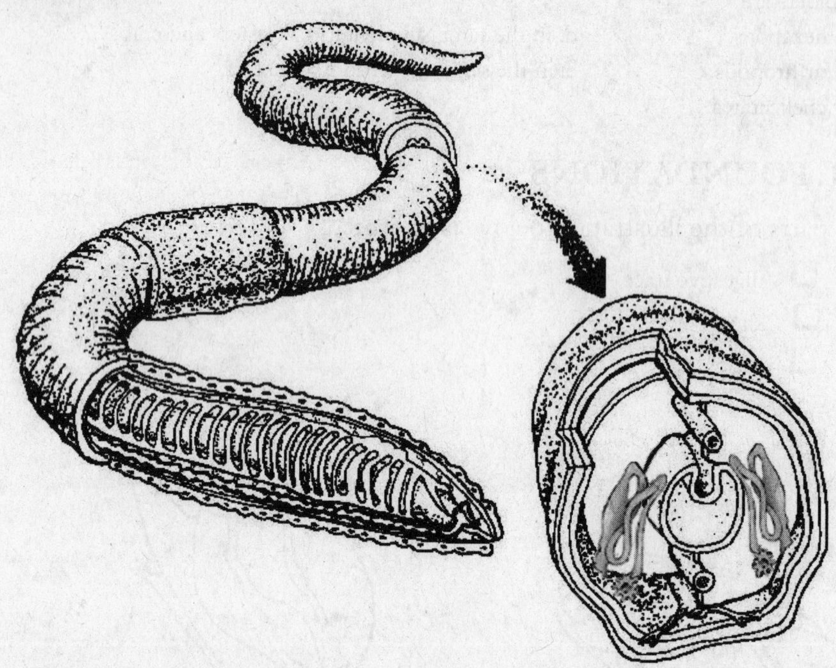

Color the parts of the illustration below as indicated.

RED	❑	heart
GREEN	❑	Malpighian tubules
YELLOW	❑	brain, nerve cord
BLUE	❑	ovary
ORANGE	❑	digestive gland
BROWN	❑	intestine

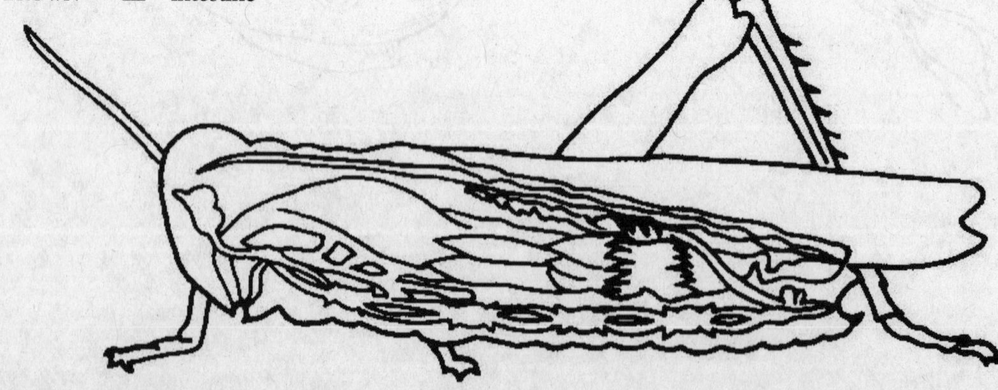

CHAPTER 32

The Deuterostomes

This chapter discusses the deuterostomes, animals with coeloms in which the second opening that develops in the embryo becomes the mouth, and the first becomes the anus. The deuterostomes include the echinoderms, hemichordates, and the chordates. The echinoderms and the chordates are by far the more ubiquitous and dominant deuterostomes. The echinoderms all live in the sea. They have spiny skins, a water vascular system, and tube feet. The chordates include animals that live both in the sea and on land. Chordates share four derived characters which distinguish them from other animals; a notochord, a dorsal tubular nerve cord, a post anal tail, and an endostyle or thyroid gland, at least at some time in their life cycle. At some point in their life cycle, chordates also have pharyngeal gill slits. There are three groups of chordates: the tunicates, the lancelets, and the vertebrates. The vertebrates are characterized by a vertebral column, a cranium, pronounced cephalization, a differentiated brain, muscles attached to an endoskeleton, and two pairs of appendages. There are 10 classes of living vertebrates: 6 which are fishes and 4 which contain animals with four limbs. The fishes include the jawless, cartilaginous, and bony fishes, and, the four-limbed vertebrates include the amphibians, reptiles, birds, and mammals. Mammals are classified first on the basis of whether embryonic development occurs within an egg, a maternal pouch, or utilizes an organ of exchange, i.e. a placenta.

REVIEWING CONCEPTS

Fill in the blanks.

INTRODUCTION

1. Scientists speculate that the last common ancestor of the deuterostomes was an animal that was a ________________ feeder.
2. The two major phyla assigned to the deuterostomes are the (a)________________________ and the (b) ____________________.

WHAT ARE DEUTEROSTOMES?

3. Deuterostomes evolved during the ________________ eon.
4. Deuterostomes are related through the (a) ________________________ characters present in their most recent common ancestor. Deuterostomes undergo a type of cleavage which is (b) ______________ and (c) ________________, meaning that the fate of their cells is fixed later in development. Cleavage in protostomes is (d) __________ and (e) __________________.

ECHINODERMS

5. Echinoderms have a true (a) ______________ which houses the internal organs. The larvae have (b) __________________ symmetry while the adults have (c) __________________________ symmetry.

6. Echinoderms have an (a) ______________ skeleton. The (b)____________________ on the body surface help to keep it clean.
7. The ____________________________ system of canals and chambers is unique to echinoderms.
8. Echinoderms are thought to have evolved during the ______________________ period.

Feather stars and sea lilies are suspension feeders

9. Sea lilies and feather stars are placed in class (a) __________________. Unlike other echinoderms, the oral surface of this class is located on the (b) _____________ surface.

Many sea stars capture prey

10. Sea stars are in the class (a) _____________________. They have a (b) _______________ from which radiate five or more arms or rays.
11. The arms or rays on sea stars contain hundreds of small ______________ connected to the radial canal.
12. As a means of obtaining food, most sea stars are either (a)______________________, or (b)_______________.

Basket stars and brittle stars makeup the largest group of echinoderms

13. Basket stars and brittle stars are placed in class __________________.
14. This class uses their arms primarily for (a) ________________ and their tube feet for (b) ______________________________.

Sea urchins and sand dollars have movable spines

15. Sea urchins and sand dollars are in the class (a) __________________. In this class the (b) ___________ are absent. They have a solid shell called a (c) _________, and their body is covered with (d) __________.

Sea cucumbers are elongated, sluggish animals

16. Sea cucumbers belong to class (a) ____________________. Their tentacles are modified (b) ________________.
17. It is not uncommon when environmental conditions are less than favorable for sea cucumbers to eject body parts in a process called __________________.

THE CHORDATES: MAJOR CHARACTERISTICS

18. Chordates are in the phylum __________________.
19. The four characteristics that distinguish chordates from all other groups are (a) _________________, (b) ______________________, (c) ________________________, and (d) ___________________.

INVERTEBRATE CHORDATES

Tunicates are common marine animals

20. Larval tunicates have typical chordate characteristics and superficially resemble ______________________.
21. Adult tunicates develop a protective covering called a (a) ____________ composed of (b) ______________________.

Lancelets clearly exhibit chordate characteristics

22. Most cephalochordates belong to the genus ________________.
23. Lancelets collect food by ______________________________.

Systematists debate chordate phylogeny

24. Cephalochordates, urochordates, and vertebrates all evolved from a chordate with an __________________________ body.

INTRODUCING THE VERTEBRATES

The vertebral column is a derived vertebrate character

25. Subphylum Vertebrata includes animals with a backbone or (a)______________________________, a braincase or (b)_____________, and a concentration of nerve cells and sense organs in a definite head, a phenomenon known as (c) _______________.
26. Early vertebrates were (a) __________________ animals that used muscles for feeding.

Vertebrate taxonomy is a work in progress

27. The extant vertebrates are currently assigned to (a) _______ (#?) classes of fishes and (b) _______ (#?) of tetrapods. The tetrapods are classified as (c) __________________, and (d) __________________.

JAWLESS FISHES

28. Some of the earliest known vertebrates, the ________________, consisted of several groups of small, armored, jawless fishes that lived on the ocean floor and strained their food from the water.
29. Hagfishes are in the class (a) ________________. Hagfish have no (b) ________________ but they do have a (c) ______________ for longitudinal support.
30. Class Petromyzontida contains the ______________________.
31. Contemporary hagfishes and lampreys have neither jaws nor paired ________.

EVOLUTION OF JAWS AND LIMBS: JAWED FISHES AND TETRAPODS

32. Jaws evolved from a portion of the (a)______________________________. This feature along with the development of fins allowed the jawed fishes to become (b)____________________.

Most cartilaginous fish inhabit marine environments

33. Cartilaginous fishes belong to class (a) ________________. This class include the (b) ____________, (c) ____________, and (d) __________.
34. The skin of cartilaginous fish contains numerous ________________, toothlike structures composed of layers of enamel and dentine.
35. (a) __________________________ are sensory grooves along the sides of all fish that contain cells capable of perceiving water movements. Similarly, the (b) __________________________ on the shark's head can sense very weak electrical currents, like those generated by another animal's muscle contractions.
36. Sharks may be (a) ____________________, that is, lay eggs; or (b) ____________________, that is the eggs are incubated and hatch internally; or (c) ____________, nutrients are transferred between the maternal and fetal blood and there is a live birth.

The ray-finned fishes gave rise to modern bony fishes

37. Bony fishes and cartilaginous fishes share many characteristics including __________ replacement.
38. In the Devonian period, bony fishes diverged into two groups the (a)____________________ fish, or actinopterygians, and the (b)____________________ with their fleshy, lobed fins and lungs.
39. The actinopterygians gave rise to the modern bony fish whose lungs became modified as ______________.
40. Activity of the _____ gene may have contributed to the diversity of bony fishes observed today

Tetrapods evolved from sarcopterygian ancestors

41. The three living groups of sarcopterygians are the (a) ____________, (b) ____________________, and (c) ____________________.
42. Ancestors of the ________________ gave rise to the tetrapods.
43. Molecular data support the hypothesis that tetrapod limbs evolved from __________________________.
44. (a) ____________________, considered to be a fish, is believed to be the transitional form between fishes and tetrapods. While it was considered a fish because it had scales and fins it also had the tetrapod features of (b) ____________________ and (c)______________________________.

Amphibians were the first successful land vertebrates

45. The three orders of modern amphibians are (a) ________________, amphibians with long tails, including salamanders, mud puppies, and newts; the (b) ________________, tailless frogs and toads; and (c) ______________________, the wormlike caecilians.

46. Some salamanders retain larval characteristics even when sexually mature. This phenomenon is called ________________________.
47. Amphibians use lungs and their (a) ____________ for gas exchange. They have a (b) ___________ (#?)-chambered heart with systemic and pulmonary circulations.

AMNIOTES: TERRESTRIAL VETREBRATES

48. The ___________________ allows terrestrial vertebrates to complete their life cycle on land.
49. The body covering of terrestrial vertebrates is typically ____________, a water-insoluble protein.

Our understanding of amniote phylogeny is changing

50. Diapsid amniotes have (a) __________ pairs of openings in the temporal bones whereas synapsid amniotes have (b) __________ pair.
51. When it comes to maintaining body temperature, fossil evidence indicates that at least some dinosaurs were ____________________.

Reptiles have many terrestrial adaptations

52. Reptiles produce and egg with a ________________________.
53. Amniotes hearts have (a) __________ (#) atria and the ventricle is either (b) _______________ or (c) _______________________.
54. Metabolically fishes, amphibians, and reptiles are ______________.

Biologists assign reptiles to several major clades

55. The order (a) __________________ (turtles, terrapins, tortoises), the order (b) ________________ (snakes, lizards, amphisbaenians, order (c)______________ (tuataras), order (d) ____________ (crocodiles, alligators, caimans, gavials) and the birds which are in order (e) _____________.

Turtles have protective shells

56. Turtles do not have teeth, but a ________________ covers the jaws.

Lizards and snakes are common modern reptiles

57. Lizards and snakes have scales that ______________ to form a continuous, flexible armor coating.
58. Vipers have a specialized ________________________ on each side of the head that is used to detect heat emitted by prey.

Tuataras superficially resemble lizards

59. In their outward appearance, tuataras look like ________________.

Crocodilians have an elongated skull

60. Crocodiles are distinguished from alligators and caimans by two major features: (a) _______________________ and the (b)______________________________ in the bottom jaw visible when the mouth is closed.

How do we know that birds are really dinosaurs?

61. Modern birds use ____________________ feathers for flight.
62. Insulating feathers could have contributed to the evolution of ________________________ which would have allowed animals to be more active.
63. Some extinct theropods as well as modern birds have furcula (wishbone) which is formed when ______________________.

Modern birds are adapted for flight

64. Ostriches and emus are called ratites. This means that there sternum does not have a ridge for the ______________________.
65. Modern birds resemble reptiles in that they (a) ______________ and have (b) ____________ on their legs.
66. Birds have a (a) __________ (#) chambered heart and very efficient lungs which combined allows them to maintain a (b)______________________ rate.
67. Birds excrete nitrogenous wastes in the form of _____________.
68. The birds digestive system has an expanded, saclike portion called the (a) ___________ where food is stored and a stomach divided into a (b)_________which secretes gastric juice and a (c) ____________ which is very muscular and grinds the food.

Mammals have hair and mammary glands

69. Derived characters of mammals include (a) ________________, (b)________________, (c)________________________, (d) _______________ and (e) ______________.
70. Fertilization is internal and the majority of mammals are (a)________________ and develop a (b) ______________ which serves as an exchange organ between the mother and the fetus.

New fossil discoveries are changing our understanding of the early evolution of mammals

71. Mammals probably evolved from (a) ______________, a reptilian group, about 200 million years ago in the (b) ____________ period.
72. The fact that early mammals were (a) _________________ (tree-dwelling) and (b) ___________________ (active at night) helped them to survive during the reign of the reptiles.

Modern mammals are assigned to three subclasses

73. Mammals are classified in two clades. The egg laying mammals, (a)___________ and mammals that bear live young, (b) _________. This last group is further divided into (c) _____________ (the marsupials or pouched mammals) and (d) _____________ (mammals that are more developed at birth and often called the placental mammals).

BUILDING WORDS

Use combinations of prefixes and suffixes to build words for the definitions that follow.

Prefixes	The Meaning	Suffixes	The Meaning
a-	without	-ata	characterized by having
caud-	tail	-chord	cord
chondr(o)-	cartilage	-derm	skin
cephal-	head	-gnath(an)	jaw
endo-	within	-ichthy(es)	fish
echino-	"spiny"	-ization	process of
noto-	back	-pod(al)	foot, footed
tetra-	four	-therm	heat
ampull-	flask	-ura	tail
caud-	tail		

Prefix	Suffix	Definition
____________	____________	1. A fish without jaws (jawless fish); a member of the class of vertebrates including lampreys and hagfishes.
an-	____________	2. An order of amphibia with legs but no tail; tailless frogs or toads.
____________	____________	3. Pertains to those amphibians with no feet, i.e., the wormlike caecilians.
____________	____________	4. The class comprising the cartilaginous fishes.
____________	____________	5. A vertebrate with limbs.
____________	-ata	6. The order that comprises the amphibians with long tails, e.g., the salamanders, mudpuppies, and newts.
____________	____________	7. A spiny-skinned animal.
____________	-skeleton	8. Bony and cartilaginous-supporting structures within the body that provide support from within.
____________	____________	9. A dorsal, longitudinal rod of cartilages.
____________	-a	10. A rounded muscular sac at the base of an echinoderm's foot.
____________	____________	11. A concentration of sensory structures on the anterior (front end) of an animal.
____________	____________	12. An animal that maintains its own body temperature through metabolism and behavior.

MATCHING

Terms:

a. Amnion
b. Amphibian
c. clasper
d. Ctylosaur
e. Cotylosaur
f. crop
g. gizzard
h. Labyrinthodont
i. Marsupium
j. Neoteny
k. Notochord
l. Oviparous
m. Ovoviviparous
n. Placenta
o. Placoderm
p. Squ;amata
q. Therapsid
r. Vertebrate

For each of these definitions, select the correct matching term from the list above.

____ 1. The dorsal, longitudinal rod that serves as an internal skeleton in the embryos of all, and in the adults of some, chordates.
____ 2. The pouch in which a marsupial's young develop.
____ 3. Animals that are egg-layers.
____ 4. An extra-embryonic membrane that forms a fluid-filled sac for the protection of the developing embryo.
____ 5. An extinct jawed fish.
____ 6. A member of the group of mammal-like reptiles of the Permian period that gave rise to the mammals.
____ 7. An organ of exchange between developing embryo and mother in eutherian mammals.
____ 8. A chordate possessing a bony vertebral column.
____ 9. Order containing lizards and snakes.
____10. A sac-like structure in which food is temporarily stored.
____11. Used by male sharks to transfer sperm to a female.

MAKING COMPARISONS

Fill in the blanks.

Vertebrate Class	Representative Animals	Heart	Skeletal Material	Other Characteristics
Myxini	Hagfish	Two-chambered heart	Cartilage	Jawless, most primitive vertebrates, gills,
Chondrichthyes	#1	Two-chambered heart	#2	#3
#4	Salmon, tuna	Two-chambered heart	#5	Gills, swim bladder
Amphibia	#6	#7	Bone	Tetrapods, aquatic larva metamorphoses into terrestrial adult, moist skin and lungs for gas exchange,

Vertebrate Class	Representative Animals	Heart	Skeletal Material	Other Characteristics
				ectothermic
#8	Birds	#9	#10	Amniotes with feathers, many adaptations for flight, endothermic, high metabolic rate
#11	Monotremes, marsupials, placentals	Four-chambered heart	#12	#13

MAKING CHOICES

Place your answer(s) in the space provided. Some questions may have more than one correct answer.

____ 1. Humans are members of the taxon(a)
a. Lagomorpha.
b. Rodentia.
c. that has chisel-like incisors that grow continually.
d. that includes placental mammals.
e. Cetacea.

____ 2. The phylum(a) that many biologists believe had a common ancestry with our own phylum is/are
a. Mollusca.
b. Annelida.
c. Arthropoda.
d. Echinodermata.
e. Chordata.

____ 3. The group(s) of animals that maintain a constant internal body temperature is/are
a. Reptilia.
b. Mammalia.
c. Aves.
d. Amphibia.
e. Pisces.

____ 4. The group(s) of animals with a part of the stomach that secretes gastric juices and a separate part that grinds food is/are
a. Reptilia.
b. Mammalia.
c. Aves.
d. Amphibia.
e. Pisces.

____ 5. A turtle is a member of the taxon(s)
a. Reptilia.
b. Chelonia.
c. Urodela.
d. Amphibia.
e. Anura.

____ 6. Mammals probably evolved from
a. amphibians.
b. a type of fish.
c. dog-like carnivores.
d. monotremes.
e. therapsids.

____ 7. The lateral line organ is found
a. in sharks.
b. in bony fish.
c. in agnathans.
d. in all fishes.
e. in all fish except placoderms.

____ 8. Members of the phylum Chordata all have
a. an embryonic notochord.
b. embryonic pharyngeal gill slits.
c. a dorsal, tubular nerve cord.
d. well-developed germ layers.
e. bilateral symmetry.

____ 9. The "spiny-skinned" animals include
a. sea stars.
b. sand dollars.
c. snails.
d. lancets.
e. marine worms.

____10. The acorn worm
a. has a notochord.
b. is a deuterostome.
c. is in the same subphylum as *Amphioxus*.
d. is in the phylum Chordata.
e. is a hemichordate.

____11. The group(s) of animals with both four-chambered hearts and a double circuit of blood flow is/are the
a. earthworms.
b. birds.
c. mammals.
d. order Aves.
e. amphibians.

____12. An animal that traps food in mucus secreted by cells of the endostyle is a/an
a. holothuroidean.
b. enchinoderm.
c. urochordate.
d. sea cucumber.
e. tunicate.

____13. The first successful land vertebrates were
a. tetrapods.
b. reptiles.
c. chondrichthyes.
d. amphibians.
e. in the superclass Pisces.

____14. Cats are members of the taxon(a)
a. Lagomorpha.
b. Cephalochordata.
c. with chisel-like incisors.
d. that includes placental mammals.
e. Cetacea.

____15. Unique features of the echinoderms include
a. radial larvae.
b. ciliated larvae.
c. endoskeleton composed of $CaCO_3$ plates.
d. gas exchange by diffusion.
e. water vascular system.

____16. A frog is a member of the taxon(a)
a. Reptilia.
b. Chelonia.
c. Urodela.
d. Amphibia.
e. Anura.

____ 17. Features of echinoderms that distinguish them from other animals include
a. tube feet.
b. radial symmetry.
c. a coelom.
d. pedicellariae.
e an endoskeleton.

____18. Included in the five extant classes of echinoderms are
a. crinoids.
b. sea lilies.
c. feather stars.
d. asteroids.
e. sand dollars.

____19. In a cladogram of deuterostomes, shared derived characters would include
a. pharyngeal gill slits.
b. determinate cleavage
c. a cranium.
d. radial cleavage.
e. a notochord.

____20. Terms associated with tunicates include
a. endostyle.
b. siphon.
c. nerve cord.
d. urochordates.
e. class Ascidiacea.

____ 21. Vertebrates are distinguished from other chordates
a. by their notochord.
b. by having a postanal tail.
c. by having amniotic eggs.
d. by their pronounced cephalization.
e. by the presence of a vertebral column.

____ 22. Terms associated with lamprey include
a. conodont.
b. ostracoderm.
c. jaws.
d. myxini.
e. parasite.

____ 23. Reproductively, sharks may be
a. hermaphrodites.
b. egg layers.
c. monoecious.
d. viviparous.
e. ovoviviparous.

VISUAL FOUNDATIONS

Color the parts of the illustration below as indicated. Also label the postanal tail.

RED ❑ heart
GREEN ❑ mouth
YELLOW ❑ brain and dorsal, hollow nerve tube
BLUE ❑ notochord
ORANGE ❑ pharyngeal gill slits
PINK ❑ pharynx, intestine
TAN ❑ muscular segments

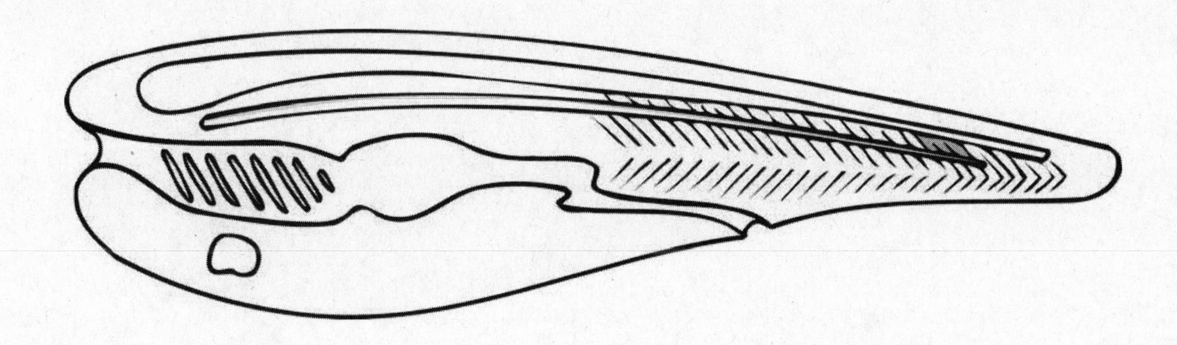

CHAPTER 33

Plant Structure, Growth, and Development

This is the first of six chapters addressing the integration of plant structure and function. This chapter examines the external structure of the flowering plant body; the organization of its cells, tissues, and tissue systems; and its basic growth patterns. The plant body is organized into a root system and a shoot system. Plants are composed of cells that are organized into tissues, and tissues that are organized into organs. Each organ performs a single function or group of functions, and is dependent on other organs for its survival. All vascular plants have three tissue systems: the dermal tissue system, the vascular tissue system, and the ground tissue system. Plant tissues may be simple or complex. Plants grow by increasing both in girth and length. Unlike growth in animals, plant growth is localized in areas of unspecialized cells, and involves cell proliferation, elongation, and differentiation.

REVIEWING CONCEPTS

Fill in the blanks.

INTRODUCTION

1. (a)____________________ plants grow, reproduce and die in a year or less; (b)____________________ plants take two years to complete their life cycle; and (c)____________________ plants can often live for more than two years.
2. (a)____________________ perennials shed their lives in fall and are dormant over winter but (b)____________________ shed their leaves over an extended period of time which means that some leaves are always present.

THE PLANT BODY

3. The body of vascular plants is organized into a (a) ____________________ system and a (b) ____________________ system.

The plant body consists of cells and tissues

4. A (a) ____________________ is a group of cells that form a structural and functional unit. (b) ____________________ are composed of only one kind of cell whereas (c) ____________________ have two or more kinds of cells.
5. Roots, stems, leaves, flower parts, and fruits are ____________________ because each is composed of all three tissue systems.

The ground tissue system is composed of three simple tissues

6. The three tissues composing the ground tissue system of herbaceous plants are (a) ______________________________, (b)____________________________, and (c) ____________________________.
7. The three primary functions of parenchyma tissue include (a)____________________________________, (b)__________________________, and (c) ____________________________.
8. Parenchyma cells have the ability to ________________________ into other kinds of cells, for example when a plant has been injured.
9. Collenchyma tissue is composed of living cells that function primarily to _____________ the plant.
10. The two types of sclerenchyma cells are (a)______________________ and (b) _________________.
11. Cellulose microfibrils are cemented together by a matrix of (a)_____________________ and (b) ____________________.
12. An important component of the cell walls of wood is __________________ which may comprise up to 35% of the dry weight of the secondary cell wall.

The vascular tissue system consists of two complex tissues

13. Xylem conducts (a) _________________, and (b)__________________________ from roots to stems and leaves. Xylem also contains (c) ____________________________ for storage and (d) ____________________ for support.
14. Having undergone (a) ____________________, mature tracheids and vessel cells are dead. They are also (b) ______________________ which contributes to their conducting properties.
15. Phloem is a complex tissue that functions to conduct (a) _______________ throughout the plant through the (b)_______________________________ which are among the most specialized plant cells.
16. ____________________________ are cytoplasmic connections through which cytoplasm extends from one cell to another.

The dermal tissue system consists of two complex tissues

17. The dermal tissue system, the (a) _______________and (b)_____________, provides a protective covering over plant parts. Epidermal cells secrete a waxy layer called the (c) _______________ that restricts water loss. (d)______________ are openings in this layer through which gases diffuse.
18. The epidermis contains special outgrowths, or hairs, called ___________________, which occur in many sizes and shapes and have a variety of functions.

19. The periderm replaces the epidermis in the stems and roots of older woody plants. Its primary function is for (a) ________________. It is composed mainly of (b) ____________ cells whose walls are coated with suberin, a waterproofing substance.

PLANT MERISTEMS

20. Plant growth is localized in regions called (a) ________________.
21. Meristematic cells do not differentiate which means that plants and not most animals are able to______________________ throughout their lifespan.
22. Plants have two kinds of meristematic growth: (a)____________________ growth, an increase in length; and (b)____________________ growth, an increase in the girth of the plant.

Primary growth takes place at apical meristems

23. The root apical meristem consists of three zones (areas), which, in order from the root cap inward, are the a)______________________________, (b)_____________________________, and (c)_____________________________.
24. (a)____________________(developing leaves) and (b)____________________ (developing buds) arise from the shoot apical meristem.

Secondary growth takes place at lateral meristems

25. There are two lateral meristems responsible for secondary growth (increase in girth), the (a) ______________________, a layer of meristematic cells that forms a long, thin, continuous cylinder within the stem and root, and the (b) ________________, a thin cylinder or irregular arrangement of meristematic cell in the outer bark.
26. The outermost covering over woody stems and roots is the ________.

DEVELOPMENT OF FORM

The plane and symmetry of cell division affect plant form

27. The ________________________ appears just prior to mitosis and determines the plane in which the cells will divide.

The orientation of cellulose microfibrils affects the direction of cell expansion

28. Growth occurs in plants when cells (a) ____________________ and (b) ____________________.

Cell differentiation depends in part on a cell's location

29. ______________________________________ is responsible for variations in chemistry, behavior, and structure among plant cells.

Morphogenesis occurs through pattern formation

30. Depending upon their location, cells are exposed to (a)________________________________ that specify positional information.

BUILDING WORDS

Use combinations of prefixes and suffixes to build words for the definitions that follow.

Prefixes	The Meaning	Suffixes	The Meaning
bi-	two	-derm(is)	skin
cut-	skin	-(i)cle	little
decidu-	falling off	-ome	mass
epi-	on, upon	-ous	pertaining to
stom-	mouth	-ennial	pertaining to a year
trich-	hair		

Prefix	Suffix	Definition
________	________	1. A hair or other special outgrowth growing out from the epidermis of plants.
________	________	2. A plant that takes two years to complete its life cycle.
________	________	3. Along with the periderm, provides a protective covering on the surface of plants.
________	-a	4. A small pore ("mouth") in the epidermis of plants.
________	________	5. The shedding of leaves at the end of the growing season.
________	________	6. A waxy covering of plant epidermal cells that prevents water loss.

MATCHING

Terms:

a.	Apical meristem	f.	Meristem	k.	Sclerenchyma
b.	Companion cell	g.	Parenchyma	l.	Shoot system
c.	Dermal tissue	h.	Perennial	m.	Tracheid
d.	Ground tissue	i.	Periderm	n.	Vascular cambium
e.	Lateral meristem	j.	Root system	o.	Xylem

For each of these definitions, select the correct matching term from the list above.

____ 1. Vascular tissue that conducts water and dissolved minerals through the plant.

____ 2. The chief type of water-conducting cell in the xylem of gymnosperms.

____ 3. An area of dividing tissue located at the tips of roots and shoots in plants.

____ 4. Secondary meristem that produces the secondary xylem and secondary phloem.

____ 5. The type of tissue system that gives rise to the epidermis.

____ 6. Plant cell that has thick secondary walls, is dead at maturity, and functions in support.

____ 7. A nucleated cell in the phloem responsible for loading and unloading sugar into the sieve tube member.

____ 8. A plant that lives longer than two years.

____ 9. General term for all localized areas of mitosis and growth in the plant body.

____10. Plant cells that are relatively unspecialized, are thin walled, and function in photosynthesis and in the storage of nutrients.

MAKING COMPARISONS

Fill in the blanks.

Tissue	Tissue System	Function	Location in Plant Body
Parenchyma	Ground tissue	Photosynthesis, storage, secretion	Throughout plant body
Collenchyma	#1	Flexible structural support	#2
#3	Ground tissue	#4	Throughout plant body; common in stems and certain leaves, some nuts and pits of stone fruit
#5	#6	Conducts water, dissolved minerals	Extends throughout plant body
Phloem	Vascular tissue	#7	#8
Epidermis	#9	#10	#11
#12	Dermal tissue	#13	Covers body of woody plants

MAKING CHOICES

Place your answer(s) in the space provided. Some questions may have more than one correct answer.

____ 1. Plants with the potential for living more than two years are called
a. annuals.
b. biennials.
c. triennials.
d. perennials.
e. polyannuals.

____ 2. The most common type of cell and tissue found throughout the plant body is the
a. parenchyma.
b. sclerenchyma.
c. collenchyma.
d. vascular tissue.
e. xylem.

____ 3. The kind of growth that results in an increase in the girth of the plant is known as
a. differentiation.
b. elongation.
c. apical meristem growth.
d. primary growth.
e. secondary growth.

____ 4. All plant cells have
a. cell walls.
d. primary cell walls.

b. secondary growth. e. secondary cell walls.
c. the capacity to form a complete plant.

____ 5. Hairlike outgrowths of plant epidermis are called
a. periderm. d. trichomes.
b. fiber elements. e. companion cells.
c. lateral buds.

____ 6. The types of cells in phloem include
a. tracheids. d. sieve tube elements.
b. vessel elements. e. fiber cells.
c. parenchyma.

____ 7. The cell type found throughout the plant body that often functions in photosynthesis, secretion, and storage is called
a. sclerenchyma. d. companion cells.
b. collenchyma. e. a tracheid.
c. parenchyma.

____ 8. Increase in the girth of a plant is due to growth of the
a. vascular cambium. d. area of cell elongation.
b. cork cambium. e. lateral meristems.
c. area of cell maturation.

____ 9. Localized areas of cell division resulting in plant growth are called
a. primordia. d. primary growth centers.
b. mitotic zones. e. secondary growth centers.
c. meristems.

____10. The types of cells in xylem include
a. tracheids. d. sieve tube members.
b. vessel elements. e. fiber cells.
c. parenchyma.

____ 11. The soft tissue part of a plant and those that are often edible consist of
a. sclerenchyma. d. collenchyma.
b. parenchyma. e. lignin.
c. periderm.

____ 12. Terms associated with collenchymal cells include
a. storage. d. support.
b. dead. e. thick primary cell walls.
c. secretion.

____ 13. The most abundant polymer in the world is
a. lignin. d. pectin.
b. cellulose. e. lectin.
c. suberin.

____ 14. In plants, the vascular tissue system
a. is embedded in the ground tissue. d. includes the xylem and phloem.
b. vessel elements. e. has parts composed of dead cells.
c. transports carbohydrates.

____ 15. Plasmodesmota are
- a. connect cells of the phloem.
- b. are cytoplasmic connections.
- c. are located in the periderm.
- d. connect cells of the xylem.
- e. companion cells.

____ 16. Stomata allow the passage of
- a. carbohydrates.
- b. water vapor.
- c. glucose.
- d. carbon dioxide.
- e. oxygen

____ 17. Cork cambium
- a. is composed of meristematic cells.
- b. is part of the vascular cambium.
- c. contains storage cells.
- d. is located in the outer bark.
- e. is part of the periderm.

VISUAL FOUNDATIONS

Color the parts of the illustration below as indicated. Also label the area of cell division and the area of cell maturation.

RED ❑ apical meristem

GREEN ❑ area of cell elongation

YELLOW ❑ root cap

ORANGE ❑ root hairs

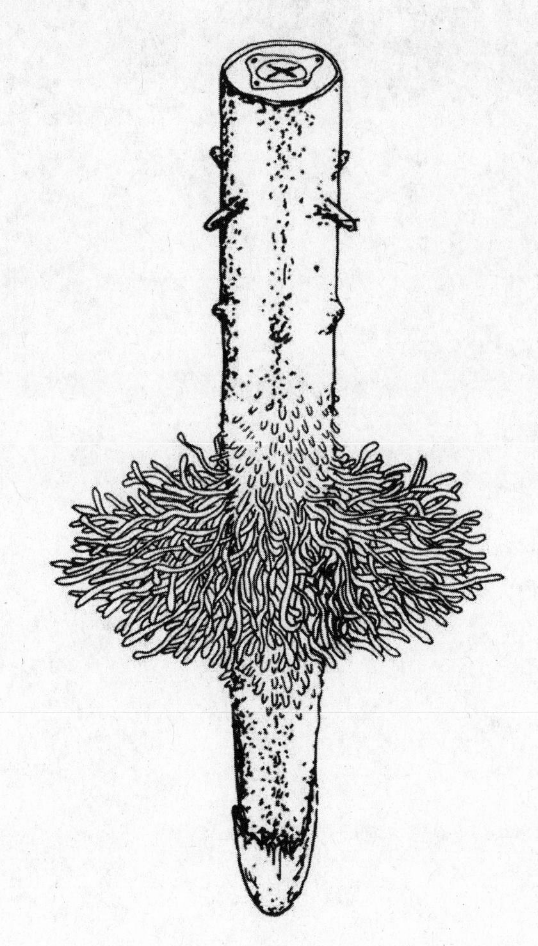

Color the parts of the illustration below as indicated.

RED	❑	outer bark (periderm)
GREEN	❑	inner bark (secondary phloem)
YELLOW	❑	vascular cambium
Brown	❑	wood (secondary xylem)3

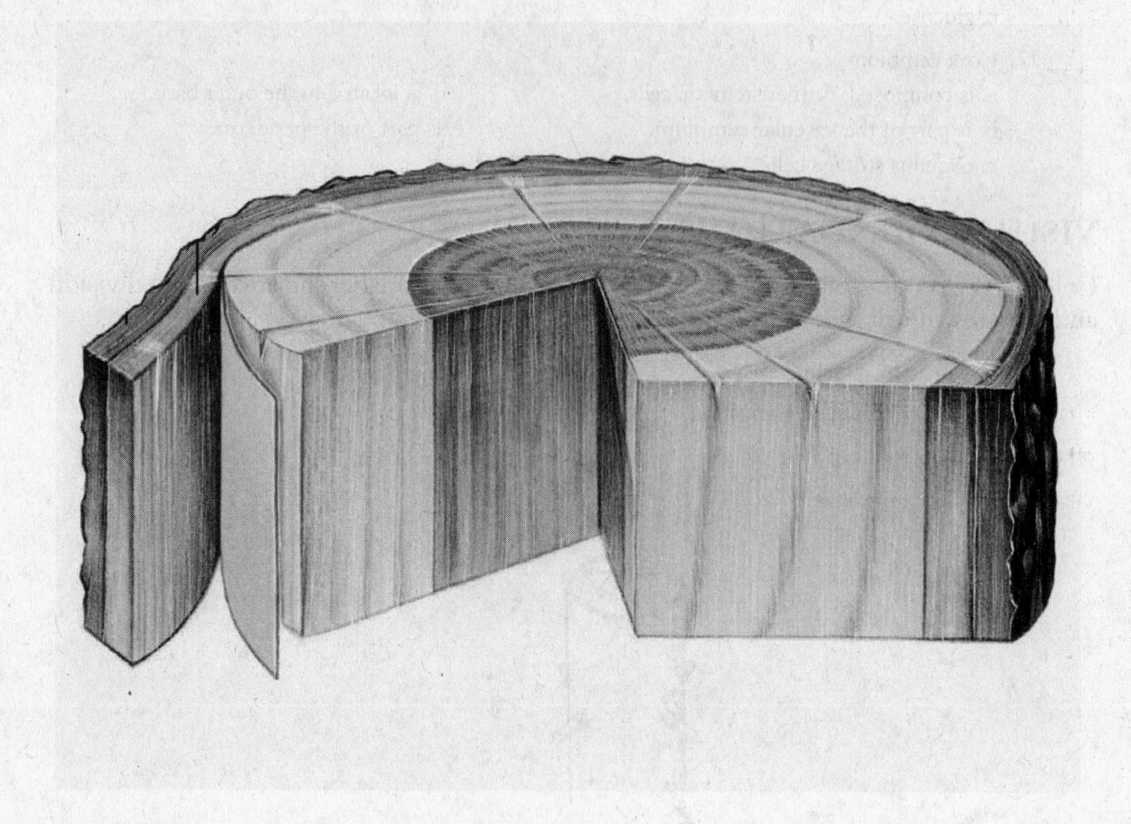

CHAPTER 34

Leaf Structure and Function

This chapter, the second of six addressing the integration of plant structure and function, discusses the structural and physiological adaptations of leaves. Leaves, as the principal photosynthetic organs in plants, are highly adapted to collect radiant energy, convert it into the chemical bonds of carbohydrates, and transport the carbohydrates to the rest of the plant. They are also adapted to permit gas exchange, and to receive water and minerals transported into them by the plant vascular system. Water loss from leaves is controlled in part by the cell walls of the epidermal cells, by presence of a surface waxy layer secreted by epidermal cells, and by tiny pores in the leaf epidermis that open and close in response to various environmental factors. Most of the water absorbed by land plants is lost, either in the form of water vapor or as liquid water. Survival at low temperatures is facilitated in some plants by the loss of leaves. The leaves of many plants are modified for functions other than photosynthesis.

REVIEWING CONCEPTS

Fill in the blanks.

INTRODUCTION

1. The shape of most leaves allows optimal absorption of (a)____________________ and the efficient diffusion of gases including (b) ______________ and (c) _____________.
2. Some leaf features that optimize photosynthesis actually promote ___________________________.

LEAF FORM AND STRUCTURE

3. The broad, flat portion of a leaf is the (a) _________________; the stalk that attaches the blade to the stem is the (b) _________________. Some leaves also have (c) ____________________, which are leaflike outgrowths usually present in pairs at the base of the petiole.
4. Leaves may be (a) __________________, having a single blade, or (b)__________________, having a blade divided into two or more leaflets.
5. Leaves are arranged on a stem in one of three possible ways. These are (a)____________________________, with one leaf at each node; (b)_____________________________, with two leaves at each node; and (c)________________________________, with three or more leaves at each node.
6. Leaf blades may have (a) ______________________ venation in which the strands of vascular tissue run parallel to one another. This venation is generally found in (b) ____________________. Venation may be (c) ___________________ in which the veins form a

meshwork. This venation is generally found in (d) ________________.

Leaf structure is adapted for maximum light absorption

7. The (a) ________________ forms the surface tissue of the leaf. Mosat cells in this layer lack (b) ________________.
8. The surface tissue of the leaf secretes a waxy (a) ________________ composed of (b) ________________ which helps to reduce water loss.
9. Guard cells change shape to regulate the stomatal opening when water and ions flow into them from the ________________ cells.
10. The photosynthetic ground tissue of the leaf is called the (a)________________. When this tissue is divided into two regions, the upper layer, nearest to the upper epidermis, is called the (b)________________, and the lower portion is called the (c) ________________.
11. The (a) ________________ in veins of a leaf conduct water and essential minerals, while the (b) ________________ in veins conducts sugar produced by photosynthesis. Veins may be surrounded by a (c)________________, consisting of parenchyma or sclerenchyma cells.
12. Light can penetrate into the body of the leaf because the leaf epidermis is (a) ________________. Light penetration is necessary for (b) ________________ to occur in complexes present in the (c) ________________.
13. (a) ________________, a raw material of photosynthesis, diffuses into the leaf through stomata, and the (b)________________ produced during photosynthesis diffuses rapidly out of the leaf through stomata.

STOMATAL OPENING AND CLOSING

14. A guard cell's shape is determined by the amount of water it contains which contributes to its (a) ________________. When water leaves a guard cell is becomes (b) ________________.

Blue light triggers stomatal opening

15. Any plant response to light must involve a (a) ________________ molecule which absorbs (b) ________________.
16. The pigment involved in the opening and closing of stomata is (a)________________ which absorbs (b) ________________.
17. Blue light triggers the synthesis of (a) ________________ and the hydrolysis of (b) ________________.
18. Blue light also triggers the activation of (a) ________________ located in the plasma membrane of guard cells. An electrochemical gradient is formed when (b) ________ is pumped out of the guard cells.

19. The electrochemical gradient drives the movement of (a) ______ ions into the guard cells. (b)____________ ions and (c)____________ ions also accumulate inside the guard cell vacuoles which changes the (d)__________________ concentration and water enters the cells.
20. As evening approaches, the concentration of (a) ____________ in the guard cell declines as it is converted to (b) ________________ which is a less osmotically active molecule and the stomata close.

Additional factors affect stomatal opening and closing

21. The opening and closing of stomata is triggered by (a) __________, and (b) ______________ along with several other environmental factors including (c) ____________________, (d) ________________, and (e) ________________.

TRANSPIRATION AND GUTTATION

22. Transpiration helps to move water from (a) ________ and throughout the plant.
23. Two benefits of transpiration are (a) ____________________ and (b) ____________________.
24. Transportation is like ______________ in humans.

Some plants exude liquid water

25. Loss of liquid water from leaves by force is known as ________.

LEAF ABSCISSION

26. Abscission is the process of ______________________.
27. Leaf abscission is a complex process that involves many physiological changes, all initiated and orchestrated by changing levels of plant hormones, particularly ____________________.
28. The brilliant colors found in autumn landscapes in temperate climates are due to the various combinations of (a) ____________ and (b)______________.

In many leaves, abscission occurs at an abscission zone near the base of the petiole

29. The area where a petiole detaches from the stem is a structurally distinct area called the (a) ____________________; it is a weak area because it contains relatively few strengthening (b) ________.
30. The "cement" that holds the primary cell walls of adjacent cells together is called the ______________.

MODIFIED LEAVES

31. Leaves are variously modified for a plethora of specialized functions. For example, the hard, pointed __________ on a cactus are leaves modified for protection.
32. ____________ are specialized leaves that anchor long, climbing vines to the supporting structures on which they are growing.

Modified leaves of carnivorous plants capture insects

33. The traps of the pitcher plant are (a) ________________, whereas those of the Venus flytrap are (b) ________________.

BUILDING WORDS

Use combinations of prefixes and suffixes to build words for the definitions that follow.

Prefixes	The Meaning	Prefixes	The Meaning
abscis-	cut off	-ation	the process of
circ-	around	-iole	little
gutt-	tear	-ion	process of
mes(o)-	middle	-ome(s)	mass
trans-	across, beyond	-phyll	leaf
trich-	hair		

Prefix	Suffix	Definition
________	________	1. The separation and falling away from the plant stem of leaves, fruit and flowers.
________	-adian	2. Pertains to something that cycles at approximately 24-hour intervals.
________	________	3. The photosynthetic tissue of the leaf sandwiched between (in the middle of) the upper and lower epidermis.
________	-piration	4. The loss of water vapor from the plant body across leaf surfaces.
________		5. The loss of liquid water which is forced out at specialized regions on leaves.
pet-	________	6. The stalk that attaches the blade of a leaf to the stem.
________	________	7. The epidermal outgrowth that covers many leaves.

MATCHING

Terms:

a.	Blade	f.	Petiole	k.	Trichome
b.	Bundle sheath	g.	Phloem	l.	Vascular bundle
c.	Cuticle	h.	Spine	m.	Vein
d.	Guard cell	i.	Stipule	n.	Xylem
e.	Guttation	j.	Tendril		

For each of these definitions, select the correct matching term from the list above.

____ 1. A ring of cells surrounding the vascular bundle in monocot and dicot leaves.

____ 2. A leaf or stem that is modified for holding or attaching to objects.

____ 3. Vascular tissue that transports sugars produced by photosynthesis.

____ 4. A waxy covering over the epidermis of the above-ground portion of plants.

____ 5. A hard, pointed leaf that is modified for protection.

____ 6. The part of a leaf that attaches to a stem.

____ 7. One of two cells that collectively form a stoma.

____ 8. In plants, vascular bundles in leaves.

____ 9. Leaflike outgrowth at the base of the peticule.

___ 10. The flat portion of a leaf.

MAKING COMPARISONS

Fill in the blanks.

Structure of Leaf	Function of Leaf Structure
Thin, flat shape	Maximizes light absorption, efficient gas diffusion
Ordered arrangement on stem	#1
Waxy cuticle	#2
#3	Allow gas exchange between plant and atmosphere
Relatively transparent epidermis	#4
#5	Allows for rapid diffusion of CO_2 to mesophyll cell surface
#6	Provide support to prevent leaf from collapsing
Xylem in veins	#7
#8	Transports sugar to other plant parts

MAKING CHOICES

Place your answer(s) in the space provided. Some questions may have more than one correct answer.

____ 1. Loss of water by evaporation from aerial plant parts is called
a. transpiration.
b. guttation.
c. abscission.
d. activation.
e. aspiration.

____ 2. The "cement" that holds the primary cell walls of adjacent cells together is called the
a. abscission zone.
b. leaf scar.
c. adhesion layer.
d. middle lamella.
e. glial substance.

____ 3. The portion of mesophyll that is usually composed of loosely and irregularly arranged cells is
a. the palisade layer.
b. the spongy layer.
c. toward the leaf's underside.
d. toward the leaf's upperside.
e. an area of photosynthesis.

____ 4. The process by which plants secrete water as a liquid is
a. evaporation.
b. transpiration.
c. guttation.
d. the potassium ion mechanism.
e. found only in monocots.

____ 5. The leaves of eudicots usually have
a. bulliform cells.
b. netted venation.
c. differentiated palisade and spongy tissues.
d. special subsidiary cells.
e. bean-shaped guard cells.

____ 6. Trichomes are found on/in the
a. palisade layer.
b. spongy layer.
c. entire mesophyll.
d. epidermis.
e. guard cells.

____ 7. Facilitated diffusion of potassium ions into guard cells
a. requires ATP.
b. closes stomates.
c. causes guard cells to shrink and collapse.
d. occurs more in daylight than at night.
e. indirectly causes pores to open.

____ 8. In general, leaf epidermal cells
a. are living.
b. lack chloroplasts.
c. protect mesophyll cells from sunlight.
d. are absent on the lower leaf surface.
e. have a cuticle.

____ 9. Factors that tend to affect the amount of transpiration include
a. the cuticle.
b. wind velocity.
c. ambient temperature.
d. relative humidity.
e. amount of light.

____10. Guard cells generally
a. have chloroplasts.
b. form a pore.
c. are found in the epidermis.
d. are found only in monocots.
e. are found only in dicots.

____11. Red water-soluble pigments in leaves are
a. carotenoids.
b. xanthophylls.
c. anthocyanins.
d. rhodophylls.
e. found only in monocots.

____12. Proton pumps in the plasma membranes of guard cells are activated by
a. blue light.
b. ultraviolet light.
c. red light.
d. wavelengths in the 800-1,000 nm range.
e. wavelengths in the 400-500 nm range.

____13. Water diffuses into guard cells by means of
a. voltage-activated ion channels.
b. proton pumps.
c. active transport.
d. facilitated diffusion.
e. osmosis.

____ 14. Monocot leaves
a. a palisade layer.
b. have parallel veins.
c. have evenly spaced veins on cross section.
d. have a spongy layer.
e. may have guard cells shaped like dumbbells.

____ 15. Terms associated with a leaf include

a. axillary bud.
b. blade.
c. petiole.
d. stipules.
e. veins.

____ 16. Venation patterns may be

a. simple.
b. opposite.
c. parallel.
d. alternate.
e. pinnately netted.

____ 17. The bundle sheath is composed of

a. sclerenchyma cells.
b. cambium.
c. cork cambium.
d. mesohyl.
e. parenchyma cells.

____ 18. Water required for photosynthesis is obtained from

a. reverse transpiration.
b. soil.
c. guard cells.
d. subsidiary cells.
e. the interstitial space.

____ 19. Carnivorous plants

a. generally grow in poor soil deficient in nitrogen.
b. have leaves modified as bracts.
c. may trap prey either passively or actively.
d. have leaves adapted for digesting prey.
e. have stems specialized to be tendrils.

____ 20. Modified leaves

a. may store water.
b. may form bulbs.
c. may from spheres to conserve water.
d. may be involved in sexual reproduction.
e. .may form spines.

____ 21 Stomatal opening and closing is affected by

a. O_2 levels.
b. CO_2 levels.
c. red light.
d. blue light.
e. follow a circadian rhythm.

VISUAL FOUNDATIONS

Color the parts of the illustration below as indicated. Also label the location of stomata. Using brackets, indicate the structures on the list below that would be components of the dermal tissue system, the ground tissue system, and the vascular tissue system.

RED	❑	stoma
YELLOW	❑	upper epidermis
ORANGE	❑	lower epidermis
VIOLET	❑	cuticle
BLUE	❑	palisade mesophyll
GREEN	❑	spongy mesophyll
PINK	❑	xylem
PURPLE	❑	phloem
BROWN	❑	bundle sheath

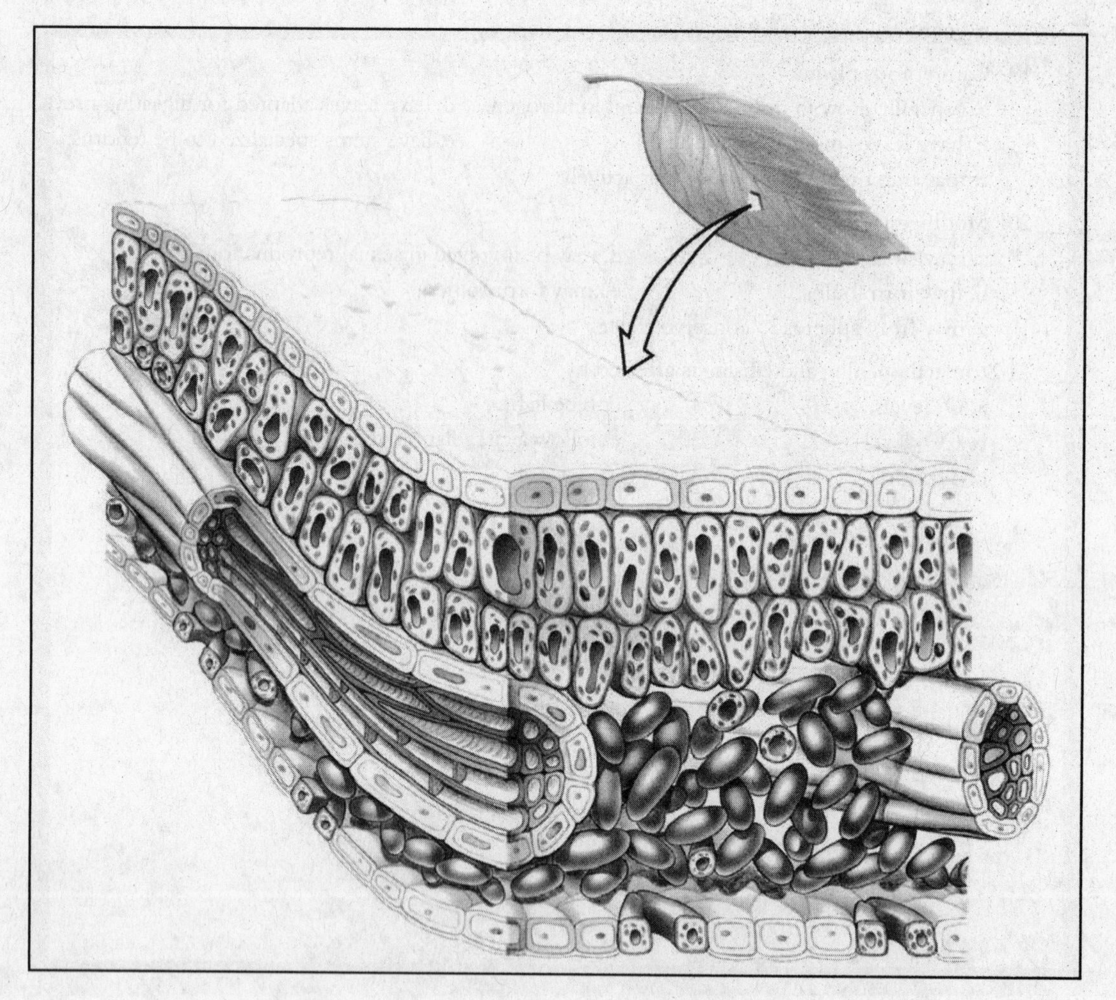

CHAPTER 35

Stem Structure and Transport

This chapter, the third of six addressing the integration of plant structure and function, discusses the structural and physiological adaptations of stems. Stems function to support leaves and reproductive structures, conduct materials absorbed by the roots and produced in the leaves to other parts of the plant, and produce new tissue throughout the life of the plant. Specialized stems have other functions as well. Although all herbaceous, or nonwoody, stems have the same basic tissues, the arrangement of the tissues in the monocot stem differs from that in the eudicot stem. The stems of herbaceous plants increase only in length. The stems of all gymnosperms and some eudicots increase in both length and girth. Increases in the length of plants result from mitotic activity in apical meristems, and increases in the girth of plants result from mitotic activity in lateral meristems. Simple physical forces are responsible for the movement of food, water, and minerals in multicellular plants.

REVIEWING CONCEPTS

Fill in the blanks.

INTRODUCTION

1. The three main parts of a vegetative vascular plant are (a) ____________, (b) ____________, and (c) ____________.
2. The primary functions of stems are to (a) ____________, (b) ____________, and (c) ____________.

STEM GROWTH AND STRUCTURE

3. All plants have ____________ growth.

Herbaceous eudicot and monocot stems differ in internal structure

4. A sunflower is a representative ____________ stem.
5. Vascular tissues in herbaceous eudicots are located in ____________.
6. Each vascular bundle contains two vascular tissues, the (a) ____________, and (b) ____________, and in some herbaceous stems a single layer of cells, the (b) ____________, sandwiched between them.
7. A cross section of the vascular bundle of a monocot stem would have (a) ____________ toward the inside and (b) ____________ toward the outside of the bundle.
8. Monocot stems do not possess ____________ which give rise to secondary growth in woody stems.

Woody plants have stems with secondary growth

9. In woody plants, cells in the (a) ______________________________, give rise to secondary xylem and phloem.
10. Cells of the outer meristem are called ______________________.
11. Cork cambium and the tissues it produces are called the __________ or outer bark.
12. The vascular cambium produces secondary (a) ________________ and secondary (b) ________________ to replace the primary conducting and supporting tissues.
13. The woody stem increases in diameter when cells in the vascular cambium divide ______________________________.
14. Secondary tissues eventually replaces the function of primary tissues as a result of the primary tissues being __________________ by the mechanical pressure of secondary growth.
15. Lateral transport takes place in (a) ____________, which are chains of (b) ____________________ cells that radiate out from the center of the woody stem or root.
16. Cork cambium produces (a) ____________________ which is the functional replacement for the (b) ____________________.
17. Variation in cork cambia and their rats of division explain the differences in the __________________ of different tree species.
18. Cork cambium cells divide to form new tissues toward the inside and the outside. The layer toward the outside consists of (a)________________, heavily waterproofed cells that protect the plant. To the inside, cork cambium forms the (b)____________________________ that stores water and starch.
19. In a woody twig the apical meristem of the terminal bud is dormant and covered by an outer layer of (a) ____________________ which are actually modified (b) ________________.
20. The number of ____________________ on a termperate-zone woody twig indicates its age.
21. When looking at a woody stem during the winter the pattern of ____________________________ indicates the leaf arrangement on the stem during the summer season.
22. The older wood in the center of a tree is called (a) ______________, and the younger, lighter-colored, more peripheral wood is (b)____________.
23. Botanically speaking (a) ______________ wood is the wood of flowering plants and (b) ____________ wood is the wood of conifers.
24. The wood of conifers typically lacks (a) __________________ and (b) ____________________ which accounts for the major differences between the wood of flowering plants and conifers.

25. Annual rings are composed of two types of cells arranged in alternating concentric circles, each layer appropriately named for the season in which it developed. The (a) ________________ has large-diameter conducting cells and few fibers, whereas the (b)______________________ has narrower conducting cells and numerous fibers.

WATER TRANSPORT

26. In contrast to internal circulation in animals where a pumping heart moves the fluids, the movement of fluids and materials throughout a plant is driven largely by ____________________.

Water and minerals are transported in xylem

27. Water moves in plants as result of being either (a) ____________ or (b) ________________. Current evidence indicates that most water is transported by being (c) ______________.

Water movement can be explained by a difference in water potential

28. One of the principal forces behind water movements through plants is a function of a cell's ability to absorb water by osmosis, also known as the "free energy of water," or the (a) __________________. Water containing solutes has (b) _________ (more or less?) free energy than pure water. Water moves from a region of (c) __________________ water potential to a region of (d) ________________ water potential.

29. Under normal conditions, the water potential of the root is more _________________ (negative or positive?) than the water potential of the soil. Thus water moves by osmosis from the soil into the root.

According to the tension-cohesion model, water is pulled up a stem

30. Water is pulled up the plant as a result of (a) ____________________ occurring at the top of the plant.

31. This upward pull is possible only as long as the column of water in the xylem remains unbroken. The two forces working to maintain this unbroken column are (a) ______________ and (b) ____________.

Root pressure pushes water from the root up a stem

32. Root pressure occurs when (a) ______________ are actively absorbed and pumped into the xylem. This movement causes a (b) _________ (decrease, increase) in the water potential of the xylem.

TRANSLOCATION OF SUGAR IN SOLUTION

The pressure-flow model explains translocation in phloem

33. The movement of sugar in the phloem occurs as the result of a pressure gradient which exists between the (a) ____________ where sugar is loaded into the phloem and the (b) ____________ where the sugar is removed from the phloem.
34. The accumulation of sugar in the sieve tube element causes a (a)____________ (decrease, increase) in the water potential which causes water to move by (b) ____________ and (c) ____________ (increase, decrease) in the turgor pressure inside the sieve tube elements.

BUILDING WORDS

Use combinations of prefixes and suffixes to build words for the definitions that follow.

Prefixes	The Meaning	Suffixes	The Meaning
adher(s)-	to stick to	-al	pertaining to
apic-	tip	-derm	skin
cohes-	to stick together	-ion	the process of
inter-	between, among	-sis	the process of
osmo-	pushing		
peri-	around		
trans-	across, beyond		

Prefix	Suffix	Definition
____________	-location	1. The movement of materials (across distances) in the vascular tissues of a plant.
____________	____________	2. Layers of cells covering the surface of woody stems and roots (i.e., the outer bark); the "skin" of woody dicots and cone-bearing gymnosperms.
____________	-node	3. The region of a stem between two successive nodes.
____________	____________	4. The binding of water molecules to the walls of the xylem cells.
____________	____________	5. The binding of water molecules to each other.
____________	____________	6. The diffusion of water from an area of higher to an area of lower concentration.
____________	____________	7. The area of plant located at the tip of roots and shoots.

MATCHING

Terms:

a.	Apical meristem	f.	Lenticels	k.	Terminal bud
b.	Cork cambium	g.	Periderm	l.	Vascular cambium
c.	Ground tissue	h.	Pith	m.	Xylem
d.	Lateral bud	i.	Ray		
e.	Lateral meristem	j.	Root pressure		

For each of these definitions, select the correct matching term from the list above.

____ 1. An area of dividing tissue located at the tips of plant stems and roots.
____ 2. Technical term for the outer bark of woody stems and roots.
____ 3. Large, thin-walled parenchyma cells found in the innermost tissue in many plants.
____ 4. A lateral meristem in plants that produces cork cells and cork parenchyma.
____ 5. The cortex and the pith are parts of this tissue system.
____ 6. Lateral meristem that gives rise to secondary vascular tissues.
____ 7. The vascular tissue responsible for transporting water and dissolved minerals in plants.
____ 8. The positive pressure in the root tissues of plants.
____ 9. A chain of parenchyma cells that functions for lateral transport of food, water, and minerals in woody plants.
____10. Another name for axillary bud.
____ 11. The embryonic shoot located at the tip of the stem.

MAKING COMPARISONS

Fill in the blanks.

Tissue	Source	Location	Function
Secondary xylem	Produced by vascular cambium	Wood	Conducts water and dissolved minerals
#1	Produced by vascular cambium	Inner bark	#2
Cork parenchyma	#3	Periderm	#4
#5	Produced by cork cambium	#6	Replacement for epidermis
#7	A lateral meristem produced by procambium tissue	Between wood and inner bark in vascular bundles	#8
Cork cambium	#9	In epidermis or outer cortex	#10

MAKING CHOICES

Place your answer(s) in the space provided. Some questions may have more than one correct answer.

____ 1. Secondary xylem and phloem are derived from
a. cork parenchyma.
b. cork cambium.
c. vascular cambium.
d. periderm.
e. epidermis.

____ 2. The outer portion of bark is formed primarily from
a. cork parenchyma.
b. cork cambium.
c. vascular cambium.
d. periderm.
e. epidermis.

____ 3. If soil water contains 0.1% dissolved materials and root water contains 0.2% dissolved materials, one would expect
a. a negative water potential in soil water.
b. a negative water potential in root water.
c. less water potential in roots than in soil.
d. water to flow from soil into root.
e. water to flow from root into soil.

____ 4. When water is plentiful, wood formed by the vascular cambium is
a. springwood.
b. summerwood.
c. late summerwood.
d. composed of large diameter conducting cells.
e. composed of thick-walled vessels.

____ 5. One daughter cell from a mother cell in the vascular cambium remains as part of the vascular cambium, the other divides to form
a. secondary xylem or phloem.
b. wood or inner bark.
c. outer bark.
d. secondary tissue.
e. primary xylem and phloem.

____ 6. If a plant is placed in a beaker of pure distilled water, then
a. water will move out of the plant.
b. water will move into the plant.
c. water pressure in the plant is positive relative to water in the beaker.
d. water potential in the beaker is zero.
e. water potential in the plant is less than zero.

____ 7. The pull of water up through a plant is due in part to
a. cohesion.
b. adhesion.
c. the low water potential in the atmosphere.
d. transpiration.
e. a water potential gradient between the soil and the root.

____ 8. Which of the following is/are true of monocot stems?
a. stem covered with epidermis
b. vascular tissues embedded in ground tissue
c. stem has distinct cortex and pith
d. vascular bundles scattered through the stem
e. vascular bundles arranged in circles

____ 9. Which of the following is/are true of eudicot stems?
a. stem covered with epidermis
b. vascular tissues embedded in ground tissue
c. stem has distinct cortex and pith
d. vascular bundles scattered through the stem
e. vascular bundles arranged in circles

____10. The area on a stem where each leaf is attached is called the
a. bud scale.
b. node.
c. lateral bud.
d. leaf scar.
e. lenticel.

____11. Sites of loosely arranged cells along the bark of a woody twig that allow gases to diffuse into the stem are called
a. bud scales.
b. nodes.
c. lateral buds.
d. leaf scars.
e. lenticels.

____12. Terms associated with secondary growth include
a. an increase in girth.
b. apical meristerm.
c. vascular cambium.
d. lateral meristem.
e. woody stems.

____13. The ground tissue at the center of a herbaceous eudicot stem is called
a. cortex.
b. cork.
c. periderm.
d. epidermis.
e. pith.

____14. Monocot stems lack
a. sclerencyma
b. vascular cambium.
e. an epidermis.
d. lateral meristems.
e. cork cambium.

____15. As a woody stem increases in circumference, the number of cells in the vascular cambium
a. increase.
b. decrease.

____16. The area of a leaf where the leaf is attached is called the
a. internode.
b. axillary node.
c. bundle.
d. lateral node.
e. node

VISUAL FOUNDATIONS

Color the parts of the illustration below as indicated.

RED	❑	vascular cambium
GREEN	❑	secondary phloem
YELLOW	❑	pith
BLUE	❑	primary phloem
ORANGE	❑	secondary xylem
BROWN	❑	primary xylem
PINK	❑	epidermis
PURPLE	❑	cortex
TAN	❑	periderm

Color the parts of the illustration below as indicated. Also label xylem and phloem.

RED	❑	path of sucrose
YELLOW	❑	companion cell
BLUE	❑	path of water
ORANGE	❑	sink
BROWN	❑	sieve tube element
PINK	❑	source

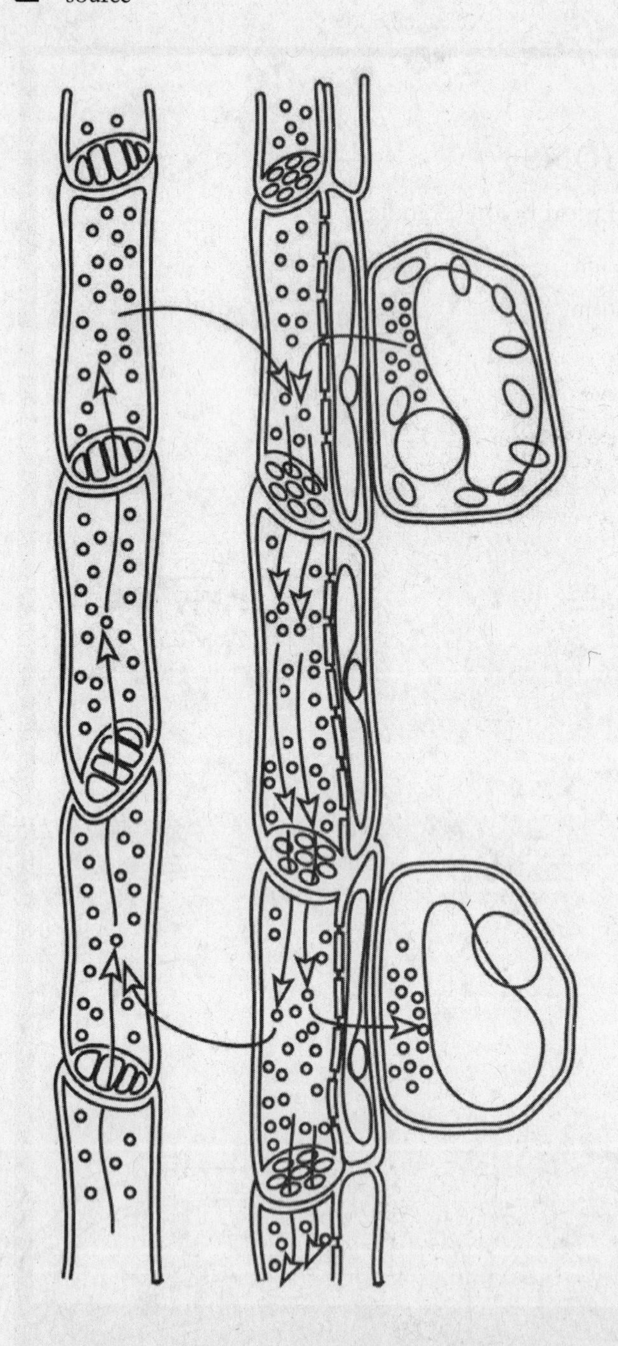

CHAPTER 36

Roots and Mineral Nutrition

The previous two chapters discussed the structural and physiological adaptations of the plant's shoot system, i.e., the leaves and stems. This chapter addresses the structural and physiological adaptations of the remaining portion of the plant, i.e., the root system. The roots of most plants function to anchor plants to the ground, to absorb water and dissolved minerals from the soil, and, in some cases, to store food. Additionally, the roots of some plants are modified for support, aeration, and/or photosynthesis. Although herbaceous roots have the same tissues and structures found in stems, they also have other tissues and structures as well. Soil is composed of inorganic minerals, organic matter, soil organisms, soil atmosphere, and soil water. Plants require at least 19 essential nutrients for normal growth, development, and reproduction. Some human practices, such as harvesting food crops, depletes the soil of certain essential elements, making it necessary to return these elements to the soil in the form of organic or inorganic fertilizer.

REVIEWING CONCEPTS

Fill in the blanks.

INTRODUCTION

1. The primary functions of roots include (a) ________________, absorption of (b) ________________ and (c) ________________, and (d) ________________.

ROOT STRUCTURE AND FUNCTION

2. Two types of root systems occur in plants; they are the (a)________________ root system and the (b) ________________ root system.
3. The embryonic root in the seed gives rise to the (a) ________________ root system and the (b) ________________ root system develops from the stem tissue.
4. Roots that arise from the stem and not preexisting roots are called ________________ roots.

Roots have root caps and root hairs

5. Each root tip is covered by a (a) ________________, which protects the root (b) ________________.
6. Short-lived tubular extensions of epidermal cells located just behind the growing root tip are called ________________.

The arrangement of vascular tissues distinguishes the roots of herbaceous eudicots and monocots

7. The eudicot root is composed primarily of (a) ____________ and lacks supporting (b) ____________.
8. Cytoplasmic bridges connecting root cells are called (a)____________. The connected, living cytoplasm is called the (b) ____________.
9. The inner layer of cortex in a eudicot root is the (a) ____________, the cells of which possess a bandlike region on their radial and transverse walls. This "band," or (b) ____________ as it's called, contains suberin.
10. ____________ are integral membrane proteins that facilitate the rapid movement of water across membranes.
11. The stele at the center of the eudicot primary root is composed of a outer layer the (a) ____________ and an inner region the (b) ____________. The (c) ____________ is located in patches around the xylem.
12. Horizontal movement of water from the soil into the center of the root can be summarized as moving from root hair/epidermis → (a)____________ → (b) ____________ → (c)____________ → root xylem.
13. Since secondary growth is absent in virtually all monocots, monocot roots lack a ____________.

Woody plants have roots with secondary growth

14. Plants that produce stems with secondary growth also produce ____________ with secondary growth.
15. As growth occurs in the roots of woody plants the epidermis is replaced by (a) ____________ which is composed of cork cells and (b) ____________.

Some roots are specialized for unusual functions

16. Adventitious roots often arise from the ____________ of stems.
17. Roots that develop from branches or a vertical stem and provide support to the plants such as corn are called (a) ____________. The swollen bases that help to support trees in an upright position are called (b) ____________.
18. (a) ____________ are plants that grow attached to other plants. They use (b) ____________ roots to anchor themselves to bark, branches, and other surfaces.

ROOT ASSOCIATIONS AND INTERACTIONS

Mycorrhizae facilitate the uptake of essential minerals by roots

19. Subterranean associations between roots and soil fungi are known as ____________.

Rhizobial bacteria fix nitrogen in the roots of leguminous plants

20. Plants and nitrogen-fixing bacteria use ______________________ to establish contact and develop nodules.

THE SOIL ENVIRONMENT

21. Soils area generally formed from _____________ continually being fragmented into smaller particles.

Soil comprises inorganic minerals, organic matter, air, and water

22. The four distinct components of soil are (a) _______________, (b) ______________________, (c) _________________________, and (d) ______________________.
23. The largest soil particles are called (a) ______________________, the medium sized particles are (b) __________________ and the smallest particles are called (c) ________________________.
24. Roots secrete (a) _____________ which are exchanged for positively charged mineral ions adhering to the surface of the soil molecules in a process called (b) ______________________.
25. Good, loamy agricultural soil contains about 40% each of (a) _______, and (b) ________ and about 20% of (c) __________. The spaces between soil particles are filled with (d) _______, and (e) _________.
26. Partly decayed organic matter called _______________ contributes to the water holding ability of soils.
27. When water drains from large pores in the soil it draws in ________.
28. Water moving downward through soil spaces carries dissolved minerals with it in a process called (a) ________________________. The deposition of the dissolved minerals in lower layers of the soil is called (b) ______________________.

Soil organisms form a complex ecosystem

29. Bits of soil that have passed through the gut of an earthworm are called ______________________.

Soil pH affects soil characteristics and plant growth

30. Air pollution in which sulfuric and nitric acids produced by human activities fall to the ground as acid rain, sleet, snow, or fog is known as ________________________.

Soil provides most of the minerals found in plants

31. More than (a) ______ (#) elements have been found on earth and over (b) ______ (#) of them have been found in plant tissues.
32. Elements required in large quantities are called (a)______________________________ while those needed only in small amounts are called (b) ____________________________.
33. In plants, potassium remains (a) _______________ and plays and important role in maintaining (b) _______________ of cells. Potassium is also involved in the opening and closing of (c) ________.

34. In order to identify elements that are essential for plant growth, investigators must reduce the number of variables to a minimum. To do this plants are grown in aerated water containing known quantities of elements, a process known as ________________.

Soil can be damaged by human mismanagement

35. The three elements that most often limit plant growth are (a)________________, (b)________________, and (c)________________.
36. The wearing away or removal of soil from the land is known as ________________.
37. Irrigation sometimes causes salt to accumulate in the soil, a process called ________________.

BUILDING WORDS

Use combinations of prefixes and suffixes to build words for the definitions that follow.

Prefixes	The Meaning	suffixes	The Meaning
adventit-	foreign	-derm(is)	skin
endo-	within	-(i)ous	pertaining to
epi-	upon	-phyt(e)	plant
hydro-	water	-rrhiz(a)	root
hum-	soil	-ule	small
macro-	large, long, great, excessive	-us	thing
micro-	small		
myc(o)-	fungus		
nod-	knob		

Prefix	Suffix	Definition
________	-ponics	1. Growing plants in water (not soil) containing dissolved inorganic minerals.
________	-nutrient	2. An essential element that is required in fairly large amounts for normal plant growth.
________	________	3. Plants that grow attached to other plants.
________	________	4. The innermost layer of the cortex in the plant root.
________	________	5. An association between roots and certain fungi.
________	________	6. Swelling on roots that houses nitrogen-fixing bacteria.
________	________	7. Organs that occur in an unusual location such as roots appearing on a stem.
________	________	8. The partly decayed organic portion of soil.

MATCHING

Terms:

a. Adventitious root
b. Apoplast
c. Casparian strip
d. Endodermis
e. Fibrous root system
f. Graft
g. Humus
h. Leaching
i. Mycorrhizae
j. Pneumatophore
k. Prop root
l. Root cap
m. Root hair
n. Stele
o. Taproot system

For each of these definitions, select the correct matching term from the list above.

____ 1. Mutualistic associations of fungi and plant roots that aid in the plant's absorption of essential minerals from the soil.
____ 2. Organic matter in various stages of decomposition in the soil.
____ 3. The innermost layer of the cortex in the plant root.
____ 4. A root that arises in an unusual position on a plant
____ 5. A band of waterproof material around the radial and transverse walls of endodermal root cells.
____ 6. An adventitious root that arises from the stem and provides additional support for plants.
____ 7. Aerial "breathing roots" of some plants in swampy and tidal environments that assist in getting oxygen to submerged roots.
____ 8. The type of root system present in a plant that has several roots of the same size developing from the end of the stem, with lateral roots of various sizes branching off these roots..
____ 9. An extension of an epidermal cell in roots, which increases the absorptive capacity of the roots.
____10. A covering of cells over the root tip that protects delicate meristematic tissue beneath it.
____11. At the center of an eudicot primary root.
____12. Root system formed from the seedling's enlarging radical.
____13. The effect of water transporting dissolved minerals deeper into the soil.

MAKING COMPARISONS

Fill in the blanks.

Root Structure	Function of the Root Structure
Root cap	Protects the apical meristem, may orient the root downward
Root apical meristem	#1
#2	Absorption of water and minerals
Epidermis	#3
Cortex	#4
#5	Controls mineral uptake into root xylem
#6	Gives rise to lateral roots and lateral meristems
#7	Conducts water and dissolved minerals
Phloem	#8

MAKING CHOICES

Place your answer(s) in the space provided. Some questions may have more than one correct answer.

____ 1. Structures found in primary roots that are also found in stems include the
a. cortex.
b. cuticle.
c. vascular tissues.
d. apical meristem cap.
e. epidermis.

____ 2. The three groups of organisms in soil that are the most important in decomposition are
a. insects.
b. fungi.
c. soil protozoa.
d. algae.
e. bacteria.

____ 3. The origin of multicellular branch roots is the
a. cambium.
b. Casparian strip.
c. pericycle.
d. parenchyma.
e. cortex.

____ 4. When water first enters a root, it usually
a. enters parenchymal cells.
b. is absorbed by cellulose.
c. moves from a water negative potential to a positive water potential.
d. moves along cell walls.
e. enters cells in the Casparian strip.

____ 5. The function(s) performed by all roots is/are
a. absorption of water.
b. absorption of minerals.
c. food storage.
d. anchorage.
e. aeration.

____ 6. The principal function(s) of the root cortex is/are
a. conduction.
b. storage.
c. production of root hairs.
d. water absorption.
e. mineral absorption.

____ 7. Essential macronutrients include
a. phosphorus.
b. potassium.
c. iron.
d. hydrogen.
e. magnesium.

____ 8. Essential micronutrients include
a. phosphorus.
b. potassium.
c. iron.
d. hydrogen.
e. magnesium.

____ 9. In general, the vascular tissues in monocot roots
a. form a solid cylinder.
b. are absent.
c. are in bundles arranged around the central path.
d. contain a vascular cambium.
e. continue into root hairs.

____10. The inorganic materials in soil come from
a. fertilizers.
b. water runoff.
c. weathered rock.
d. the atmosphere.
e. percolating water.

____11. Roots produced in unusual places on the plants, often as aerial roots, are called ___________ roots.
a. secondary
b. enhancement
c. adventitious
d. contractile
e. aerial water-absorbing

____12. Sugar produced during photosynthesis are transported in the
a. epidermis.
b. xylem.
c. cambium.
d. radical.
e. phloem.

____13. Root hairs are modified
a. endodermis.
b. sclerenchyma
c. epidermis.
d. parenchyma.
e. collenchymas.

____14. Oxygen needed by roots
a. enters from spaces in the soil.
b. is produced by the breakdown of glucose.
c. is transported by the xylem.
d. is transported by the phloem.
e. is less available when it rains.

____15. The Casparian strip
a. contain suberin.
b. is waterproof.
c. forms to prevent sugar loss from the root.
d. is associated with the endodermis.
e. forms when leaves prepare to fall from the plant.

____16. Aquaporins
a. are intergral membrane proteins.
b. facilitate solute movement into roots.
c. contribute to the pericyle.
d. can be found in roots.
e. are located in the epidermis.

____17. Aerial breathing roots are also known as
a. prop roots.
b. nodules.
c. rhizobia.
d. arbuscules.
e. pneumatophores.

____18. When soils are low in nitrogen, legume roots
a. secrete flavinoids.
b. secrete nitrogenase compounds.
c. attract rhizobial bacteria.
d. initiate cell signaling.
e. enlarge their nodules.

____19. The plant hormone _______ triggers cell division and is involved in root nodule formation.
a. giberellin.
b. cytokinin.
c. abscisic acid.
d. ethylene.
e. auxin.

VISUAL FOUNDATIONS

Color the parts of the illustration below as indicated.

RED	❑	primary xylem
GREEN	❑	cortex
BLUE	❑	vascular cambium
ORANGE	❑	secondary xylem
YELLOW	❑	epidermis
BROWN	❑	periderm
TAN	❑	pericycle
PINK	❑	secondary phloem
VIOLET	❑	primary phloem

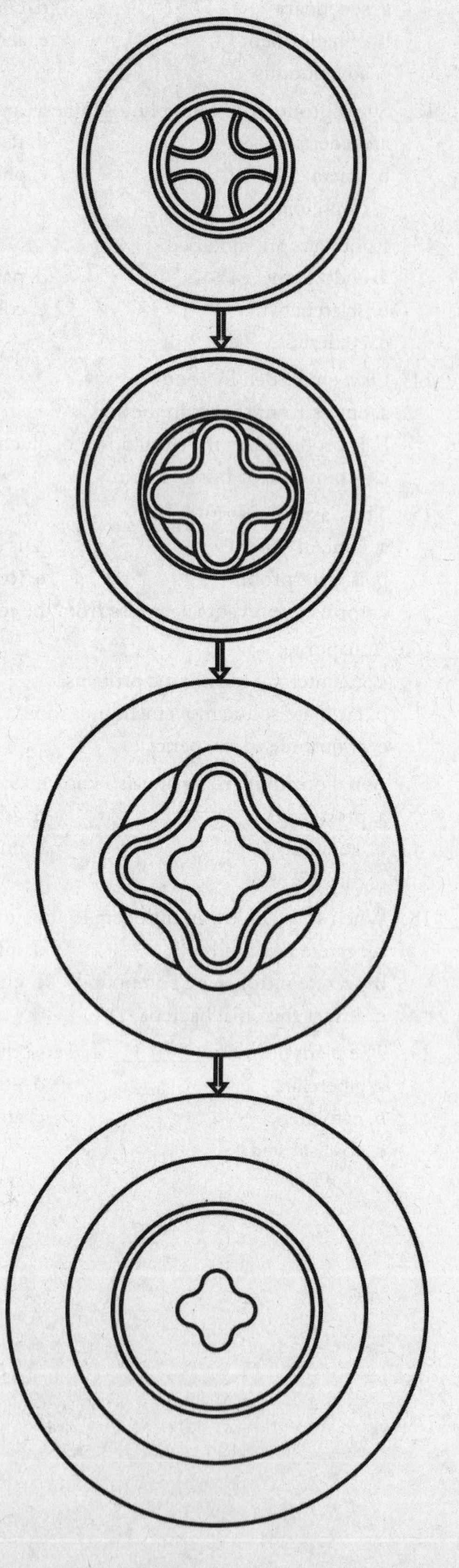

CHAPTER 37

Reproduction in Flowering Plants

This chapter, the fifth in a series of six addressing the integration of plant structure and function, discusses reproduction in flowering plants. As you recall from the first chapter in the series, flowering plants are the largest and most successful group of plants. All flowering plants reproduce sexually; some also reproduce asexually. Sexual reproduction involves flower formation, pollination, fertilization within the flower ovary, and seed and fruit formation. The offspring resulting from sexual reproduction exhibit a great deal of individual variation, due to gene recombination and the union of dissimilar gametes. Although sexual reproduction has some disadvantages, it offers the advantage of new combinations of genes that might make an individual plant better suited to its environment. Asexual reproduction involves only one parent; the offspring are genetically identical to the parent and each other. Asexual reproduction, therefore, is advantageous when the parent is well adapted to its environment. The stems, leaves, and roots of many flowering plants are modified for asexual reproduction. Also, in some plants, seeds and fruits are produced asexually without meiosis or the fusion of gametes.

REVIEWING CONCEPTS

Fill in the blanks.

INTRODUCTION

1. The union of gametes is called (a) ________________. Sexual reproduction provides an advantage to the offspring because they are genetically (b) ________________ from either parent.
2. Many flowering plants reproduce (a) ________________.
   Genetically these offspring are (b) ________________ to their parents.

THE FLOWERING PLANT LIFE CYCLE

3. Plants that spend a portion of their life cycle in a multicellular haploid stage and part in a multicellular diploid stage are said to undergo a reproductive cycle called the (a) ________________________.
   The haploid part is the (b) ____________________________
   and the diploid portion is called the (c)______________________.

Flowers develop at apical meristems

4. The Flowering Locus C gene codes for a transcription factor that represses ________________________________.

Each part of a flower has a specific function

5. Flowers are reproductive shoots, usually consisting of four kinds of organs, (a) ________________, (b) ______________________, (c) ____________________, and (d) ________________.
6. The collective term for all the sepals of a flower is (a) ____________.
   The collective term for all the petals of a flower is (b) ____________.

7. Carpels bear ________________.
8. Pollen sacs within the anther contain numerous diploid cells called (a)____________________, each of which undergoes meiosis to produce four haploid cells called (b) ____________________, each of which divides mitotically to produce an immature male gametophyte called a (c) ______________. The pollen grain becomes mature when its generative cell divides to form two nonmotile (d) ________________.

POLLINATION

9. The transfer of pollen grains from (a) __________ to (b) __________ is known as pollination.

Many plants have mechanisms that prevent self-pollination

10. The mating of genetically similar individuals is known as (a)____________________, and the mating of dissimilar individuals is called (b) ____________________.
11. A genetic condition in which the pollen is ineffective in fertilizing the same flower or other flowers on the same plant is known as __________.

Flowering plants and their animal pollinators have coevolved

12. Two things flowers have evolved to attract pollinators are (a)________________, and (b) ______________.
13. Plants pollinated by insects often have (a) ____________, or (b)__________ (color?) petals, but usually not (c) __________ petals.
14. Bees see ultraviolet light as a color referred to as ______________.
15. Flowers pollinated by birds are usually colored (a) ____________, (b) ____________, or (c) ____________ because birds see well in this range of (d) ____________________.
16. Flowers pollinated by bats bloom (a) ____________ and have (b) ____________________ petals. They usually smell like (c)____________.
17. Some plants have evolved flowers resembling specific shapes. The flower of one species of orchid actually resembles an insect, a ______________.
18. Scientists think humming-bird pollinated flowers arose from ________________ pollinated flowers.

Some flowering plants depend on wind to disperse pollen

19. Wind-pollinated plants produce many small, inconspicuous ____________.

FERTILIZATION AND SEED AND FRUIT DEVELOPMENT

20. Once pollen grains have been transferred from anther to stigma, the tube cell grows a pollen tube down through the (a) ____________ and into an (b) ____________ in the ovary.

A unique double fertilization process occurs in flowering plants

21. The tissue with nutritive and hormonal functions that surrounds the developing embryonic plant in the seed is called ____________.

Embryonic development in seeds is orderly and predictable

22. A seed contains an (a) ____________ and stores (b) ____________.
23. The tissue that anchors the developing embryo and aids in nutrient uptake from endosperm is called the ____________.

The mature seed contains an embryonic plant and storage materials

24. The mature embryo within the seed consists of a short embryonic root, or (a) ____________; an embryonic shoot; and one or two seed leaves, or (b) ____________.
25. The shoot apex or terminal bud is called the ____________.

Fruits are mature, ripened ovaries

26. The four basic types of fruits are (a) ____________, (b) ____________, (c) ____________, and (d) ____________.
27. One type of fruit, the (a) ____________ fruit, develops from a single ovary. It may be fleshy or dry. Fruits that have soft tissues throughout, like those in tomatoes and grapes, are called (b)____________, while fleshy fruits that have a hard, stony pit surrounding a single seed are known as (c) ____________.
28. Fruits that split along one suture, like the milkweed are an example of a (a) ____________. Those that split along just two sutures are an example of a (b) ____________, and those that split along two or more (multiple) sutures are an example of a (c) ____________.
29. (a)____________ fruits result from the fusion of several developing ovaries in a single flower. The raspberry is an example. Similarly, (b) ____________ fruits form from the fusion of several ovaries of many flowers that grow in proximity on a common floral stalk. The pineapple is an example.
30. (a)____________ fruits contain plant tissue in addition to ovary tissue. When one eats a strawberry, for example, a person is eating the fleshy (b) ____________, and when eating apples and pears, one is consuming the (c) ____________ that surrounds the ovary.

Seed dispersal is highly varied

31. Flowering plant seeds and fruits are adapted for various means of dispersal, among which four common means are (a) ____________, (b) ____________, (c) ____________, and (d) ____________.

GERMINATION AND EARLY GROWTH

32. Dry seeds take up water by a process called ____________.

Some seeds do not germinate immediately

33. The seeds of desert plants contain ____________ which inhibits germination until conditions are favorable.

34. A process of scratching or scarring the seed coat before sowing it is called ________________________________.

Eudicots and monocots exhibit characteristic patterns of early growth

35. Corn and grasses have a special sheath of cells called a ____________________ that surrounds and protects the young shoot.

ASEXUAL REPRODUCTION IN FLOWERING PLANTS

36. Various vegetative structures may be involved in asexual reproduction. For example, (a) ______________, are horizontal underground stems. When these stems are fleshy it is an indication that they are storing (b) ______________. (c) ______________, are underground stems greatly enlarged for food storage, like those in potatoes. (d)__________, are short underground stems with fleshy storage leaves, as in onions and tulips. (e) __________, are thick underground stems covered with papery scales; and (f) ____________ or horizontal, aboveground stems with long internodes, like those in strawberries.
37. Some plants reproduce asexually by producing ______________, aboveground shoots that develop from adventitious buds on roots.

Apomixis is the production of seeds without the sexual process

38. The advantage of apomixis over other methods of asexual reproduction is that the seeds and fruits produced can be ________________ by methods associated with sexual reproduction.

A COMPARISON OF SEXUAL AND ASEXUAL REPRODUCTION

Sexual reproduction has some disadvantages

39. The many adaptations of flowers for different modes of pollination represent one cost of ________________________________.

BUILDING WORDS

Use combinations of prefixes and suffixes to build words for the definitions that follow.

Prefixes	The Meaning	Suffixes	The Meaning
angio-	enclosed	-ancy	state of
co-	with, together, in association	-ate	to make
cole(o)-	sheath	-cotyl	cup
dorm-	to sleep	-ome	mass
endo-	within	-phyt(e)	plant
germin-	to sprout	-ptil(e)	feather
hypo-	under, below	-sperm	seed
plum-	feather	-ule	small
rhiz-	root		
spor(o)-	spore		

Prefix	Suffix	Definition
________	________	1. Nutritive tissue within seeds.
________	________	2. The part of a plant embryo or seedling below the point of attachment of the cotyledons.

_______________ -evolution 3. Evolutionary changes that result from close interaction and reciprocal adaptations between two species.

_______________ _______________ 4. The shoot above the cotyledon.

_______________ _______________ 5. A temporary state of arrested physiological activity.

_______________ _______________ 6. When a seed beings to develop and the embryo resumes growth.

_______________ _______________ 7. The sheath of cells that protects and surrounds a young shoot such as in grasses.

_______________ _______________ 8. A horizontal underground stem.

_______________ _______________ 9. Another name for flowering plants.

_______________ _______________ 10. The diploid portion of a plant's life cycle.

MATCHING

Terms:

a.	Aggregate fruit	f.	Drupe	k.	Seed
b.	Apomixis	g.	Fruit	l.	Sepal
c.	Carpel	h.	Ovary	m.	Stolon
d.	Corolla	i.	Ovule		
e.	Cotyledon	j.	Pollen		

For each of these definitions, select the correct matching term from the list above.

____ 1. A collective term for the petals of a flower

____ 2. A ripened ovary.

____ 3. The outermost parts of a flower, usually leaf-like in appearance, that protect the flower as a bud.

____ 4. A type of reproduction in which fruits and seeds are formed asexually.

____ 5. A fruit that develops from a single flower with many separate carpels, such as a raspberry.

____ 6. A plant reproductive body composed of a young embryo and nutritive tissue.

____ 7. The male gametophyte in plants.

____ 8. An above-ground, horizontal stem with long internodes.

____ 9. The female reproductive unit of a flower.

____10. A simple fleshy fruit that contains a hard and stony pit surrounding a single seed; such as a plum or peach.

____11. A structure with the potential to develop into a seed.

____12. The fist leaf to be produced by seeds, monocots produce a single one and eudicots produce two.

MAKING COMPARISONS

Fill in the blanks.

Examples of Fruit	Main Type of Fruit	Subtype	Description
Bean	Simple	Dry: opens to release seeds; legume	Splits along the 2 sutures
Acorn	Simple	#1	#2

Examples of Fruit	Main Type of Fruit	Subtype	Description
Corn	Simple	Dry: does not open; grain	#3
Tomato	#4	#5	Soft and fleshy throughout, usually with few to many seeds
Strawberry	#6	None	Other plant tissues and ovary tissue comprise the fruit
Pineapple	#7	None	#8
Raspberry	#9	None	#10
Peach	#11	#12	#13

MAKING CHOICES

Place your answer(s) in the space provided. Some questions may have more than one correct answer.

____ 1. A modified underground bud in which fleshy storage leaves are attached to a stem is a
a. corm.
b. stolon.
c. bulb.
d. fruit.
e. capsule.

____ 2. A simple, dry fruit that develops from two or more fused carpels is a
a. corm.
b. stolon.
c. bulb.
d. fruit.
e. capsule.

____ 3. A short, erect underground stem surrounded by a few papery scales is a
a. corm.
b. stolon.
c. bulb.
d. fruit.
e. capsule.

____ 4. A horizontal, above ground stem with long internodes is a
a. corm.
b. stolon.
c. bulb.
d. fruit.
e. capsule.

____ 5. Which of the following is/are somehow related to simple dry fruits?
a. drupe
b. legume
c. grain
d. single carpel
e. green bean

____ 6. The process by which an embryo develops from a diploid cell in the ovary without fusion of haploid gametes is called
a. lateral bud generation.
b. dehiscence.
c. asexual reproduction by means of suckers.
d. apomixis.
e. spontaneous generation.

____ 7. The tomato is actually a
a. drupe.
b. berry.
c. fleshy lateral bud.
d. fruit.
e. mature ovary.

____ 8. The "eyes" of a potato are
a. diploid gametes.
b. parts of a stem.
c. capable of producing complete plants.
d. axillary buds.
e. rhizomes.

____ 9. Horizontal, asexually reproducing stems that run above ground are
a. diploid gametes.
b. rhizomes.
c. tubers.
d. corms.
e. stolons.

____10. Raspberries and blackberries are examples of
a. follicles.
b. achenes.
c. drupes.
d. aggregate fruits.
e. multiple fruits.

____11. A fruit that forms from many separate carpels in a single flower is a/an
a. follicle.
b. achene.
c. drupe.
d. aggregate fruit.
e. accessory fruit.

____12. A fruit composed of ovary tissue and other plant parts is a/an
a. follicle.
b. achene.
c. drupe.
d. aggregate fruit.
e. accessory fruit.

____13. In flowering plants, the fusion of two gametes
a. produces the gametophyte generation.
b. produces the sporophyte generation.
c. produces a haploid generation.
d. produces a diploid generation.
e. is called fertilization.

____14. In flowering plants, haploid spores are produced by
a. the sporophyte generation.
b. the gametophyte generation.
c. meiosis.
d. gametes.
e. self fertilization.

____15. The outermost and lowest whorl on a floral shoot is made up of
a. anthers.
b. carpels.
c.. sepals.
d. petals.
e. peduncels.

____16. Terms associated with anthers include
a. pollen grains.
b. pollen tube.
c. carpel.
d. generative cell.
e. tube cell.

____17. Pollen sacs in an anther
a. form from pistols.
b. are collectively call the calyx.
c. contain diploid cells.
d. are the site of male gamete formation.
e. contain microsporocytes.

____18. Flowers pollinated by insects are often colored
a. blue.
b. indigo.
c. yellow.
d. red.
e. violet.

____19. Flowers pollinated by birds are often colored
a. pink.
b. orange.
c. purple.
d. red.
e. yellow.

____20. Features that plants have evolved to attract pollinators include
a. pollen.
b. nectar.
c. coloration.
d. scent.
e. petals.

____21. Flowers pollinated by bats are often colored
a. pink.
b. white.
c. purple.
d. red.
e. yellow.

VISUAL FOUNDATIONS

Color the parts of the illustration below as indicated.

RED ❑ ovule
GREEN ❑ sepal
YELLOW ❑ ovary (and derived from ovary)
BLUE ❑ stamen
ORANGE ❑ style
BROWN ❑ stigma
TAN ❑ seed
PINK ❑ floral tube (and derived from floral tube)
VIOLET ❑ petal

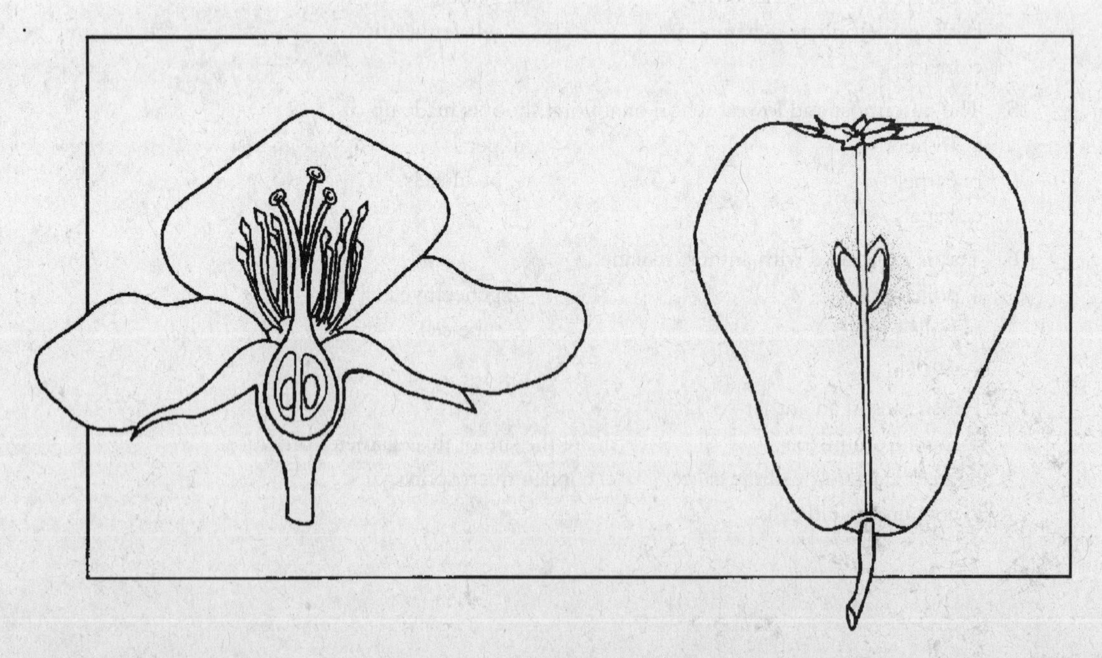

CHAPTER 38

Plant Developmental Responses to External and Internal Signals

This is the last chapter in a series of six addressing the integration of plant structure and function. How a particular gene is expressed is determined by a variety of factors, including signals from other genes and from the environment. Environmental cues, such as temperature, light, growth, and touch, influence plant growth and development. All aspects of plant growth and development are affected by hormones, organic compounds present in very low concentrations that act as highly specific chemical signals between cells. Hormones interact in complex ways with one another to produce a variety of responses in plants. Auxins are involved in cell elongation, phototropism, gravitropism, apical dominance, and fruit development. Giibberellins are involved in stem elongation, flowering, and seed germination. Cytokinins promote cell division and differentiation, delay senescence, and interact with auxins in apical dominance. Ethylene has a role in the ripening of fruits, apical dominance, leaf abscission, and wound response. Abscisic acid is involved in stomatal closure due to water stress, and bud and seed dormancy. Brasinosteroids are plant steroid hormones and they are involved in several aspects of plant growth and development. Other chemicals have been implicated in plant growth and development, and are the subject of ongoing research.

REVIEWING CONCEPTS

Fill in the blanks.

INTRODUCTION

1. Growth and development of a plant are controlled by its ____________________, organic compounds present in low concentration in the plant's tissues that act as chemical signals between cells.
2. Environmental cues exert an influence on (a)____________________________ and hormone production.

TROPISMS

3. Tropism is a ________________________.
4. Growth or movement initiated by light is called a (a) ______________ response and it is triggered by (b) ___________ light. Phototropins are light-activated (c) ______________.
5. For an organism to have a biological response to light, it must contain a light-sensitive substance, called a __________________________, to absorb the light.
6. Tropisms are categorized according to the stimulus that causes them to occur. A response to gravity is (a) ________________, and movement caused by a mechanical stimulus is (b) _____________.

PLANT HORMONES AND DEVELOPMENT

7. The six major classes of hormones that regulate responses in plants are (a) ____________, (b) ____________, (c) ____________, (d) ____________, (e) ____________, and (f) ____________.

PLANT HORMONES ACT BY SIGNAL TRANSDUCTION

8. Many plant hormones bind to (a)____________, located in the plasma membrane where they trigger (b) ____________ reactions.

AUXINS PROMOTE CELL ELONGATION

9. Bending toward light occurs below the tip of the ____________.
10. ____________ is the most common and physiologically important auxin.
11. The movement of auxin is (a) ____________ because it always moves in one direction, specifically from the (b)____________ toward the (c) ____________.
12. Plants that grow almost entirely at the apical meristem rather than from axillary buds are exhibiting ____________.
13. Auxins produced by developing seeds stimulates the development of the ____________.

GIBBERELLINS PROMOTE STEM ELONGATION

14. A disease seen in rice seedlings causes them to grow extremely tall, fall over, and die. It is caused by a ____________ that produces gibberellin.
15. In addition to influencing stem elongation, gibberellins are also involved in (a) ____________, and (b)____________.

CYTOKININS PROMOTE CELL DIVISION

16. Cytokinins are structurally similar to (a)____________and mainly promote (b) ____________and (c)____________.
17. They also delay ____________ (aging).

ETHYLENE PROMOTES ABSCISSION AND FRUIT RIPENING

18. Ethylene regulates developmental response to a mechanical stimulus. The plant response to a mechanical stimulus is known as ____________.
19. Leaf abscission is influenced by two antagonistic hormones, (a)____________ and (b)____________.

ABSCISIC ACID PROMOTES SEED DORMANCY

20. ______________________ is the temporary state of arrested physiological activity in flowering plants.
21. Abscisic acid is an ______________________ hormone.
22. In seeds, the level of abscisic acid decreases during the winter, and the level of (a) ______________________ increases. (b)______________________ are also involved in breaking dormancy.

Brassinosteroids are plant steroid hormones

23. *Arabidopsis* mutants that cannot synthesize brassinosteroids develop as __________ plants.

Identification of a universal flower-promoting signal remains elusive

24. __________ has been identified as a flower-promoting substance.

LIGHT SIGNALS AND PLANT DEVELOPMENT

25. ______________________ is any response of a plant to the relative lengths of daylight and darkness.
26. Short-day plants are also called ______________________.
27. (a)______________________ when they detect the lengthening nights of late summer and fall; (b) ______________________ when they detect the shortening nights of spring and early summer; (c)______________________ plants do not flower when night length is either too long or too short; (d) ______________________ plants do not initiate flowering in response to seasonal changes in the period of daylight and darkness but instead respond to some other type of stimulus, external or internal.

Phytochrome detects day length

28. Phytochrome is a family of five (a) ______________________ pigments, the main photoreceptor for (b) ______________________ and many other light-initiated plant responses.

Competition for sunlight among shade-avoiding plants involves phytochrome

29. Plants tend to grow taller when closely surrounded by other plants, a phenomenon known as ______________________.

Phytochrome is involved in other responses to light, including germination

30. Seeds with a light requirement must be exposed to light containing (a)________ wavelengths to convert (b) ______ to (c) ________ before germination occurs.

Phytochrome acts by signal transduction

31. Red light causes (a) ______________ channels to open in all cell membranes involved in plant movements caused by changes in (b)______________. Far-red light causes these channels to close.
32. The active form of phytochrome moves from the cytoplasm into the nucleus, where it activates a ______________________.

Light influences circadian rhythms

33. Internal cycles, known as circadian rhythms, help organisms detect ____________________.

RESPONSES TO HERBIVOREES AND PATHOGENS

34. Plants have an (a) ____________________ immune system. Infections trigger a (b) ____________________ response by the plant.
35. Defense signals travel thorough out the plant via the ____________________.

Jasmonic acid activates several plant defenses

36. Jasmonic acid is structurally similar to (a) ____________________ in animals. Jasmonic acid triggers the production of enzymes that increase the plant's resistance to (b) ____________________.

Methyl salicylate may induce systemic acquired resistance

37. Salicyclic acid was first extracted from (a) ____________________. It is chemically related to (b) ____________________.

BUILDING WORDS

Use combinations of prefixes and suffixes to build words for the definitions that follow.

Prefixes	The Meaning	Suffixes	The Meaning
amyl(o)-	starch	-ancy	state of being
aux-	enlarge	-ence	the condition of
dorm-	to sleep	-chrom(e)	color
senes(c)-	to grow old	-ism	the process of
photo-	light	-plast	membrane
phyto-	plant	-tropism	turn, turning
promo-	to move forward		
trop-	turn		

Prefix	Suffix	Definition
__________	__________	1. The growth response of an organism to light; usually the turning toward or away from the light source.
gravi-	__________	2. The growth response of an organism to gravity; usually the turning toward or away from the direction of gravity.
__________	__________	3. A blue-green, proteinaceous pigment involved in photoperiodism and a number of other light-initiated physiological responses of plants.
__________	-periodism	4. The physiological response of organisms to variations of light and darkness.
__________	__________	5. The natural ageing process that occurs in most cells.

________ -ter		6. A sequence of DNA that is a binding site for RNA polymerase.
________	________	7. Directional growth.
________	________	8. Part of specialized ells in the root cap that collect toward the bottom of the cells in response to gravity.
________ -in		9. A group of natural plant hormones which commonly promote cell elongation.
________	________	10. A temporary state of arrested physiological activity.

MATCHING

Terms:

a. Abscisic acid
b. Apical dominance
c. Auxin
d. Coleoptile
e. Cytokinin
f. Ethylene
g. Florigen
h. Gibberelin
i. Imbibition
j. Nastic movements
k. Necrotic
l. Proteostome
m. Senescence
n. Statolith
o. Thigmotropism
p. Tropism

For each of these definitions, select the correct matching term from the list above.

____ 1. The inhibition of axillary bud growth by the apical meristem.
____ 2. A plant hormone involved in dormancy and responses to stress.
____ 3. The process of aging.
____ 4. A protective sheath that encloses the stem in certain monocots.
____ 5. A plant hormone involved in apical dominance and cell elongation.
____ 6. A plant hormone that promotes fruit ripening.
____ 7. Plant growth in response to contact with mechanical stimuli, such as a solid object.
____ 8. A plant hormone that promotes rapid cell division and is involved in other aspects of plant growth and development.
____ 9. A change in the position of a plant root or stem.
___ 10. A hormone that promotes stem elongation.
___ 11. Thought to be a flower-promoting substance.
___ 12. Describes dead areas in plant tissue.

MAKING COMPARISONS

Fill in the blanks.

Principal Action	Hormone(s)
Regulates growth by promoting cell elongation	Auxin and gibberellin
Promotes apical dominance, stem elongation, root initiation, fruit development	#1
Delays leaf senescence, inhibition of apical dominance, embryo development	#2
Promotes seed germination, stem elongation, flowering,	#3

Principal Action	Hormone(s)
fruit development	
Fruit ripening, seed germination, root initiation, abscission	#4
Promotes cell division	#5
Promotes seed dormancy	#6

MAKING CHOICES

Place your answer(s) in the space provided. Some questions may have more than one correct answer.

____ 1. Growth in response to gravity is known as
a. a tropism.
b. phototropism.
c. gravitropism.
d. thigmotropism.
e. turgor.

____ 2. A hormone involved in rapid stem elongation just prior to flowering is
a. auxin.
b. gibberellin.
c. cytokinin.
d. ethylene.
e. abscisic acid.

____ 3. Which of the following is/are correct about hormones?
a. They are effective in very small amounts.
b. They are organic compounds.
c. For the most part, their effects occur near the area where they are produced.
d. The effects of different hormones overlap.
e. Each plant hormone has multiple effects.

____ 4. The hormone(s) principally responsible for cell division and differentiation is/are
a. auxin.
b. gibberellin.
c. cytokinin.
d. ethylene.
e. abscisic acid.

____ 5. The hormone(s) principally responsible for the growth of a coleoptile toward light is/are
a. auxin.
b. gibberellin.
c. cytokinin.
d. ethylene.
e. abscisic acid.

____ 6. A plant that touches your house, continues to grow toward and attach itself to the house, is exhibiting
a. tropism.
b. phototropism.
c. gravitropism.
d. thigmotropism.
e. turgor movement.

____ 7. The plant hormone(s) about which Charles Darwin gathered information is/are
a. auxins.
b. gibberellin.
c. cytokinin.
d. ethylene.
e. abscisic acid.

____ 8. The rapid elongation of a floral stalk during the initiation of flowering is known as
a. acceleration.
b. bolting.
c. blooming.
d. floral enhancement.
e. internodal elongation.

____ 9. The aging process is called
a. dormancy.
b. bolting.
c. abscissionation.
d. systemination.
e. senescence.

____10. Environmental signals that trigger a response in plants include
a. temperature.
b. touch.
c. hours of darkness.
d. light.
e. hours of daylight.

____11. The site of gravity perception in roots is the
a. meristem.
b. F-box.
c. thigmotrope.
d. root cap.
e. amyloplast.

____12. Plant hormones may bind to
a. interstitial receptors.
b. enzyme-linked receptors.
c. cytosolic receptors.
d. intranuclear receptors.
e. T1R1 receptors.

____13. The hormone derived from the Greek word meaning to enlarge or increase is
a. cytokinin.
b. abscisic acid.
c. brassinolide.
d. gibberellin.
e. auxin.

____14. Apical dominance
a. promotes growth at the apical meristem.
b. involves ubiquinylated proteins.
c. inhibits lateral bud growth.
d. is driven by the levels of indolacetic acid.
e. leads to phototropism.

____15. The auxin signaling pathway shares similarities with the
a. abscisic acid pathway.
b. brassinosteroid pathway.
c. ethylene pathway.
d. gibberellin pathway.
e. cytokinin pathway.

____16. Plants that flower when the night length is equal to or greater than some critical period are
a. intermediate-day plants.
b. short-day plants.
c. long-night plants.
d. day-neutral plants.
e. long-day plants.

____17. Circadian rhythms
a. affect sleep movements in plants.
b. the rate of photosynthesis.
c. stomata opening and closing.
d. affect gene expression in plants.
e. seasonal reproduction in plants.

____18. The hormone involved in fruit ripening is
a. auxins
b. gibberellin.
c. cytokinin.
d. ethylene.
e. abscisic acid.

____19 Brassinosteroids
a. are produced in leaves.
b. are produced in seeds.
c. are produced in fruits.
d. are involved in vascular devlopment.
e. regulate light-mediated gene expression.

____20. Many plant hormones bind
a. enzyme-linked receptors.
b. TIR1 receptors.
c. cytosolic receptors.
d. nuclear receptors.
e. F-box receptors.

VISUAL FOUNDATIONS

Color the parts of the illustration below as indicated.

RED	❑ auxin receptor	GREY	❑ enzyme complex
VIOLET	❑ repressor protein	YELLOW	❑ transcription activator
TAN	❑ auxin response element	BLUE	❑ DNA
ORANGE	❑ auxin response gene	BROWN	❑ newly formed protein

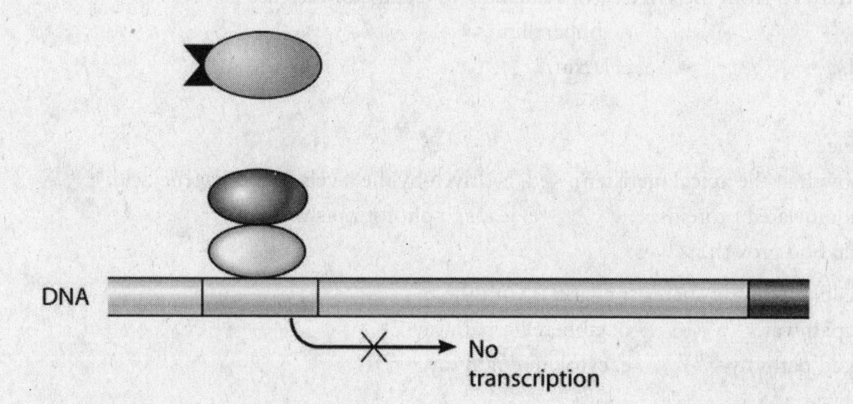

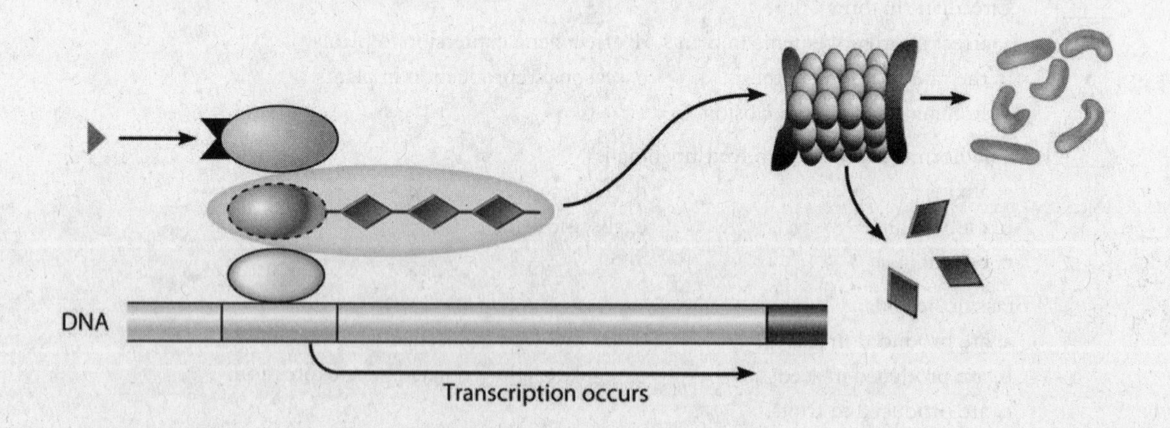

CHAPTER 39

❑

Animal Structure and Function: An Introduction

This is the first in a series of chapters that examine the structural, functional, and behavioral adaptations that help animals meet environmental challenges. This chapter examines the architecture of the animal body. In most animals, cells are organized into tissues, tissues into organs, and organs into organ systems. The principal animal tissues are epithelial, connective, muscular, and nervous. Epithelial tissues are characterized by tight-fitting cells and the presence of a basement membrane. Covering body surfaces and lining cavities, they function in protection, absorption, secretion, and sensation. Connective tissue joins together other tissues, supports the body and its organs, and protects underlying organs. There are many different types of connective tissues, consisting of a variety of cell types. Muscle tissue is composed of cells that are specialized to contract. There are three major types of muscle tissue — cardiac, smooth, and skeletal. Nervous tissue is composed of cells that are specialized for conducting impulses and those that support and nourish the conducting cells. Organs are comprised of two or more kinds of tissues. Complex animals have many organs and organ systems. Organ systems work together to maintain the body's homeostasis.

REVIEWING CONCEPTS

Fill in the blanks.

INTRODUCTION

1. A cell's size is limited by the ____.
2. New cells formed by cell division remain associated in multicellular animals. The size of an animal is determined by the (a) ____________ of cells that make up its body, not their (b) ______________.
3. Bacteria and protists can be small because they rely on diffusion and do not require ________________________________.

TISSUES, ORGANS, AND ORGAN SYSTEMS

4. A group of cells that carry out a specific function is called a (a)____________, and these groups associate to form (b)____________, which in turn are grouped into the (c)__________________ of the body.

Epithelial tissues cover the body and line its cavities

5. Epithelial cells form a (a) ______________ layer of (b)____________________ cells that are attached on one surface to a noncellular (c) _____________________________.
6. Four major functions of epithelial cells are (a) ______________________, (b) ______________________, (c) ______________________, and (d) ______________________.
7. Epithelial cells can be distinguished on the basis of their shape. (a)________________ cells are flat, (b) ________________ cells resemble dice, and (c) ___________________ cells are tall, slender cells shaped like cylinders.
8. Epithelial tissues also vary in the number of cell layers composed of each shape. For example, (a) ________________ epithelium is made up of only one layer of cells and (b) ________________ epithelium has two or more cell layers.
9. An epithelial tissue in which the cells appear to be layered but actually form a single layer is said to be ____________________________.

Glands are made of epithelial cells

10. A gland is one location where secretory epithelial cells may be located. The two types of glands are (a) ________________, for example sweat glands, and (b) ______________________ which produce hormones.

Epithelial cells from membranes

11. (a) ____________________ membranes line body cavities that open to the outside of the body and (b) ____________________ membranes line body cavities that do not open to the outside of the body.

Connective tissues support other body structures

12. Characteristically, connective tissues contain very few cells, an (a)____________________________________ in which the cells are embedded and (b) ____________________________ in a matrix the cells secrete.
13. The nature and function of each kind of connective tissue is determined by the structure and properties of the ____________________________.
14. There are three types of fibers in connective tissues. Collagen fibers are numerous, strong fibers composed of the protein (a) ____________. Elastic fibers are composed of (b) ________________ and can stretch. Reticular fibers, composed of (c)_________________________________, form delicate networks of connective tissue.

15. Fibroblast cells produce (a) _____________, (b) _____________, and (c) _____________ present in the matrix of connective tissue.
16. _____________ function as the body's scavenger cells.
17. The most widely distributed connective tissue in the vertebrate body is (a) _____________ which allows the body parts it connects to (b) _____________.
18. Dense connective tissues are predominantly composed of _____________ fibers and are less flexible
19. _____________ is the supporting skeleton in the embryos of all vertebrates.
20. All vertebrates have an internal supporting structure called the _____________ that is composed of cartilage and/or bone.
21. Cartilage cells are called (a) _____________ secrete (b)_____________ that strengthen the extracellular matrix. As the secreted matrix forms, the cells become isolated in holes called (c)_____________.
22. Bone cells are called (a) _____________ and, like chondrocytes, they become isolated in (b) _____________. Unlike cartilage, bone is highly (c) _____________, a term referring to the abundant supply of blood vessels in bone.
23. Osteocytes communicate with one another through small channels called _____________.
24. Compact bone consists of spindle-shaped units called (a) _____________. Osteocytes are arranged in concentric layers called (b) _____________. The lamellae surround central microscopic channels known as (c)_____________, through which blood vessels and nerves pass.
25. The noncellular component of blood is the _____________.
26. Red blood cells function to _____________.
27. White blood cells function to _____________.
28. Small cell fragments that originate in bone marrow and are involved in blood clotting are called _____________.

Muscle tissue is specialized to contract

29. Each muscle cell is called a _____________.
30. Contractile proteins called (a) _____________, and (b) _____________ are contained within the elongated (c) _____________ of muscle cells.
31. Vertebrates have three kinds of muscles: (a) _____________ muscle that attaches to bones and causes body movements; (b)_____________ muscle occurring in the walls of the digestive tract, uterus, blood vessels, and many other internal organs; and heart or (c) _____________ muscle.

Nervous tissue controls muscles and glands

32. Nervous tissue is composed of cells that conduct impulses, (a)__________, and support cells which are called (b) __________ cells.
33. Neurons generally contain three functionally and anatomically distinct regions: the (a) ________________, which contains the nucleus; the (b) ________________, which receive incoming impulses; and the (c) __________, which carry impulses away from the cell body.
34. Neurons communicate with one another and with glands and muscles at cellular junctions called ________________.
35. A ____________ is a collection of neurons bound together by connective tissue.

Tissues and organs make up the organ systems of the body

36. Mammals have 11 organ systems. They are (a)______________________, (b)______________________, (c)______________________, (d)______________________, (e)______________________, (f)______________________, (g)______________________, (h)______________________, (i)______________________, (j)______________________, and (k)______________________.

REGULATING THE INTERNAL ENVIRONMENT

37. A balanced internal environment is referred to as ______________.
38. __________________ are changes in either the internal or external environment that affects normal body conditions.
39. Animals that maintain a relatively constant internal environment despite changes occurring in the external environment are called __________________.

Negative feedback systems restore homeostasis

40. In a negative feedback system, a (a) ______________ detects a change from the normal condition and an (b) ______________ activates hoemostatic mechanisms to restore the steady state.
41. The response from the integrator will be ______________ from the output of the sensor.

A few positive feedback systems operate in the body

42. Unlike what occurs in a negative feedback system, the response of a positive feedback system ______________ the changing condition detected by the sensor.

REGULATING BODY TEMPERATURE

Ectotherms absorb heat from their surroundings

43. An ectotherm's metabolic rate tends to change with the weather; therefore, they have a much __________________ daily energy expenditure than endotherms.

Endotherms derive heat from metabolic processes

44. Endotherms can have a metabolic rate as much as __________ (#?) times as high as that of ectotherms.
45. In mammals receptors involved in temperature regulation are located in the ______________________________.

Many animals adjust to challenging temperature changes

46. When stressed by cold, many animals sink into (a) ______________, a short-term decrease in body temperature below normal levels. (b)______________ is long-term torpor in response to winter cold and scarcity of food. (c) __________________ is a state of torpor caused by lack of food or water during periods of high temperature.

BUILDING WORDS

Use combinations of prefixes and suffixes to build words for the definitions that follow.

Prefixes	The Meaning	Suffixes	The Meaning
chondro-	cartilage	-blast	embryo
dendr-	tree	-cyte	cell
fibro-	fiber	-ial	pertaining to
homeo-	similar, "constant"	- ite	part of
inter-	between, among	-phage	eat, devour
interstit	to stand between	-sis	process of
macro-	large, long, great, excessive	-stasis	equilibrium
multi-	many		
myo-	muscle		
nich-	nest		
osteo-	bone		
pseudo-	false		
synap-	a union		

Prefix	Suffix	Definition
__________	-cellular	1. Composed of many cells.
__________	-stratified	2. An arrangement of epithelial cells in which the cells falsely appear to be stratified.
__________	__________	3. A cell, especially active in developing ("embryonic") tissue and healing wounds, that produces connective tissue fibers.
__________	-cellular	4. Situated between or among cells.
__________	__________	5. A large cell, common in connective tissues, that phagocytizes ("eats") foreign matter including bacteria.
__________	__________	6. A cartilage cell.
__________	__________	7. A bone cell.
__________	-fibril	8. A thin longitudinal contractile fiber inside a muscle cell.
__________	__________	9. Maintaining a constant internal environment.
__________	__________	10. An extension of a neuron that transmits information to the cell body.

__________ __________ 11. The junction where nerves communicate with other nerves, glands, or muscles.

__________ -e 12. The functional role of a species in a community.

__________ __________ 13. The type of fluid found between cells.

MATCHING

Terms:

a. Acclimitization
b. Adipose tissue
c. Cardiac muscle
d. Cartilage
e. Collagen
f. Elastic
g. Endocrine gland
h. Exocrine gland
i. Gland
j. Glial cell
k. Homeostasis
l. Matrix
m. Organ
n. Organ system
o. Osteon
p. Recticular
q. Skeletal muscle
r. Stressor
s. Striations

For each of these definitions, select the correct matching term from the list above.

____ 1. Striated (voluntary) muscle.
____ 2. Tissue in which fat is stored, or the fat itself.
____ 3. Spindle-shaped unit of bone composed of concentric layers of osteocytes.
____ 4. Supporting skeleton in the embryonic stages of all vertebrates.
____ 5. A specialized structure made up of tissues and adapted to perform a specific function or group of functions.
____ 6. Glands that secrete products directly into the blood or tissue fluid instead of into ducts.
____ 7. Tissue characterized by the presence of intercalated discs.
____ 8. A cell that supports and nourishes neurons.
____ 9. General term for a body cell or organ specialized for secretion.
____10. A protein in connective tissue fibers.
____11. Thin, branched fibers that from delicate networks.
____12. Microscopically helps to identify skeletal and cardiac muscle from smooth muscle.
____13. Adjustment by animals to seasonal changes.

MAKING COMPARISONS

Fill in the blanks.

Principle Tissue	Tissue	Location	Function
Epithelial tissue	Stratified squamous epithelium	Skin, mouth lining, vaginal lining	Protection; outer layer continuously sloughed off and replaced from below
Epithelial tissue	#1	#2	Allows for transport of materials, especially by diffusion
#3	Pseudostratified epithelium	#4	#5
#6	Adipose tissue	Subcutaneous layer, pads	#7

Principle Tissue	Tissue	Location	Function
		certain internal organs	
Connective tissue	#8	Forms skeletal structure in most vertebrates	#9
#10	Blood	#11	#12
Muscle tissue	Cardiac muscle	#13	Contraction of the heart
Muscle tissue	#14	Attached to bones	Movement of the body
#15	Nervous tissue	Brain, spinal cord, nerves	Respond to stimuli, conduct impulses

MAKING CHOICES

Place your answer(s) in the space provided. Some questions may have more than one correct answer.

____ 1. The nucleus of a neuron is typically found in the
a. cell body.
b. synapse.
c. dendrite.
d. axon.
e. glial body.

____ 2. The main types of animal tissues include
a. epithelial.
b. nervous.
c. muscular.
d. skeletal.
e. connective.

____ 3. Myosin and actin are the main components of
a. cartilage.
b. blood.
c. collagen.
d. muscle.
e. adipose tissue.

____ 4. The basement membrane
a. lies beneath epithelium.
b. contains polysaccharides.
c. consists of cells that lie beneath the epithelium.
d. is synonymous with plasma membrane.
e. is noncellular.

____ 5. Thin, flattened epithelial cells are called ________ cells
a. columnar.
b. cuboidal.
c. squamous.
d. stratified.
e. endothelial.

____ 6. Chondrocytes are
a. part of cartilage.
b. found in bone.
c. immature osteocytes.
d. found in lacunae.
e. embryonic osteocytes.

____ 7. Large muscles attached to bones are
a. skeletal muscles.
b. smooth muscles.
c. composed largely of actin and myosin.
d. nonstriated.
e. multinucleated.

____ 8. Platelets are
a. bone marrow cells.
b. fragments of large cells.
c. a type of white blood cells.
d. a type of RBCs.
e. derived from bone marrow cells.

____ 9. Blood is one type of
a. endothelium.
b. collagen.
c. mesenchyme.
d. connective tissue.
e. cardiac tissue.

____10. Epithelium located in the ducts of glands would most likely be
a. simple columnar.
b. simple cuboidal.
c. simple squamous.
d. stratified squamous.
e. stratified cuboidal.

____11. The portion of the neuron that is specialized to receive a nerve impulse is the
a. cell body.
b. synapse.
c. dendrite.
d. axon.
e. glial body.

____12. Collagen is
a. part of blood.
b. part of bone.
c. part of connective tissue.
d. composed of fibroblasts.
e. fibrous.

____13. Connective tissue matrix may be
a. noncellular.
b. mostly lipids.
c. a gel.
d. a polysaccharide.
e. fibrous.

____14. Haversian canals
a. contain nerves.
b. run through cartilage.
c. run through bone.
d. connect lacunae.
e. connect canaliculi.

____15. Adipose tissue is a type of
a. cartilage.
b. epithelium.
c. connective tissue.
d. modified blood tissue.
e. marrow.

____16. Cells that nourish and support nerve cells are known as
a. neurons.
b. glial cells.
c. neuroblasts.
d. synaptic cells.
e. neurocytes.

____17. A group of closely associated cells that carries out a specific function is a/an
a. organ.
b. system.
c. organ system.
d. tissue.
e. clone.

____18. Cells lining internal cavities that secrete a lubricating mucus are
a. goblet-cells.
b. pseudostratified.
c. exocrine glands.
d. epithelial.
e. part of connective tissue.

____19. Cells that make up the outer layer of skin are
a. the columnar layer.
b. the cuboidal layer.
c. stratified squamous epithelium.
d. basement membrane cells.
e. endothelium.

____20. You would expect the outer layer of skin on the soles of your feet to be
a. cuboidal.
b. simple.
c. connected to a basement membrane.
d. stratified.
e. squamous.

____21. A serous membrane would be expected to
a. line a body cavity not open to the outside.
b. line a body cavity not open to the outside.
c. secrete fluid.
d. line the abdominal cavity.
e. contain epithelial cells.

____22. Both bone and cartilage
a. are vascularized.
b. have canaliculi.
c. have lamellae.
d. have cells present in lacunae.
e. are connective tissue.

____23. Thermoregulation
a. occurs in ectotherms.
b. may involve hormones.
c. occurs in invertebrates.
d. occurs in endotherms.
e. may involve hibernation.

____24. Torpor is or can be associated with the term(s)
a. hibernation.
b. acclimatization.
c. change in temperature.
d. estivation.
e. lack of food.

____25. A balanced internal environment
a. occurs during homeostasis.
b. describes a state of standing safe.
c. .is seldom achieved.
d. occurs during homostasis.
e. is affected by stressors.

____22. Negative feedback systems
a. involve integration.
b. intensify the condition.
c. are common body responses.
d. reinforce the sensor outuput.
e. respond in a way opposite of the sensor output.

____22. Endotherms
a. will be mammals.
b. have adaptations for insulation.
c. allow their body temperature to rise and fall with the environment.
d. have high energy costs.
e. may be birds.

VISUAL FOUNDATIONS

Color the parts of the illustration below as indicated. Also label compact bone and spongy bone.

- RED ❑ blood vessel
- GREEN ❑ lacuna
- YELLOW ❑ osteon
- BLUE ❑ osteocyte
- ORANGE ❑ Haversian canal
- VIOLET ❑ cytoplasmic extensions

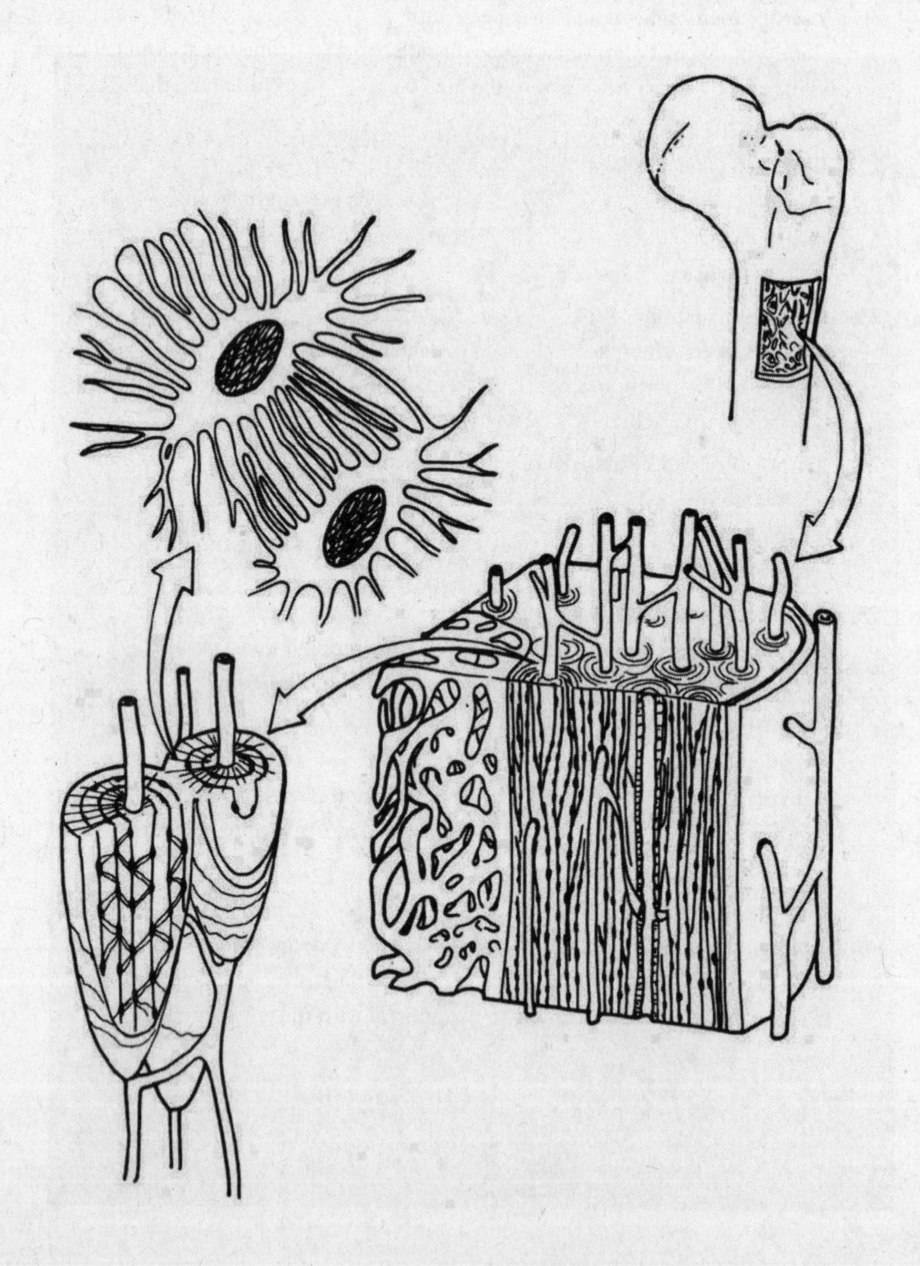

CHAPTER 40

Protection, Support, and Movement

The next several chapters discuss how animals carry out life processes and how the organ systems of a complex animal work together to maintain homeostasis of the organism as a whole. This chapter focuses on epithelial coverings, skeleton, and muscle – systems that are closely interrelated in function and significance. Epithelium covers all external and internal body surfaces. In invertebrates, the external epithelium may contain secretory cells that produce a protective cuticle, secrete lubricants or adhesives, produce odorous or poisonous substances, or produce threads for nests or webs. In vertebrates, specifically humans, the external epithelium (skin) includes nails, hair, sweat glands, oil glands, and sensory receptors. In other vertebrates, it may include feathers, scales, mucous, and pigmentation. The skeleton supports and protects the body and transmits mechanical forces generated by contractile cells. Among the invertebrates are found hydrostatic skeletons and exoskeletons. All exoskeletons are composed of nonliving material above the epidermis. In many invertebrates it prevents growth and requires periodic molting. In invertebrates such as bivalves, clams and oysters, the exoskeleton is added to by secretions from the underlying epithelium as the animal grows and it is not shed. Endoskeletons on the other hand, extensive in echinoderms and chordates, are composed of living tissue that can grow. All animals have the ability to move. Muscle tissue is found in most invertebrates and all vertebrates.

REVIEWING CONCEPTS

Fill in the blanks.

INTRODUCTION

1. The epithelium, skin, of *Rana ridibanda* produces secretions that kill the antibiotic resistant strain of *S. aureus* known as ____________________.
2. Both the epithelium and the skeletal system provide an animal with ____________________.
3. Muscle can contract. When it is anchored to the ________________, muscle contraction causes movement (locomotion).

EPITHELIAL COVERINGS

4. The structure and functions of the external epithelium are adapted to the animal's (a) ______________________ and (b) __________________.

Invertebrate epithelium may secrete a cuticle

5. In insects, the outer epithelium secretes a protective and supportive outer covering of nonliving material called the ____________________.
6. In many species of animals, the epithelium contains cells that secrete (a) ___________________________ and (b) ____________________.

Vertebrate skin functions in protection and temperature regulation

7. The integumentary system of vertebrates includes the (a) __________ and (b) ____.
8. Structures derived from the epithelium can provide insulation. Examples are the (a) ____________ of birds and the (b) __________ of mammals.
9. Examples of structures derived from the epithelium in mammals are (a)________________________, (b) ________________________, (c) ________________________, (d) ________________________, and (e) ________________________.
10. The outer layer of skin is called the (a) ______________. It consists of several sublayers, or (b) ________.
11. Cells in the ____________________ continuously divide. As they are pushed upward, these cells mature, produce keratin, and eventually die.
12. Epithelial cells produce an insoluble, elaborately coiled protein called (a) ___________, which functions in the skin for (b) __________, (c) ____________________, and (d) ____________________.
13. The dermis consists of dense, fibrous (a)____________________ composed mainly of (b)________________ fibers and rests on a layer of (c)__________________ tissue composed largely of fat.
14. Exposure to ultraviolet light causes the skin epidermis to (a)______________ and pigment cells to produce (b) __________. However, UV radiation also damages (c) __________ which can lead to (d) ______________.

SKELETAL SYSTEMS

15. A skeletal system transmits and transforms ____________________ generated by muscle contractions resulting in movement.

In hydrostatic skeletons body fluids transmit force

16. Hydrostatic skeletons are made of ____.
17. The hydrostatic skeleton transmits forces generated by ______________.
18. Many invertebrates (e.g., hydra) have a (a)________________________ in which fluid is used to transmit forces generated by contractile cells or muscle. In these animals, contractile cells are arranged in two layers, an outer (b)____________________ oriented layer and an inner (c)________________ arranged layer. When the (d) __________ (outer or inner?) layer contracts, the animal becomes shorter and thicker, and when the (e) __________ (outer or inner?) contracts, the animal becomes longer and thinner.
19. Annelid worms have sophisticated hydrostatic skeletons. The body cavity is divided by transverse partitions called __________, creating isolated, fluid-filled segments that can operate independently.

Mollusks and arthropods have nonliving exoskeletons

20. The support system in some organisms is an ________________ consisting of a non-living substance overlying the epidermis.
21. Arthropod exoskeletons, composed mainly of the polysaccharide (a)__________. This nonliving skeleton prevents growth, necessitating periodic (b) ________________.

Internal skeletons are capable of growth

22. Internal skeletons are also called (a) ____________________. They are extensively developed only in the echinoderms and the (b)____________.
23. The endoskeletons of echinoderms is composed of spines and plates composed of ____________.
24. The vertebrate skeleton provides support, protection, and it ____________________.

The vertebrate skeleton has two main divisions

25. The two main divisions of the vertebrate endoskeleton are the (a)________________ skeleton along the long axis of the body and the (b) ________________ skeleton which includes the bones of the limbs and girdles.
26. Components of the axial skeleton include the (a) ____________, which consists of cranial and facial bones; the (b)________________________, made up of a series of vertebrae; and the (c)________________, consisting of the sternum and ribs.
27. Components of the appendicular skeleton include the (a)__________________, consisting of clavicles and scapulas; the (b) __________________, which consists of large, fused hipbones; and the (c) __________, each of which terminates in digits.

A typical long bone amplifies the motion generated by muscles

28. The long bones are covered by a connective tissue layer called the (a)____________________ to which (b) ______________ and (c) ______________ attach.
29. The ends of long bones are called the (a) ______________, and the shaft is the (b) ______________. A cartilaginous "growth center" in children called the (c) ______________ becomes an (d)______________ in adults. (e) ______________, made up of osteons, forms a dense, and hard outer shell.

Bones are remodeled throughout life

30. Bones are built by (a) ______________ which become isolated in the matrix and become (b) ______________. Bone is absorbed by (c) ____________________ prior to remodeling.

Joints are junctions between bones

31. Junctions between bones are called (a) ____________. Joints are classified according to the degree of their movement: (b)________________, such as sutures, are tightly bound by fibers;

(c) ____________________, like those between vertebrae; and the most common type of joint, the (d) ____________________.

32. Freely movable joints are enclosed by a (a) ____________________ and are lubricated by (b) ____________________.

MUSCLE CONTRACTION

33. All eukaryotic cells contain the contractile protein (a) __________. In many cells, it functions in association with the contractile protein (b)__________.

Invertebrate muscle varies among groups

34. Bivalves mollusks have (a) ____________________ muscle which they use to shut the shell quickly and (b) ____________________ muscle which they use to hold the shell tightly closed for long periods of time.

Insect flight muscles are adapted for rapid contraction

35. ______________________________ contractions are muscle contractions that are not synchronized with signals from motor neurons.

Vertebrate skeletal muscles act antagonistically to one another

36. Skeletal muscles pull on cords of connective tissue called ______________, which are attached to bones and then pull on them.

37. The movement on one muscle can be reversed by the movement of another. This means that muscles are acting ____________________.

38. The muscle that contracts to produce a particular movement is called the (a) ____________________ and the muscle that produces the opposite movement is called the (b) ____________________.

A vertebrate muscle may consist of thousands of muscle fibers

39. A muscle fiber is actually ______________________________.

40. Within the sacroplasm of each muscle fiber are large, threadlike structures called (a) ____________________ that run lengthwise through the fiber. These larger structures are made up of smaller structures called (b) ____________________.

41. Actin is the structural unit of (a) ____________________ and myosin is the structural unit of (b) ____________________.

42. The basic unit of skeletal muscle contraction is called a __________.

Contraction occurs when actin and myosin filaments move past one another

43. A motor neuron releases (a) ____________________ into the synaptic cleft between the motor neuron and each muscle fiber, where it binds with receptors on the surface of the muscle fiber, depolarizing the sarcolemma and initiating an (b)______________________________.

44. Depolarization of the T tubules stimulates calcium release from the (a)______________________________. Calcium then binds to (b) ____________________ which undergoes a conformational change exposing active sites on the (c) ____________________ filaments. These sites are also called (d) ______________________________sites.

45. During the power stroke of muscle contraction the actin filament is pulled toward the ________________ of the sarcomere.

ATP powers muscle contraction

46. __________ is the immediate energy source for muscle contraction.
47. A medical examiner can estimate the time of death based upon the amount of cadaver ____________________observed.
48. _________________________ is the backup energy storage compound in muscle cells.
49. __________ is the chemical energy stored in muscle fibers.

The type of muscle fibers determines strength and endurance

50. (a) _________________________ fibers are well-adapted for endurance activities such as swimming and running. They derive most of their energy from (b) ____________________. These fibers are rich in (c) __________________ which enhances oxygen movement from blood into muscle.
51. (a) _________________________ fibers contract rapidly and have an intermediate level of fatigue.
52. (a) ______________________ fibers generate power and carry out rapid movements by they can only sustain this activity for a short period of time. These fibers obtain most of their energy from (b)________________.

Several factors influence the strength of muscle contraction

53. A motor neuron and the fibers to which it is functionally connected is called a ________________.
54. The single, quick contraction of a skeletal muscle is a ________________.
55. Even when not moving skeletal muscles remain in a partially contracted state called ____________________________.

Smooth muscle and cardiac muscle are involuntary

56. (a)________________ muscle is capable of slow, sustained contractions and (b)________________ muscle contracts rhythmically.

BUILDING WORDS

Use combinations of prefixes and suffixes to build words for the definitions that follow.

Prefixes	The Meaning	Suffixes	The Meaning
ax-	center line	-blast	embryo, "formative cell"
chit-	tunic	-chondr(al)	cartilage
endo-	within	-dermis	skin
epi-	upon, over, on	-ial	pertaining to
kerat-	horn	-in	chemical substance
myo-	muscle	-lemma	sheath
osteo-	bone	-mere	a part of
peri-	about, around, beyond	-oste(um)	bone
sarc(o)-	muscle		
tend-	to stretch		

Prefix	Suffix	Definition
________	________	1. The outermost layer of skin, resting on the dermis.
________	________	2. Connective tissue membrane on the surface of bone that is capable of forming bone.
________	________	3. Pertains to the occurrence or formation of "something" within cartilage.
________	________	4. A bone-forming cell.
________	-clast	5. Cells that break down bone.
________	-filament	6. Subunit of myofibril consisting of either actin or myosin.
________	________	7. Serves as a diffusion barrier in the skin.
________	________	8. A polysaccharide contained in exoskeletons.
________	________	9. The central skeleton of the body consisting of the skull, vertebral column, ribs, and sternum.
________	________	10. The contractile unit of skeletal muscles.
________	________	11. The plasma membrane of a skeletal muscle fiber.
________	-on	12. Connects muscle to bone.

MATCHING

Terms:

a. Actin
b. Axial skeleton
c. Endoskeleton
d. Endosteum
e. Exoskeleton
f. Keratin
g. Ligament
h. Lacunae
i. Motor unit
j. Oxygen debt
k. Periosteum
l. Sarcomere
m. Sebum
n. Skeletal muscle
o. Stratum basale
p. Tendon
q. T-tubule
r. Vertebral column

For each of these definitions, select the correct matching term from the list above.

____ 1. A motor neuron and the muscle fibers it stimulates.
____ 2. A water-insoluble, elaborately coiled protein manufactured by epidermal cells that gives skin mechanical strength and flexibility.
____ 3. A segment of a striated muscle cell that serves as the basic unit of muscle contraction.
____ 4. The deepest layer of the epidermis, consisting of cells that continuously divide.
____ 5. A band of connective tissue that connects bones and limits movement at the joints.
____ 6. The skull, vertebral column, sternum, and ribs.
____ 7. The hard exterior covering of certain invertebrates.
____ 8. Tough cords of connective tissue that anchor muscles to bone.
____ 9. The oxygen necessary to metabolize the lactic acid produced during strenuous exercise.
____10. The protein composing one type of myofilament.
____11. A mixture of fats and waxes.
____12. The connective tissue layer covering the outer surface of a bone.
____13. Small cavities in bone containing mature osteocytes.
____14. An inward extension of the sacrolemma.

MAKING COMPARISONS

Fill in the blanks.

Organism	Skeletal Material	Kind of Skeleton	Characteristics
Many invertebrates	Fluid	Hydrostatic	Contractile tissue generates force that moves the body
Arthropods	#1	#2	#3
#4	#5	Endoskeleton	Internal shell, some have spines that project to the outer surface
Chordates	#6	#7	#8

Characteristics	Slow-Oxidative fibers	Fast-Glycolytic fibers	Fast-Oxidative Fibers
Contraction speed	Slow	#9	#10
Rate of fatigue	#11	#12	Intermediate
Major pathway for ATP synthesis	#13	Glycolysis	#14
Intensity of contraction	#15	#16	Intermediate

MAKING CHOICES

Place your answer(s) in the space provided. Some questions may have more than one correct answer.

____ 1. A major component of the arthropod skeleton is the polysaccharide
a. keratin.
b. melanin.
c. chitin.
d. actin.
e. myosin.

____ 2. Actin filaments contain
a. actin.
b. tropomyosin.
c. troponin complex.
d. myosin.
e. keratin.

____ 3. The H zone in a sarcomere consists of
a. actin.
b. tropomyosin.
c. troponin complex.
d. myosin.
e. keratin.

____ 4. The human skull is part of the
a. cervical complex.
b. girdle.
c. axial skeleton.
d. appendicular skeleton.
e. atlas.

____ 5. Most of the mechanical strength of bone is due to
a. spongy bone.
b. osteocytes.
c. intramembranous bone.
d. endochondral bone.
e. marrow.

____ 6. You perceive touch, pain, and temperature through sense organs in your
a. epidermis.
b. stratum basale.
c. stratum corneum.
d. epithelium.
e. dermis.

____ 7. Myofilaments are composed of
a. myofibrils.
b. fibers.
c. actin.
d. myosin.
e. sarcoplasmic reticulum.

____ 8. The outer layer of a vertebrate's skin is the
a. epidermis.
b. stratum basale.
c. stratum corneum.
d. epithelium.
e. dermis.

____ 9. The tissue around bones that lays down new layers of bone is the
a. metaphysis.
b. epiphysis.
c. marrow.
d. periosteum.
e. endosteum.

____10. The human axial skeleton includes the
a. ulna.
b. shoulder blades.
c. skull.
d. femur.
e. breastbone.

____11. Cartilaginous growth centers in children are called the
a. metaphysis.
b. epiphysis.
c. marrow.
d. periosteum.
e. endosteum.

____12. The human appendicular skeleton includes the
a. ulna.
b. shoulder blades.
c. centrum.
d. femur.
e. breastbone.

____13. Vertebrate appendages are connected to
a. the cervical complex.
b. girdles.
c. the axial skeleton.
d. the appendicular skeleton.
e. the atlas.

____14. Skeletons that operate entirely by hydrostatic pressure are found in
a. annelids.
b. echinoderms.
c. hydra.
d. lobsters and crayfish.
e. vertebrates.

____15. An opposable digit is found in
a. lobsters.
b. great apes.
c. crayfish.
d. some insects.
e. humans.

____16. Internal skeletons are found in
a. annelids.
b. echinoderms.
c. hydra.
d. lobsters and crayfish.
e. chordates.

____17. External skeletons are found in
a. annelids.
b. echinoderms.
c. hydra.
d. insects.
e. chordates.

____18. The "space" between the terminal axon of one neuron and a dendrite of the next neuron is called the
a. synaptic cleft.
b. chemotrasmitter zone.
c. cross bridge.
d. Z line.
e. binding site.

____19. Melanocytes
a. are located in the dermis.
b. produce keratin.
c. produce melanin.
d. are continuously replaced.
e. are located in the stratum basale.

____20. Epithelial derivatives include
a. antlers.
b. claws.
c. glands.
d. sebum.
e. melanin.

____21. Terms directly associated with the dermis include
a. blood vessels.
b. collagen.
c. sensory receptors.
d. cuticle.
e. dense connective tissue.

____22. Exoskeletons
a. are rigid and inflexible.
b. are a one-piece sheath.
c. are found only in invertebrates.
d. transmit forces.
e. must be shed in animals that have them.

____23. The vertebrate vertebral column has
a. 7 cervical vertebrae.
b. 6 sacral vertebrae.
c. 5 lumbar vertebrae.
d. 14 thoracic vertebrae.
e. 6 fused vertebrae making up the coccygeal region.

____24. When a sarcomere contracts
a. the myofilaments shorten.
b. the I-band shortens.
c. the A-band gets longer.
d. the Z-lines get closer together.
e. the H-zone shortens.

____25. Skeletal muscles use stored ____ as a backup energy supply even though it is present in very limited quantities.
a. myoglobin
b. ATP
c. creatine phosphate
d. glycogen
e. glucose

____26. Exposure to UV radiation
a. stimulates melatonin production.
b. affects dark-skinned individuals the least.
c. .causes the skin to become inflamed.
d. causes the epidermis to thicken.
e. stimulates keratin production.

____ 27. Hydrosatic skeletons
a. are found in many soft-bodied animals.
b. can be found in vertebrates with internal skeletons.
c. are found in shelled invertebrates.
d. are found in roundworms.
e. transmit force.

____ 28. An osteon contains
a. lamellae.
b. lacunae.
c. osteoblasts.
d. spongy bone.
e. canaliculi.

____ 29. Strength of muscle contraction
a. is affected by the amount of tension developed by each fiber.
b is determined by the number of fibers contracting.
c. is affected by the frequency of fiber stimulation.
d. the amount of muscle tone.
e. is affected by the amount of summation that occurs

VISUAL FOUNDATIONS

Color the parts of the illustration below as indicated. Also label H zone, A band, I band, sarcomere, and muscle fiber.

RED ❑ Z line
GREEN ❑ mitochondria
YELLOW ❑ sarcoplasmic reticulum
BLUE ❑ T tubule
ORANGE ❑ myofibrils
TAN ❑ sarcomere
PINK ❑ sarcolemma
VIOLET ❑ nucleus

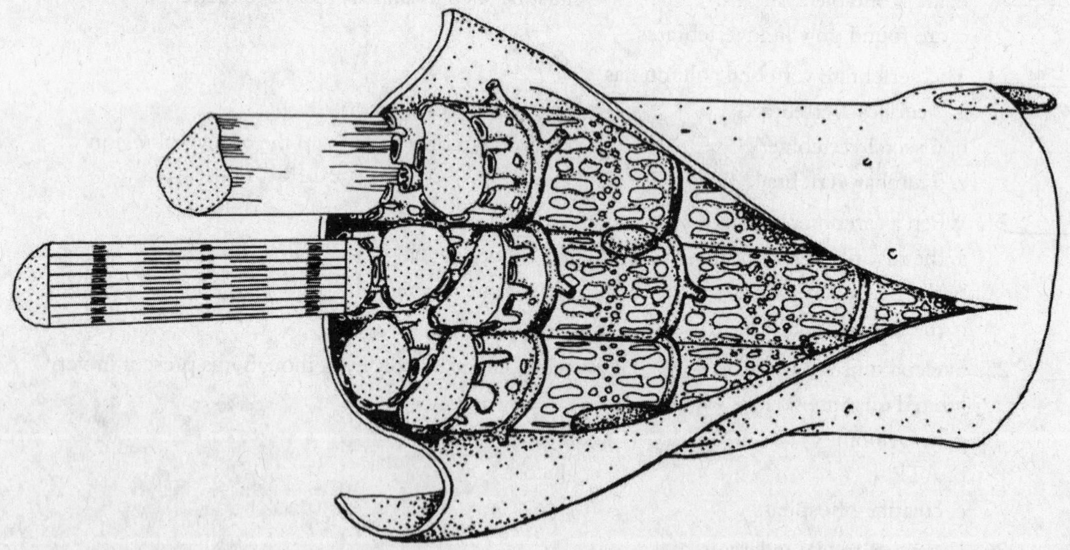

CHAPTER 41

Neural Signaling

Behavior and physiological processes in animals are regulated by the endocrine and nervous systems. Endocrine regulation is generally slow and long-lasting, whereas nervous regulation is typically rapid and brief. Changes within an animal or in the outside world are detected by receptors and transmitted as electrical signals to the central nervous system where the information is sorted and interpreted. An appropriate response is then sent to muscles or glands where the response occurs. The nervous system is composed mainly of two types of specialized cells, glial cells and neurons. Glial cells help support, maintain, and regulate the nervous system and neural function. The neuron is the structural and functional unit that carries electrical impulses. This chapter focuses on the role of the neuron in nervous regulation and the importance of glial cells. The typical neuron consists of a cell body, dendrites, and an axon. The cell body houses the nucleus and has branched cytoplasmic projections called dendrites that receive stimuli. The axon is a single, long structure that transmits impulses to a neuron, muscle, or gland. Transmission involves changes in ion distribution between the inside and outside of the neuron. Transmission from one neuron to the next generally involves the release of chemicals into the space between neurons, the synapse. These chemicals diffuse across the space and bring about a response in the adjacent neuron, muscle, or gland. Hundreds of messages may arrive at a single neuron at the same time and must be integrated prior to a given response. Integration may occur at the level of the neuronal cell body, in the spinal cord, or in the brain. Following integration, the neuron may or may not initiate an impulse along its axon.

REVIEWING CONCEPTS

Fill in the blanks

INTRODUCTION

1. The ability of an organism to survive is largely dependent on its ability to detect and respond appropriately to environmental referred to as ____________________.
2. In animals, the principal regulatory system is the (a)____.
3. The formation of new nerves is called __________________________.

NERUAL SIGNALING: AN OVERVIEW

4. Information flow through the nervous system begins with the detection of a stimulus by a "sensory device," in a process called
5. A signal passes from a receptor along an (a)____________________________ and is then transmitted to an (b)____________________________ which sorts and interprets incoming sensory information before a response goes out from the

CNS along an (c) ______________________. Responses sent to muscles travel along (d) ______________________.

6. Signals are carried from the central nervous system (CNS) along efferent neurons to ____________ which may be a muscle or a gland.

NEURONS AND GLIAL CELLS

Neurons receive stimuli and transmit neural signals

7. Neurons are highly specialized cells with a distinctive structure consisting of a cell body and two types of cytoplasmic extensions. Numerous, short (a) ______________________ receive impulses and conduct them to the (b) ______________ which integrates impulses. Once integrated, the impulse is conducted away by the (c) ______________, a long process that terminates at another neuron or effector.
8. Branches at the ends of axons, called (a) ______________, end in tiny (b) ______________________ that release a transmitter chemical called a (c) ______________________.
9. Many axons outside of the vertebrate CNS are surrounded by a series of (a) ______________________ that form an insulating covering called the (b) ______________________. Gaps in this insulating covering, called (c) ______________________, occur between successive Schwann cells.

Certain regions of the CNS produce new neurons

10. ______________________ is the production of new neurons.

Axons aggregate to form nerves and tracts

11. A (a) ______________ is a bundle of axons outside the CNS, whereas a (b) ______________ is a bundle of axons within the CNS.
12. A mass of cell bodies outside the CNS are called (a) ______________, whereas a mass of cell bodies within the CNS are called (b) ________.

Glial cells play critical roles in neural function

13. Collectively, glial cells make up the ______________________.
14. There are several types of glial cells. (a) ______________________ are star-shaped glial cells that provide physical support and nutrients for neurons. (b) ______________________ form the myelin sheath in the CNS. (c) ______________________ are ciliated and line internal cavities of the CNS, and (d) ______________________ mediate responses to injury or disease in the CNS.

TRANSMITING INFORMATION ALONG THE NEURON

15. The voltage measured across the plasma membrane of a cell is called the ______________________.

Ion channels and pumps maintain the resting potential of the neuron

16. The membrane potential in a non-excited neuron or muscle cell is its (a)____________________________. A typical value for this membrane potential is (b) ____________.
17. Two main factors that determine the magnitude of the membrane potential are
    (a)____
    and (b)____.

Ions cross the plasma membrane by diffusion through ion channels

18. The three types of channels found in neurons are
    (a)____,
    (b)____, and
    (c)____.
19. __________________________ channels are the most common type of passive ion channels in the plasma membrane.
20. When a neuron is not conducting an electrical impulse, the charge in the extracellular fluid relative to the intracellular fluid is (a)________________ (negative, positive) and the intracellular fluid relative to the extracellular fluid is (b) ________________ (negative, positive).
21. In most cells, potassium ion concentration is highest (a)____________________ (inside or outside?) the cell and sodium ion concentration is highest) (b)__________________ (inside or outside?) the cell.

Ion pumping maintains the gradients that determine the resting potential

22. The ionic balance across the plasma membrane of neurons is a result of several factors. The (a) ____ is an active transport system that moves (b) ______________ (#) sodium ions out of the cell and (c) ________________ (#) potassium ions into the cell.

Graded local signals vary in magnitude

23. Depolarization is described as an (a) ________________ action because it brings the neuron closer to (b) __________________________________.
24. Hyperpolarization is described as an (a) __________________ action because it (b) __________________ a neurons ability to conduct an impulse.
25. A _______________________________________ occurs in a relatively small region of the plasma membrane and it varies in magnitude based upon stimulus strength.

Axons transmit signals called action potentials

26. If a stimulus is strong enough, it depolarizes the membrane to the (a)________________________________, and it will provoke a response from the neuron called an (b) ____.

27. All cells can generate graded potentials, but action potentials are only generated by (a) ________________, (b) ________________ and a few cells of the endocrine and immune systems.
28. Local anesthetics bind to ________________________ which blocks the transmission of an action potential.

An action potential is generated when the voltage reaches threshold level

29. Action potentials are initiated when (a) ________________ in the plasma membrane of the neuron open in response to a critical level of voltage change in the membrane potential allowing (b)__________ to enter the neuron. This response is the (c)________________________ phase of an action potential.
30. Depolarization is an example of a ______________ (positive, negative) feedback system.

The neuron repolarizes and returns to a resting state

31. After a certain period (a) ________________ close and the (b)____________________ open which is the (c)______________________ phase of the action potential.
32. As the action potential moves down the axon, ______________ occurs behind it.
33. The time in which a new action potential cannot be generated is called the ________________________ period.

The action potential is an all-or-none response

34. An action potential either occurs or it does not. This is called an ________________________ response.

An action potential is self-propagating

35. An action potential is a ________________________ that ravels down the length of the axon.

Several factors determine the velocity of an action potential

36. In ummyelinated axons the process of conduction is __________.
37. In myelinated axons the process of conduction is (a) __________, because the action potential skips along the axon from one (b)__________________ to the next. This results in a nerve impulse being conducted (c) ________ (#) times faster than the fastest conductance in an unmyelinated axon.

TRANSMITTING INFORMATION ACROSS SYNAPSES

38. A junction between two neurons, or between a neuron and an effector, is called a ______________.

39. The neuron that terminates at a synapse is known as the (a)_____________________ neuron, and the neuron that begins at that synapse is called the (b) _____________________ neuron.

Signals across synapses can be electrical or chemical

40. In (a) _____________________, the pre- and postsynaptic neurons occur very close together and form gap junctions. Most synapses are (b) _____________________ that involve release of (c) _____________________ that conducts the neural signal across the synapse.

Neurons use neurotransmitters to signal other cells

41. Cells that release (a) _____________________ are called cholinergic neurons. This neurotransmitter is released by some neurons in the (b) ___________ and (c) _____________________.
42. _____________________ neurons release norepinephrine.
43. The three neurotransmitters that are called catecholamines are (a)_____________________, (b)_____________________, and (c)_____________________.
44. Catecholmines and the neurotransmitters (a) _____________________ and (b) _____________________ are called biogenic amines.
45. The neuropeptides (a) _____________________ and (b)_____________________ function mainly as neuromodulators, molecules that stimulate long-term changes.
46. The body's endogenous opioids are (a) _____________________, and (b) _____________________.
47. _____________________is a gas that acts as a retrograde messenger, transmitting information from the postsynaptic to the presynaptic neuron.

Neurotransmitters bind with receptors on postsynaptic cells

48. Neurotransmitters are stored in the synaptic terminals within small membrane-enclosed sacs called (a) _____________________. An increase in intracellular (b) ___________ levels induces the vesicles to fuse with the synaptic membrane and release their contents.
49. Many neurotransmitter receptors are _____________________ channels.

Activated receptors can send excitatory or inhibitory signals

50. A depolarization of the postsynaptic membrane that brings the neuron closer to firing is called an (a) _____________________. A hyperpolarization of the postsynaptic membrane that reduces the probability that the neuron will fire is called an (b)_____________________.

NEURAL INTEGRATION

Postsynaptic potentials are summed over time and space

51. Local responses in the postsynaptic membrane that vary in magnitude, fade over distance, and can be summated are called ______________. (EPSPs and IPSPs are examples.)
52. Graded potentials can be added together in a process called (a)______________. When a second EPSP occurs before the depolarization caused by the first EPSP has decayed, (b)______________ occurs. In (c)______________, several synapses in the same area generate EPSPs simultaneously. Adding EPSPs together can bring the neuron to threshold.

NEURAL CIRCUITS: COMPLEX INFORMATION SIGNALING

53. In a (a) ______________ circuit a single neuron is controlled by synapses with two or more presynaptic neurons. In a (b) ______________ circuit a single presynaptic neuron stimulates may postsynaptic neurons

BUILDING WORDS

Use combinations of prefixes and suffixes to build words for the definitions that follow.

Prefixes	The Meaning	Suffixes	The Meaning
hyper-	above	-glia	glue
inter-	between, among	-neuro(n)	nerve
multi-	many, much, multiple		
neuro-	nerve		
post-	behind, after		
pre-	before, prior to, in advance of		
synap-	a union		

Prefix	Suffix	Definition
________	________	1. A nerve cell that carries impulses from one nerve cell to another, and is between a sense receptor and an effector.
________	________	2. Cells providing support and protection for neurons.
________	-transmitter	3. Substance used by neurons to transmit impulses across a synapse.
________	-polar	4. Pertains to a neuron with more than two (often many) processes or projections.
________	-synaptic	5. Pertains to a neuron that begins after a specific synapse.
________	-synaptic	6. Pertains to a neuron that ends before a specific synapse.

____________	-polarized	7. The state of the membrane as a result of too many potassium ions entering the cell at the end of repolarization.
____________	-se	8. A junction between two nerves or a nerve and an effector.

MATCHING

Terms:

a. Action potential
b. Axon
c. Convergence
d. Dendrite
e. Facilitation
f. Ganglion
g. Integration
h. Myelin sheath
i. Nerve
j. Neuron
k. Refractory period
l. Schwann cell
m. Summation
n. Synapse
o. Threshold level

For each of these definitions, select the correct matching term from the list above.

____ 1. One of many projections from a nerve cell that conducts a nerve impulse toward the cell body.

____ 2. A conducting cell of the nervous system that typically consists of a cell body, dendrites, and an axon.

____ 3. A mass of neuron cell bodies located outside the central nervous system.

____ 4. The long extension of the neuron that transmits nerve impulses away from the cell body.

____ 5. A large bundle of axons wrapped together in connective tissue that are located outside the CNS.

____ 6. The brief period of time that must elapse after the response of a neuron or muscle cell, during which it cannot respond to another stimulus.

____ 7. The electrical activity developed in a muscle or nerve cell during activity.

____ 8. An insulating covering around axons of certain neurons.

____ 9. The junction between two neurons or between a neuron and an effector.

____10. The sorting and interpreting of incoming sensory information and determining a response.

____11. A combination of EPSPs and IPSPs.

____12. The necessary critical value before an action potential is initiated.

MAKING COMPARISONS

Fill in the blanks.

Kind of Potential	mV	Membrane Response	Ion Channel Activity
Membrane potential	-70mV	Stable	Sodium-potassium pump active, passive sodium and potassium channels open
Resting potential	#1	#2	#3
#4	From -50 to -55mV	Depolarization	#5
Action potential	#6	#7	Voltage-activated sodium ion channels open during depolarization and potassium ion channels open during repolarization

Kind of Potential	mV	Membrane Response	Ion Channel Activity
EPSP	Increasingly positive	#8	#9
IPSP	Increasingly negative	#10	Neurotransmitter-receptor combination may open potassium ion channels or chloride ion channels

MAKING CHOICES

Place your answer(s) in the space provided. Some questions may have more than one correct answer.

____ 1. The inner surface of a resting neuron is generally ____________ compared with the outside.
a. positively charged
b. negatively charged
c. polarized
d. 70 mV
e. –70 mV

____ 2. The part of the neuron that transmits an impulse from the cell body to an effector cell is the
a. dendrite.
b. axon.
c. Schwann body.
d. collateral.
e. hillock.

____ 3. Saltatory conduction
a. occurs between nodes of Ranvier.
b. occurs only in the CNS.
c. is more rapid than the continuous type.
d. requires less energy than continuous conduction.
e. involves depolarization at nodes of Ranvier.

____ 4. The nodes of Ranvier are
a. on the cell body.
b. on the dendrites.
c. on the axon.
d. gaps between adjacent Schwann cells.
e. insulated with myelin.

____ 5. A neuron that begins at a synapse is called a
a. presynaptic neuron.
b. postsynaptic neuron.
c. synapsing neuron.
d. neurotransmitter.
e. acetylcholine releaser.

____ 6. Which of the following correctly expresses the movement of ions by the sodium-potassium pump?
a. sodium out, potassium in
b. sodium in, potassium out
c. sodium and potassium both in and out, but in different amounts
d. more sodium out than potassium in
e. less sodium out than potassium in

____ 7. A nerve pathway or tract in the CNS is
a. one neuron.
b. a group of nerves.
c. a bundle of axons.
d. a ganglion.
e. a bundle of cell bodies.

____ 8. An axon cannot transmit an action potential no matter how great a stimulus is applied when it is
a. hyperpolarized.
b. depolarized.
c. in the absolute refractory period.
d. in the relative refractory period.
e. in the resting state.

____9. If a neurotransmitter hyperpolarizes a postsynaptic membrane, the change in potential is referred to as
a. spatial summation.
b. temporal summation.
c. a threshold level impulse.
d. IPSP.
e. EPSP.

____10. Aside from the sodium-potassium pump, the resting potential is due mainly to
a. inward diffusion of chloride ions.
b. inward diffusion of sodium ions.
c. outward diffusion of potassium ions.
d. open sodium channels in the membrane.
e. large protein anions inside the cell.

____11. Branchlike extensions of the cell body involved in receiving stimuli are
a. dendrites.
b. axons.
c. Schwann bodies.
d. collaterals.
e. hillocks.

____ 12. The first thing that needs to happen in the nerve signaling pathway for transmission of a nerve impulse to occur is
a. transmission of the signal.
b. perception.
c. depolarization.
d. integration.
e. reception.

____ 13. __________ are phagocytic glial cells.
a. Oligodendrocytes
b. Ependymal cells
c. Neuroglia
d. Micoglia
e. Astrocytes

____ 14. The word nucleus is used to refer to
a. the part of the eukaryotic cell containing DNA.
b. masses of nerve cell bodies outside the CNS.
c. the site where nerve impulses are integrated.
d. a region of the CNS containing nerve cell bodies.
e. the region of a nerve where dendrites attach.

____ 15. In neurons, potassium ions
a. move by passive diffusion.
b. move through voltage-gated channels.
c. are in greater concentration outside the cell.
d. are pumped into the cell.
e. contribute to the equilibrium potential.

____ 16. A graded potential
a. is transmitted for short distances.
b. varies with stimulus strength.
c. is a localized response to a stimulus.
d. is determined by the resting potential.
e. results in hyperpolarization of a neuron.

____ 17. During depolarization of a neuron
a. voltage-gated sodium channels close.
b. the neuron is in a relative refractory period.
c. the voltage-gated potassium channels are closed.
d. potassium channels open
e. the transmembrane potential becomes more positive.

____ 18. ________ are cells that myelinate axons.
a. Node cells
b. Oligodendrocytes
c. Ependymal cells
d. Schwann cells
e. Astrocytes

____ 19. Chemical synapses
a. are found at gap junctions.
b. are the majority of synapses.
c. are involved in the escape response of animals.
d. occur at clefts about 20 nm wide.
e. are used by many interneurons to communicate.

____ 20. Adrenergic neurons
a. bind acetylcholine.
b. release norepinephrine.
c. release acetylcholine.
d. form electrical synapses.
e. bind norepinephrine.

____ 21. Neuropeptides that act as neurotransmitters include
a. substance P
b. gamma-aminobutyric acid (GABA).
c. norepinephrine.
d. serotonin.
e. dopamine.

____ 22. Neuropeptides
a. act as neuromodulators.
b. inhibit neurons in the brain.
c. inhibit neurons in the spinal cord.
d. are excitatory molecules on skeletal muscles.
e. enhance the action of GABA.

____ 23. Postsynaptic potentials
a. may be summed temporally.
b. may be summed spatially.
c. may involve convergent circuits.
d. may be inhibitory.
e. may occur at a subthreshold level.

____ 24. Which cells guide neurons during embryonic development?
a. oligodendrocytes
b. microglia.
c. No response answers the question.
d. ependymal cells
e. astrocytes

VISUAL FOUNDATIONS

Color the parts of the illustration below as indicated. Also label nodes of Ranvier and Schwann cell.

RED ❑ synaptic terminals
GREEN ❑ terminal branches
YELLOW ❑ myelin sheath
BLUE ❑ axon
ORANGE ❑ cellular sheath
PINK ❑ cell body and dendrite
VIOLET ❑ nucleus of neuron

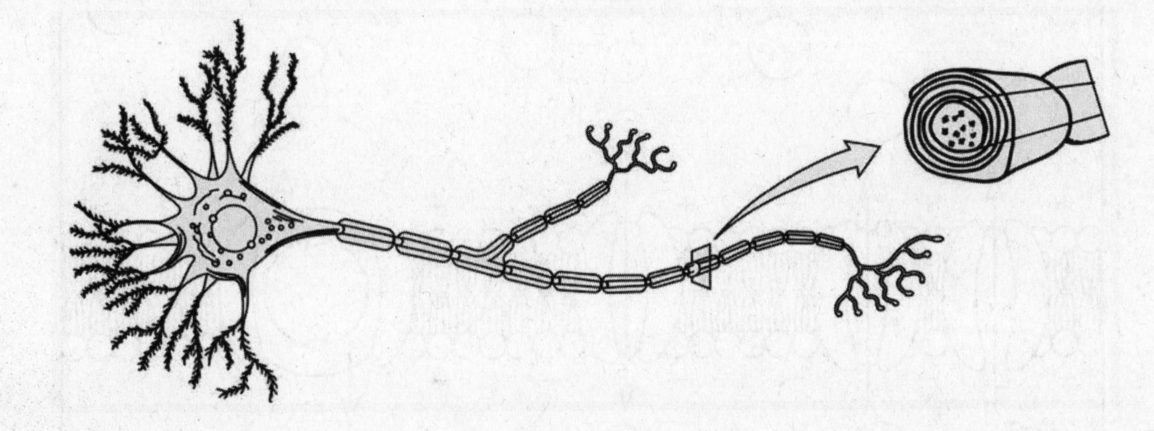

Color the parts of the illustration below as indicated. Also label extracellular fluid and cytoplasm.

RED ❑ sodium ion
GREEN ❑ potassium ion
YELLOW ❑ plasma membrane
BLUE ❑ arrows indicating diffusion into the cell
ORANGE ❑ sodium channel
BROWN ❑ potassium channel
TAN ❑ large anions
PINK ❑ sodium-potassium pump
VIOLET ❑ arrows for diffusion out of the cell

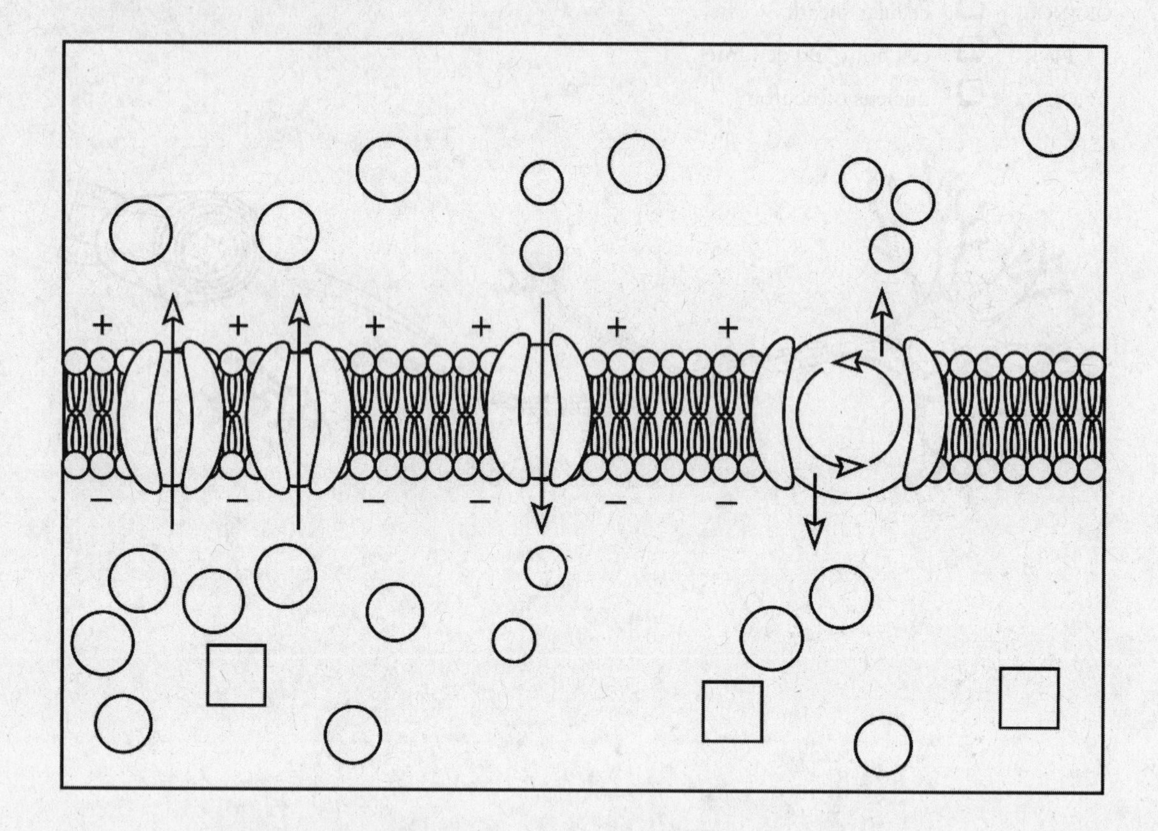

CHAPTER 42

❑

Neural Regulation

This chapter compares several animal nervous systems, and then examines the vertebrate nervous system, with emphasis on the human brain. Also examined are the cellular mechanisms of memory and learning. The simplest organized nervous system is the nerve net found in cnidarians. A nerve net consists of nerve cells scattered throughout the body; there is no central control organ and no definite nervous pathways. The nervous system of echinoderms is more complex, with a nerve ring and nerves that extend into various parts of the body. Bilaterally symmetric animals have more complex nervous systems. The vertebrate nervous system has two main divisions: a central nervous system (CNS) consisting of a complex tubular brain that is continuous with a tubular nerve cord, and a peripheral nervous system (PNS). The CNS provides centralized control, integrating incoming information and determining appropriate responses. The PNS consists of sensory receptors and cranial and spinal nerves. It functions to regulate the organism's internal environment and to help it adjust to its external environment. Learning is a function of the brain, involving the storage of information and its retrieval. Many drugs affect the nervous system, some by changing the levels of neurotransmitters within the brain.

REVIEWING CONCEPTS

Fill in the blanks.

INTRODUCTION

INVERTEBRATE NERVOUS SYSTEMS: TRENDS IN EVOLUTION

1. In general, the lifestyle of an animal is closely related to the design and complexity of its nervous system. For example, the simplest nervous system, called a (a) ______________, is found in (b) ______________, which includes sessile animals such as Hydra. This system of interconnecting neurons has no (c) ____________________________.
2. Echinoderms have a modified nerve net. In this system, a ______________________ surrounds the mouth, and large, radial nerves extend into each arm. Branches from these nerves coordinate the animal's movements.
3. Planarian flatworms have a (a) ________________________ nervous system with nerve cords and a concentrations of nerve cells in the head called (b) __________________; these serve as a primitive brain.
4. Annelids have a solid ventral nerve cord with the cell bodies of many of the neurons massed into (a) ______________. The earthworm system is controlled by a (b) ______________.
5. Some arthropods have cerebral ganglia that have specific ______________________________________.

6. Cephalopod mollusks have complex nervous systems that include well-developed (a) ____________________ organs. Cephalopods are considered to be the most (b) ______________________________ invertebrates.
7. Five major trends can be observed in the evolution of nervous systems. These trends are
(a)____,
(b) ____,
(c) ____,
(d) ____
and, (e) ____.

AN OVERVIEW OF THE VERTEBRATE NERVOUS SYSTEM

8. The human brain contains about (a) ______________________ (#?) neurons. How these neurons are organized is critical in determining an individual's (b) ____________________________.
9. The vertebrate nervous system is divided into a (a)____________________ system consisting of the brain and spinal cord, and a (b) ____________________ system consisting of sensory receptors and nerves.
10. The PNS system is divided into the (a) ________________ division, which is concerned with changes in the external environment, and the (b) ______________ division, which regulates the internal environment. The autonomic division has two efferent pathways made up of (c) ______________ and (d) ______________________ nerves.

EVOLUTION OF THE VERTEBRATE BRAIN

11. The (a) ______________________ of the vertebrate embryo differentiates anteriorly into the (b) ______________ and posteriorly into the (c) ______________________.

The hindbrain develops into the medulla, pons, and cerebellum

12. The (a) ____________________, which coordinates muscle activity, and the (b) ____________, which connects the spinal cord and medulla with upper parts of the brain, are formed from the (c)____________________. The (d) ________________, made up largely of nerve tracks, is formed from the (e)____________________________.
13. The brainstem is made up of the (a) ______________, (b)______________ and (c) ______________.
14. The cerebellum is responsible for (a) ______________________, (b) ________________________, and (c) ____________________.

The midbrain is prominent in fishes and amphibians

15. The midbrain is the most prominent part of the brain in fish and amphibians. It is their main ______________________ area, linking sensory input and motor output.

16. In mammals, the midbrain consists of the (a)______________________ which regulates visual reflexes, and the (b) ______________________, which regulate some auditory reflexes. It also contains the (c) ______________________ that integrates information about muscle tone and posture.

The forebrain gives rise to the thalamus, hypothalamus, and cerebrum

17. The forebrain, or proencephalon, differentiates into the diencephalon and the telencephalon. The diencephalon gives rise to the (a)__________________ and (b) ______________________. The telencephalon gives rise to the (c) ______________________.and in most vertebrate groups to the (d) ______________________.
18. The cerebrum is divided into right and left cerebral (a) ______________ within each lies a (b) ______________________.
19. The cerebrum is made up of an inner layer of (a) ______________ matter which is predominantly comprised of (b) ______________ axons that connect various parts of the brain. The outer layer of (c)______________matter contains cell bodies and dendrites.
20. Some reptiles and all mammals have a type of cerebral cortex called the ______________________ which consists primarily of association areas.
21. The surface area of the human cerebral cortex is increased by folds called (a) ______________________. The furrows between the tissue are called (b)______________________ if they are shallow and (c) ________________ if they are deep.

THE HUMAN CENTRAL NERVOUS SYSTEM

22. The human central nervous system consists of the brain and spinal cord. Both are protected by bone and three meninges, the innermost (a) ______________, the middle (b) ______________________, and the outermost (c) ____________. The CNS is bathed by cerebrospinal fluid produced in a specialized capillary bed called the (d)______________________.

The spinal cord transmits impulses to and from the brain

23. The spinal cord extends from the base of the brain to the ______________________________ vertebra.
24. Grey matter in a cross section of the spinal cord is shaped like the letter "H." It is surrounded by (a) ______________________ that contains nerve pathways, or (b) ______________________. In the center of the cord is the (c) ______________________.
25. The spinal cord also controls certain involuntary responses to stimuli. These responses are called ______________________.

The most prominent part of the human brain is the cerebrum

26. The human cerebral cortex consists of three functional areas. (a)______________ areas receive incoming sensory information, (b)______________ areas control voluntary movement, and (c)______________________ areas link the other two areas.

27. Each hemisphere of the human cerebrum is divided into lobes. The (a) ____________________ lobes contain motor and association areas and are separated from the primary sensory areas in the (b)________________ lobes by a groove called the (c)____________________. The (d) __________________lobes contain centers for hearing and the (e) ________________ lobe contains the visual centers.

Axons in white matter of the cerebrum connect parts of the brain

28. Deep in the white matter of the cerebrum lie the paired ____________________ that are important in coordinating movement.
29. The neurotransmitter (a) ____________________ found in the substantia nigra helps balance (b) ________________ and (c)__________________ of neurons involved in motor functions.
30. In Parkinson's disease, neurons producing the neurotransmitter ____________________ die over time resulting in a decrease in the amount of available neurotransmitter.

The body follows a circadian cycle of sleep and wakefulness

31. The (a) ____________________ is an endocrine gland located in the hypothalamus. It produces (b) ________________ a hormone that plays a role in regulating the sleep-wake cycle.
32. The hypothalamus, thalamus and brain stem are all involved in regulating the ______________________________.
33. The reticular activating system (RAS) is a nerve pathway responsible for maintaining consciousness. It is located within the ____________________________.
34. Electrical activity in the form of brain waves can be detected by a device that produces brain wave tracings called an (a)__________________________. The (b) _______ waves are associated with times of relaxation, (c) ________________ waves have a fast-frequency and are associated with heightened mental activity. (d) ____________ and (e) ____________ are slow waves that are associated with sleep.
35. There are two main stages of sleep: (a)____________________________ sleep, which is associated with dreaming, and (b) __________________ sleep, which is characterized by theta and delta waves. About 20% of sleep time is spent in (c) ____________ sleep.

The limbic system affects emotional aspects of behavior

36. The limbic system affects the emotional aspects of behavior, evaluates rewards, and is important in motivation. It consists of parts of the ____________ as well as parts of the thalamus, and hypothalamus, several nuclei in the midbrain, and the neural pathways that connect these structures.

37. The ______________________ filters incoming sensory information and interprets it in the context of emotional needs and survival.
38. The mesolimbic dopamine pathway is a major ______________ circuit.

Learning and memory involve long-term changes at synapses

39. Memory is the process of (a) _______________, (b) ____________, and (c) ________________ information. (d) ____________________ is unconscious memory for perceptual and motor skills and (e)______________________ involves factual knowledge of people, places, or objects, and requires conscious recall of the information.
40. _______________________________ describes the ability of synaptic connections to change in response to experience.
41. The __________________________ functions in the formation and retrieval of declarative memories.
42. The increases in strength of a synaptic connection as a result of a high rate of action potential transmission is called (a)______________________________________. The low-frequency stimulation of neurons results in a long-lasting decrease in the strength of synaptic connections. This decrease is called (b)____.
43. The establishment of long-term memory involves (a)______________________ and (b) _______________________.

Language involves comprehension and expression

44. Wernicke's area is located in the ______________________________ and is an important center for language comprehension.
45. ____________________ located in the left frontal lobe controls our ability to speak.

THE PERIPHERAL NERVOUS SYSTEM

The somatic division helps the body adjust to the external environment

46. The somatic nervous system consists includes (a)______________________ that detect changes in the external environment. (b) __________________________ transmit information to the CNS, and (c) ____________________________ that adjust the positions of the skeletal muscles that help maintain the body's posture and balance.
47. Neurons are organized into nerves. The 12 pairs of (a) ______________ nerves connect to the brain and are largely involved with sense receptors. Thirty-one pairs of (b) ___________ nerves connect to the spinal cord.

The autonomic division regulates the internal environment

48. The efferent component of the autonomic division is divided into two systems. The (a) ________________ system frequently operates to stimulate organs and to mobilize energy. The

(b)_________________ system influences organs to conserve and restore energy.

49. The autonomic system uses two neurons between the CNS and the effector. The first neuron is called the (a)_________________. Its cell body and dendrites lie within in the CNS its axon synapses with the second neuron called the (b) _________________ whose dendrite synapses with the effector.

EFFECTS OF DRUGS ON THE NERVOUS SYSTEM

50. About _______% of all prescribed drugs are taken to alter psychological conditions.
51. The most commonly abused drug is _________________.
52. Habitual use of almost any mood-altering drug can result in _________________, in which the user becomes emotionally dependent on the drug.
53. Drug _________________ has occurred when a decrease in response to a drug has occurred and greater amounts are required to obtain the same results.
54. The neurobiological mechanisms for drug addiction involve the _________________ pathway.

BUILDING WORDS

Use combinations of prefixes and suffixes to build words for the definitions that follow.

Prefixes	The Meaning	sufixes	The Meaning
cerebr-	brain	-cle	small
hypo-	under, below	-ellum	little
limb-	border	-ic	pertaining to
para-	beside, near	-ory	related to
post-	behind, after	-um	structure
pre-	before, prior to, in advance of		
sens-	perceive		
som	body		
ventr-	belly		

Prefix	Suffix	Definition
__________	-thalamus	1. Part of the brain located below the thalamus; principal integration center for the regulation of the viscera.
__________	-ganglionic	2. Pertains to a neuron located distal to (after) a ganglion.
__________	-ganglionic	3. Pertains to a neuron located proximal to (before) a ganglion.
__________	-vertebral	4. Pertains to structures located beside the vertebral column.

___________ ___________ 5. Nerve that receive impulses about the internal and external environment and transmit them to the CNS.

___________ ___________ 6. A system in the brain that influences the emotional aspects of behavior.

___________ ___________ 7. One of four chambers in the brain where cerebral spinal fluid flows.

___________ ___________ 8. The largest part of the vertebrate brain which is divided into hemispheres and lobes.

___________ ___________ 9. The second largest part of the vertebrate brain and the part responsible for muscle tone, posture and equilibrium.

MATCHING

Terms:

a.	Autonomic nervous system	f.	Gyrus	k.	Pons
b.	Central nervous system	g.	Limbic system	l.	Sensory areas
c.	Cerebellum	h.	Meninges	m.	Somatic nervous system
d.	Cerebral cortex	i.	Parasympathetic nervous system	n.	Temporal
e.	Corpus callosum	j.	Peripheral nervous system	o.	Thalamus

For each of these definitions, select the correct matching term from the list above.

____ 1. An action system of the brain that plays a role in emotional responses.

____ 2. The outer layer of the cerebrum, composed of gray matter and consisting of densely-packed nerve cells.

____ 3. The receptors and nerves that lie outside of the central nervous system.

____ 4. The folds of brain tissue located in the cerebrum.

____ 5. The connective tissue that envelops the brain and spinal cord.

____ 6. The portion of the peripheral nervous system that controls the visceral functions of the body.

____ 7. A large bundle of nerve fibers interconnecting the two cerebral hemispheres.

____ 8. The bulge on the anterior surface of the brainstem between the medulla and midbrain; connects various parts of the brain.

____ 9. The nervous system consisting of the brain and spinal column.

____10. A division of the autonomic nervous system concerned primarily with storing and restoring energy.

____11. A relay center in the brain for motor and sensory messages.

____12. Lobe of the brain located just above the ear.

MAKING COMPARISONS

Fill in the blanks.

Human Brain Structure	Subdivision or Location of Structure	Function of Structure
Medulla	Myelencephalon	Contains vital centers and other reflex centers
#1	Metencephalon	Connects various parts of the brain
#2	Mesencephalon	#3
Thalamus	#4	#5
#6	#7	Controls autonomic functions; links nervous and endocrine systems; controls temperature, appetite, and fluid balance; involved in some emotional and sexual responses
Cerebellum	#8	Responsible for muscle tone, posture, and equilibrium
#9	Telencephalon	Complex association functions
#10	Brain stem and thalamus	Arousal system
#11	Certain structures of the cerebrum, diencephalon	Affect emotional aspects of behavior, motivation, sexual behavior, autonomic responses, and biological rhythms

MAKING CHOICES

Place your answer(s) in the space provided. Some questions may have more than one correct answer.

____ 1. The largest part of the human brain is the
a. cerebrum.
b. cerebellum.
c. myelencephalon.
d. medulla oblongata.
e. midbrain.

____ 2. The part of the human brain that is the center for intellect, memory, and language is the
a. cerebrum.
b. cerebellum.
c. myelencephalon.
d. medulla oblongata.
e. midbrain.

____ 3. Skeletal muscles are controlled by the
a. parietal lobes.
b. frontal lobes.
c. central sulcus.
d. temporal lobes.
e. mesolimbic ganglia.

____ 4. Cerebrospinal fluid (CSF)
a. is produced by choroid plexi.
b. circulates through ventricles.
c. is between the arachnoid and pia mater.
d. is within layers of the meninges.
e. is in the subarachnoid space.

____ 5. In vertebrates, the cerebrum is typically
a. divided into two hemispheres.
b. mostly gray matter.
c. mainly cell bodies and some sensory neurons.
d. mostly white matter.
e. mainly axons connecting parts of the brain.

____ 6. Under ordinary circumstances regulation of body temperature is under the control of the
a. sympathetic n.s.
b. autonomic n.s.
c. forebrain.
d. cranial nerve VI.
e. dorsal root ganglion.

____ 7. When you are eating one of your favorite foods, the cranial nerve(s) directly involved in your perception of this pleasurable experience is/are
a. facial.
b. VII.
c. X.
d. XIX.
e. XI.

____ 8. The regulation of body temperature, appetite, and fluid balance is mainly controlled by the
a. cerebellum.
b. cerebrum.
c. thalamus.
d. hypothalamus.
e. red nucleus.

____ 9. In non-REM sleep, as compared to REM sleep, there is/are
a. higher-amplitude delta waves.
b. faster breathing.
c. lower blood pressure.
d. more dream consciousness.
e. more norepinephrine released.

____10. The brain and spinal cord are wrapped in connective tissue called
a. gray matter.
b. meninges.
c. gyri.
d. sulcus.
e. neopallium.

____11. The part of the mammalian brain that integrates information about posture and muscle tone is the
a. cerebrum.
b. cerebellum.
c. myelencephalon.
d. medulla oblongata.
e. midbrain.

____12. The part(s) of the brain that coordinate(s) muscular activity is/are the
a. cerebrum.
b. cerebellum.
c. myelencephalon.
d. medulla oblongata.
e. midbrain.

____13. If you are having strong sexual feelings one minute and enraged over little or nothing the next minute, chances are very good that you have just had a very active
a. frontal lobe.
b. limbic system.
c. thalamus.
d. reticular activating system (RAS).
e. cerebellum.

____14. If only one temporal lobe is damaged, one might expect
a. blindness in one eye.
b. total blindness.
c. decrease in hearing acuity in both ears.
d. partial loss of hearing in both ears.
e. inability to detect taste.

____15. The limbic system
a. is important in motivation.
b. is involved in biological rhythms.
c. is present in all mammals.
d. affects sexual behavior.
e. plays a role in autonomic responses.

____16. Flatworms have
a. eyespots.
b. bilateral symmetry.
c. nerve cords.
d. cerebral ganglia.
e. a primitive brain.

____17. Nerve nets
a. connect to primitive brains.
b. connect to ganglia.
c. are found in cnidarians.
d. are found in some species of flatworms.
e. coordinate movements in echinoderms.

____18. The forebrain gives rise to the
a. myelencphalon.
b. diencephalon.
c. metaencephalon.
d.mesencephalon.
e. telecephalon.

____19. The medulla
a. contains a space, the third ventricle.
b. contains centers that regulate respiration.
c. contains nuclei that rely information between the cerebrum and cerebellum.
d. consists primarily of nerve tracts.
e. arises from the metancephalon.

____20. The hypothalamus
a. controls body temperature.
b. contains visual reflex centers.
c. regulates appetite.
d. connects to olfactory centers.
e. contains the neocortex.

____21. Terms associated with the cerebrum include
a. corpus callosum.
b. balance and coordination.
c. prominent in amphibians.
d. highly developed association functions.
e. sulci.

____22. Visual centers are located in the
a. insular lobe.
b. parietal lobe.
c. frontal lobe.
d. temporal lobe.
e. occipital lobe.

____23. The left and right hemispheres are connected by the
a. fornix.
b. corpus callosum.
c. substantia nigra.
d. paleocortex.
e. amygdale.

____24. Heightened mental activity is associated with an increase in
a. alpha waves.
b. delta waves.
c. low frequency brain waves.
d. theta waves.
e. beta waves.

____25. Implicit memory
a. is unconscious memory for motor skills.
b. is another name for short term memory.
c. uses dopamine as the neurotransmitter.
d. is another name for declarative memory.
e. involves factual knowledge.

____26. Formulating a meaningful word order for a sentence involves
- a. involves the parietal lobe.
- b. activation of Broca's area.
- c. involves the right hemisphere of the brain.
- d. activation of Wernicke's area.
- e. the left hemisphere of the brain.

VISUAL FOUNDATIONS

Color the parts of the illustration below as indicated. Also label sensory neuron, interneuron, and motor neuron.

RED	❑	muscle
YELLOW	❑	receptor
ORANGE	❑	location of cell bodies of sensory neuron
TAN	❑	central nervous system

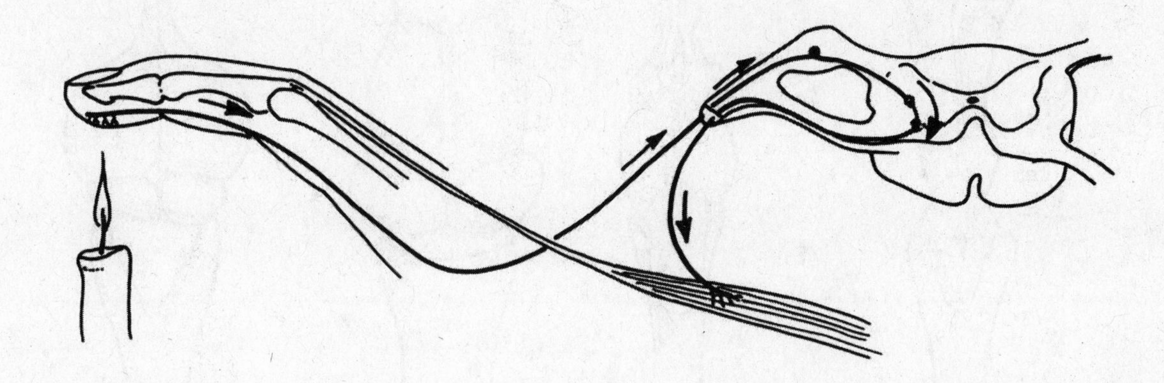

Color the parts of the illustration below as indicated. Also label sulcus. Indicate the vertebrate class represented by each brain illustration.

RED	❑	cerebrum
GREEN	❑	olfactory bulb, tract, and lobe
YELLOW	❑	corpus striatum
BLUE	❑	cerebellum
ORANGE	❑	optic lobe
BROWN	❑	epiphysis
TAN	❑	medulla
VIOLET	❑	diencephalon

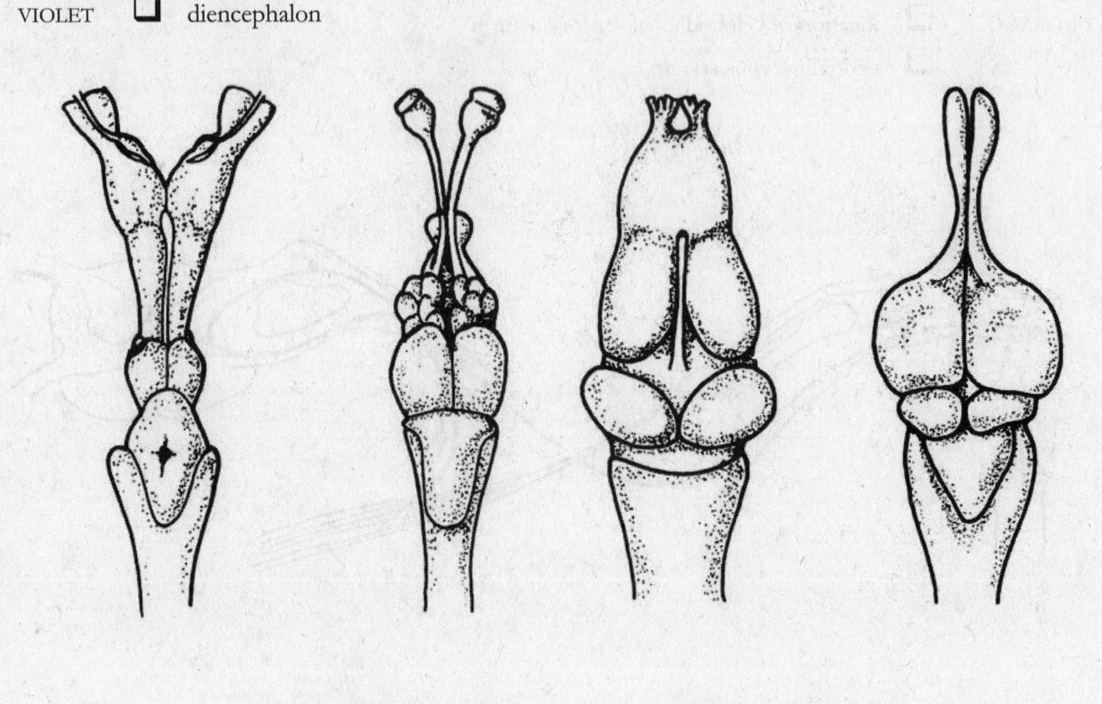

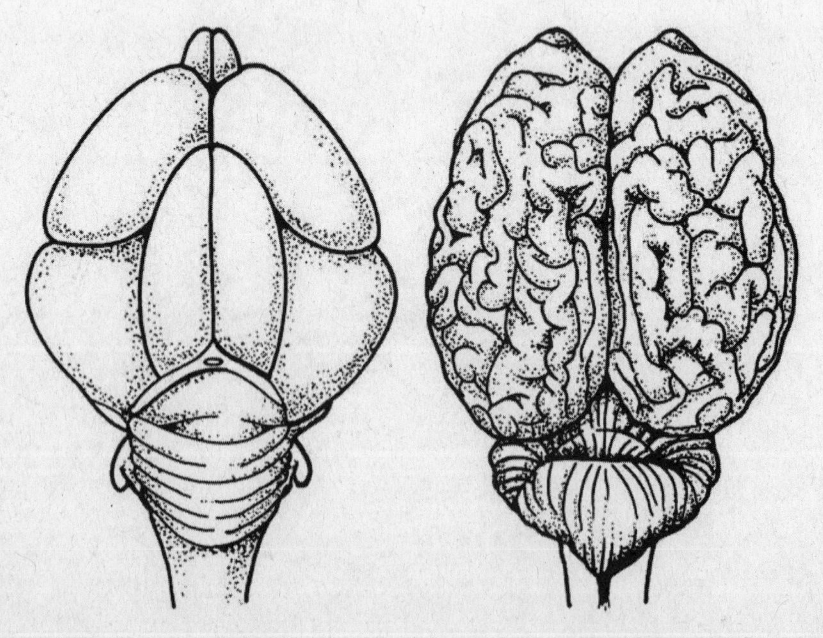

Color the parts of the illustration below as indicated. Also label diencephalon and midbrain.

RED	❑	pituitary
GREEN	❑	thalamus
YELLOW	❑	spinal cord
BLUE	❑	hypothalamus
ORANGE	❑	medulla and pons
BROWN	❑	cerebrum
TAN	❑	cerebellum
PINK	❑	corpus callosum
VIOLET	❑	pineal body

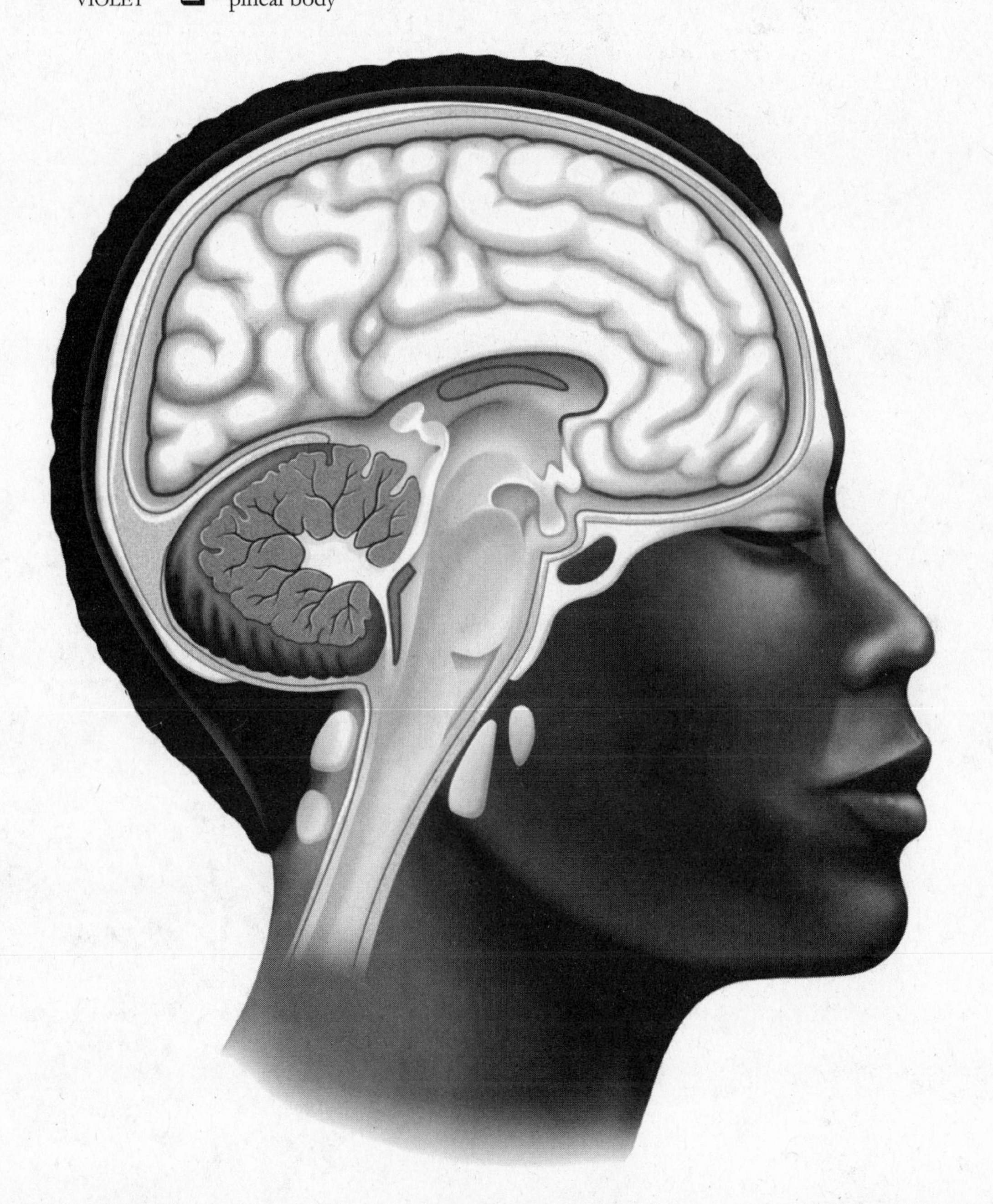

CHAPTER 43

❑

Sensory Systems

Sense organs are specialized structures with receptor cells that detect changes in the internal or external environment and transmit the information to the nervous system. Receptor cells may be neuron endings or specialized cells in contact with neurons. Sensory receptors can be classified according to the location of the stimuli to which they respond or to the type of energy they transduce. They work by absorbing energy and converting it into electrical energy. Depending upon which ion channels are affected, a receptor cell will either depolarize or hyperpolarize. When the state of depolarization reaches a threshold level, an impulse is generated in the axon, thereby transmitting the information to the CNS. Sense organs that respond to heat and cold provide important cues about body temperature, and help some animals locate a warm-blooded host or prey. Those that detect electrical energy are found in some fish where they can detect electric currents in water. Some fish species can generate a shock for defense or to stun prey. Sense organs that detect magnetic fields are used for orientation and migration, and those that detect physical force are responsible for pain reception. Sense organs that respond to mechanical energy include the various tactile receptors in the skin, the lateral line organs in fish, the receptors that respond continuously to tension and movement in muscles and joints, the receptors that respond to body position, and the receptors responsible for hearing. Sense organs that respond to chemical energy are responsible for the sense of taste and smell, and those that respond to light energy are responsible for vision.

REVIEWING CONCEPTS

Fill in the blanks.

INTRODUCTION

1. Dolphins, bats, and a few other vertebrates detect distant objects by ______________________, sometimes called biosonar.

HOW SENSORY SYSTEMS WORK

Sensory receptors receive information

2. In the process of ______________________, sensory receptors absorb a small amount of energy from some stimulus.

Sensory receptors transduce energy

3. The energy to which a receptor is responding must be (a)__________________ into (b) __________________ energy which is the form of information used by the nervous system. The process is called (c) ______________________.
4. A change in the transmembrane potential of a receptor cell is called a (a)______________________ which is a (b)______________________ because the magnitude of the change depends upon the strength of the stimulus.

5. If a receptor potential (a) _______________ the sensory neuron to threshold, it initiates an (b) ________________________ that travels along the (c) ____________ neuron to the central nervous system.

Sensory input is integrated at many levels

6. A decrease in the frequency of action potentials in a sensory neuron, even though the stimulus is maintained at the same intensity, is called ____________________________.
7. Sensation is determined by the type of (a) _____________________ that is activated and the part of (b) _________ that receives the message from the sense organ.
8. Impulses from a sensory receptor may differ in (a)________________________________, (b)________________________________, (c)________________________________, and (d)________________________________.
9. Interpretation of a message and the type of sensation depends on which _______________ receive the message.
10. The process of selecting, interpreting, and organizing sensory information is called (a) _______________________ and it occurs in the (b) _____________.

We can classify sensory receptors based on location of stimuli or on the type of energy they transduce

11. Receptors that detect stimuli in the external environment are called (a)__________________; and (b) ________________ detect changes in such things as pH, osmotic pressure, body temperature, and the chemical composition of the blood.
12. Receptors are classified according to the energy they transduce. (a)____________________________ respond to changes in temperature, (b) ______________________________ respond to magnetic fields, (c) ______________________________ respond to pain, (d) ______________________________ respond to chemical changes and (e) ______________________________ respond to light energy.

THERMORECEPTORS

13. Three types of snakes, the (a) _____________ and (b) __________, and (c) _______________ use thermoreceptors to locate prey.
14. In mammals, free nerve endings in the (a) ____________ and (b)________________detect temperature changes in the external environment.
15. Thermoreceptors that detect internal temperature changes in mammals are found in the ______________________ of the brain.

ELECTRORECEPTORS AND MAGNETIC RECEPTION

16. Electroreceptors are found in predatory species of (a) ___________, (b) ____________________, and (c) ____________________which use them to detect muscle activity in their prey.
17. Electric ells and electric rays use electric organs to ___________ their prey.

NOCICEPTORS

18. Nociceptors (pain receptors) respond to pain from different stimuli. The three types of nociceptors are (a) ____________________, (b) ____________________, and (c) ____________________
19. Nociceptors transmit signals through sensory neurons to interneurons in the (a) ____________________. The sensory neurons release the neurotransmitters (b) ____________________ and (c)____________________.
20. In mammals, certain interneurons release (a) ____________________ and (b) ____________________ that inhibit the release of substance P and block the sensation of pain.

MECHANORECEPTORS

21. Mechanoreceptors are activated when they ____________________ in response to a stimulus.

Tactile receptors are located in the skin

22. The tactile receptors of many animals are located at the base of either a (a) ___________, or a (b) ___________.
23. In mammalian skin, ____________________ forms discs in the deep epidermis that are sensitive to light touch.
24. (a) ____________________ located in the upper dermis are sensitive to light touch and vibration.
    (b)____________________ located in the dermis detect heavy, continuous pressure and stretching of the skin.
    (c)____________________ located in the deep dermis are sensitive to deep pressure that causes rapid movement of tissues, especially vibrations.

Proprioceptors help coordinate muscle movement

25. Vertebrates have three categories of proprioceptors:
    (a)____________________, which detect muscle movement;
    (b)____________________, which respond to tension in contracting muscles and in the tendons that attach muscle to bone; and
    (c) ____________________, which detect movement in ligaments.

Many invertebrates have gravity receptors called statocysts

26. Statocysts are one type of receptor found in invertebrates. A statocyst, which is an infolding of the epidermis, is lined with (a)_______________ that are stimulated when tiny granules called (b) _______________ are pulled upon by gravity.

Hair cells are characterized by stereocilia

27. A ____________________________ is a true cilium with a 9+2 arrangement of microtubules.
28. Sterocilia are not really cilia but are ________________ that contain actin filaments.

Lateral line organs supplement vision in fishes

29. Lateral line organs detect ____________________ in the water.
30. Lateral line organs consist of a (a) __________ lined with (b)______________ that runs the length of the animal's lateral surface. Water disturbances move the (c) ____________ secreted by the hair cells (receptor cells), generating an electrical response.

The vestibular apparatus maintains equilibrium

31. The (a) __________________ and the (b)______________ are gravity detectors in inner ear. They have sensory hair cells covered with a (c)__________________________ in which are embedded (d)_______________ composed of calcium carbonate. This structural arrangement and the detection of gravity are similar to that in (e)______________________.
32. Turning movements, referred to as (a) _______________________, cause movement of a fluid called (b) _______________ in the semicircular canals, which in turn stimulate the hair cells of the (c)____________. There are no (d) _______________ associated with these receptors.

Auditory receptors are located in the cochlea

33. Auditory receptors are located in an inner ear structure of birds and mammals, called the (a) ____________. These receptors are classified as (b) ___________________________ because they contain hair cells that detect pressure waves.
34. In terrestrial vertebrates, sound waves first initiate vibrations in the eardrum, or (a) ______________________. Three tiny ear bones, the (b) ___________, (c) ______________, and (d) ______________ transmit the vibration to fluids in the inner ear through a membrane covered opening called the (e) ____________________.
35. As fluid waves cause movement in the basilar membrane the stereocilia of the (a) __________________________________ rub against the (b) __________________ membrane opening ion channels in the membrane. This results in an action potential being sent along the (c)________________________ to the brain.
36. The _________________________ helps maintain equal pressure between the middle ear and the atmosphere.

37. When detecting sound, pitch is dependent upon the (a)________________ of sound waves and loudness is dependent upon the (b) ________________ of sound waves.

CHEMORECEPTORS

38. Throughout the animal kingdom, two special senses (a)________________ and (b) ________________ allow animals to detect chemical substances in food, water, and air.

Taste receptors detect dissolved food molecules

39. Traditionally four basic tastes have generally recognized. They are (a)________________, (b) ________________, (c)________________, and (d) ________________. A fifth taste known as (e) ________________ has now been identified and another basic taste (f) ________________ has been proposed.

The olfactory epithelium is responsible for the sense of smell

40. Most invertebrates depend on ________________, or the detection of odors, as their main sensory modality.
41. In terrestrial vertebrates, the sense of smell occurs in the ________________.
42. The olfactory cortex is part of the ________________ in the brain.

Many animals communicate with pheromones

43. Pheromones are ________________ that are secreted into the environment.

PHOTORECEPTORS

44. Cephalopod mollusks, arthropods, and vertebrates all have photosensitive pigments called ________________ in their eyes.

Invertebrate photoreceptors include eyespots, simple eyes, and compound eyes

45. Some flatworms have ________________, which are photoreceptive organs capable of differentiating the intensity of light.
46. Effective image formation generally requires a ________________ to concentrate light on groups of photoreceptors.
47. The compound eye in insects and crustaceans consists of individual visual units called ________________, which collectively form a mosaic image.

Vertebrate eyes form sharp images

48. The tough outer coat of the mammalian eye, called the (a)________________, helps maintain the (b) ________________ of the eyeball. The anterior, transparent part of this coat, which allows the entry of light, is called the (c) ________________.
49. The viscous fluid located behind the lens of the eye is called ________________.
50. The ciliary processes secrete ________________ between the cornea and the lens.

51. The amount of light entering the eye is regulated by the __________.

The retina contains light-sensitive rods and cones

52. Rods and cones function as ________________________.
53. The (a) ______________, a specific region of the retina, has the highest density of (b) _________ and therefore is the region of (c)________________.
54. The retina has five main types of neurons. They are (a)____________________, (b) ______________, (c)____________________, (d) ______________, and (e) ________________________________
Light activates rhodopsin

55. Chemically, rhodposin consists of (a) ______________ and (b)__________.
56. ______________ occurs when enzymes restore rhodopsin to a form in which it can respond to light after a person has been out in bright light.

Color vision depends on three types of cones

57. The three types of cones found in primates are (a) __________, (b) ________, and (c) __________. Each cone is so named because that is the wavelength that is most strongly absorbed. All cones respond to a (d) ______________.

Integration of visual information begins in the retina

58. Optic nerves cross in the floor of the (a) ________________ forming an X-shaped structure called the (b) ______________.

BUILDING WORDS

Use combinations of prefixes and suffixes to build words for the definitions that follow.

Prefixes	The Meaning	Suffixes	The Meaning
chemo-	chemical	-chiasm	cross
endo-	within	-cyst	cell
inter(o)-	between, among	-ic	pertaining to
olfac-	smell	-ile	pertaining to
opt-	eye	-lith	stone
oto-	ear	-(t)ory	place for
proprio-	one's own		
stat-	standing still		
tact-	to touch		
thermo-	heat, warm		

Prefix	Suffix	Definition
____________	-ceptor	1. Sense organs in muscles, tendons, and joints that enable the animal to perceive the position of its own body parts.
____________	____________	2. Calcium carbonate "stone" in the inner ear of vertebrates.
____________	-lymph	3. Fluid within the semicircular canals of the vertebrate ear.
____________	-receptor	4. A sense organ or sensory cell that responds to chemical stimuli.
____________	-receptor	5. A sensory receptor that provides information about body temperature.
____________	-ceptor	6. A sensory receptor within (among) body organs that helps maintain homeostasis.
____________	____________	7. Receptors that are sensitive to touch.
____________	____________	8. Simplest organs of equilibrium that sense positional changes relative to gravity.
____________	____________	9. Receptor type involved in detecting odorants.
optic	____________	10. The x-shaped structure formed by the crossing over of axons from the retina.
____________	____________	11. The nerve that .is formed by the union of axons from ganglion cells in the eye.

MATCHING

Terms:

a.	Chemoreceptor	f.	Exteroceptor	k.	Pacinian corpuscle
b.	Cochlea	g.	Fovea	l.	Rhodopsin
c.	Cone	h.	Interoceptor	m.	Statocyst
d.	Cornea	i.	Iris	n.	Thermoreceptor
e.	Electroreceptor	j.	Mechanoreceptor	o.	Tympanic membrane

For each of these definitions, select the correct matching term from the list above.

____ 1. A light-sensitive pigment found in rod cells of the vertebrate eye.
____ 2. The structure of the inner ear of mammals that contains the auditory receptors.
____ 3. A sensory receptor that responds to mechanical energy.
____ 4. The structure of the vertebrate eye that regulates the size of the pupil.
____ 5. Sensory receptor that provides information about body temperature.
____ 6. A mechanoreceptor that is sensitive to deep pressure touches.
____ 7. The "ear drum."
____ 8. The transparent anterior covering of the eye.
____ 9. A sense organ or sensory cell that responds to chemical stimuli.
____10. A conical photoreceptive cell of the retina that is particularly sensitive to bright light, and, by light of various wave lengths, mediates color vision.
____11. Area of the retina with the greatest density of receptor cells.
____12. A gravity receptor found in invertebrates.

MAKING COMPARISONS

Fill in the blanks.

Stimuli	Receptor Classification: Type of Stimuli	Receptor Classification: Location	Example
Light touch	Mechanoreceptor	Exteroceptor	Meissner's corpuscle in skin
Electrical currents in water	#1	#2	Organ in skin of some fish
#3	#4	Exteroceptor	Organ of Corti in birds
Change in internal body temperature	Thermoreceptor	#5	#6
#7	Mechanoreceptor	#8	Muscle spindle in human skeletal muscle
Heat from prey	#9	Exteroceptor	#10
#11	#12	Exteroceptor	Ocelli of flatworms
Food dissolved in saliva	#13	Exteroceptor	#14
Pheromones	#15	#16	Vomeronasal organ in the epithelia of terrestrial vertebrates

MAKING CHOICES

Place your answer(s) in the space provided. Some questions may have more than one correct answer.

____ 1. The process by which sensory receptors change the form of energy of a stimulus to a form of energy that can be transmitted along neurons is called
a. sensory adaptation.
b. conversion.
c. energy transduction.
d. receptor potential.
e. integration.

____ 2. A change in ion distribution that causes a change in the voltage across the membrane of a sensory receptor is a(n)
a. sensory adaptation.
b. conversion.
c. transduction.
d. receptor potential.
e. integration.

____ 3. If the release of a neurotransmitter from a presynaptic terminal decreases during a constant input of action potentials, ____________ is said to have occurred.
a. sensory adaptation
b. conversion
c. transduction
d. receptor potential
e. integration

____ 4. In humans, visual stimuli are interpreted in the
a. brain.
b. photoreceptors.
c. postsynaptic photoganglia.
d. retina.
e. optic nerves.

____ 5. The state of depolarization or hyperpolarization in a receptor neuron that is caused by a stimulus is the
a. depolarization potential.
b. resting potential.
c. receptor potential.
d. action potential.
e. threshold potential.

____ 6. Variations in the quality of sound are recognized by the
a. number of hair cells stimulated.
b. pattern of hair cells stimulated.
c. frequency of nerve impulses.
d. intensity of stimulation.
e. amplitude of response.

____ 7. The membrane at the opening of the inner ear that is in contact with the stapes is the
a. tectorial membrane.
b. basilar membrane.
c. oval window.
d. eardrum.
e. round window.

____ 8. Receptors within muscles, tendons, and joints that perceive position and body orientation are
a. exteroceptors.
b. proprioceptors.
c. interoceptors.
d. mechanoreceptors.
e. electroreceptors.

____ 9. Movement in ligaments is detected by
a. muscle spindles.
b. joint receptors.
c. Golgi tendon organs.
d. proprioceptors.
e. tonic sense organs.

____10. Sense organs that detect changes in pH, osmotic pressure, and temperature within body organs are
a. exteroceptors.
b. proprioceptors.
c. interoceptors.
d. mechanoreceptors.
e. electroreceptors.

____11. Eyespots that detect light but do not form clear images are called
a. simple eyes.
b. ommatidia.
c. facets.
d. ocelli.
e. ciliary bodies.

____12. In the human eye, the "posterior cavity," between the lens and the retina, is filled with
a. aqueous fluid.
b. aqueous humor.
c. rods and cones.
d. vitreous humor.
e. the vitreous body.

____13. Accommodation involves
a. changing the shape of the lens.
b. changing the shape of the fovea.
c. flattening of the retina.
d. ciliary muscle contractions.
e. redistribution of photoreceptors in the retina.

____14. Sensory perception
a. occurs at the level of the receptor.
b. is influenced by past experiences.
c. occurs when light hits the retina.
d. occurs when impulses reach the brain.
e. involves the interpretation of a sensation.

____15. Thermorecptors are important to species survival in
a. ticks.
b. mosquitoes.
c. blood-sucking arthropods.
d. snakes.
e. fleas.

____16. Words associated with pain include
a. nociceptor.
b. dopamine.
c. glutamate.
d. free nerve ending.
e. substance P

____17. __________ are receptors that are sensitive to light touch.
a. Ruffini endings.
b. Pacinian corpuscles.
c. Meissner corpuscles.
d. Merkel cells.
e. hair cells.

____18. A structural feature common to both a statocyst and a utricle is the presence of
a. otoliths.
b. hair cells.
c. motor nerves.
d. calcium carbonate.
e. a gelatinous matrix.

____19. Lateral line organs
a. can be found in amphibians.
b. can be found in reptiles.
c. can be found in lizards.
d. can be found in aquatic organisms.
e. share structural similarities with the cupula of the inner ar.

____20. The organ of Corti
a. is located in the labyrinth.
b. is surrounded by endolymph.
c. is stimulated when receptor cells come in contact with the basilar membrane.
d. rests on the tectorial membrane.
e. is stimulated by mechanical forces.

____21. Cranial nerve I
a. is involved in hearing.
b. is involved in sight.
c. connects to the brain stem.
d. connects to the olfactory bulb.
e. is also known as the olfactory nerve.

VISUAL FOUNDATIONS

Color the parts of the illustrations below as indicated. Also label the locations for endolymph, perilymph, tympanic canal, and vestibular canal.

RED	❑	hair cells	ORANGE	❑	basilar membrane
GREEN	❑	cilia	BROWN	❑	tympanic canal and vestibular canal
YELLOW	❑	cochlear nerve	TAN	❑	bone
BLUE	❑	tectorial membrane	PINK	❑	cochlear duct

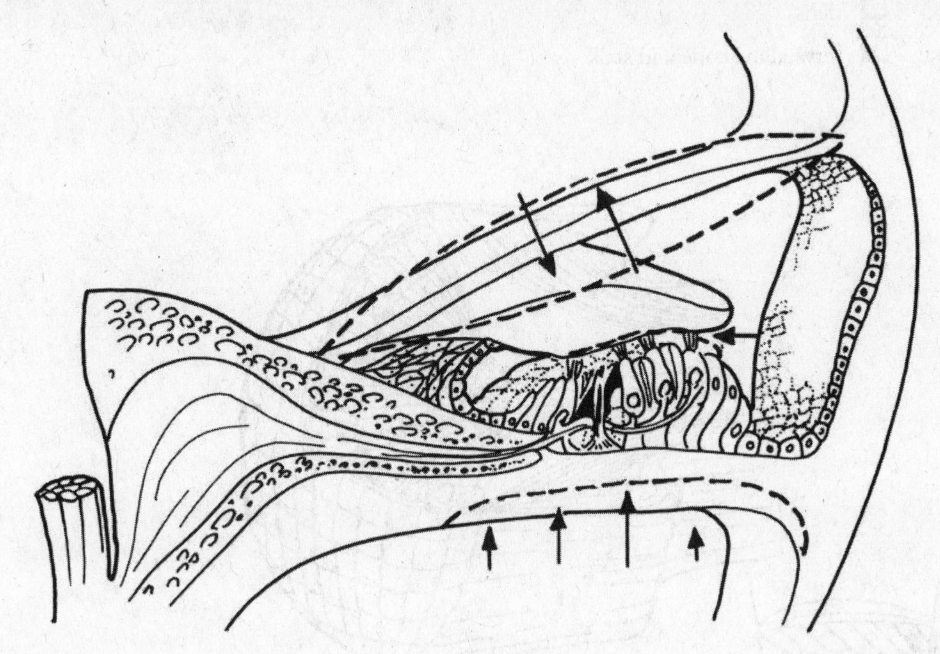

Color the parts of the illustrations below as indicated. Also label the light rays and circle the optic nerve fibers.

RED	❑	rod cell
GREEN	❑	cone cell
ORANGE	❑	bipolar cell
WHITE	❑	ganglion cell
VIOLET	❑	optic nerve fibers
TAN	❑	choroid layer and sclera
PINK	❑	horizontal cell
BLUE	❑	amacrine cell
YELLOW	❑	vitreous body
BROWN	❑	pigmented epithelium

Color the parts of the illustrations below as indicated.

- RED ❑ ommatidia
- GREEN ❑ facets
- YELLOW ❑ optic nerve and ganglion
- BLUE ❑ retinular cell
- ORANGE ❑ rhabdome
- BROWN ❑ iris cell with pigment
- TAN ❑ lens
- PINK ❑ crystalline cone and stalk

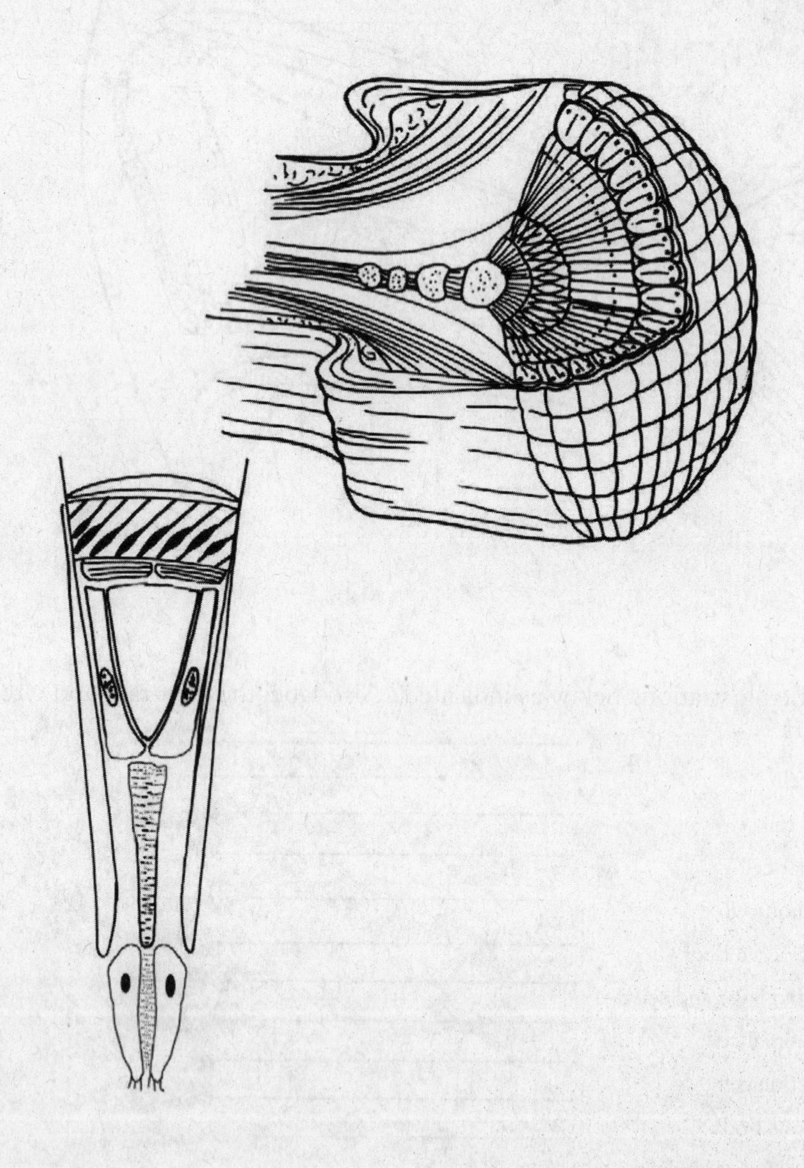

CHAPTER 44

Internal Transport

This chapter discusses how materials get to and from the body cells of animals. Some animals are so small that diffusion alone is effective at transporting materials. Larger animals, however, require a circulatory system. Some invertebrates have an open circulatory system in which blood is pumped from a heart into vessels that have open ends. Blood spills out of the vessels into the body cavity and baths the tissues directly. The blood then passes back into the heart, either directly through openings in the heart (arthropods) or indirectly, passing first through open vessels that lead to the heart (some mollusks). Other animals have a closed circulatory system in which blood flows through a continuous circuit of blood vessels. The walls of the smallest vessels are thin enough to permit exchange of materials between the vessels and the extracellular fluid that baths the tissue cells. A heart can be a muscular organ (vertebrates) or a pulsating vessel (annelids). The muscular heart consists of one or two chambers that receive blood, and one or two that pump blood into the arteries. The lymphatic collects extracellular fluid and returns it to the blood.

REVIEWING CONCEPTS

Fill in the blanks.

INTRODUCTION

1. High-density lipoproteins play a protective role, removing excess ______________ from the blood and tissues.

TYPES OF CIRCULATORY SYSTEMS

2. (a) ______________, (b) ______________, (c) ______________, (d) ______________, (e) ______________ are animal groups that have no specialized circulatory structures.
3. Some animals have a gastrovascular cavity which serves two different organ functions: It serves as a (a) ______________ organ and as a (b) ______________ organ.
4. The movement and distribution of nutrients and gases or even wastes requires a ______________ in which the materials can circulate.
5. Components of a circulatory system include the (a) ________, that is usually pumped by a (b) ________ through a system of (c) ________ or (d) ______________.

Many invertebrates have an open circulatory system

6. An open circulatory system consists of a heart connected to (a)_____________ vessels from which the blood flows into (b)__________.
7. Blood is called (a) ______________________ in animals with open circulatory systems because it has been mixed with (b)________________________ fluid.
8. (a) __________________ and (b) ________________________ have an open circulatory system.
9. Most mollusks have an open circulatory system and a heart that has ________ (#) of chambers.
10. (a) ______________, a blood pigment that imparts a bluish color to hemolymph in some invertebrates, contains the metal (b) _________.
11. In insects, hemolymph mainly distributes (a) ______________ and (b) ________________. Oxygen is supplied to cells through a system of tubules.

Some invertebrates have a closed circulatory system

12. The three groups of inverbetrats having a closed circulatory system are (a) ______________________, (b) ______________________, and (c) ______________________.
13. Annelids have a (a) __________ blood vessel that conducts blood anteriorly, a (b) ___________ blood vessel that conducts blood posteriorly, and, in the anterior part of the worm, (c) _________ (#?) pairs of contractile blood vessels connecting the two.
14. Hemoglobin is found in the _____________ of earthworm blood.

Vertebrates have a closed circulatory system

15. All vertebrates have a _____________ located heart
16. The vertebrate circulatory system primarily transports (a) _________, (b) ______________________, (c) ______________________, and (d) ______________________.

VERTEBRATE BLOOD

17. Human blood consists of fluid, (a) __________________, in which (b) ______________________, (c) ______________________, and (d) ________________ are suspended.

Plasma is the fluid component of blood

18. Plasma is in dynamic equilibrium with (a) __________________ fluid, which bathes cells, and with (b) ______________________ fluid.
19. The major groups of proteins found in plasma are (a) ____________, a protein involved in blood clotting, (b) _________________, some of which are involved in immunity, and (c) _______________.

Red blood cells transport oxygen

20. RBCs are produced in (a) ___________________, contain the respiratory pigment (b) _____________, and live about (c) ________ days.

21. A deficiency in hemoglobin and/or a decrease in red blood cells are two situations that can result in a condition known as (a)________________. Two other causes might be either (b)____________________ or (c)__________________.

White blood cells defend the body against disease organisms

22. White blood cells are also known as (a) ____________________.
23. The two major groups of white blood cells are (a) ________________ and (b) ________________.
24. The largest white blood cells are called (a) __________________. Some of these cells differentiate into (b) __________________ which are phagocytic cells often found in the extracellular space.
25. The three types of granular leukocytes are (a) __________________ (b) ____________________, and (c) ____________________.
26. Two types of agranular leukocytes are (a) __________________ which secrete antibodies, and (b) ____________________.

Platelets function in blood clotting

27. The blood cells responsible for blood clotting in vertebrates other than mammals are called (a) ____________________.
28. In mammals, pieces of cells called (b) ____________________ perform the same function.

VERTEBRATE BLOOD VESSELS

29. (a) ______________ carry blood away from the heart; (b) ______________ return blood to the heart.
30. The exchange of nutrients and waste products takes place across the thin wall of ____________________.
31. Blood pressure is affected by constriction and relaxation of smooth muscle associated with the wall of the ____________________.

EVOLUTION OF THE VERTEBRATE CARDIOVASCULAR SYSTEM

32. Chambers in the heart that pump blood into arteries are called (a)________________, and chambers that receive blood from veins are called (b) ____________.
33. Fish hearts have (a) ________ (#?) atrium(ia) and (b) ________ ventricle(s).
34. The amphibian heart has (a) ________ (#?) chambers, (b)____________ (#?) atrium(ia) and (c) __________________ ventricle(s). Oxygen-poor blood returning from the systemic circulation is pumped into the right atrium by the (d) ____________________.
35. The three groups of animals with a four chambered heart in which the chambers are completely separated are (a) ____________________, (b) ____________________, and (c) ____________________.

36. Having a heart with four completely separated chambers allows birds and mammals to maintain different blood pressures in the systemic and pulmonary circuits. The pressure in the systemic circulation is ________________ (higher/lower) than in the pulmonary circulation.

THE HUMAN HEART

37. The human heart is able to vary the amount of blood it pumps per minute from (a) ______ liters to (b) ______ liters.
38. The human heart is enclosed by a connective tissue sac called the (a)______________________ which forms a (b)_______________ sac around the heart that reduces friction when the heart beats.
39. The (a) ______________________ septum separates the two ventricles, and the (b) ______________________ separates the two atria in a four-chambered heart.
40. (a) ________ (#?) valves control the flow of blood in the human heart. The (b) ______________________________ prevents backflow into the atria during ventricular contractions. The valve on the right is also known as the (c) ______________________, and the valve on the left is known as the (d) ____________________. The (e) ____________________ "guard" the exits from the heart.

Each heartbeat is initiated by a pacemaker

41. The conduction system of the heart contains a pacemaker called the (a) ______________, an (b) ______________________________ which links the atria to the (c) ______________________________, which divides, sending a branch into each ventricle.
42. The portion of the cardiac cycle in which contraction occurs is called (a) ______________, and the portion in which relaxation occurs is called (b) ________________.

The cardiac cycle consists of alternating periods of contraction and relaxation

43. Of the two main heart sounds, the (a) __________ occurs first and is associated with closure of the (b)__________.
44. The (a) ______ sound is produced by the closure of the (b)__________________, which marks the beginning of ventricular diastole.

The nervous system regulates the heart rate

45. The heart beats independently, but the rate can be regulated by the (a)______________ and the (b) ______________ systems.

Stroke volume depends on venous return

46. According to ____, if veins deliver more blood to the heart, the heart pumps more blood.

Cardiac output varies with the body's need

47. (a) ______________________________ is the volume of blood pumped by one ventricle in one minute. It is determined by multiplying the number of ventricular beats per minute times the amount of blood pumped by one ventricle per beat, a value referred to as the (b)__________________.
48. The cardiac output of a resting adult is about ______________________.

BLOOD PRESSURE

49. High blood pressure is called (a) ________________________. It can be caused by an (b) ________ (increase or decrease?) in blood volume such as frequently occurs with a high dietary intake of (c) ________.
50. The most important factor in determining peripheral resistance to blood flow is the ______________________.

Blood pressure varies in different blood vessels

51. Flow rate can be maintained in veins at low pressure because they are ______________________________ vessels.

Blood pressure is carefully regulated

52. ______________________________ are sense organs in the walls of some arteries.
53. Sense organs in arteries detect (a) ____________________ and send the perceived information to (b) ______________________ in the medulla of the brain.

THE PATTERN OF CIRCULATION

54. A double circulatory system consists of a (a) ____________________ connecting the heart and lungs, and (b) ____________________ connecting the heart and body.

The pulmonary circulation oxygenates the blood

55. Pulmonary veins carry oxygen-(a) ________ (rich or poor?) blood to the (b) ________ atrium of the heart.

The systemic circulation delivers blood to the tissues

56. Arteries in the systemic circuit branch from the (a) __________, the largest artery in the body, and serve major body areas. For example, the carotid arteries feed the (b) ________, subclavian arteries supply the (c) ______________, and iliac arteries feed the (d) ________.
57. Veins returning blood from the head and neck empty into the large (a)______________________________ vein, while venous return from the lower body empties into the (b) __________________________ vein.

THE LYMPHATIC SYSTEM

58. The lymphatic system is considered to be an accessory (a)____ in vertebrates.
59. The lymphatic system has three main functions. It collects (a)______________________ to be returned to the blood, it defends the body by initiating (b) ______________________________ and it surrounds the digestive tract where it (c) ________________.

The lymphatic system consists of lymphatic vessels and lymph tissue

60. The fluid contained within lymphatic vessels is called __________.
61. Lymph tissue is organized into masses called ________________.
62. The (a) ____________________, (b) ______________________, and (c) ____________________ are organs that consist mainly of lymph tissue and are consider to be part of the lymphatic system.
63. Lymphatic vessels from all over the body join the circulatory system at the base of the (a) ______________________ by way of ducts: the (b) ___________________________ duct on the left and the (c)______________________ duct on the right.

The lymphatic system plays an important role in fluid homeostasis

64. The tendency for plasma to leave the blood at the arterial end of a capillary and enter the interstitial fluid is referred to as the (a)__________________________pressure. Three forces contribute to this pressure. The main occurs as the result of blood pressure against the capillary wall and is the (b) __________________ pressure. This pressure is augmented by the (c) _________________ pressure of the interstitial fluid. Principle opposing force is the (d) ________________ pressure of the blood.

CARDIOVASCULAR DISEASE

65. (a) _______________________ is a disease in which the walls of affected arteries are damaged, inflamed, and narrowed as a result of (b)_________ deposits.
66. (a) ____________________ is said to occur when tissues lack an adequate blood supply as the result of a decrease in the diameter or blockage of a blood vessel. If this situation occurs in the heart chest pains may occur during exercise. This pain is a characteristic of a condition called (b) ___________________.
67. A clot that forms in a blood vessel is called a _______________.
68. (a) ______________________________ is the insertion of a small balloon into a blocked coronary artery in order to insert a (b)___________ which will be left in the artery to hold it open.

BUILDING WORDS

Use combinations of prefixes and suffixes to build words for the definitions that follow.

Prefixes	The Meaning	Suffixes	The Meaning
baro-	pressure	-ary	pertaining to
capill-	hairy	-cardium	heart
erythro-	red	-coel	cavity
hemo-	blood	-cyte	cell
leuk(o)-	white (without color)	-emia	condition of
neutro-	neutral	-(i)cle	small
peri-	about, around, beyond	-lunar	moon
semi-	half	-phil	loving, friendly, lover
sin-	hollow	-us	thing
vaso-	vessel		
ventr-	belly		

Prefix	Suffix	Definition
________	________	1. The blood cavity that comprises the open circulatory system of arthropods and some mollusks.
________	-cyanin	2. The copper-containing blood pigment in some mollusks and arthropods.
________	________	3. Red blood cell.
________	________	4. General term for all of the body's white blood cells.
________	________	5. The principal phagocytic cell in the blood that has an affinity for neutral dyes.
eosino-	________	6. WBC with granules that have an affinity for eosin.
baso-	________	7. WBC with granules that have an affinity for basic dyes.
________	-emia	8. A form of cancer in which WBCs multiply rapidly within the bone marrow.
________	-constriction	9. Constriction of a blood vessel.
________	-dilation	10. Relaxation of a blood vessel.
________	________	11. The tough connective tissue sac around the heart.
________	________	12. Shaped like a half-moon.
________	-receptor	13. Pressure receptor.
________	________	14. A fluid filled space such as that found in invertebrates which is filled with hemolymph.
________	________	15. The cavity in the mammalian heart from which blood enters the systemic circulation.
________	________	16. The smallest blood vessels in a closed circulatory system.
________	________	17. A cancer caused by the rapid replication of white blood cells in the bone marrow.

MATCHING

Terms:

a. Aorta
b. Arterioles
c. Artery
d. Atrium
e. AV node
f. Blood pressure
g. Interstitial fluid
h. Lymph
i. Lymphatic system
j. Mitral valve
k. Parasympathetic
l. Plasma
m. Pulse
n. SA node
o. Systole
p. Sympathetic
q. Vein

For each of these definitions, select the correct matching term from the list above.

____ 1. Network of fluid-carrying vessels and associated organs that participate in immunity and in the return of tissue fluid to the main circulation.
____ 2. The heart valve located between the left atrium and the left ventricle.
____ 3. The fluid in lymphatic vessels.
____ 4. The fluid portion of the blood consisting of a pale, yellowish fluid.
____ 5. A contracting chamber of the heart that forces blood into the ventricle.
____ 6. The rhythmic expansion and recoil of an artery that may be felt with the finger.
____ 7. The smallest arteries, which carry blood to the capillary beds.
____ 8. Contraction of the heart muscle, especially that of the ventricle, during which the heart pumps blood into arteries.
____ 9. Largest blood vessel in the body through which blood leaves the heart and enters the systemic circulation.
____10. A blood vessel that carries blood from the tissues toward the heart.
____11. Nerves that release acetylcholine to slow the heart rate.
____12. In the electrical conductance system of the heart, it connects to the AV bundle.

MAKING COMPARISONS

Fill in the blanks.

Specific Protein or Cell Type	Blood Component	Function
Fibrinogen	Plasma	Involved in blood clotting
#1	#2	Transport fats, cholesterol
Erythrocytes	#3	#4
Thrombocytes	Cell Component	#5
Monocytes	#6	#7
#8	#9	Help regulate the distribution of fluid between plasma and interstitial fluid
Neutrophils	Cell Component	#10

MAKING CHOICES

Place your answer(s) in the space provided. Some questions may have more than one correct answer.

____ 1. The pericardium surrounds
a. atria only.
b. ventricles only.
c. blood vessels.
d. the heart.
e. clusters of some blood cells.

____ 2. Very small vessels that deliver oxygenated blood to capillaries are called
a. arterioles.
b. venules.
c. arteries.
d. veins.
e. setum vessels.

____ 3. One would expect an ostrich to have a heart most like a
a. lizard.
b. snail.
c. earthworm.
d. human.
e. guppy.

____ 4. Fibrinogen is
a. a plasma lipid.
b. in serum.
c. a protein.
d. a gamma globulin.
e. involved in clotting.

____ 5. Three chambered hearts generally consist of the following number of atria/ventricles:
a. 2/1.
b. 1/2.
c. 1/1 and an accessory chamber.
d. 0/3.
e. 3/0.

____ 6. Blood enters the right atrium from the
a. right ventricle.
b. inferior vena cava.
c. superior vena cava.
d. pulmonary artery.
e. jugular vein.

____ 7. In general, blood circulates through vessels in the following order:
a. veins, capillaries, arteries, arterioles.
b. veins, capillaries, arterioles, arteries.
c. veins, arteries, arterioles, capillaries.
d. arteries, arterioles, capillaries, veins.
e. capillaries, arteries, arterioles, veins.

____ 8. Prothrombin
a. is produced in liver.
b. is a precursor to thrombin.
c. catalyzes conversion of fibrinogen to fibrin.
d. requires vitamin K for its production.
e. is a toxic by-product of metabolism.

____ 9. Erythrocytes are
a. also called RBCs.
b. spherical.
c. produced in bone marrow.
d. one kind of leucocyte.
e. carriers of oxygen.

____10. Monocytes
a. are RBCs.
b. are WBCs.
c. are produced in the spleen.
d. are large cells.
e. can become macrophages.

____11. The blood-filled cavity of an open circulatory system is called the
a. lymphocoel.
b. hemocoel.
c. atracoel.
d. sinus.
e. ostium.

____12. Pulmonary arteries carry blood that is
a. low in oxygen.
b. low in CO_2.
c. high in oxygen.
d. high in CO_2.
e. on its way to the lungs.

____13. The semilunar valve(s) is/are found between a ventricle and
a. an atrium.
b. another ventricle.
c. the aorta.
d. the pulmonary vein.
e. the pulmonary artery.

____14. Phenomena that tend to be associated with hypertension include
a. obesity.
b. increased vascular resistance.
c. increased workload on the heart.
d. decrease in ventricular size.
e. deteriorating heart function.

____15. A heart murmur may result when
a. individuals contract rheumatic fever.
b. systole is too high.
c. a semilunar valve is injured.
d. diastole is too low.
e. diastole is either too low or too high.

____16. Ventricles receive stimuli for contraction directly from the
a. SA node.
b. AV node.
c. Purkinje fibers.
d. atria.
e. intercalated disks.

____17. If a person's heart rate is 65 and one ventricle pumps 75 ml with each contraction, what is that person's cardiac output in liters?
a. 3.575
b. 4.000
c. 4.525
d. 4.875
e. 5.350

____18. An open circulatory system will
a. have a pump to move the blood.
b. have a capillary bed.
c. be found in cephalopods.
d. circulate fluid called hemolymph.
e. have arteries.

____19. Hemocyanin
a. contains copper.
b. is found in earthworms.
c. contains iron.
d. circulated oxygen in flatworms.
e. transports oxygen in arthropods.

____20. A closed circulatory system is found in
a. some arthropods.
b. in annelids.
c. is some reptiles.
d. some mollusks.
e. in echinoderms.

____21. Functions of a closed circulatory system include
a. maintaining fluid balance.
b. transporting carbon dioxide.
c. transporting oxygen to the gills from the tissues.
d. transporting hormones.
e. transporting red blood cells.

____22. Plasma consists of about ____% water and _____% proteins.
a. 95/5
b. 92/7
c. 97/2
d. 85/15
e. 98/1

____23. Serum lacks
a. globulins.
b. hormones.
c. albumins.
d. antibodies.
e. clotting proteins.

____24. The cell group granularcytes includes
a. lymphocytes.
b. basophils.
c. neutrophils.
d. monocytes.
e. eosinophils.

____25. Red blood cell(s)
a. are produced in yellow marrow.
b. last about 60 days.
c. are also called leukocytes.
d. are red because they contain hemocyanin.
e. formation is regulated by erythropoietin.

____26. In a normal individual the greatest number of white blood cells are
a. lymphocytes.
b. neutrophils.
c. basophils.
d. monocytes.
e. eosinophils.

____27. In a normal individual the white blood cells present in the lowest number are
a. lymphocytes.
b. neutrophils.
c. basophils.
d. monocytes.
e. eosinophils.

____28. For a normal blood clot to form
a. vitamin K is required.
b. magnesium ions are required.
c. fibrinogen is converted to soluble fibrin.
d. calcium ions are required.
e. thrombin must be converted to prothrombin.

____29. A closed circulatory system with two completely separated circuits
a. provides a higher pressure to the pulmonary circuit.
b. provides more rapid deliver of oxygen to the tissues.
c. allows birds to have a high metabolic rate.
d. is found in some amphibians.
e. is found in annelids.

____30. Parasympathetic neurons affect heart rate by
a. opening calcium channels.
b. activating a G protein.
c. by activating a protein kinase.
d. releasing norepinephrine.
e. opening potassium channels.

____31. An increase in blood pressure
a. is reversed by the hormone rennin.
b. is reversed by angiotensin II.
c. is reversed by inhibitory nerve impulses sent to sympathetic nerves.
d. causes signals to be sent to centers in the pons.
e. is detected by baroreceptors.

____32. Net filtration will occur in a capillary bed
a. if blood hydrostatic pressure exceeds blood osmotic pressure.
b. if interstitial osmotic pressure exceeds blood osmotic pressure.
c. if blood osmotic pressure exceeds interstitial osmotic pressure.
d. if blood osmotic pressure exceeds blood hydrostatic pressure.
e. None of these conditions would result in net filtration in the capillary bed.

VISUAL FOUNDATIONS

Color the parts of the illustration below as indicated. Also circle the capillary bed.

RED ❑ artery and arrows indicating flow of blood away from the heart

PINK ❑ arteriole

VIOLET ❑ capillaries

PURPLE ❑ venuole

BLUE ❑ vein and arrows indicating flow of blood to the heart

GREEN ❑ lymphatic, lymph capillaries

YELLOW ❑ lymph node

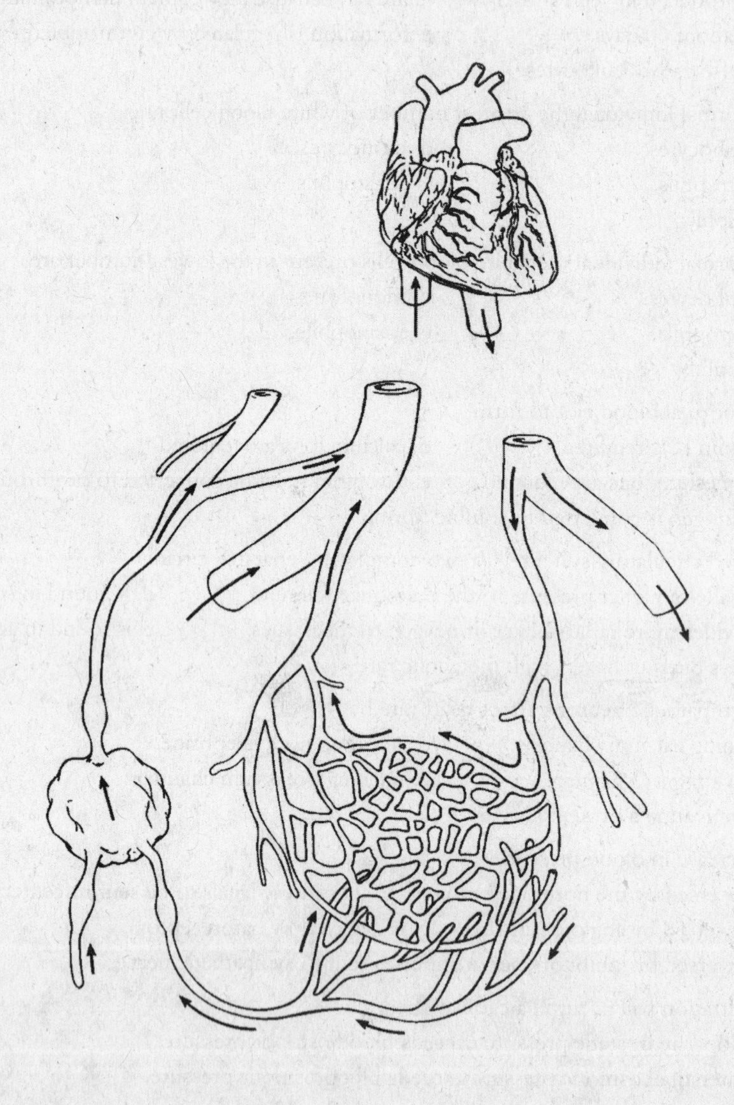

Color the parts of the illustration below as indicated. Label artery, vein, and capillary

RED	❑	smooth muscle
GREEN	❑	outer coat (connective tissue)
YELLOW	❑	endothelium

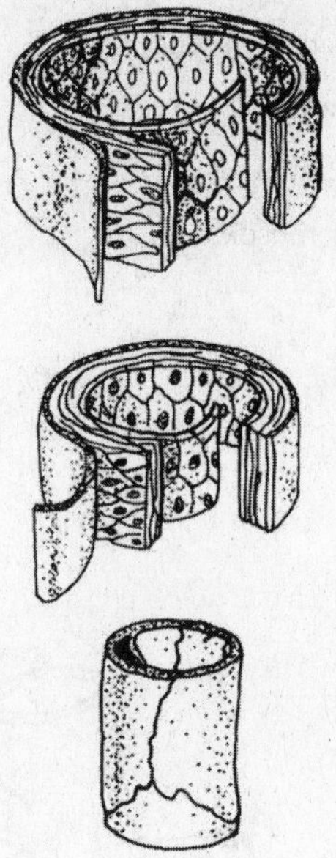

Color the parts of the illustration below as indicated.

RED	❑	aorta and pulmonary artery	ORANGE	❑	valve
GREEN	❑	ventricle	BROWN	❑	partition
YELLOW	❑	atrium	TAN	❑	conus
BLUE	❑	vein from body and pulmonary vein	PINK	❑	sinus venosus

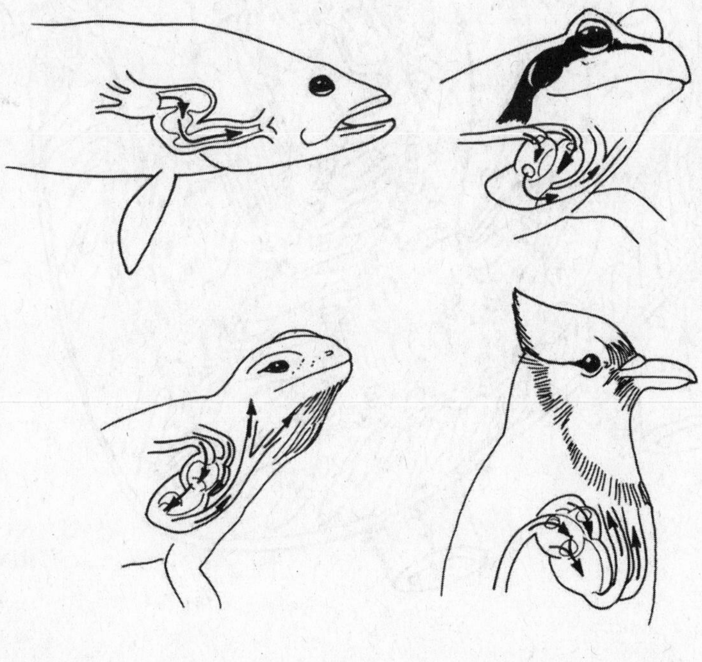

Color the parts of the illustration below as indicated. Also label interventricular septum, superior vena cava, inferior vena cava, pulmonary veins, aorta, and pulmonary artery.

RED	❑	chordae tendineae
GREEN	❑	tricuspid valve
YELLOW	❑	pulmonary semilunar valve
BLUE	❑	mitral valve
ORANGE	❑	aortic semilunar valve
BROWN	❑	papillary muscle
TAN	❑	atrium
PINK	❑	ventricle

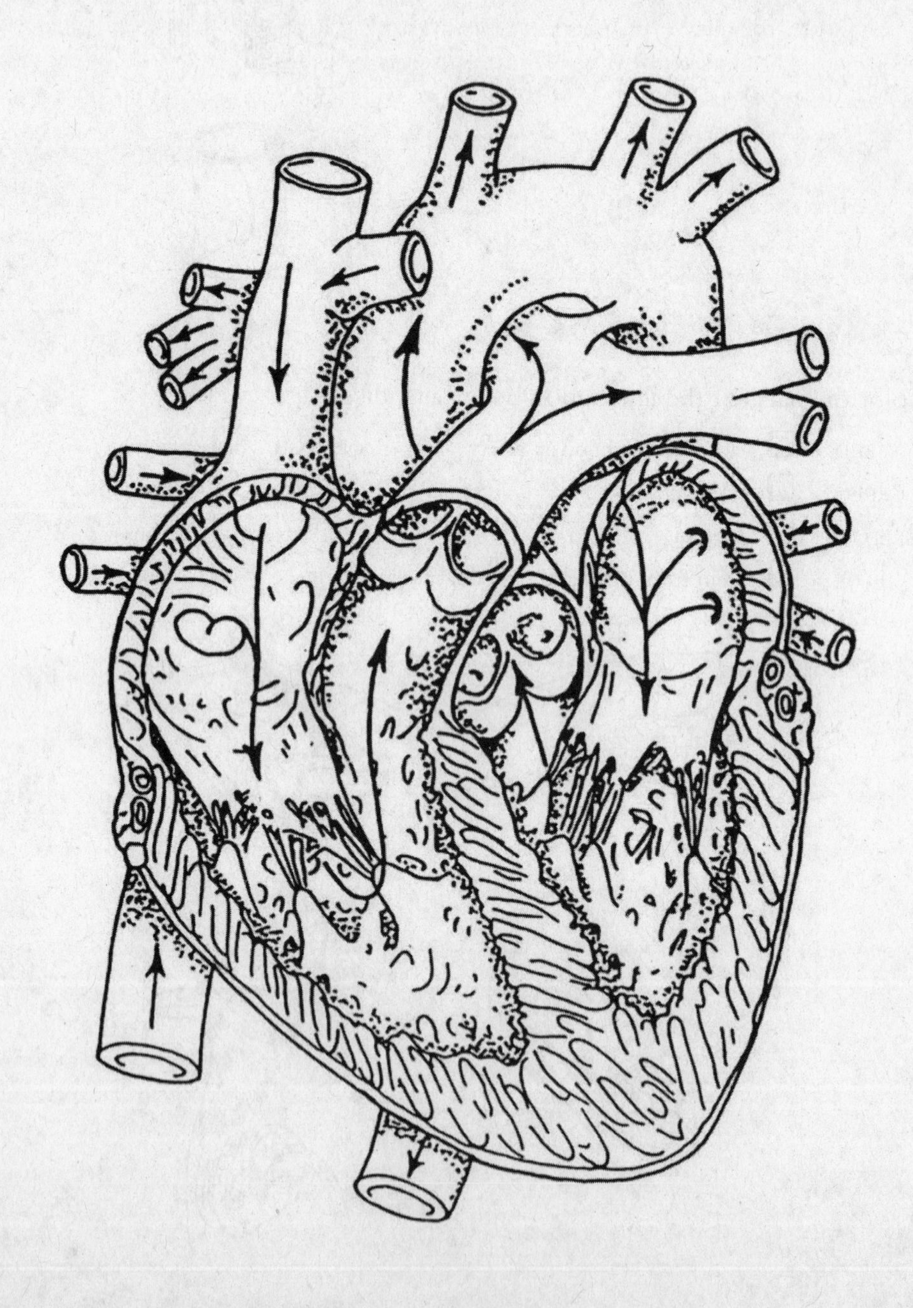

CHAPTER 45

The Immune System: Internal Defense

All animals have the ability to prevent disease-causing microorganisms from entering their body. Because these external defense mechanisms sometimes fail, most animals have developed internal defense mechanisms as well. Before an animal can attack an invader, it must be able to recognize its presence and to distinguish between its own cells and the invader. Whereas most animals are capable of nonspecific responses, such as phagocytosis, it is primarily vertebrates that are capable of specific responses, such as the production of antibodies that target specific pathogens. This type of internal defense is characterized by a more rapid and intense response the second time the organism is exposed to the pathogen. Resistance to many pathogens can be induced artificially by either the injection of the pathogen in a weakened, killed, or otherwise altered state, or the injection of antibodies produced by another person or animal. The immune system is constantly surveying the body for virally-infected cells and tumor cells. It destroys them whenever they arise; failure results in cancer. This same surveillance system is responsible for graft rejection by reading the foreign cells as nonself cells and destroying them. Sometimes the system falsely reads self cells as nonself and destroys them, resulting in an autoimmune disease. Allergies are another example of abnormal immune responses. There are also times that pathogens evade the immune system and cause disease.

REVIEWING CONCEPTS

Fill in the blanks.

INTRODUCTION

1. Disease causing organisms are called ________________________.
2. ________________ is the study of internal defense systems.
3. Immune cells and tissues are scattered throughout the body and it is important that they are able to ____________ one another.

EVOLUTION OF IMMUNE RESPONSES

4. An immune response is a two-step process that involves both ____ and reacting to them.
5. There are the two main types of immune responses (a) ____________, and (b) ____________________.
6. A substance that the immune system specifically recognizes as foreign is called an ________________.
7. ________________________ are highly specific proteins that recognize and bind to specific antigens.

Invertebrates launch nonspecific immune responses

8. The first line of defense is an animal's ________________________.
9. Distinguishing self from nonself relies upon the presence of ________________________ receptors in the plasma membrane.
10. All animals species studied have ________________________, cells that move throughout the body to protect them from foreign matter and bacteria.
11. All animals produce an ________________________ response when they have an infection.
12. Annelids are the only invertebrates known to have ________________________ cells, a type of cell that destroys tumor cells and virally infected cells.
13. ________________________ peptides inactivate or kill pathogens.
14. The primary immune response in invertebrates is a ________________________ response.

Vertebrates launch both innate and adaptive immune responses

15. The lymphatic system that evolved in ________________________ includes lymphocytes and lymph nodes.

INNATE IMMUNE RESPONSES IN VERTEBRATES

Physical barriers prevent most pathogens from entering the body

16. An animal's first line of defense is its (a) ________________________ which, by being physically intact, blocks the entry of (b) ____________.
17. ________________________ are antimicrobial peptides found in saliva.
18. ________________________ is found in tears and it attacks the cell walls of gram-positive bacteria.

Cells of the innate immune system destroy pathogens and produce chemicals

19. Innate immunity depends upon the action of several types of cells including (a) ________________________, (b)________________________, and (c) ________________________.
20. ________________________ are the main phagocytes in the body.
21. ________________________ cells are active against tumor cells and virally infected cells.
22. (a) ________________________ cells are found in all epithelial tissues that come in contact with the environment. They produce (b)________________________ which are antiviral cytokines.

Cytokines are important signaling molecules

23. Cytokines are a large group of (a) ________________________ and (b) ________________________ that are important signaling molecules and they perform regulatory functions.

24. When regulating cellular activity, cytokines can act as either (a)_______________ or (b) _______________ agents. When produced in large amounts, some cytokines act as (c) _______________.
25. _______________ interferons are produced by macrophages or fibroblasts and inhibit viral replication and activate NK cells.
26. _______________ interferons stimulate macrophages to destroy tumor cells and virally infected cells.
27. In response to gram-negative bacteria, other pathogens, and some tumor cells, _______________ stimulates immune cells to initiate an inflammatory response.
28. _______________ are cytokines that regulate interactions between white blood cells such as lymphocytes and macrophages.
29. _______________ are cytokines that act as signaling molecules to attract, activate, and direct the movement of cells of the immune system.

Complement promotes destruction of pathogens and enhances inflammation

30. The action of complement proteins are not specific. They are involved in (a)_______________, (b)_______________, (c)_______________, and (d)_______________.

Inflammation is a protective response

31. The inflammatory response includes three main processes. These are (a) _______________, (b) _______________, and (c) _______________.

ADAPTIVE IMMUNE RESPONSES IN VERTEBRATES

32. Adaptive immunity is also referred to as (a) _______________ immunity. The response is (b) _______________ for distinct macromolecules.

Many types of cells are involved in adaptive immune responses

33. The two main groups of cells that participate in the specific immune response are (a) _______________ and (b)_______________.
34. The two types of lymphocytes that function in specific immunity are (a) __________ which are responsible for cell-mediated immunity and (b) __________ which are responsible for antibody-mediated immunity.
35. All lymphocytes develop from stem cells in the _______________.
36. B cells complete development in the _______________ where they become immunocompetent.
37. Activated B cells form a clone of cells which differentiate into _______________ that produce antibodies.

38. Within the (a) ______________, T cells become (b)__________________________, that is capable of immunological response.
39. Two main types of T cells are: the (a) __________________ T cells, or killer T cells, that recognize and destroy cells with foreign antigens on their surfaces, and the (b) ______________ T cells that secrete cytokines that activate T cells, B cells and macrophages.
40. The three types of antigen-presenting cells are (a) ___________, (b) ______________________ and (c) ______________________. When they are involved in an immune response, antigen presenting cells display (d) ______________________________________ and (e)____.

The major histocompatibility complex is responsible for recognition of self

41. The MHC genes encode for the (a)__________________________, or self antigens, that differ in chemical structure, function, and tissue distribution. These antigens could be considered to be your biochemical (b) ____________________.
42. Class I MHC can be found on the surface of (a)______________________ cells and class II MHC are found in the surface of (b) _________________.

CELL-MEDIATED IMMUNITY

43. T-cells do not recognize antigens unless they are ____.
44. T_C cells act much the same way as _____________ cells which also destroy target cells.

ANTIBODY-MEDIATED IMMUNITY

45. (a) _______ cells stimulate and activate (b) __________ cells which are responsible for antibody-mediated immunity and they are . Stimulation results in the formation of a (c) ____________ of the activated cells.
46. A major difference between B cells and T cells is that B cells do not leave the ________________________.

A typical antibody consists of four polypeptide chains

47. Antibodies are also called (a) ____________________________, abbreviated as (b) _____. These molecules "recognize" specific amino acid sequences on antigens called (c)______________________. The antibody molecule is Y-shaped, the two arms functioning as (d) _______________.

Antibodies are grouped in five classes

48. Antibodies are grouped into five classes, which are abbreviated as (a)________, (b) __________, (c) ___________, (d) __________, and (e) _______.
49. In humans, about 75% of circulating antibodies are (a) _________. (b) ________ predominate in secretions, (c) _______ is an important

immunoglobulin on the B cell surface, and (d) _______ mediates the release of histamine from mast cells.

Antigen-antibody binding activates defenses

50. The antigen-antibody complex may inactivate (a) ____________ or (b) ____________ and it may stimulate (c) ____________.

The immune system responds to millions of different antigens

51. ________________________ is the process whereby a lymphocyte responds to a specific antigen by repeatedly dividing, giving rise to a clone of cells with identical receptors.

Monoclonal antibodies are highly specific

52. Monoclonal antibodies are (a) ____________ antibodies produced by cells from the clone of (b) ____________.
53. Each monoclonal antibody is specific for a single ________________________.
54. B cells can lymphoma cells can be induced to fuse forming a (a)____________. These fused cells can be used to produce (b) ________________________.

Immunological memory is responsible for long-term immunity

55. Memory B cells have a (a) ____________ gene that prevents (b)____________ which is programmed cell death that occurs in plasma cells.
56. The first exposure to an antigen stimulates a (a)____________________ response. The principal antibody synthesized in this initial response is (b) __________.
57. A second exposure to an antigen evokes a (a)____________________response, which is more rapid and more intense than that in the initial response. The principal antibody synthesized in the second exposure is (b) __________.
58. Active immunity can be either (a) ____________________ or (b)____________________ induced.
59. Passive immunity is characterized by the fact that its effects are only ________________________.
60. Passive immunity involves the injection of (a) ____________. As a result, there will be no (b) ________________ cells produced.
61. Babies who are breast-fed receive ________ antibodies in the milk.

RESPONSE TO DISEASE, IMMUNE FAILURES, AND HARMFUL REACTIONS

Cancer cells evade the immune system

62. Cancer cell antigens are classified according to their (a)________________________________ and their (b)____________.
63. ________________ inhibitors slow the development of blood vessels which helps to slow the growth of tumors.

Immunodeficiency disease can be inherited or acquired

64. The leading cause of acquired immunodeficiency in children is ______________________.
65. Lack of protein decreases T cell numbers and the ability to make antibodies. This combination results in an increased risk of ______________________ infections.
66. In DiGeorge syndrome the ______________ is absent or reduced resulting in a deficiency of T cells.

HIV is the major cause of acquired immunodeficiency in adults

67. HIV is transmitted through (a) ______________ and typically enters the body through the epithelial (b) ____________.
68. HIV enters the (a) ____ cell and is (b) ______________ to form DNA which is incorporated into the host genome.
69. Individuals infected with HIV eventually suffer from immunodepression and develop opportunistic infections such as (a)______________________, and cancers such as (b)________________ and (c) ______________.
70. If AIDS infects the nervous system, it can cause AIDS ______________________.

In an autoimmune disease, the body attacks its own tissues

71. ______________________ is the ability to recognize self and is established during lymphocyte development.
72. A lymphocyte that has the potential to be ______________________ may launch an immune response against self tissues.
73. ______________________________ is an autoimmune disease that affects the central nervous system.

Rh incompatibility can result in hypersensitivity

74. Rh incompatibility occurs when individuals produce antibodies against ______________________.
75. Rh incompatibility can cause ______________________________, which destroys red blood cells in a fetus and may cause death

Allergic reactions are directed against ordinary environmental antigens

76. In an allergic response, an allergen stimulates production of ______ type immunoglobulins.
77. Hay fever is also known as ______________________.
78. In allergic asthma, (a) __________ cells release mediators that cause the smooth muscle of the airways to (b)______________________________.
79. When a person develops an allergy to a specific drug, ______________________________________ can occur.

Graft rejection is an immune response against transplanted tissue

80. Graft rejection occurs because the ______________ of the host and graft tissue are not compatible.

BUILDING WORDS

Use combinations of prefixes and suffixes to build words for the definitions that follow.

Prefixes	The Meaning	Suffixes	The Meaning
anti-	against, opposite of	-cyte	cell
auto-	self, same	-gen	origin
comple-	to fill up	-ment	state of being
lyso-	loosening, decomposition		
mono-	alone, single, one		
path	disease		
phag(o)-	to eat		

Prefix	Suffix	Definition
______	-body	1. A specific protein that acts against pathogens and helps destroy them.
______	-histamine	2. A drug that acts against (blocks) the effects of histamine.
lympho-	______	3. A white blood cell strategically positioned in the lymphoid tissue; the main "warrior" in specific immune responses.
______	-clonal	4. An adjective pertaining to a single clone of cells.
______	-immune	5. An adjective pertaining to the situation wherein the body reacts immunologically against its own tissues (against "self").
______	-zyme	6. An enzyme that attacks and degrades the cell walls of gram-positive bacteria.
______	______	7. A cell that engulfs foreign antigens.
______	______	8. A group of proteins that works with other defensive response.
______	______	9. A disease causing organism.

MATCHING

Terms:

a.	Allergen	g.	Immune response	m.	Memory cell
b.	Antigen	h.	Immunoglobulin	n.	Passive
c.	Complement	i.	Innate	o.	Pathogen
d.	Cytokine	j.	Interferon	p.	Plasma cell
e.	Dendritic	k.	Interleukin	q.	T cell
f.	Histamine	l.	Mast cell	r.	Thymus gland

For each of these definitions, select the correct matching term from the list above.

____ 1. Any substance capable of stimulating an immune response; usually a protein or large carbohydrate that is foreign to the body.

____ 2. A gland that functions as part of the lymphatic system.

____ 3. Substance released from mast cells that is involved in allergic and inflammatory reactions.

____ 4. A type of white blood cell responsible for cell-mediated immunity.

____ 5. The process of recognizing foreign and dangerous macromolecules and responding to eliminate them.

____ 6. A protein produced by animal cells when challenged by a virus.

____ 7. Type of lymphocyte that secretes antibodies.
____ 8. An organism capable of producing disease.
____ 9. A type of cell found in connective tissue; contains histamine and is important in allergic reactions.
____10. Temporary immunity derived from the immunoglobulins of another organism.
____11. One type of antigen presenting cells.
____12. Signaling molecule that may act as either an autocrine or a paracrine agent.
____13. Nonspecific immune responses.

MAKING COMPARISONS

Fill in the blanks.

Cells	Nonspecific Immune Response	Specific Immune Response: Antibody-Mediated Immunity	Specific Immune Response: Cell-Mediated Immunity
B-lymphocytes	None	Activated by helper T cells and macrophages; multiply into a clone	None
#1	None	Secrete specific antibodies	None
#2	None	Long-term immunity	None
#3	None	None	Activated by helper T cells and macrophages; multiply into a clone
#4	None	Involved in B cell activation	Involved in T cell activation
#5	None	None	Chemically destroy cancer cells, foreign tissue grafts, and cells infected with viruses
#6	None	None	Long-term immunity
#7	Kill virus-infected cells & tumor cells	Kill virus-infected cells and tumor cells	None
#8	Phagocytic, destroy bacteria	Antigen-presenting cells, activate helper T cells	Antigen-presenting cells; stimulates cloning
#9	Phagocytic, destroy bacteria	None	None

MAKING CHOICES

Place your answer(s) in the space provided. Some questions may have more than one correct answer.

____ 1. T-cell receptors
a. bind antigens.
b. have no known function.
c. are identical on all T cells.
d. are found on killer T cells.
e. stimulate antibody production.

____ 2. Complement
a. is a system of several proteins.
b. is highly antigen-specific.
c. is stimulated into action by an antibody-antigen complex.
d. helps destroy pathogens.
e. is an antibody.

____ 3. The first cells to be involved in a second response to the same antigen are
a. killer T cells.
b. memory cells.
c. plasma cells.
d. macrophages.
e. helper T cells.

____ 4. Active immunity can be artificially induced by
a. transfusions.
b. injecting vaccines.
c. passing maternal antibodies to a fetus.
d. injecting gamma globulin.
e. stimulating macrophage growth.

____ 5. B cells
a. are granular.
b. are lymphocytes.
c. clone after contacting a targeted antigen.
d. are derived from plasma cells.
e. include many antigen-binding forms.

____ 6. Which of the following is true of AIDS?
a. HIV infects helper T cells.
b. AIDS is not spread by casual contact.
c. Both heterosexuals and homosexuals are at risk.
d. HIV can be transmitted by sharing needles.
e. There is currently no cure for AIDS.

____ 7. T cells
a. are lymphocytes.
b. are called LGLs.
c. are also called T lymphocytes.
d. are platelets
e. are involved in cell-mediated immunity.

____ 8. An allergic reaction involves
a. killer T cells.
b. mast cells.
c. interaction between an allergen and mast cells.
d. production of IgE.
e. histocompatibility antigens.

____ 9. Which of the following is/are true of Ig?
a. They are antibodies.
b. They contain a C region.
c. They are produced in response to specific antigens.
d. They contain an antigenic determinant.
e. They are also called immunoglobulins.

____10. Nonspecific immune responses in vertebrates include
a. skin.
b. acid secretions.
c. inflammation.
d. phagocytes.
e. antibody-mediated immunity.

____11. The histocompatibility complex is
a. found in cell nuclei.
b. called HLA in humans.
c. different in each individual.
d. the same in individuals comprising a species.
e. a group of closely linked genes.

____12. Pathogens include
a. viruses.
b. bacteria.
c. fungi.
d. worms.
e. protests.

____13. The most powerful antigens are
a. lipids.
b. polysaccharides.
c. proteins.
d. pathogens.
e. bacteria.

____14. Animals that have nonspecific imuunity include
a. annelids.
b. humans.
c. cnidarians.
d. grasshoppers.
e. mollusks.

____15. The most ancient animals to have differentiated white blood cells are
a. echinoderms.
b. annelids.
c. arthropods.
d. jawed vertebrates.
e. tunicates.

____16. A specific immune response is thought to have evolved first in
a. jawed vertebrates.
b. annelids.
c. mollusks.
d. primates.
e. nematodes.

____17. Macrophages develop from
a. neutrophils.
b. eosinophils.
c. monocytes.
d. basophils.
e. dendritic cells.

____18. NK cells
a. have their origin in bone marrow.
b. will destroy abnormal body cells.
c. use specific processes to destroy target cells.
d. are agranular lymphocytes.
e. use nonspecific processes to destroy target cells.

____19. Dendritic cells can be found in
a. skin.
b. urinary epithelium.
c. digestive epithelium.
d. subcutaneous tissue.
e. the epithelium lining the airways.

____20. Cytokines and complement
a. stimulate cell growth and repair.
b. lyse bacteria and viruses.
c. have been associated with the development of atherosclerosis.
d. stimulate white blood cells.
e. are involved in nonspecific immune responses.

____21. Cell-mediated immunity involves
a. helper T cells.
b. cytokines.
c. the antigen-MHC class II complex.
d. B cells.
e. the antigen-MHC class I complex.

____22. Helper T cells
- a. express the CD_8 molecule on their surface.
- b. are also known as PAMP cells.
- c. secrete cytokines.
- d. can be antigen presenting cells.
- e. activate T_c cells.

____23. Antibodies
- a. inactivate antigens.
- b. consist of four polypeptide chains.
- c. are present in tears.
- d. may stimulate phagocytosis.
- e. serve as the B cell receptor for antigen.

____24. The Rh system
- a. was first identified in monkeys.
- b. results in individuals naturally producing antibodies against the foreign blood type.
- c. identifies Rh negative individuals as being homozygous recessive.
- d. includes antigen D on the surface of red blood cells.
- e. includes over 40 antigens.

VISUAL FOUNDATIONS

Color the parts of the illustration below as indicated. Label variable region and constant region, and circle the antigen-antibody complex.

- RED ❑ antigenic determinants
- GREEN ❑ antigen
- YELLOW ❑ antibody-heavy chain
- BLUE ❑ binding sites
- ORANGE ❑ antibody-light chain

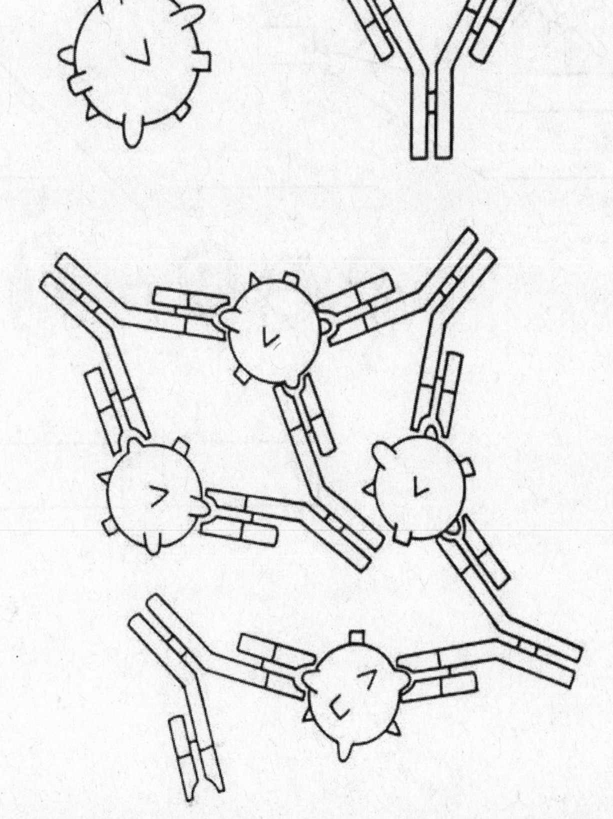

Color the parts of the illustration below as indicated. Label cell-mediated immunity and antibody-mediated immunity. Also label antigen presentation, cooperation, and migration to lymph nodes.

RED	❑	bone marrow	VIOLET	❑	B cell
BROWN	❑	thymus	PINK	❑	plasma cell
YELLOW	❑	T cell	BROWN	❑	natural killer cell
ORANGE	❑	helper T cell	TAN	❑	monocyte, macrophage, and dendritic cell
GREY	❑	cytotoxic T cell			
BLUE	❑	memory cell			

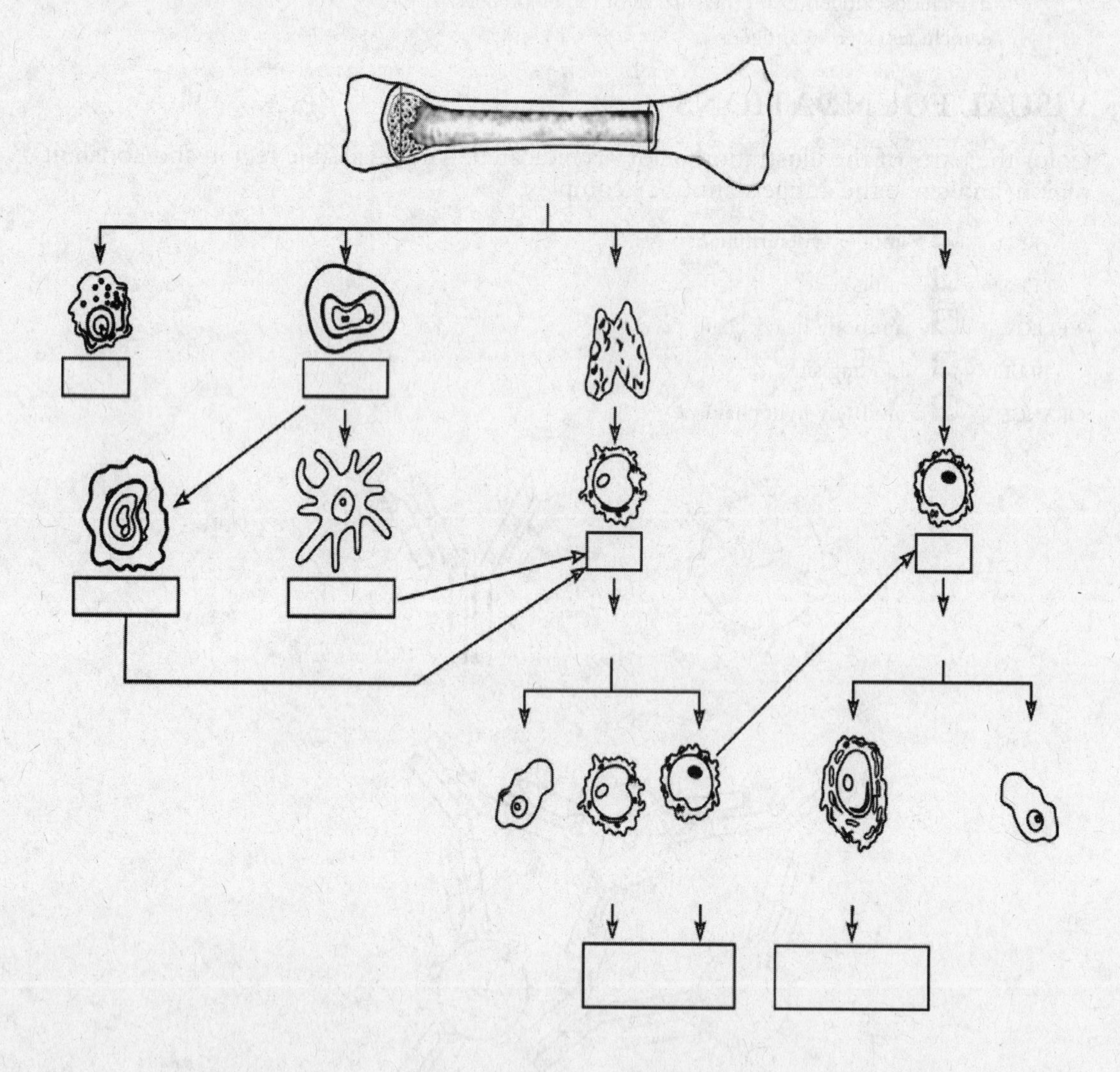

CHAPTER 46

❑

Gas Exchange

In an earlier chapter you studied cellular respiration, the intracellular process that uses oxygen as the final electron acceptor in the mitochondrial electron transport chain and produces carbon dioxide as a waste product. This chapter discusses organismic respiration, that is, how animal cells take up oxygen from the environment in order for it to be made available to the mitochondria, and how the carbon dioxide generated in the mitochondria gets out of the animal body. Oxygen exchange with air is more efficient than with water. This is because air contains a higher molecular concentration of oxygen than water, and oxygen diffuses more rapidly in air than in water. In small, aquatic organisms, gases diffuse directly between the environment and all body cells. In larger, more complex organisms, specialized respiratory structures, such as tracheal tubes, gills, and lungs increase the surface area for gas exchange to occur. Respiratory pigments, such as hemoglobin and hemocyanin, greatly increase respiratory efficiency by increasing the capacity of blood to transport oxygen.

REVIEWING CONCEPTS

Fill in the blanks.

INTRODUCTION

1. The exchange of gases between an organism and the medium in which it lives is called ______________________.
2. The metabolic waste product of respiration is ____________________.

ADAPTATIONS FOR GAS EXCHANGE IN AIR OR WATER

3. Air contains (a) _______ (more or less?) oxygen than water, and oxygen diffuses (b) _______ (faster or slower?) in air than in water.
4. A mammal uses almost _______ (%) less energy to breathe on land than a fish does to breathe in water.

TYPES OF RESPIRATORY SURFACES

5. All respiratory surfaces need to be (a) __________ and have (b)_______________..
6. The four main types of respiratory surfaces are (a) ____________, (b) ____________________, (c) ______________________, and (d) _______________________.
7. Most respiratory systems actively move air or water over their respiratory surfaces, a process known as ________________________.

The body surface may be adapted for gas exchange

8. In aquatic animals, the body surface is kept moist by the (a)____ and in some

terrestrial organisms the respiratory surface is kept moist by
(b)______________________________.

Tracheal tube systems deliver air directly to the cells

9. Arthropods have a respiratory system consisting of a network of (a)_____________________________; air enters the system through small openings called (b) _______________. In large insects (c)_______________________ pump air in and out of these openings.
10. The respiratory network of an arthropod terminates in small fluid-filled _______________________ where gas exchange takes place.

Gills are the respiratory surfaces in many aquatic animals

11. Sea stars have (a) _____________ gills. Gas exchange occurs between the (b) _____________ and the (c) _______________.
12. In bivalves water is moved over the gills by the movement of the (a)__________ and gas exchange also occurs across the surface of the (b) ______________.
13. The gills of bony fish have many ___________________ that extend out into the water and within which are numerous capillaries to exchange gases with the passing water.
14. In bony fish, blood flows through respiratory structures in a direction opposite to water movement, an arrangement called a (a)____.
This system maximizes gas exchange because blood is always coming in contact with water that has a (b) __________________ amount of oxygen. This arrangement results in more than (c) ________% of the available oxygen diffusing into the blood.

Terrestrial vertebrates exchange gases through lungs

15. Lungs are respiratory structures that develop as ingrowths of the (a)____________________________ or from the wall of a (b)____________________________ such as the pharynx.
16. The ____________________ of spiders are within an inpocketing of the abdominal wall.
17. Modern bony fishes almost all have a homologous structure, a ____________________ that allows them to control their buoyancy.
18. Most amphibian gas exchange occurs _______________________.
19. In reptiles, gas exchange is not efficient enough to sustain ____.
20. Birds have the most efficient ________________________ of any living vertebrate group.
21. In birds, air flows in (a) ____________________ and is renewed during (b)__________ cycles of ventilation. The air flows first to the (c)_________________________ and from there is flows into (d)______________________. From here the air flows into the (e)_______________________ and from there out of the body.

22. In birds the blood flows at right angles to the parabronchi where gas exchange occurs. This arrangement of airflow and blood flow is describes as ____________________________.

THE MAMMALIAN RESPIRATORY SYSTEM

The airway conducts air into the lungs

23. The nasal cavities are lined with a ______________________________ epithelium rich in blood vessels.
24. Air flows from the nasal cavity into the (a) ____________. From here air passes into the (b) ______________ and then the (c) ____________.
25. Food and water are prevented from entering the larynx by the ____________________.

Gas exchange occurs in the alveoli of the lungs

26. The lungs are located in the (a) _____________ cavity. The right lung has (b) ____ (#?) lobes, and the left lung has (c) ___ (#?) lobes.
27. Each lung is covered by a (a) ________________________ and enclosed in a fluid filled (b) ____________________________. Gas exchange acurrs across the surface of the (c) ____________________.

Ventilation is accomplished by breathing

28. During inspiration, the volume of the thoracic cavity (a) ___________ as a result of the diaphragm (b) __________________. During exhalation, the volume (c) __________________ as a result of the diaphragm (d) _____________________.
29. Forced inhalation also involves contraction of the (a)______________________________________ muscles which has the effect of (b) ____________________ the volume of the thoracic cavity.
30. The work of breathing is reduced by ____________________________ secreted by specialized epithelial cells lining the surface of the alveoli.

The quantity of respired air can be measured

31. (a) ____________________ is the volume of air that is inhaled and exhaled with each normal resting breath. It averages about (b)________ml. The volume of air remaining in the lungs after maximal expiration is the (c) ___________________________ and the maximum amount of air that can be inhaled following a maximum exhalation is the (d) _______________________.

Gas exchange takes place in the alveoli

32. The factor that determines the direction and rate of diffusion of a gas across a respiratory surface is the ____________________ of that gas.
33. ______________________________________ states that the total pressure in a mixture of gases is the sum of the pressures of the individual gases.
34. Fick's law of diffusion says the amount of oxygen or carbon dioxide that diffuses across the alveolar membrane depends on the difference in (a) ____________________ of the gases on the two sides of the

membrane and on the amount of (b) ____________________ of the membrane.

Gas exchange takes place in the tissues

35. The partial pressure of oxygen in arterial blood is about (a)____________ compared to about (b) ____________ in the tissues. Therefore, oxygen diffuses (c) ____________________.

Respiratory pigments increase capacity for oxygen transport

36. ________________________ are copper-containing respiratory pigments found in the hemolymph of mollusks and arthropods.
37. The most common respiratory pigments found in animals are (a)____________________ and (b) ____________________.
38. Hemoglobin is a general term for a group of compounds which consist of an (a) ____________________ or heme which is bound to a protein called (b) ________________.
39. Oxygen combines with the element (a) ________ in the (b) ________ group of hemoglobin. This "association" can be illustrated as Hb + O2 $\rightarrow HbO_2$ known as (c) ____________________. The result is a compound that carries oxygen and releases it where it is in lower concentrations.
40. HbO_2 is prone to dissociate in a/an (a) ____________(acidic or basic) environment. A change in the normal HbO_2 dissociation curve caused by a change in pH is known as the (b) ____________.

Carbon dioxide is transported mainly as bicarbonate ions

41. Carbon dioxide is transported in the blood in three forms. 10% is transported as (a) ____________________, 30% is (b) ____________________ and 60% is transported as (c) ____________________, the formation of which is catalyzed by an enzyme in RBCs called (d) ____________________.
42. Any condition that interferes with the removal of carbon dioxide by the lungs can lead to ____________________ because carbon dioxide is allowed to build up in the blood.

Breathing is regulated by respiratory centers in the brain

43. Respiratory centers are located in the (a) ____________ and the neurons in the (b) ____________regulate the rhythm of ventilation. Respiratory centers in the (c) ____________ regulate the transition from inspiration to exhalation.
44. Changes in arterial carbon dioxide levels are sensed by chemoreceptors in the (a) ____________________, (b) ____________________, and (c) ____________________. When stimulated they send signals that increase (d) ____________________.
45. A decrease in pH leads to (a) ____________ (faster or slower) breathing and an increase in pH leads to (b) ____________ (faster or slower) breathing.
46. Changes in ____________ (carbon dioxide, pH, or oxygen) have the least affect on breathing.

47. Shallow breathing lead to ___________________, a deficiency in oxygen.
48. Irreversible brain damage occurs within about ______ minutes of the start of oxygen deprivation.

Hyperventilation reduces carbon dioxide concentration

49. A certain concentration of carbon dioxide is needed in the blood to maintain normal ________________________________.

High flying or deep diving can disrupt homeostasis

50. Regardless of the altitude the concentration of oxygen in the air remains at (a) ____%. However the (b) ____________________ of oxygen does change with altitude.
51. A sudden decrease in environmental pressure may cause the release of dissolved gases in the blood in the form of bubbles that block capillaries, causing a very painful syndrome which is commonly called "the bends," and in the scientific community is called ____.

Some mammals are adapted for diving

52. The diving reflex is present to some extent in humans. Several physiological changes occur that can lead to the survival of cold water near drowning victims. These changes are:
(a)____,
(b)____,
(c)____, and
(d)____.

BREATHING POLLUTED AIR

53. When dirty air is breathed into the lungs the bronchial tubes narrow as part of a response called ______________________________.

BUILDING WORDS

Use combinations of prefixes and suffixes to build words for the definitions that follow.

Prefixes	The Meaning	Suffixes	The Meaning
alveol-	small cavity	-al	pertaining to
derm-	skin	-(a)tion	the process of
dia-	across	-cle	little
diffuse-	to pour out	-ion	the process of
hyper-	over	-ox(ia)	containing oxygen
hyp(o)-	under	-phragm	fence
opercula	a lid	-um	structure
ox(y)-	containing oxygen	-us	thing
respir-	to breathe		
spira-	to breathe		
ventila-	to fan		

Prefix	Suffix	Definition
________	-ventilation	1. Excessive rapid and deep breathing; a series of deep inhalations and exhalations.
________	________	2. Oxygen deficiency.
________	________	3. The process whereby animals actively move air or water over their respiratory surfaces.
________	-hemoglobin	4. A complex of oxygen and hemoglobin.
________	________	5. The combined process of inhalation and exhalation.
________	________	6. The movement of gas from an area of higher to lower concentration.
________	________	7. The type of gills found in echinoderms.
________	________	8. An individual air sac in the lung.
________	________	9. An individual opening leading into the tracheal tubes of insects.
________	________	10. An external bony plate that covers the gills of bony fishes.
________	________	11. Dome-shaped muscle that forms the floor of the thoracic cavity.

MATCHING

Terms:

a. Alveolus
b. Bronchioles
c. Bronchus
d. Diaphragm
e. Epiglottis
f. Gill
g. Larynx
h. Lung
i. Operculum
j. Pharynx
k. Pleural cavity
l. Pleural membrane
m. Respiratory center
n. Trachea

For each of these definitions, select the correct matching term from the list above.

____ 1. "Windpipe."
____ 2. An air sac of the lung through which gas exchange with the blood takes place.
____ 3. A type of respiratory organ of aquatic animals.
____ 4. The membrane that lines the thoracic cavity and envelopes the lungs.
____ 5. One of the branches of the trachea and its immediate branches within the lung.
____ 6. The throat region in humans.
____ 7. Tiny air ducts of the lung that branch to form the alveoli.
____ 8. .Respiratory organ of air breathers.
____ 9. .Contains an individual lung.
____10. The organ at the upper end of the trachea that contains the vocal cords.
____11. Closes to prevent food and water from entering the larynx.
____12. Located in the medulla.

MAKING COMPARISONS

Fill in the blanks.

Organism	Type of Gas Exchange Surface
Amphibians	Body surface and lungs
Fish	#1
Spiders	#2
Reptiles	#3
Birds	#4
Nudibranch mollusks	#5
Insects	#6
Sea stars	#7
Clams	#8
Mammals	#9

MAKING CHOICES

Place your answer(s) in the space provided. Some questions may have more than one correct answer.

____ 1. A cockroach obtains oxygen for tissues located deep in its body by means of
a. book gills.
b. pseudolungs.
c. circulation of hemolymph.
d. a countercurrent exchange system.
e. branching tracheal tubes.

____ 2. Air passes through structures of a mammal's respiratory system in the following order:
a. trachea, pharynx, bronchi, bronchioles.
b. larynx, trachea, bronchioles, bronchi.
c. pharynx, larynx, bronchi, bronchioles.
d. larynx, trachea, bronchi, bronchioles.
e. trachea, larynx, bronchi, bronchioles.

____3. The ability of oxygen to be released from oxyhemoglobin is affected by
a. temperature.
b. pH.
c. concentration of carbon dioxide.
d. oxygen concentration in tissues.
e. ambient oxygen concentration.

____ 4. If a patient is found to have inelastic air sacs, trouble expiring air, an enlarged right ventricle, obstructed air flow, and large alveoli, he/she probably
a. has lung cancer.
b. has emphysema.
c. has chronic obstructive pulmonary disease.
d. lives at a high altitude.
e. experiences chronic bronchitis.

____ 5. If you are SCUBA diving for a lengthy time at 200 feet, you might get the bends if
a. you surface quickly.
b. you are using a helium mixture.
c. nitrogen in your blood is rapidly absorbed by your tissues.
d. you come to the surface very slowly.
e. you stay at 200 feet even longer.

____ 6. The diaphragm and rib or intercostal muscles alternately contract and relax, these actions result in
a. inspiration and expiration.
b. inhalation and exhalation.
c. expiration and inspiration.
d. exhalation and inhalation.
e. oxygen intake and carbon dioxide output.

____ 7. Characteristics that all respiratory surfaces share in common include
a. moist surfaces.
b. thin walls.
c. specialized structures such as tracheal tubes, gills, or lungs.
d. large surface to volume ratio.
e. alveoli or spiracles.

____ 8. If the PO_2 in the tissue of your pet dog is 10, atmospheric PO_2 is 150 mm Hg, and arterial PO_2 is 110 mm Hg, you might expect the dog to
a. function normally.
b. accumulate carbon dioxide.
c. have a serious, but not lethal, oxygen deficit.
d. die.
e. become dizzy from too much oxygen.

____ 9. Vital capacity
a. is about 4500 ml in humans.
b. includes the tidal volume.
c. is one measure of the functional capacity of lungs.
d. is the maximum amount of air to be inhaled.
e. is about 500 ml in humans.

____ 10. Most of the mucus produced by epithelial cells in the nasal cavities is disposed of by means of
a. a large handkerchief.
b. reabsorption.
c. evaporation.
d. absorption in surrounding lymph ducts.
e. swallowing.

____ 11. Gas exchange in air
a. requires lungs.
b. requires a moist surface.
c. is easier because there is a higher molecular concentration of oxygen.
d. uses more energy than in water.
e. can lead to desiccation.

____ 12. Ventilation
a. is an active process.
b. occurs only in animals with lungs.
c. is the exchange of gases between the blood and the cells.
d. in sponges utilizes flagella.
e. occurs only in land dwelling animals.

____ 13. Words or phrases that can be associated with tracheal tubes include
a. tracheole.
b. respiratory system.
c. spider.
d. grasshopper.
e. spiracle

____ 14. Gills can be found in animals that are
a. mollusks.
b. chordates.
c. crustaceans.
d. vertebrates.
e. annelids.

____15. A countercurrent exchange system
a. allows blood high in oxygen to come in contact with water high in oxygen.
b. allows blood high in oxygen to come in contact with water that is low in oxygen.
c. allows blood low in oxygen to come in contact with water that is high in oxygen.
d. allows blood low in oxygen to come in contact with water that is low in oxygen.
e. occurs in birds.

____16. In amphibians
a. most gas exchange occurs across the body surface.
b. the lungs are long, simple sacs.
c. the lungs are better developed than in reptiles.
d. the lungs have a spongy texture.
e. the lungs have filaments.

____17. The bird respiratory system
a. connects to air spaces in some bones.
b. is less efficient than reptiles.
c. is similar to the human system.
d. has a crosscurrent arrangement.
e. ends in tiny, thin-walled parabronchi open at both ends.

____18. In the human respiratory system
a. inhaled air contains about 15% oxygen.
b. exhaled air contains about 50 times more carbon dioxide than inhaled air.
c. the concentration of carbon dioxide is lower in the cells than the blood.
d. pressure of a gas determines its movement.
e. the majority of carbon dioxide is transported bound to hemoglobin.

____19. An increase in ventilation rate
a. would lead to respiratory alkalosis.
b. can be triggered by receptors in the jugular vein.
c. would decrease the amount of oxygen bound to hemoglobin.
d. would lead to respiratory acidosis.
e. would increase mitochondrial activity.

____20. If a person become submerged in very cold water
a. metabolic rate increases.
b. oxygen is shunted to internal organs.
c. oxygen demand goes up.
d. blood pressure increases.
e. heart rate increases.

____21. Sponges and flatworms
a. exchange gases by diffusion.
b. have no specialized respiratory structures.
c. exchange gases by diffusion so are less than 1 mm thick..
d. use active transport for gas diffusion.
e. rely on body cells being in direct contact with the environment for gas exchange..

____22. ________ are a respiratory surface that is rich in blood vessels.
a. tracheal tubes
b. bronchi.
c. alveoli.
d. tracheae.
e. dermal gills.

____23. Terms that are associated with gills include
a. operculum.
b. cilia.
c. trapping food.
d. filament.
e. countercurrent.

____24. The amount of oxygen delivered to the tissues is affected by
a. Fick's law of diffusion.
b. the amount of atmospheric carbon dioxide.
c. the amount of lactic acid in the tissues.
d. the Bohr effect.
e. Dalton's law of partial pressures.

VISUAL FOUNDATIONS

Color the parts of the illustration below as indicated.

RED	❑	lungs
GREEN	❑	gills
YELLOW	❑	body surface used for gas exchange
BLUE	❑	tracheal tube
ORANGE	❑	book lung

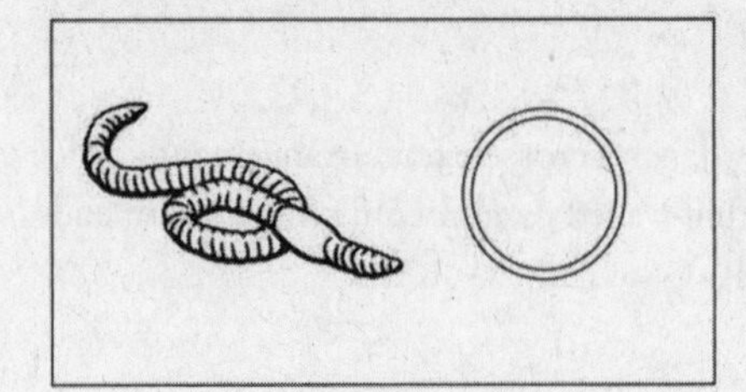

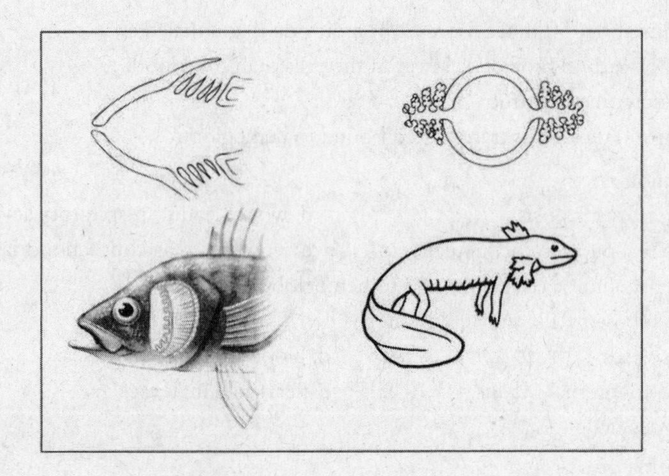

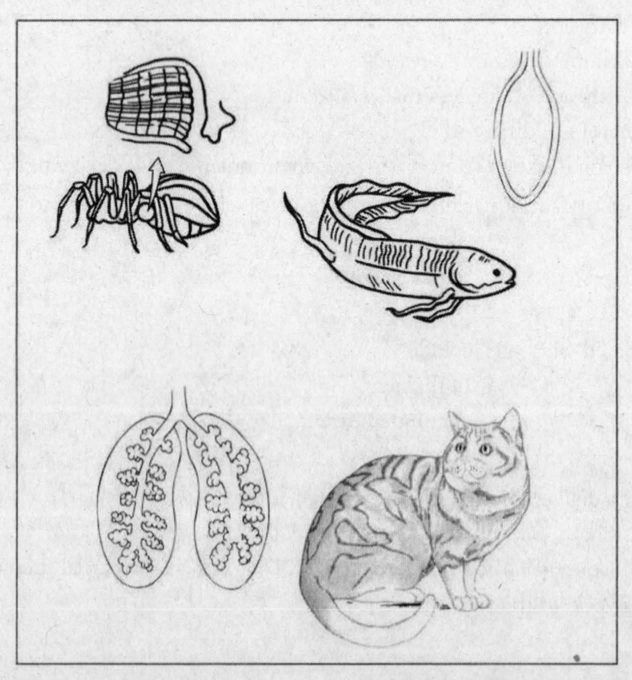

Color the parts of the illustration below as indicated.

RED	❑	red blood cell
BROWN	❑	bronchiole
YELLOW	❑	epithelial cell of alveolus
BLUE	❑	macrophage
PINK	❑	capillary

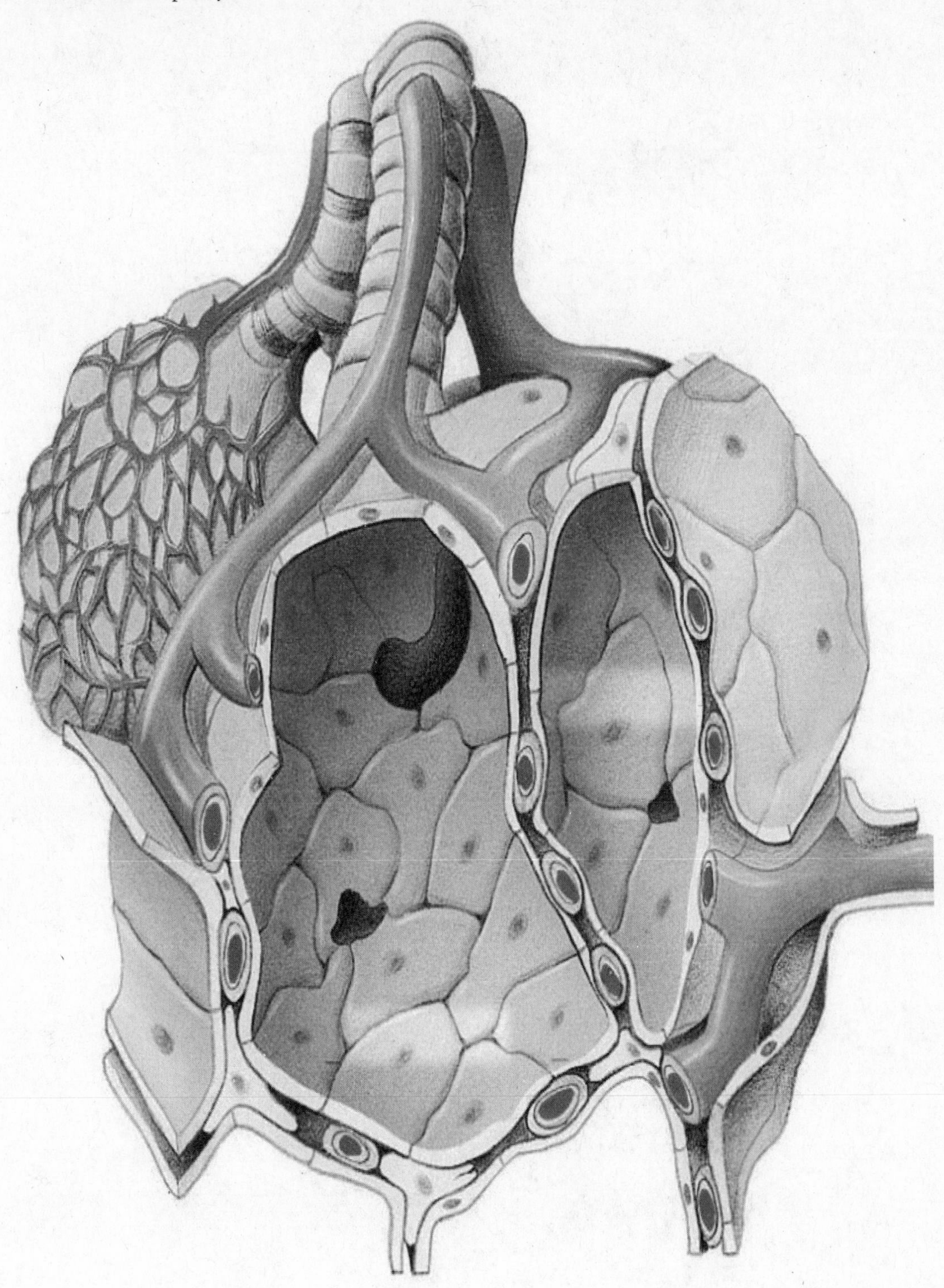

CHAPTER 47

Processing Food and Nutrition

The processing of food involves several steps — taking food into the body, breaking it down into its constituent nutrients, absorbing the nutrients, and eliminating the material that is not broken down and absorbed. Animals eat either plants, animals, or both. Some animals have no digestive systems, with digestion occurring intracellularly within food vacuoles. Others have incomplete digestive systems with only a single opening for both food to enter and wastes to exit. Still other animals have complete digestive systems, in which the digestive tract is a complete tube with two openings, a mouth where food enters and an anus where waste is expelled. The human digestive system has highly specialized structures for processing food. All animals require the same basic nutrients — carbohydrates, lipids, proteins, vitamins, and minerals. Carbohydrates are used by the body as fuel. Lipids are also used as fuel, and additionally, as components of cell membranes, and as substrates for the synthesis of steroid hormones and other lipid substances. Proteins serve as enzymes and as structural components of cells. Vitamins and minerals are needed for many biochemical processes. Serious nutritional problems result from eating too much food, eating too little food, or not eating a balanced diet.

REVIEWING CONCEPTS

Fill in the blanks.

INTRODUCTION

1. Organisms that obtain their main source of energy from organic molecules synthesized by other organisms are known as ____________________.
2. Substances in the diet that are used as energy sources are known as ____________________.
3. With slight variations all animals require the same basic nutrients: (a)____________________, (b) ____________________, (c) ____________________, (d) ____________________, and (e) ____________________.

NUTRITIONAL STYLES AND ADAPTATIONS

4. Feeding is the (a) ____________________, (b) ____________________. and (c) ____________________ of food.
5. (a) ____________________ is the process of taking in food, (b)____________________ is the process of breaking down food, (c)____________________ is the process of nutrients passing from the digestive tract into the blood, and (d) ____________________ is the discharge of undigested material from the digestive tract.

Animals are adapted to their mode of nutrition

6. Animals that feed directly on producers are called (a) ______________ or (b) ______________________________.
7. Animals cannot digest the (a) _______________ of plant cell walls. Therefore many herbivores have developed a (b) _________________ relationship with microorganisms.
8. Cattle, sheep, and deer are an example of a group of animals called ___________________.
9. Carnivores are (a) _______________________ or higher-level consumers and are therefore (b) _________________. They have well-developed (c) ____________ teeth for stabbing prey.
10. Consumers of plant and animal material are called _____________.
11. Animals that obtain food as suspended particles in water are called ____________________.
12. ____________________ feeders are animals that consume nutrients in soil or sediments.

Some invertebrates have a digestive cavity with a single opening

13. Examples of animals in which a gastrovascular cavity can be found are (a) ________________________ and (b) ______________________.

Most animal digestive systems have two openings

14. The vertebrate digestive system is a complete tube extending from the (a) __________________ to the (b) _____________________.
15. __________________________ is responsible for the movement of ingested food along the digestive tract.

THE VERTEBRATE DIGESTIVE SYSTEM

16. The specialized portions of the vertebrate digestive tube, in order, are the: mouth → (a) ____________________ → (b) ________________ → stomach → (c) ________________________ → large intestine → anus.
17. The wall of the digestive tract has four layers. The innermost is the (a)__________ that lines the lumen and consists of a layer of epithelial tissue and the underlying connective tissue. Next is a connective tissue layer the (b) ______________ in which are blood vessels, lymphatic vessels, and nerves. It is surrounded by the muscle layer. The fourth layer, the outer connective tissue layer, is the (c)___________________.

Food processing begins in the mouth

18. Ingestion and the beginning of mechanical and enzymatic breakdown of food take place in the mouth. The teeth of mammals perform varied functions: (a) _______________ are designed for biting, (b)_______________ for tearing, and (c) _____________, and (d)____________ crush and grind food. Each tooth is covered by a

coating of very hard (e) ____________, under which is the main body of the tooth, the (f) ____________, which resembles bone. The (g) ____________ contains blood vessels and nerves.

19. The salivary glands of terrestrial vertebrates moisten food and release ____________, an enzyme that initiates carbohydrate (starch) digestion.

The pharynx and esophagus conduct food to the stomach

20. A lump of food passing through the esophagus is called a ____________.
21. During swallowing the ____________ closes the opening to the airway.
22. Waves of muscular contractions called ____________ move a lump of food through the esophagus into the stomach.

Food is mechanically and enzymatically digested in the stomach

23. The stomach is able to increase and decrease in volume because the lining is folded into pleats called ____________.
24. Pits in the stomach contain (a) ____________ which are composed of different types of cells. (b) ____________ cells secrete hydrochloric acid and (c) ____________ cells secrete pepsinogen, an enzyme precursor that is converted to (d) ____________.
25. Open sores, peptic ulcers, are caused by the bacterium ____________.
26. Food that enters the stomach is digested and mechanically mixed for about ______ hours.

Most enzymatic digestion takes place in the small intestine

27. The three regions of the small intestine are (a) ____________, (b) ____________, and (c) ____________. Most chemical digestion of food takes place in the (d) ____________ portion.
28. The surface area of the small intestine is increased by small fingerlike projections called (a) ____________, and by (b) ____________, which are projections of the plasma membrane of the simple columnar epithelial cells of the villi.

The liver secretes bile

29. The function of bile which is secreted by the liver is ____________.
30. Bile is composed of water, bile (a) ____________ and bile (b)____________, (c)____________, salts and (d)____________.
31. Bile produced by the liver is stored in the (a) ____________ where it is also (b) ____________.

The pancreas secretes digestive enzymes

32. Pancreatic juice contains multiple enzymes: (a) ________________ and (b) ________________________________ digest polypeptides (c) ________________________________ digests fats, (d)________________________________ digests most carbohydrates and ribonuclease and deoxyribonuclease which digest nucleic acids.

Nutrients are digested as they move through the digestive tract

33. Carbohydrate digestion is completed by enzymes produced by cells lining the small intestine. Maltase digests maltose into (a)________________________________ and dipeptidases digest small peptides into (b) ______________________.
34. Bile released from the gall bladder into the duodenum (a)____________________ fats and exposing them to the action of pancreatic lipase which break fats down into (b)______________________, (c)______________________, (d)______________________, and (e)______________________.

Nerves and hormones regulate digestion

35. Neuropeptides act on digestive tract smooth muscle. (a)______________________ stimulated contraction and (b)__________________________ inhibits it.
36. Hormones such as secretin, gastrin, CCK, and GIP are polypeptides secreted by ______________________ in the mucosa of the digestive tract.

Absorption takes place mainly through the villi of the small intestine

37. Most digested nutrients are absorbed from the intestinal lumen by passing through the epithelial cells of the ____________________ lining the small intestine and then into the blood or lymph.
38. Absorbed amino acids and glucose are transported directly to the (a)______________ by (b) ______________________________..
39. Absorbed fatty acids and monacyglycerols, along with absorbed cholesterol and phospholipids, are packaged into protein-covered fat droplets called (a) ____________________ which enter the (b)________________ of the villus.
40. Following digestion and absorption, material remaining at the end of the small intestine passes into the large intestine through the ____________________.

The large intestine eliminates waste

41. The large intestine absorbs sodium and water, cultures bacteria, and eliminates wastes. It is made up of seven regions, which beginning at the junction of the small and large intestine are in the following order: the (a) ______________, a blind pouch near the junction of the small and large intestines; the (b) ______________________________; the (c) ____________________; the (d)

_______________________; the (e) ____________________; the (f) __________, the last portion of the tube; and the (g) _______, the opening at the end of the tube.

42. Intestinal bacteria benefit their host by producing vitamin (a) ____ and some (b) _____ vitamins that can be absorbed and used by the host.
43. (a) _______________ is the process of getting rid of metabolic wastes, while (b) _________________ is the process of getting rid of digestive wastes that never participated in metabolism.

REQUIRED NUTRIENTS

44. According to the most recent guidelines, (a) _____ to _____% of our calories should come from carbohydrates, (b) _____ to _____% from fat, and (c) ____ to ____% from protein.
45. Food energy is expressed in terms of (a) ____________________ which are equivalent to (b) ________________________.

Carbohydrates provide energy

46. Most carbohydrates are ingested in the form of (a) ______________ and (b) ________________.
47. Polysaccharides are referred to as _______________________.
48. When we eat an excess of carbohydrate-rich food, the liver cells become packed with (a) ______________ and convert excess glucose to (b) ______________ and (c) _______________. After forming triacylglycerols, the liver sends them to (d) ________________ for storage.
49. A rapid rise in blood glucose concentration stimulates the pancreas to release (a) ____________________ which lowers blood sugar levels.
50. Dietary fiber helps to decrease blood cholesterol levels. This fiber is mainly a mixture of (a) _____________________, and (b)____.

Lipids provide energy and are used to make biological molecules

51. About 98% of ingested dietary lipids are in the form of __________________.
52. The three polyunsaturated fatty acids that are essential fatty acids and must be obtained from food are (a) _______________________, (b) ____________________, and (c) _______________________.
53. The recommended daily dietary amount of cholesterol is __________mg.
54. Macromolecular complexes of cholesterol and triacylglycerols bound to proteins are called (a) ___________________________. Two types of these important complexes are the (b) __________________________, which apparently decrease the risk of heart disease by transporting excess cholesterol to the liver; and (c)___________________________________, which transport cholesterol to the cells and have been associated with coronary artery disease.

55. In general, animal foods are rich in both (a) ________________ and (b) ____________________, whereas most plant foods are rich in (c) ____________________ and no (d) ____________.

56. (a) ________________ fatty acids decrease LDL levels and help to decrease the risk for coronary artery disease while (b) __________ fatty acids increase blood levels of LDL and decrease levels of HDL.

Proteins serve as enzymes and as structural components of cells

57. There are approximately (a) ________ (#?) amino acids important in nutrition. Of these about (b) ______ (#) that cannot be synthesized in adult humans. These are called (c) ____________________ and need to be consumed in the diet.

58. Complete proteins, those with the best distribution of amino acids are found in certain foods including (a) __________, (b) ____________, (c) ____________, (d) ____________, and (e) ____________.

59. Liver cells (a) ____________________ amino acids and in the process produce (b) ________________.

Vitamins are organic compounds essential for normal metabolism

60. Vitamins are divided into two broad groups: the (a)________________ vitamins such as A, D, E, and K, and the (b)____________________ vitamins that include vitamins B and C..

61. Surpluses from overdoses of the __________________ vitamins can accumulate to harmful levels.

Minerals are inorganic nutrients

62. Major minerals are required by the body in amounts of 100 mg. These minerals include (a)____________________________,
(b)____________________, (c)____________________________,
(d)________________,
(e)____________________,
(f)____________________, (g)______________________,
and (h)________________.

63. Minerals such as rioin, copper, iodide, fluoride and selenium are required in amounts of less than 100 mg per day are known as ____________.

Antioxidants protect against oxidants

64. Antioxidants destroy __________________________, molecules with one or more unpaired electrons.

65. Vitamins (a) _____, (b) _____, and (c) ____ have strong antioxidant activity.

Phytochemicals play important roles in maintaining health

66. Diets rich in (a) ____________, and (b) ______________ may be more important for health than diets with a low intake of (c) _______.

ENERGY METABOLISM

67. (a) ______________________________ is a measure of the rate of energy used during resting conditions.
(b)______________________________ is the sum of an individuals' BMR and the energy needed to carry out daily activities.
68. Body mass index is an index of (a) ______________ in relation to (b) ____________________.

Energy metabolism is regulated by complex signaling

69. The (a) ____________________ regulates energy metabolism and food intake. It also produces two hormones involved in regulating eating. (b) ____________________ stimulates appetite and (c) ____________________suppress appetite.
70. When the stomach is empty, it secretes the peptide hormone (a)____________________ which stimulates appetite. It stimulates the hypothalamus to produce (b) ______. The small and large intestine secrete the peptide hormone (c) __________ which signals the hypothalamus to stop secreting (d) ______.
71. ____________ a hormone released by adipose tissue, is important in the regulation of body weight

Obesity is a serious nutritional problem

72. Approximately (a) _______% of U.S. adults are overweight and (b)_______% are obese.
73. An individual has a steady state called a __________ around which body weight is regulated.

Undernutrition can cause serious health problems

74. Individuals suffering from undernutrition generally are consuming a diet that is most often deficient in
____________________________.
75. Severe protein malnutrition results in a condition known as (a)______________________________. Severe calorie deficiency along with protein deficiency results in a form of malnutrition called (b) ____________________.

BUILDING WORDS

Use combinations of prefixes and suffixes to build words for the definitions that follow.

Prefixes	The Meaning	Suffixes	The Meaning
cec-	blind end	-al	pertaining to
epi-	upon, over, on	-amine	of chemical origin
intrins-	internally	-ic	pertaining to
jejun-	empty	-itis	inflammation
micro-	small	-micro(n)	small, "tiny"
omni-	all	-um	structure
pariet-	wall	-vore	eating
sub-	under, below		
vit-	life		

Prefix	Suffix	Definition
herbi-	________	1. An animal that eats plants.
carni-	________	2. An organism that eats flesh.
________	________	3. An organism that eats both plants and animals.
________	-mucosa	4. A layer of connective tissue below the mucosa that binds it to the muscle layer beneath.
periton-	________	5. Inflammation of the peritoneum.
________	-glottis	6. A flap of tissue over the airway that prevents food and drink from entering the airway when swallowing.
________	-villi	7. Small projections of the cell membrane that increase the surface area of the cell.
chylo-	________	8. Tiny droplets of lipid that are absorbed from the intestine into the lymph circulation.
________	________	9. A factor needed for adequate absorption of vitamin B_{12}.
________	________	10. Cells in the gastric pit that secrete HCl.
________	________	11. The region of the small intestine right after the duodenum.
________	________	12. A blind pouch formed where the small and large intestines meet.
________	________	13. Organic compound required in small amounts and it is often a component of coenzymes.

MATCHING

Terms:

a. absorption	f. gastrin	k. peristalsis
b. adventitia	g. mineral	l. rugae
c. bile	h. mucosa	m. stomach
d. digestion	i. pancreas	n. villus
e. elimination	j. pepsin	o. vitamin

For each of these definitions, select the correct matching term from the list above.

____ 1. A cell layer that lines the digestive tract and secretes a lubricating layer of mucous.

____ 2. The ejection of waste products, especially undigested food remnants, from the digestive tract.

____ 3. The chief enzyme of gastric juice; hydrolyses proteins.

____ 4. A digestive gland located in the vertebrate abdominal cavity, having both exocrine and endocrine functions.

____ 5. An organic compound necessary in small amounts for the normal metabolic functioning of a given organism; usually acts as a coenzyme.

____ 6. Folds in the stomach wall, giving the inner lining a wrinkled appearance.

____ 7. The taking up of a substance by the lining of the digestive tract.

____ 8. Produced by the liver and stored in the gallbladder.

____ 9. Powerful, rhythmic waves of muscular contraction and relaxation in the walls of hollow tubular organs that aid in the movement of substances through the tube.

____10. A minute "finger-like" projection from the surface of a membrane.

____11. Inorganic nutrient absorbed in the form of salts dissolved in food and water.

____12. One of several enzymes that helps regulate the digestive system.

MAKING COMPARISONS

Fill in the blanks.

Enzyme	Function	Site of Production
Salivary amylase	Enzymatic digestion of starch	Salivary glands
Maltase	#1	#2
#3	Proteins to polypeptides	#4
#5	Polypeptides to dipeptides	#6
Ribonuclease	#7	#8
#9	Degrades fats	Pancreas
#10	Splits small peptides to amino acids	#11
Lactase	#12	Small intestine

MAKING CHOICES

Place your answer(s) in the space provided. Some questions may have more than one correct answer.

____ 1. Some important functions of minerals include their role in/as
a. cofactors.
b. nerve impulses.
c. neurotransmitters.
d. maintaining fluid balance.
e. digestive hormones.

____ 2. The relative amounts of nutrient types consumed by impoverished societies, as compared to affluent societies, would likely be
a. more protein, less carbohydrate.
b. more lipid, less carbohydrate.
c. more carbohydrate, less protein.
d. more carbohydrate, less lipid.
e. more vegetable matter, less animal matter.

____ 3. Most absorption of macromolecular subunits occurs in the
a. stomach.
b. rectum.
c. large intestine.
d. small intestine.
e. colon.

____ 4. Products of complete fat digestion include
a. fatty acids.
b. amino acids.
c. chylomicrons.
d. glycerol.
e. triacylglycerol.

____ 5. Nerves and blood vessels in teeth are located in the
a. pulp.
b. enamel.
c. dentin.
d. cementum.
e. canines only, not other teeth.

____ 6. A herbivore is a/an
a. plant consumer.
b. primary consumer.
c. carnivore.
d. omnivore.
e. mutualistic symbiote.

____ 7. Which of the following concerning human nutrition is *true*?
a. Water is essential.
b. Excess nutrients are converted to fat.
c. Some fatty acids are required nutrients.
d. Proteins cannot be used for energy.
e. The liver deaminates amino acids.

____8. The principal function(s) of intestinal villi is/are to
a. absorb nutrients.
b. secrete enzymes.
c. increase surface area.
d. stimulate digestion.
e. secrete hormones.

____ 9. All animals are
a. omnivores.
b. carnivores.
c. primary consumers.
d. consumers.
e. heterotrophs.

____10. Lipids are used to
a. as fuel in cellular respiration.
b. build membranes.
c. provide energy.
d. synthesize steroid hormones.
e. synthesize bile salts.

____11. Bile is principally involved in the digestion of
a. carbohydrates.
b. lipids.
c. starch.
d. nucleic acids.
e. proteins.

____12. Because it enters the citric acid cycle, the most important single, intermediate molecule in the metabolism of lipids is
a. keto acid.
b. ATP.
c. acetyl CoA.
d. NAPH.
e. glucose.

____13. A lacteal is a/an
a. lymph vessel.
b. capillary bed.
c. digestive gland.
d. absorptive surface.
e. structure found in the intestine.

____14. Peristalsis originates in the
a. submucosa.
b. mucosa.
c. adventitia.
d. muscle layer.
e. peritoneum.

____15. Which of the following regarding cellulose in the human diet is *true*?
a. It's harmful.
b. It's a major source of protein.
c. We can't digest it.
d. It's an important carbohydrate nutrient.
e. It's an important source of fiber.

____16. Pepsin is principally involved in the digestion of
a. carbohydrates.
b. lipids.
c. starch.
d. nucleic acids.
e. proteins.

____17. The major sources of energy in the human diet are

a. proteins.
b. lipids.
c. carbohydrates.
d. starches and sugars.
e. meat.

____18. The total metabolic rate

a. includes the basal metabolic rate.
b. is less than the basal metabolic rate.
c. refers to metabolic rate after exercise.
d. includes the energy used in daily activity.
e. does not include the basal metabolic rate.

____19. Which of the following has/have a complete digestive system?

a. birds.
b. earthworms.
c. sponges.
d. jelly fish.
e. fish.

____20. A herbivore would likely have

a. well-developed claws.
b. flattened molars.
c. large, sharp canines.
d. a short gut.
e. symbiotic microorganisms.

____21. Animals that are omnivores include

a. earthworms.
b. black bears.
c. filter feeders.
d. pigs.
e. fish.

____22. Cnidarians and flatworms

a. digest prey extracellularly.
b. have a pharynx.
c. have a gastrodermis that aides in digestion.
d. have a mouth and an anus.
e. ingest whole prey.

____23. Having a complete digestive tract is more of an advantage than having a gastrovascular cavity because a complete digestive tract

a. digests a wider variety of foods.
b. is lined with absorptive cells.
c. allows the animal to take in more food before all digestive activity is complete.
d. has regions with specific activities.
e. provides better mixing of enzymes with the food.

____24. A tooth

a. is arranged enamel, dentin, pulp.
b. is arranged crown, root.
c. is held in the socket by cementum.
d. is arranged enamel, pulp, dentin.
e. has an artery, a vein, a nerve, a lymph vessel.

____25. The liver

a. stores bile.
b. stores glycogen.
c. is divided into lobes.
d. stores iron.
e. regulates blood nutrient levels.

____26. In which of the following are the two terms correctly paired?

a. amylase : digests maltose
b. pancreatic amylase : digests cellulose
c. trypsin: activated by enterokinase
d. substance P : stimulation of smooth muscle contraction
e. pancreas : production of ribonuclease

____27. Gastrin

a. is produced by the small intestine.
b. stimulates the release of bile.
c. is released when the stomach becomes distended.
d. stimulates HCl production and secretion.
e. stimulates insulin secretion.

____28. Hormones associated with the duodenal mucosa include
a.cholecystokinin
b. gastric stimulatory peptide.
c. inhibin.
d. secretin.
e. glucose-dependent insulinotropic peptide.

____29. Which hormone(s) target the pancreas?
a. gastrin
b. cholecystokinin (CCK)
c. enkephalin
d. secretin
e. glucose-dependent insulinotropic peptide (GIP)

____30. You can promote a healthy proportion of HDL to LDL by
a. exercising.
b. drinking alcohol to promote fat metabolism.
c. not smoking.
d. eating monounsaturated fats.
e. keeping a healthy body weight.

____31. Healthy sources of protein include
a. chicken.
b. peas.
c. nuts.
d. peanuts.
e. gelatin.

____32.Antioxidants include
a. vitamin E.
b. Vitamin B_{12}.
c. pantothenic acid.
d. vitamin K..
e. folic acid.

____33. Which hormone stimulates appetite?
a. neuropeptide Y
b. insulin
c. melanocortins
d. peptide YY
e. ghrelin

____34. A deficiency in which mineral(s) can result in anemia?
a. iron
b. selenium
c. zince
d. copper
e. phosphorous

____35. A source of complete proteins would be
a. nuts.
b. eggs.
c. milk.
d. fish.
e. meat.

____36. Omega-3 fatty acids
a. are found in fish.
b. decrease LDL levels.
c. are found in coconut oil.
d. are found in plant oils.
e. decrease cholesterol levels.

____37. The recommended daily maximum intake of cholesterol is ______ mg.
a. 100
b. 200
c. 300
d.400
e. 500

____38. Dietary fiber
a. includes cellulose.
b. is obtained by eating fruits.
c. affects blood cholesterol levels.
d. includes indigestible carbohydrates.
e. includes undigested proteins.

____39. Protein digesting enzymes released by the pancreas include
a. trypsin
b. dipeptidase.
c. amylase.
d. chymotrypsin.
e. pepsin.

____40. Parietal cells.

a. secrete pepsinogen.
b are part of the gastric glands..
c. absorb vitamin B_{12}.
d. are located in gastric pits.
e. secrete intrinsic factor.

VISUAL FOUNDATIONS

Color the parts of the illustration below as indicated. Label the mucosa, submucosa, muscle layer, visceral peritoneum, intestinal glands, and villi.

- RED ❑ artery
- GREEN ❑ lymph
- YELLOW ❑ nerve fiber
- BLUE ❑ vein
- VIOLET ❑ goblet cell
- PINK ❑ epithelial cell of villus

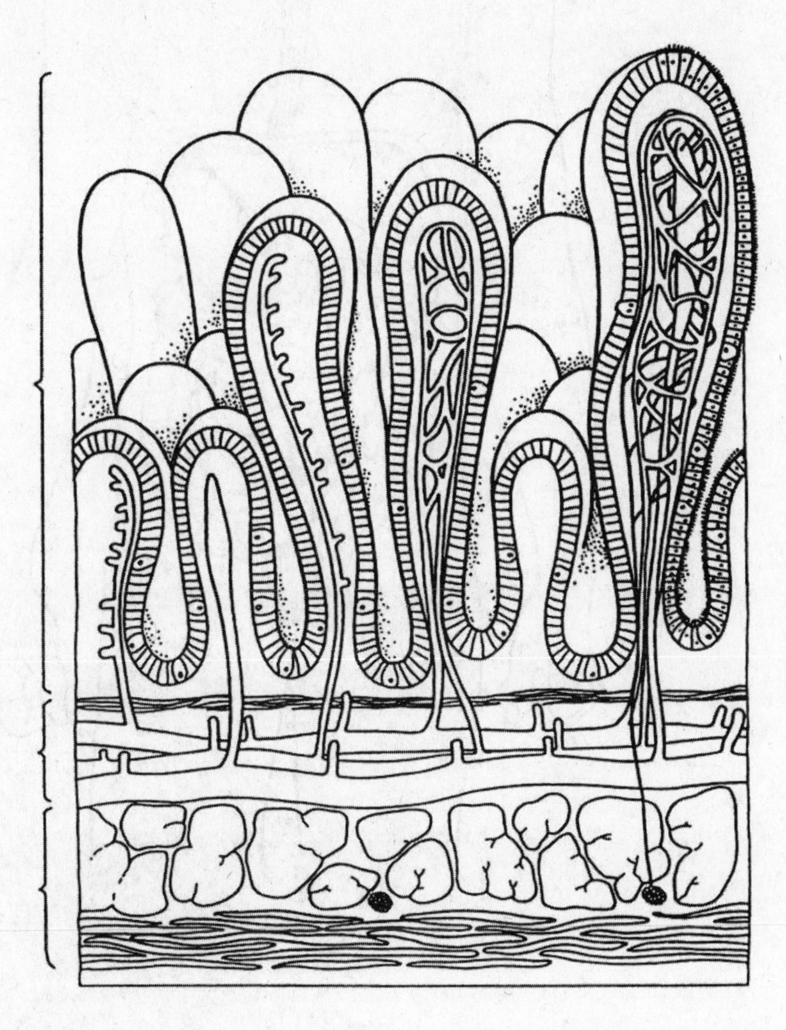

Color the parts of the illustration below as indicated.

RED	❑	liver
GREEN	❑	gall bladder
YELLOW	❑	esophagus
BLUE	❑	pancreas
ORANGE	❑	stomach
BROWN	❑	large intestine
TAN	❑	small intestine
PINK	❑	rectum and anus
VIOLET	❑	vermiform appendix

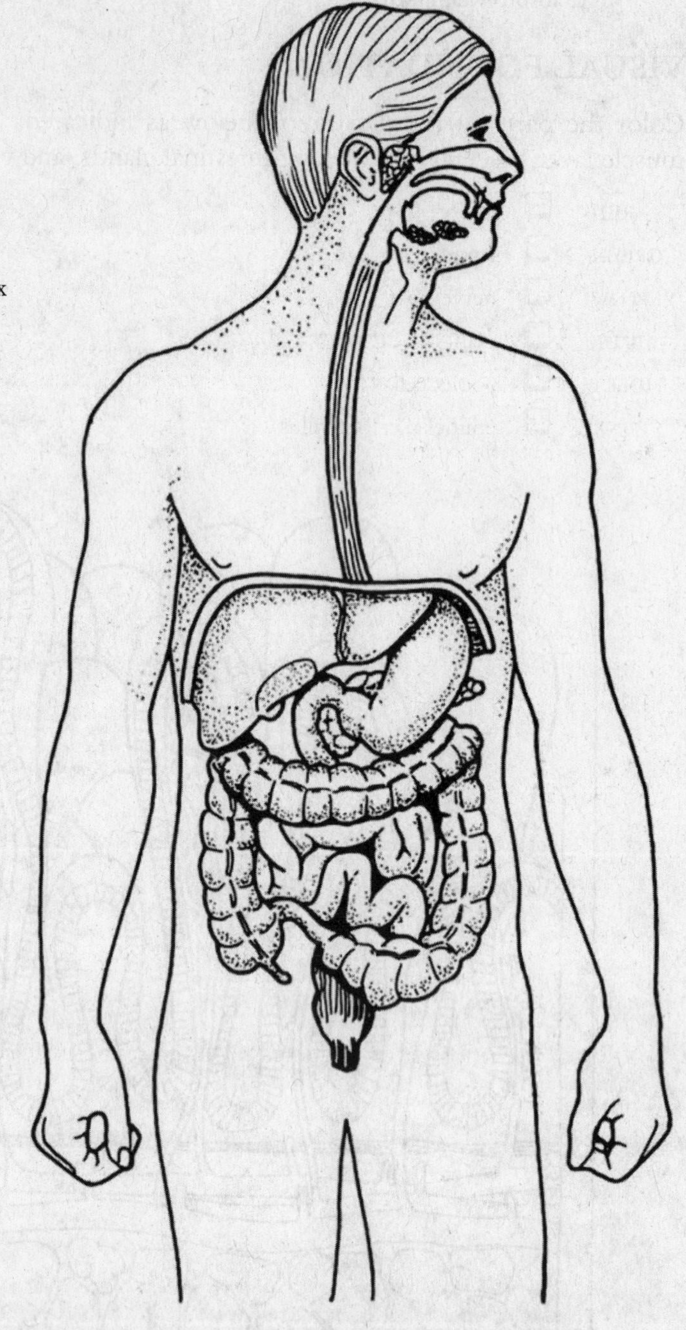

CHAPTER 48

Osmoregulation and Disposal of Metabolic Wastes

The water content of the animal body, as well as the concentration and distribution of ions in body fluids is carefully regulated. Most animals have excretory systems that function to rid the body of excess water, ions, and metabolic wastes. The excretory system collects fluid from the blood and interstitial fluid, adjusts its composition by reabsorbing from it the substances the body needs, and expels the adjusted excretory product, e.g., urine in humans. The principal metabolic wastes are water, carbon dioxide, and nitrogenous wastes. Excretory systems among invertebrates are diverse and adapted to the body plan and lifestyle of each species. The kidney, with the nephron as its functional unit, is the primary excretory organ in vertebrates.

REVIEWING CONCEPTS

Fill in the blanks.

INTRODUCTION

1. ____________ is the medium in which most metabolic reactions take place.
2. Two processes that maintain homeostasis of fluids in the body are (a)____________________ and (b)____________________________.

MAINTAINING FLUID AND ELECTROLYTE BALANCE

3. Compounds such as inorganic salts, acids, and bases that form ions in solution are called ______________________.
4. A unit of osmotic pressure equal to the molarity of the solution times the number of particles produced when the solute dissolves is an _______________.
5. ______________________________ controls the concentration of salt and water so that body fluids don't become too dilute or too concentrated.

METABOLIC WASTE PRODUCTS

6. The principal metabolic wastes in most animals are (a)___________________, (b) ____________________, and (c)__________________.
7. The catabolism of proteins leads to the production of three nitrogen containing waste products. They are (a) _____________, (b) ________, and (c) ___________________.
8. The production of (a) __________________ is an important water-conserving adaptation in many terrestrial animals, including insects,

certain reptiles, and birds. It is produced from (b) ______________ and by the (c) ____.

9. The major nitrogenous waste product produced by amphibians and mammals is ___________.

OSMOREGULATION AND EXCRETION IN INVERTEBRATES

10. The body fluid of most marine invertebrates is (a) ______________ with the sea water. As a result, these animals are said to be (b)____________________________.
11. Animals living in water environments much less stable than the ocean and those that live on land maintain an optimal salt concentration independent of the environmental conditions. These animals are called ______________.

Nephridial organs are specialized for osmoregulation and/or excretion

12. The nephridial organs of many invertebrates consist of tubes that open to the outside of the body through (a) ____________________. In flatworms, these excretory organs are called (b) ______________, and in annelids and mollusks they are known as (c) ______________.

Malpighian tubules conserve water

13. Malpighian tubules are slender extensions of the (a) ______________, and they have blind ends that lie in the (b) ________________ where they are surrounded by (c) ____________________.
14. Wastes are transferred from blood to the Malpighian tubules by either (a) __________________________ or (b) ________________.

OSMOREGULATION AND EXCRETION IN VERTEBRATES

15. The main osmoregulatory and excretory organ in most vertebrates is the (a) ______________, but the (b) ____________, (c) __________ or (d) __________, and the digestive system also help maintain fluid balance and secrete metabolic wastes.

Freshwater vertebrates must rid themselves of excess water

16. A freshwater fish is ____________ to the water around it.
17. In freshwater fish the majority of nitrogenous waste is excreted by the ________.
18. In freshwater fishes, (a) __________________ is the main nitrogenous waste; about 10% of nitrogenous wastes are excreted as (b) ______________.
19. As part of osmoregulation, both freshwater fish and amphibians produce a large amount of ____________________.

Marine vertebrates must replace lost fluid

20. Marine bony fishes are __________________ to the surrounding water.
21. To compensate for fluid loss, marine bony fishes ________________.
22. To help them maintain osmotic balance with the surrounding water, marine cartilaginous fishes accumulate (a) _________ in their tissues. This makes these fishes (b) ________________ to their environment.

23. Marine mammals need to produce a _______________ urine which helps them excrete the high amounts of urea that they produce as a result of consuming a lot of protein in their diet.

Terrestrial vertebrates must conserve water

24. Birds conserve water by excreting nitrogen as (a) _______________; Mammals excrete (b) _______________.

THE URINARY SYSTEM

25. The outer portion of the kidney tissue is the (a) _______________ and the inner portion is the (b) _______________.
26. The urinary system is the principal excretory system in human beings and other vertebrates. Urine produced in the kidneys is transported through two tubes called (a) _______________ leading to the (b)_______________ where it is stored until it passes to the outside of the body through the (c) _______________.

The nephron is the functional unit of the kidney

27. The functional units of the kidneys are the nephrons. Each nephron consists of a cup-shaped (a) _______________ and a (b)_______________. Filtrate passes through the nephron structures in the order (c) _______________ → proximal convoluted tubule (d) _______________ → distal convoluted tubule → (e) _______________
28. There are two types of nephrons in kidneys. Those that have a small glomerulus and lie almost entirely within the cortex and outer medulla are (a) _______________ nephrons. Those that have a large glomerulus and loops of Henle that extend deep into the medulla are called (b)_______________ nephrons.

Urine is produced by glomerular filtration, tubular reabsorption, and tubular secretion

29. Urine is produced by a combination of three processes (a) _______________, (b) _______________, and (c) _______________.
30. Plasma is filtered out of the glomerular capillaries and into _______________.
31. Three major factors contribute to filtration and the formation of filtrate. They are (a) _______________, (b)_______________, and (c)_______________.
32. Specialized epithelial cells called (a) _______________ surround the capillaries of the glomerulus. These cells are separated by narrow gaps called (b) _______________ which are one part of the (c) _______________ that holds back blood cells and most plasma proteins.
33. The renal tubules reabsorb about (a) ____% of the filtrate into the blood, leaving only about (b) _____ L to be excreted as urine in a 24 hour period.
34. About 65% of the filtrate is reabsorbed as it passes through the first section of the nephron tubule called the _______________.

35. The maximum rate at which a substance can be reabsorbed is its ______________________________.

36. When the circulating level of potassium ions is too high, the heart rhythm (a) ______________________. The hormone (b)________________ stimulates secretion of excess potassium by the kidney.

Urine becomes concentrated as it passes through the renal tubule

37. The osmolarity of human blood is about (a) ________ mOsm/L and the kidneys can produce a urine that is about (b) _____ mOsm/L.

38. Filtrate is concentrated as it moves downward through the (a)_________ loop of Henle as a result of (b)______________________________ and it is diluted as it moves upward through the (c) ____________ loop as a result of (d) ______________________________. The movement of the filtrate in opposite directions along the loop of Henle is, called (e) ______________________. This helps maintain a hypertonic interstitial fluid that draws water out of the collecting ducts.

39. The ________________ are capillaries that extend from the efferent arterioles of the juxtamedullary nephrons; they collect water from interstitial fluid and help to maintain the high salt concentration of the interstitial fluid.

Urine consists of water, nitrogenous wastes, and salts

40. Urine is the ____________________ consisting of water, nitrogenous wastes, salts, and traces of other substances.

41. ____________________ is the physical, chemical, and microscopic examination of urine.

Hormones regulate kidney function

42. Urine volume is regulated by the hormone (a) ________________ which is released by the (b) ____________________________. The hormone is actually produced in the (c) ________________.

43. The "thirst center" that responds to dehydration is located in the __________________.

44. If the pituitary gland routinely does not release enough ADH or if an individual develops an acquired insensitivity of the kidney to ADH, the disorder __________________________ may occur.

45. The hormone (a) __________________ secreted by the adrenal cortex stimulates sodium reabsorption in the (b) ______________ and the (c) ____________________.

46. When blood pressure falls, cells of the (a) ____________________ secrete the enzyme renin. Renin converts the plasma protein (b)__________________ into (c) ____________________. The enzyme ACE then converts (d) ____________________ into (e) ____________________ which stimulates aldosterone secretion and the constriction of blood vessels.

47. Atrial natriurectic peptide (ANP) is a hormone that is produced by the (a) ______________. ANP increases the excretion of (b)

__________ and decreases (c) __________________________.
ANP also (d)___________________ afferent arterioles which increases the glomerular filtration rate.

48. The renin-angiotensin-aldosterone pathway works __________________ with ANP to regulate fluid and electrolyte balance as well as blood pressure.

BUILDING WORDS

Use combinations of prefixes and suffixes to build words for the definitions that follow.

Prefixes	The Meaning	Suffixes	The Meaning
electr-	amber	-cyte	cell
glomerul-	little ball	-lyt(e)	dissolve
juxta-	beside, near	-us	thing
podo-	foot		
proto-	first, earliest form of		
ureth-	canal		

Prefix	Suffix	Definition
____________	-nephridium	1. The flame cell excretory organs of flatworms and nemerteans; the earliest form of specialized excretory organ.
____________	____________	2. A specialized epithelial cell possessing elongated foot processes, which cover the surfaces of most of the glomerular capillaries.
____________	-medullary	3. Pertains to nephrons situated nearest the medulla of the kidney.
____________	____________	4. A cluster of capillaries located inside of Bowman's capsule.
____________	____________	5. A compound such as an inorganic salt or certain acids and bases.
____________	-a	6. A passageway leading from the bladder to the outside of the body.

MATCHING

Terms:

a.	aldosterone	f.	glomerulus	k.	reabsorption
b.	antidiuretic hormone	g.	malpighian tubule	l.	urea
c.	bowman's capsule	h.	metanephridium	m.	ureter
d.	osmoconformer	i.	nephron	n.	urethra
e.	osmoregulator	j.	osmoregulation	o.	uric acid

For each of these definitions, select the correct matching term from the list above.

____ 1. An animal that maintains an optimal salt concentration of its body fluids despite changes in the salinity of its surroundings.

____ 2. A hormone produced by the vertebrate adrenal cortex that functions to increase sodium reabsorption.

____ 3. The ball of capillaries at the proximal end of a nephron; enclosed by the Bowman's capsule.

____ 4. The beginning of the nephron tubule which encloses the glomerulus.

____ 5. The principal nitrogenous excretory product of many terrestrial animals, including insects and certain birds and reptiles.

____ 6. The active regulation of the osmotic pressure of body fluids.

____ 7. One of the paired ducts that conducts urine from the kidney to the bladder.

____ 8. The principal nitrogenous excretory product of mammals; one of the water-soluble end products of protein metabolism.

____ 9. The functional unit of the vertebrate kidney.

____10. A hormone secreted by the posterior lobe of the pituitary that controls the rate of water reabsorption by the kidney.

____11. An animal whose internal salt composition reflects that of its environment.

MAKING COMPARISONS

Fill in the blanks.

Organism	Excretory Mechanism/Structure
Marine sponges	Diffusion
Vertebrates	#1
Insects	#2
Earthworms	#3
Flatworms	#4

MAKING CHOICES

Place your answer(s) in the space provided. Some questions may have more than one correct answer.

____ 1. The amount of needed substances that can be reabsorbed from renal tubules is a function of the
a. Tm.
b. blood pH.
c. tubular transport maximum.
d. concentration of urea in forming urine.
e. saturation of tubule receptors.

____ 2. The principal functional unit(s) in the vertebrate kidneys is/are
a. nephridia.
b. nephrons.
c. Bowman's capsules.
d. antennal glands.
e. Malpighian tubules.

____ 3. Water is reabsorbed by interstitial fluids when osmotic concentration in interstitial fluid is increased by
a. salts from filtrate.
b. urea from filtrate.
c. free ions.
d. concentrating the filtrate.
e. deamination in kidney cells.

____ 4. Urea is a principal nitrogenous waste product produced by
a. amphibians.
d. mammals.

b. insects.
c. grasshoppers.
e. the liver.

____ 5. In the kidney, most reabsorption of filtrate takes place in the
a. ureter.
b. loop of Henle.
c. proximal convoluted tubule.
d. urethra.
e. collecting duct.

____ 6. Urea is synthesized
a. from uric acid.
b. in kidneys.
c. from ammonia and carbon dioxide.
d. in the urea cycle.
e. in aquatic invertebrates.

____ 7. Relative to sea water, fluids in the bodies of marine organisms
a. are isotonic.
b. are hypotonic.
c. are hypertonic.
d. lose water.
e. gain water.

____ 8. The duct in humans that leads from the urinary bladder to the outside is the
a. ureter.
b. loop of Henle.
c. proximal convoluted tubule.
d. urethra.
e. collecting duct.

____ 9. Which of the following statements most accurately describes changes in the concentration of filtrate in the two portions of the loop of Henle?
a. decreases in both
b. increases in both
c. increases in descending/decreases in ascending
d. decreases in descending/increases in ascending
e. remains essentially the same in both

____10. A potato bug would excrete wastes by means of
a. nephridia.
b. nephrons.
c. a pair of kidney-like structures.
d. green glands.
e. Malpighian tubules.

____11. Relative to fresh water, fluids in the bodies of aquatic organisms
a. are isotonic.
b. are hypotonic.
c. are hypertonic.
d. lose water.
e. gain water.

____12. Excretion is specifically defined as
a. maintaining water balance.
b. homeostasis.
c. removal of metabolic wastes from the body.
d. the elimination of undigested wastes.
e. the concentration of nitrogenous products.

____13. The main way(s) that *any* excretory system maintains homeostasis in the body is/are to
a. excrete metabolic wastes.
b. eliminate undigested food.
c. regulate body fluid constituents.
d. regulate salt and water balance.
e. concentrate urea in urine.

____14. The group(s) of animals that has/have no specialized excretory systems include
a. insects.
b. cnidarians.
c. annelids.
d. sponges.
e. sharks and rays.

____15. The principal form(s) of nitrogenous waste products in various animal groups include(s)
a. uric acid.
d. ammonia.

b. carbon dioxide.
c. oxalic acid.
e. urea.

___16. The principal nitrogenous waste product(s) in human urine is/are
a. uric acid.
b. NH_3.
c. NH_4.
d. ammonia.
e. urea.

___17. The first step in the catabolism of amino acids
a. is conversion of ammonia to uric acid.
b. is deamination.
c. produces ammonia.
d. is removal of the amino group.
e. is accomplished with peptidases.

___18. The type(s) of excretory organ(s) found in animals that collect wastes through flame cells is/are
a. protonephridia.
b. metanephridia.
c. open to the outside through pores.
d. green glands.
e. Malpighian tubules.

___19. The nitrogenous waste requiring the largest amount of water for adequate excretion is
a. urea.
b. uric acid.
c. an amino acid.
d. uracil.
e. ammonia.

___20. Urea is synthesized in the liver from
a. uric acid and ammonia.
b. uric acid and oxygen.
c. ammonia and nitrogen.
d. ammonia and carbon dioxide.
e. ammonia and oxygen.

___21. Ammonia is produced in
a. reptiles.
b. fish.
c. birds.
d. amphibians.
e. mammals.

___22. Excretory systems
a. maintain homeostasis.
b. are found in all animals.
c. help to avoid dehydration.
d. selectively adjust the amount of salt and water.
e. evolved first in amphibians.

___23. Most amphibians
a. produce a dilute urine.
b. are semi-aquatic.
c. excrete urea.
d. have osmoregulatory systems similar to mammals.
e. produce a concentrated urine to conserve water.

___24. Salt glands can be found in some
a. sharks.
b. reptiles.
c. whales.
d. birds.
e. dolphins.

___25. Urine enters the renal pelvis from the
a. papillae.
b. cortex.
c. medullary duct.
d. collecting duct.
e. renal tubules.

____26. The kidneys produce
a. erythropoietin.
b. vitamin D_3.
c. cortical hormone.
d. renin.
e. factor K.

____27. Under normal circumstances, the total amount of blood passing through a kidney is
a. 180 L in 24 hours.
b. 5 L per minute.
c. 100 L per hour.
d. 1.2 L per minute.
e. about ¼ of the cardiac output.

____28. Each day the kidney tubules reabsorb
a. 1200 g of salt.
b. 20 L of water.
c. 180 L of water.
d. 500 g of chloride.
e. 250 g of glucose.

____29. The loop of Henle
a. is permeable to water in the ascending limb.
b. is a countercurrent mechanism.
c. is specialized to maintain a high interstitial sodium chloride concentration.
d. is permeable to salt in the descending limb.
e. connects to the collecting duct.

____30. Antidiuretic hormone
a. is released by the hypothalamus.
b. is produced by the hypothalamus.
c. stimulates water reabsorption.
d. is produced by the pituitary gland.
e. is released by the pituitary gland.

____31. Which of the following is/are correctly paired?
a. atrial natriuretic peptide : collecting ducts
b. aldosterone : afferent arteriole
c. antidiuretic hormone : collecting ducts
d. angiotensin II : adrenal glands
e. aldosterone: distal tubules

VISUAL FOUNDATIONS

Color the parts of the illustration below as indicated.

RED ❑ abdominal aorta
GREEN ❑ adrenal gland
YELLOW ❑ ureter
BLUE ❑ inferior vena cava
ORANGE ❑ urethra
BROWN ❑ kidney
TAN ❑ urinary bladder
PINK ❑ renal artery
VIOLET ❑ renal vein

Color the parts of the illustration below as indicated. Label the juxtamedullary nephron and the cortical nephron.

RED	❑	artery	ORANGE	❑	loop of Henle
GREEN	❑	renal pelvis	BROWN	❑	medulla
YELLOW	❑	collecting duct	BLUE	❑	vein
PINK	❑	glomerulus	TAN	❑	cortex
VIOLET	❑	proximal convoluted tubule and distal convoluted tubule			

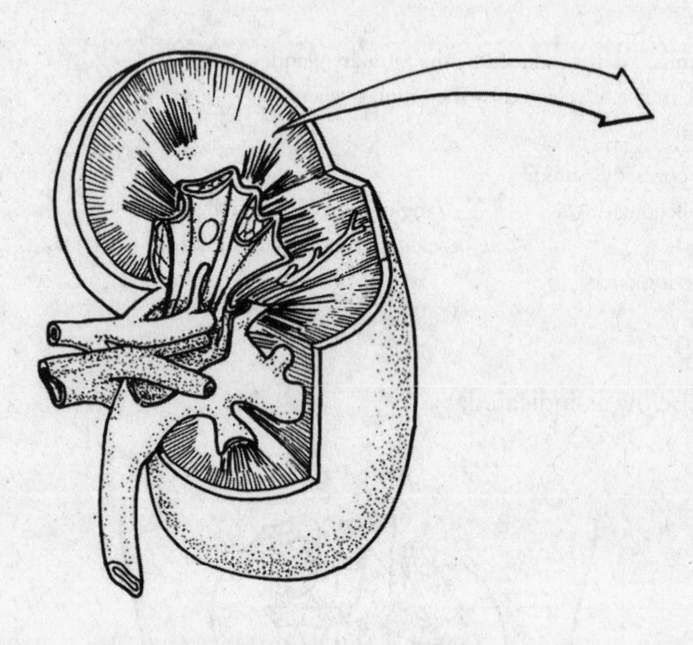

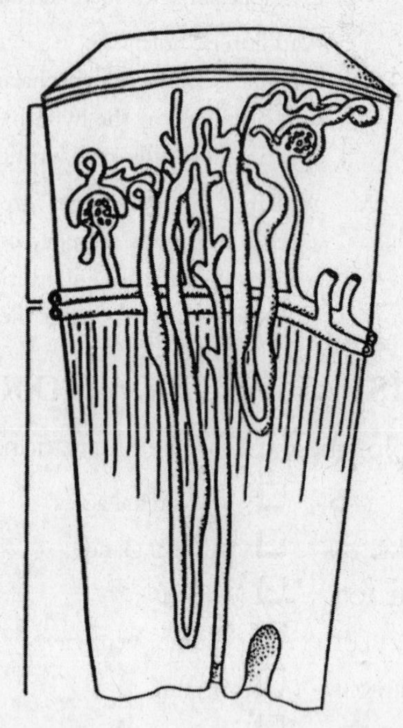

CHAPTER 49

❑

Endocrine Regulation

This chapter discusses the actions of a variety of hormones and examines how overproduction or deficiency of various hormones interferes with normal functioning. The endocrine system is a diverse collection of glands and tissues that secrete hormones, chemical messengers that signal other cells. The majority of hormones are transported by the blood to the target tissue where they stimulate a physiological change. The endocrine system works closely with the nervous system to maintain the steady state of the body. They diffuse from the blood into the interstitial fluid and then combine with receptor molecules on or in the cells of the target tissue. Hormones may be steroids, peptides, proteins, or derivatives of amino acids or fatty acids. Some hormones activate genes that lead to the synthesis of specific proteins. Others activate a second messenger that relays the hormonal message to the appropriate site within the cell. Most invertebrate hormones are secreted by neurons rather than endocrine glands. They regulate growth, metabolism, reproduction, molting, and pigmentation. In vertebrates, hormonal activity is controlled by the hypothalamus, which links the nervous and endocrine systems. Vertebrate hormones regulate growth, reproduction, salt and fluid balance, and many aspects of metabolism. Malfunction of any of the endocrine glands can lead to specific disorders.

REVIEWING CONCEPTS

Fill in the blanks.

INTRODUCTION

1. The endocrine system regulates many physiological processes including
   (a) ____,
   (b) ____,
   (c) ____,
   (d) ____,
   and (e) ____.
2. Nuclear receptors for which ligands are not yet known are called ______________ receptors.

AN OVERVIEW OF ENDOCRINE REGULATION

3. Endocrine glands differ from exocrine glands in that they have no (a)________________ and secrete their hormones (b)____.
4. About _______ (#) discrete endocrine glands have been identified.
5. Specialized hormone releasing cells have been found in no discrete endocrine glands. These hormone releasing cells are located in the (a)______________, (b) ________________, and (c)______________________ as well as many other parts of the body.

6. Hormones are typically transported by (a) ________________ and produce a response only after they (b)________________________________.

The endocrine system and nervous system interact to regulate the body

7. The endocrine and nervous systems work together to maintain ________________.
8. The endocrine system responds more ________________ (slowly, quickly) than the nervous system.
9. The response of the endocrine system lasts a ________________ (longer, shorter) time than the nervous system.

Negative feedback systems regulate endocrine activity

10. Most endocrine action is regulated by (a)________________ that act to restore (b)________________.

Hormones are assigned to four chemical groups

11. The four chemical groups of hormones are (a)________________, (b)________________, (c)________________, and (d) ________________.
12. Signaling molecules produced by neurons are called ________________.

TYPES OF ENDOCRINE SIGNALING

13. Hormones signal their target cells by (a) ________________ signaling, (b) ________________ signaling, (c) ________________ signaling, and (d) ________________ signaling.
14. Steroid and thyroid hormones are transported bound to ________________.

Neurohormones are transported in the blood

15. Neurohormones are produced by (a) ________________, transported down (b) ________________ and released into the (c)________________.

Some local regulators are considered hormones

16. There are two types of local regulation. In (a)________________ signaling a hormone acts on the very cell that produces it and in (b)________________ signaling the released hormone works on nearby cells.
17. ________________ is a signaling molecule that is stored in mast cells.
18. (a) ________________ is a gaseous signaling molecule and (b)________________ are modified fatty acids released continuously by cells of most tissues and most of them target (c)________________.

MECHANISMS OF HORMONE ACTION

19. Receptors are continuously synthesized and degraded. Their numbers can be increased or decreased by (a) ________________ and (b) ________________.

Lipid-soluble hormones enter target cells and activate genes

20. Steroid and thyroid hormones bind receptors that are located (a)________________ because these hormones are (b) ________________.
21. Many steroid hormones work through (a) ________________ and (b) ________________.

Water-soluble hormones bind to cell-surface receptors

22. Peptide hormones bind to specific receptors on the plasma membrane because they are ________________.
23. Two main types of cell-surface receptors are (a) ________________ -linked and (b) ________________-linked.
24. With respect to a hormones action, the conversion of an extracellular signal to an intracellular signal is called ________________.
25. G-protein linked receptors activate G proteins. Some activated G proteins in turn activate (a) ________________ an enzyme that converts ATP to cAMP. cAMP activates (b)________________, enzymes that phosphorylate proteins.
26. In other signaling pathways, activated G proteins activate the membrane-bound enzyme phospholipase C which interacts with its membrane lipid target to produce (a) ________________ and (b) ________________.
27. cAMP, inositol trisphosphate, and diacyglycerol all acts as ________________.
28. Most enzyme-linked receptors are ________________ receptors.

NEUROENDOCRINE REGULATION IN INVERTEBRATES

29. Most invertebrate hormones are secreted by ________________ rather than by endocrine glands. They help to regulate regeneration, molting, metamorphosis, reproduction, and metabolism.
30. Hormones control growth and development in insects. Neurosecretory cells in the brain of insects produce (a)________________ which is stored in the corpora cardiaca. When released from the corpora cardiaca, this hormone stimulates the (b) ________________ glands to produce (c) ________________ also called (d) ________________. This hormone stimulates growth and molting.

ENDOCRINE REGULATION IN VERTEBRATES

Homeostasis depends on normal concentrations of hormones

31. A target cell is over stimulated if (a) ________________ of the hormone occurs and it is under stimulated if (b) ________________ occurs.

The hypothalamus regulates the pituitary gland

32. The pituitary gland connects to the hypothalamus by the ________________.
33. The pituitary is often referred to as the ________________ because of the number of body activities it controls.

The posterior pituitary gland releases hormones produced by the hypothalamus

34. The hypothalamus produces vasopressin, also known as (a)_________________ and a second hormone (b)_____________, which stimulates milk production in nursing mothers. Both of these neurohormones are stored in and released from the (c) _________________.

The anterior pituitary gland regulates growth and other endocrine glands

35. The hypothalamus produces (a) ____________ hormones and (b) _________________ hormones that regulate the activity of the anterior pituitary. These hormones travel from the hypothalamus to the pituitary via a (c) _____________ system of blood vessels.
36. Among the hormones released by the anterior pituitary are ____________ hormones which stimulate other endocrine glands.
37. Melanocyte-stimulating hormones (MSH) are secreted by the (a)_________________ gland and certain (b)_____________ in the hypothalamus. In humans MSH causes the skin to darken, it (c) _________________ and it is involved in regulation of energy and body weight.
38. The anterior pituitary secretes growth hormone (GH), also called (a)_____________ which is referred to as an (b)_________________ because it promotes tissue growth.
39. Hypersecretion of GH during childhood may cause a disorder known as (a) _____________. In adulthood, hypersecretion can cause "large extremities," or (b) _____________.

Thyroid hormones increase metabolic rate

40. The thyroid secretes (a) _____________, also known as T4 because each molecule contains four atoms of (b) ____________, and (c) _____________, also known as T3 because each molecule contains three atoms of (d) _________.
41. Thyroid secretion is regulated by a negative feedback system between the thyroid gland and the (a) _____________ gland, which releases less (b) _____________ when thyroid hormone blood titers rise above normal.
42. Hyposecretion of thyroid hormones during infancy and childhood may result in retarded development, a condition known as ___________.
43. The most common type of hyperthyroidism is ___________ which is actually an autoimmune disease.

The parathyroid glands regulate calcium concentration

44. Calcium levels are regulated by the interaction of two hormones. Parathyroid hormone stimulates the release of calcium from (a)_________ and calcium reabsorption by the (b)_____________________. The (c) ____________

gland secretes (d) ______________, which inhibits parathyroid hormone activity and calcium reabsorption.

The islets of the pancreas regulate glucose concentration

45. Numerous clusters of cells in the pancreas, called (a)________________________, secrete hormones that regulate glucose concentration in the blood. When blood glucose levels are high, (b) ______ cells release the hormone (c) ________ which lowers blood glucose levels; when it is low, (d) ________ cells release (e) ______________which raises blood glucose levels.
46. Insulin lowers the concentration of glucose in the blood by stimulating uptake of glucose by (a) ____________, (b) ____________, and (c) ____________ cells.
47. In addition to regulating glucose levels, insulin also regulates (a)______________ and (b) ______________ metabolism.
48. Glucagon stimulates (a) __________ cells to convert glycogen to glucose. It also stimulates (b) ______________, the production of glucose from noncarbohydrates.
49. Type I diabetes is an (a) ____________________ in which (b)________________ are destroyed resulting in insulin deficiency. Type II diabetes is typically observed in (c) ____________ individuals. It is often proceeded by (d) ____________ syndrome. Type II diabetes typically begins as (e)________________________.
50. Similar metabolic disturbances occur in both types of diabetes. They are disruptions in (a) ____________, (b) ____________, and (c)____________ metabolism and (d)____________ imbalance.
51. Glucose is excreted in the urine when its ______________________________ is exceeded.

The adrenal glands help the body respond to stress

52. The adrenal medulla and the adrenal cortex secrete hormones that help the body cope with stress and regulate metabolism. The adrenal medulla secretes (a) ____________________ and (b)________________________. These hormones increase heart rate, metabolic rate, strength of muscle contraction, and reroute blood to organs that require more blood in a time of stress and are in the chemical group known as (c) ____________________.
53. The adrenal cortex synthesizes (a) ________________hormones but it only secretes three in appreciable amounts. They are (b)________________________, (c)________________________, and (d)________________________. The sex hormone precursors are converted to (e) ________________, the principal male sex hormone, and (f) ______________________, the principal female sex hormone.
54. Kidneys reabsorb more sodium and excrete more potassium in response to the mineralocorticoid hormone ________________.

55. The main function of ______________________________ is to enhance gluconeogenesis in the liver.
56. Stress stimulates the hypothalamus to secrete CRF, which is the abbreviation for (a) ______________________________. CRF stimulates the anterior pituitary to release (b)______________________________, which regulates the secretion of glucocorticoids and aldosterone.
57. __________________ disease results when the adrenal cortex produces insufficient amounts of aldosterone and cortisol.

Many other hormones help regulate life processes

58. The (a) ________________ gland in the brain produces the hormone (b) ________________, which influences biological rhythms and the onset of sexual maturation.
59. The thymus gland produces the hormone ________________ which plays a role in immunity.
60. The heart secretes ANF, which stands for ______________________________, which promotes sodium excretion and lowers blood pressure.

BUILDING WORDS

Use combinations of prefixes and suffixes to build words for the definitions that follow.

Prefixes	The Meaning	Suffixes	The Meaning
gluc-	sweet	-agon	to fight
hom(eo)-	same	-megaly	enlargement
hyper-	over	-oid	resembling
hypo-	under	-stasis	to control the process of
neuro-	nerve		
ster-	hard,solid		

Prefix	Suffix	Definition
__________	-hormone	1. A hormone secreted by certain nerve cells.
__________	-endocrine	2. Refers to a nerve cell that secretes a neurohormone.
__________	-secretion	3. An excessive secretion; over secretion.
__________	-secretion	4. A diminished secretion; under secretion.
__________	-thyroidism	5. A condition resulting from an overactive thyroid gland.
__________	-glycemia	6. An abnormally low level of glucose in the blood.
__________	-glycemia	7. An abnormally high level of glucose in the blood.
acro-	__________	8. An abnormal condition characterized by enlargement of the head, and sometimes other structures.
__________	__________	9. The balanced internal environment of the body.
__________	__________	10. A hormone with 4 carbon rings as a part of their structure. Testosterone for example.
__________	__________	11. A hormone produced by alpha cells that raises blood glucose levels.

MATCHING

Terms:

a. adrenal gland
b. aldosterone
c. beta cells
d. calcitonin
e. endocrine gland
f. glucagon
g. hormone
h. insulin
i. oxytocin
j. prostaglandin
k. thymus
l. thyroid gland
m. thyroxine
n. tropic hormone

For each of these definitions, select the correct matching term from the list above.

____ 1. A hormone produced by the hypothalamus and released by the posterior lobe of the pituitary; causes the uterus to contract and stimulates the release of milk from the mammary glands.

____ 2. A hormone secreted by the thyroid gland that rapidly lowers the calcium content in the blood.

____ 3. An endocrine gland that lies anterior to the trachea and releases hormones that regulate the rate of metabolism.

____ 4. Paired endocrine glands, each located just superior to each kidney.

____ 5. General term for an organic chemical produced in one part of the body and transported to another part where it affects some aspect of metabolism.

____ 6. Cells of the pancreas that secrete insulin.

____ 7. A gland that secretes products directly into the blood or tissue fluid instead of into ducts.

____ 8. General term for a hormone that helps regulate another endocrine gland.

____ 9. One of the hormones produced by the thyroid gland.

____10. An endocrine gland that produces thymosin; important in the development of the immune response mechanism.

MAKING COMPARISONS

Fill in the blanks.

Hormone	Chemical Group	Function
Testosterone	Steroid	Develops and maintains sex characteristics of males; promotes spermatogenesis
Aldosterone	#1	#2
Thyroxine	#3	#4
#5	Peptide	Stimulates reabsorption of water; conserves water
#6	#7	Helps body adapt to long-term stress; mobilizes fat; raises blood glucose levels
#8	#9	Affects wide range of body processes; may interact with other hormones to regulate metabolic activities
ACTH	Peptide	#10
Epinephrine	#11	#12
#13	#14	Stimulates uterine contraction

MAKING CHOICES

Place your answer(s) in the space provided. Some questions may have more than one correct answer.

____ 1. GH is referred to as an anabolic hormone because it
a. suppresses appetite.
b. stimulates other endocrine glands.
c. promotes tissue growth.
d. regulates the anterior pituitary gland.
e. affects neural membrane potentials.

____ 2. The thyroid gland is located
a. in the neck.
b. behind the trachea.
c. in front of the trachea.
d. above the larynx.
e. below the larynx.

____ 3. The hormone and target tissue involved in raising glucose concentration in blood by glycogenolysis and gluconeogenesis are
a. insulin and pancreas.
b. glucagon and liver.
c. thyroid-stimulating hormone and thyroid gland.
d. adrenocorticotropic hormone and adrenal cortex.
e. thyroxine and various metabolically active cells.

____ 4. Activity of the anterior lobe of the pituitary gland is controlled by
a. the hypothalamus.
b. ADCH.
c. epinephrine and norepinephrine.
d. releasing hormones.
e. inhibiting hormones.

____ 5. A chemical produced by one cell that has a specific regulatory effect on another cell defines a/an
a. neurohormone.
b. hormone.
c. exocrine gland secretion.
d. pheromone.
e. prohormone.

____ 6. A three-year old cretin and an adult suffering from myxedema most likely have
a. a low metabolic rate.
b. Cushing's disease.
c. Addison's disease.
d. a low level of thyroid hormones.
e. a high level of thyroid hormones.

____ 7. A person with a fasting level of 750 mg glucose per 100 ml of blood
a. is hyperglycemic.
b. is hypoglycemic.
c. is about normal.
d. probably has cells that are not using enough glucose.
e. has too much insulin.

____ 8. Increased skeletal growth results directly and/or indirectly from activity of
a. aldosterone.
b. growth hormone.
c. hormones from the hypothalamus.
d. andosterides.
e. epinephrine and/or norepinephrine.

____ 9. Diabetes in an adult with ample numbers of adequately functioning beta cells and normal concentrations of insulin
a. is type I diabetes.
b. is type II diabetes.
c. indicates that target cells are not using insulin.
d. indicates that not enough insulin is being produced.
e. is unusual since this condition usually occurs in children.

____10. Small, hydrophobic hormones that form a hormone-receptor complex with intranuclear receptors include
a. steroids.
b. cyclic AMP.
c. chemicals that activate genes.
d. thyroid hormones.
e. prostaglandins.

____11. A possible negative consequence of hyposecretion of insulin is
a. Addison's disease.
b. Cushing's disease.
c. myxedema.
d. diabetes mellitus.
e. dwarfism.

____12. Which of the following is mismatched?
a. exocrine glands : ductless
b. negative feedback : calcium regulation
c. neuropeptide : oxytocin
d. steroid hormone : G protein
e. GTP : adenylyl cyclase

____13. The two hormones involved in the regulation of calcium levels are
a. thyroid hormone and parathyroid hormone.
b. thyroid hormone and calcitonin.
c. growth hormone and thyroid hormone.
d. parathyroid hormone and calcitonin.
e. a mineralocorticoid and a glucocorticoid.

____14. Which hormone is incorrectly matched with its function?
a. oxytocin : stimulates smooth muscle contraction.
b. antidiuretic hormone : stimulates water secretion
c. melatonin : regulates biological rhythms
d. progesterone : development of male characteristics
e. epinephrine : stimulates heart rate

____15. Which gland is incorrectly matched with its hormone?
a. thyroid gland : parathyroid hormone
b. anterior pituitary : oxytocin
c. adrenal cortex : epinephrine
d. posterior pituitary : antidiuretic hormone
e. hypothalamus : prolactin

____16. Melanocyte stimulating hormones
a. are secreted by the anterior pituitary.
b. are steroid hormones.
c. are secreted by neurons in the hypothalamus.
d. are important in animal camouflage.
e. regulate secretion of anabolic hormone.

____17. Goiter
a. may result from hypersecretion of thyroid hormone.
b. may occur when there is an iodine deficiency.
c. may result from hyposecretion of thyroid hormone.
d. may be associated with TSH receptors and Graves disease.
e. is an autoimmune disease.

____18. Diabetes
a. may cause a disruption in fat metabolism.
b. may cause a disruption in protein metabolism.
c. may cause a disruption in carbohydrate metabolism.
d. involves alpha cells and insulin.
e. may cause an electrolyte imbalance.

____19. Which disease and gland is not correctly paired?
a. Addison's disease : adrenal gland
b. Cushing's syndrome : adrenal gland
c. insulin resistance : pancreas
d. myxedema : hypothalamus
e. credtinism : thyroid gland

____20. Which of the following is not correctly paired?
a. prostaglandin : fatty acid derivative
b. neuropeptide : water-soluble
c. melatonin : steroid hormone
d. thyroid hormone : amino acid derivative
e. molting hormone : steroid hormone

____21. Local regulators of cellular processes include
a. histamine.
b. growth factors.
c. some peptides.
d. dopamine.
e. nitrous oxide.

____22. Second messengers involved in signal transduction pathways include
a. calcium
b. diacylglycerol
c. phospholipase C
d. cAMP
e. inositol diphosphate

____23. Juvenile hormone
a. is secreted by the corpora allata.
b. stimulates the release of grain hormone.
c. stimulated the prothoracic glands
d. suppresses metamorphosis.
e. is secreted by the corpora cardiaca

____24. Which of the following is not correctly paired with its target tissue?
a. pineal gland : hypothalamus
b. anterior pituitary : hypothalamus
c. adrenal cortex : kidney tubules
d. anterior pituitary : thyroid gland
e. pancreas: liver

VISUAL FOUNDATIONS

Color the parts of the illustration below as indicated. Also label transcription, translation, and the target cell.

RED	❑	receptor molecule	BROWN	❑	endocrine gland cell
YELLOW	❑	hormone molecules	TAN	❑	nucleus
BLUE	❑	DNA	PINK	❑	blood vessel
ORANGE	❑	mRNA	VIOLET	❑	protein molecule

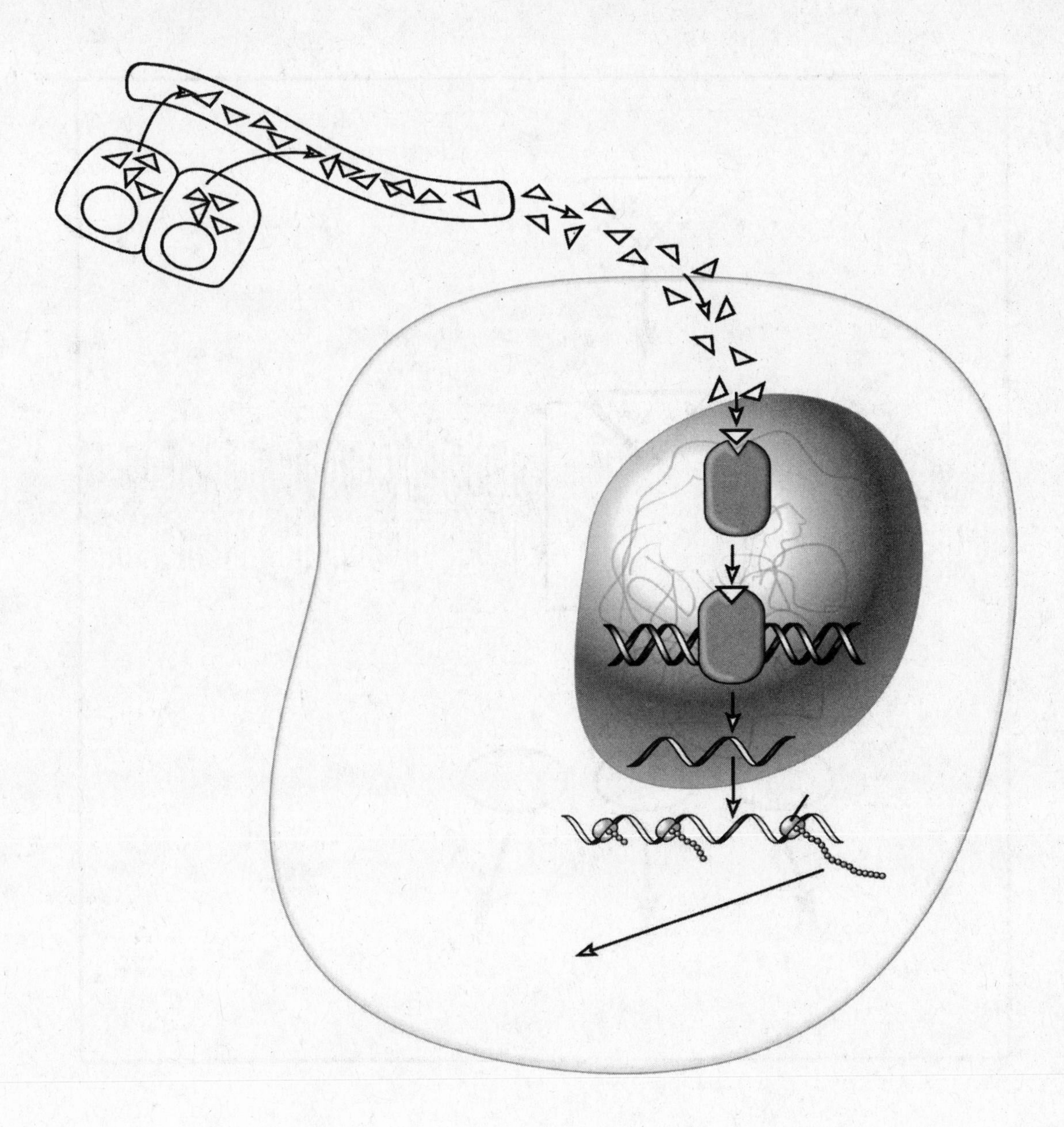

Color the parts of the illustration below as indicated.

RED	❑	receptor
GREEN	❑	hormone
YELLOW	❑	cytosol
BLUE	❑	second messengers
ORANGE	❑	extracellular fluid
BROWN	❑	plasma membrane

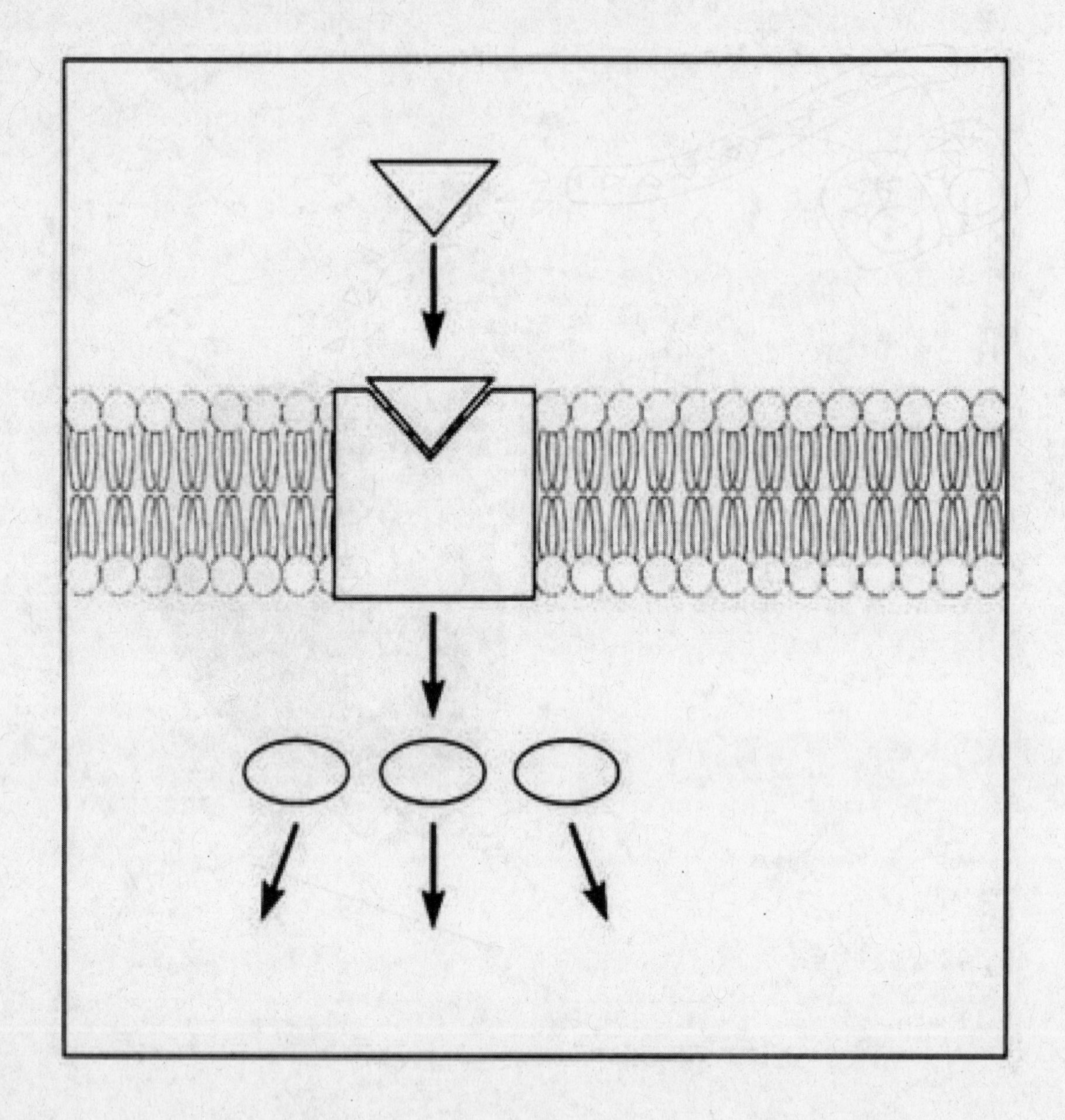

Color the parts of the illustration below as indicated.

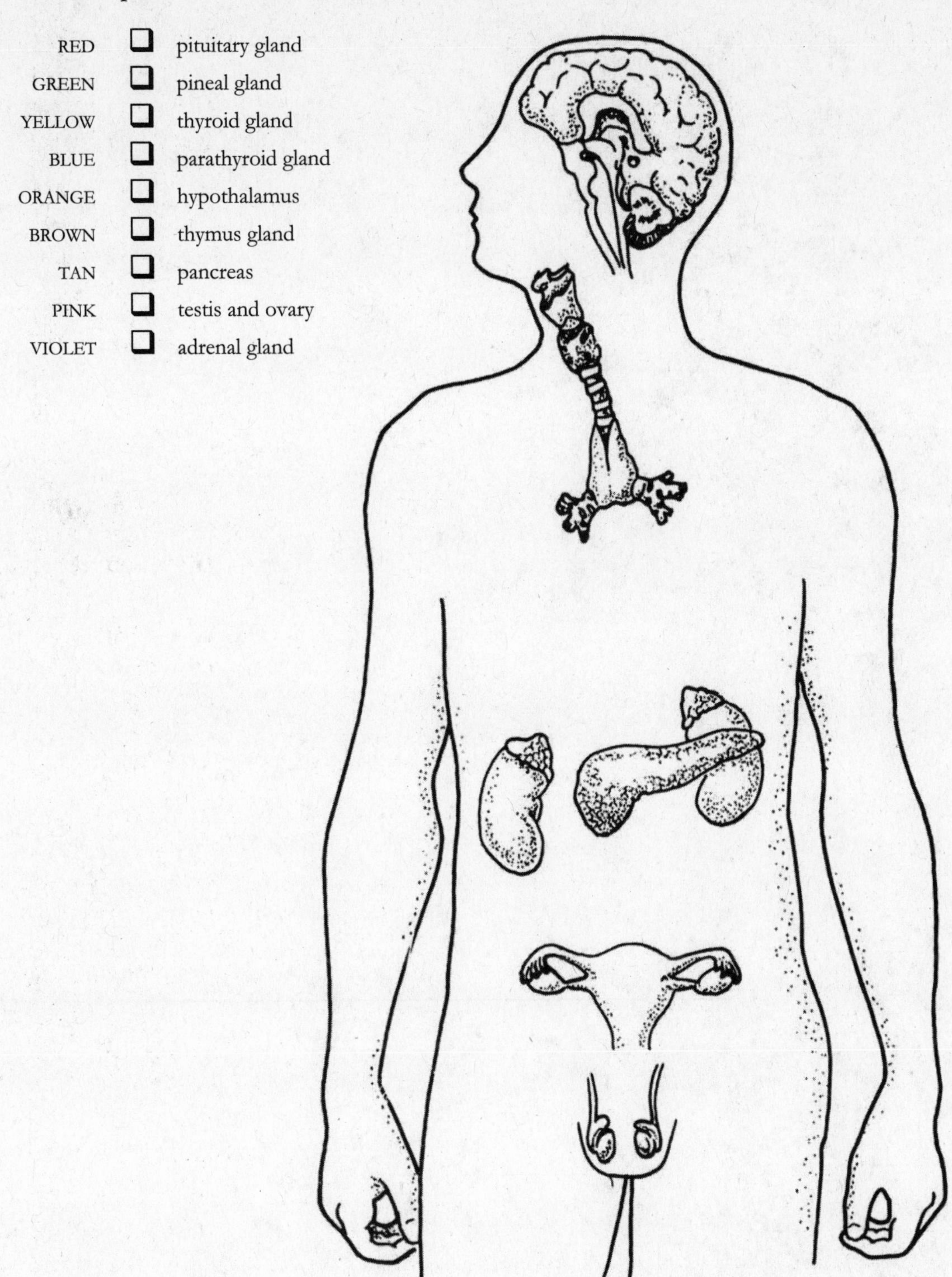

CHAPTER 50

❑

Reproduction

For a species to survive, it is essential that it replace individuals that die. Animals do this either asexually or sexually. In asexual reproduction, a single parent splits, buds, or fragments, giving rise to two or more offspring that are genetically identical to the parent. In sexual reproduction, sperm contributed by a male parent, and an egg contributed by a female parent, unite, forming a zygote that develops into a new organism. This method of reproduction promotes genetic variety among the offspring, and therefore, gives rise to individuals that may be better able to survive than either parent. Depending on the species, fertilization occurs either outside or inside the body. Some animals have both asexual and sexual stages. In some, the unfertilized egg may develop into an adult. In still others, both male and female reproductive organs may occur in the same individual. More typically, animals have only a sexual stage, fertilization is required for development to occur, and male and female reproductive organs occur only in separate individuals. The complex structural, functional, and behavioral processes involved in reproduction in vertebrates are regulated by hormones secreted by the brain and gonads. Several hormones regulate the birth process in humans. There are a variety of effective methods of birth control. Next to the common cold, sexually transmitted diseases are the most prevalent communicable diseases in the world.

REVIEWING CONCEPTS

Fill in the blanks.

INTRODUCTION

1. Hermaphrodites have (a) ________________and (b) ____________ reproductive organs and produce (c) ____________ and (d) ________.

ASEXUAL AND SEXUAL REPRODUCTION

Asexual reproduction is an efficient strategy

2. In asexual reproduction, the offspring are __________ to the parent that produces them.
3. Sponges and cnidarians can reproduce by ______________, which occurs when a group of cells separate from the "parent's" body and form a new individual.
4. _______________________ is a form of asexual reproduction whereby a "parent" breaks into several pieces, each piece giving rise to a new individual.
5. Parthenogenesis refers to the production of a complete organism from a(n) ______________________.

Most animals reproduce sexually

6. The union of an egg and a sperm cell produces a _______________.
7. Many aquatic animals practice (a) ________________________ in which the gametes meet outside the body. (b)____________________________ occurs when the sperm and egg unite inside the body.
8. ______________________ is a form of sexual reproduction in which a single individual produces both eggs and sperm.

Sexual reproduction increases genetic variability

9. The fastest and most efficient method of reproduction is by ____________________ reproduction.
10. The main benefit of sexual reproduction is that it can increase the (a)______________ of a population. Sexual reproduction is thought to remove (b) ___________________ from a population and at the same time it maintains (c) __________.

HUMAN REPRODUCTION: THE MALE

The testes produce gametes and hormones

11. The production of sperm is called (a) _______________. The initial development of a sperm begins with an undifferentiated stem cell called a (b) ________________, which enlarges to form the (c) _________________ that goes through meiosis. The resulting secondary spermatocytes undergo a second (d)__________ division forming (e) _________________.
12. The tip of a mature sperm cell is covered by the (a) ______________ which contains enzymes that help the sperm to (b)____________________________.
13. (a) _______________ cells provide nutrients to the developing sperm cells and secrete hormones. These cells form (b)______________________ and the blood-testis barrier.
14. The blood-testis barrier prevents developing sperm cells from stimulating an ________________________________.
15. The testes, are housed in an external sac called the (a) ____________. And contain the (b) ________________________ tubules in which the spermatids are formed.
16. The ______________________ is the passage way that a testis takes as it descends through the abdominal wall into the scrotum.

A series of ducts store and transport sperm

17. Sperm complete their maturation and are stored in the (a)______________, from which they are moved up into the body through the (b) _____________. The next portion of the tubule, the (c) _________________passes through the prostate gland and then opens into the (d) _________________.

The accessory glands produce the fluid portion of semen

18. Human semen contains an average of about ________ million sperm.

19. When ejaculated, sperm are suspended in fructose rich secretions that are produced by two paired glands called the (a)______________________________. These glands also produce (b) ______________ which stimulate smooth muscle contractions in males and females. The (c) ____________________ secretes an alkaline fluid into the forming semen to neutralize acidic conditions in the female vagina. Prostate secretions also include clotting enzymes and prostate-specific antigen.
20. Prior to intercourse, the _________________ glands release a mucous secretion that lubricates the penis.

The penis transfers sperm to the female

21. The penis consists of an elongated (a) __________ that terminates in an expanded portion called the (b) _______________, which is partially covered by a fold of skin called the (c) ______________. An operation that removes this "extra" skin is known as (d) ________________.
22. The penis contains three columns of erectile tissue, two (a)________________________ and one (b) ________________________. When they become engorged with (c) _______________, the penis becomes erect.

Testosterone has multiple effects

23. Testosterone is classified as an (a) _______________ and is the principal male sex hormone. In some target tissues, testosterone is converted to other (b) ______________. Interestingly, in brain cells testosterone is converted to (c) ________________, a female sex hormone.
24. Testosterone is produce by the _______________________ cells located between the seminiferous tubules.

The hypothalamus, pituitary gland, and testes regulate male reproduction

25. The two gonadotropic hormones, (a) ____ and (b) _______________________________________, along with testosterone, stimulate sperm production.
26. LH stimulates secretion of the hormone __________________, which is responsible for establishing and maintaining primary and secondary sex characteristics in the male.
27. The Sertoli cells secrete (a) __________________________________ and other signaling molecules necessary for spermatogenesis. They also secrete the hormone (b) _____________________ which regulates FSH secretion.

HUMAN REPRODUCTION: THE FEMALE

The ovaries produce gametes and sex hormones

28. Formation of ova is called (a) ___________________. In this process stem cells in the fetus, the (b) ____________________, enlarge, forming the (c) ___________________________, which begin meiosis before birth.
29. A primary oocyte and the granulosa cells immediately around it comprise the (a) ______________. At puberty, the hormone (b) _______

stimulates development of a primary oocyte, causing it to complete meiosis. Meiosis produces a large cell and two smaller (c)________________. The large cell or (d)________________, continues development to form the ovum.

30. After ovulation, the portion of the follicle that remains in the ovary develops into a temporary endocrine gland called the (a)____________ which secretes (b) ________________.

The oviducts transport the secondary oocyte

31. After ovulation, the secondary oocyte is swept into the oviduct also known as the (a) ________________tube or (b)____________ tube.
32. In a tubal pregnancy, the embryo begins to develop in the ________________ rather than the uterus.

The uterus incubates the embryo

33. The layer of the uterus facing the lumen is the (a) ____________. This lies on top of the thicker (b) ________________.
34. Embryos normally implant in the (a) ____________ of the uterus. If there is no embryo, or the embryo fails to implant, this layer is shed during (b) ____________.
35. The lower portion of the uterus, the ____________, extends slightly into the vagina.

The vagina receives sperm

36. The vaginal opening is located in the (a) ________________ region. The vagina serves as a receptacle for sperm and as a part of the (b) ________________.

The vulva are external genital structures

37. The female external genitalia are collectively known as the ________.
38. The most sensitive part in the vulva is the (a) ________ that responds to sexual arousal in the same way as the male (b) ________.

The breasts function in lactation

39. The gland cells of the breast are arranged in grapelike clusters called ________________.
40. (a) ________________ is the production of milk. Milk production is stimulated by the hormone (b) ____________. Nursing stimulates the release of (c) ________________ which promotes the ejection of milk from the alveoli.
41. Oxytocin released during breast-feeding stimulate the ____________ to contract and return to its original size.

The hypothalamus, pituitary gland, and ovaries regulate female reproduction

42. The hormone ________________ acts in females in a way similar to testosterone in males.

43. The menstrual cycle begins at puberty and occurs each month until about the age of ______.
44. The first day of the menstrual cycle is the onset of menstruation. The first two weeks of the menstrual cycle correspond with the (a)______________________ phase in the ovary. At about day 14 (b) ________________ occurs leading to the (c)____________________ phase in the final two weeks. The menstrual phase is about the first five days of the preovulatory phase.
45. During the first part of the preovulatory phase, the hypothalamus secretes (a) ______________________________ hormone which stimulates the anterior pituitary to release (b)__________________hormone which promotes follicle development and (c) __________________hormone which stimulates the theca cells to proliferate and produce androgens.
46. After the first week, one dominate follicle remains and its granulosa cells become sensitive to LH and FSH. The granulosa cells secrete estrogen which inhibits (a) _______ and (b) ___________ secretion.
47. Around day 14 or midcycle, LH levels rise again and this stimulates the final maturation of the follicle and _______________________.
48. The postovulatory phase is also called the (a) ________________ phase. During this time LH promotes the development of the corpus luteum which secretes (b) ____________________ and (c)____________________,. These hormones stimulate final preparation of the uterus for a possible pregnancy.
49. If pregnancy does not occur the corpus luteum degenerates and ________________________ begins.

Menstrual cycles stop at menopause

50. A change in the (a) ________________________ triggers menopause, a period when (b) __________ are no longer produced and a woman becomes (c) ________________________.

Most mammals have estrous cycles

51. In an estrous cycle, the endometrium is ______________________ if conception does not occur and not expelled as occurs in humans and other primates.

FERTILIZATION, PREGNANCY, AND BIRTH

Fertilization is the fusion of sperm and egg

52. Fertilization and the establishment of pregnancy together are referred to as ______________________________.
53. After being deposited in the female reproductive tract, sperm generally retain the ability to fertilize and ovum for about (a) ______ hours. but they can remain alive for up to 5 days. The ovum remains fertile for up to (b) _____ hours after ovulation.
54. Uterine and oviduct smooth muscle contractions help to move the sperm toward the ovum. These contractions are stimulated by (a)______________ secreted by the female and (b)______________

present in the semen. The secondary oocyte is also thought to release chemoattractants that attract sperm to the oocyte for fertilization.

55. While fertilization is a monspermatic event, millions are released into the female reproductive tract because many die or are phagocytized and the ________________ from many acrosomes are needed to penetrate the oocyte barriers.

Hormones are necessary to maintain pregnancy

56. Membranes that develop around an embryo secrete a peptide hormone called (a) ________________ that signals the corpus luteum to continue to function and release (b)________________ that will stimulate development of the uterine wall including the muscle and (c) ________________ that inhibits uterine contractions.

57. Membranes surrounding the embryo and uterine tissue form the ________________, an organ of exchange between the embryo and the mother. Once fully developed, it will take over the secretion of estrogen and progesterone.

The birth process depends on a positive feedback system

58. At the end of pregnancy, (a) ________________, and increased concentrations of (b) ________________, and (c) ________________ initiate uterine contractions.

59. During the first stage of labor uterine contractions move the fetus toward (a) ________________. The amnion also breaks releasing (b) ________________. A (c) ________________ cycle of uterine contractions continues to move the baby toward the birth canal.

60. During the second stage of labor the fetus is ________________.

61. During the third stage of labor, the (a) ________________ and (b) ________________ are expelled as afterbirth.

HUMAN SEXUAL RESPONSE

62. Sexual stimulation results in two basic physiological responses: (a)________________ and (b)________________.

63. The four phases of the sexual response cycle are (a)________________, (b)________________, (c)________________, and (d) ________________.

BIRTH CONTROL METHODS AND ABORTION

Many birth control methods are available

64. Any deliberate separation of sexual intercourse from reproduction is called ________________.

Most hormone contraceptives prevent ovulation

65. The most common oral contraceptives are combinations of (a)________________, and (b) ________________.

Intrauterine devices are widely used

66. The IUD is the most widely used method of (a) ________________ contraception. It is used by an estimated (b) ____________________ (#?) women around the world.

Barrier methods of contraceptive methods include the diaphragm and condom

67. The condom is the only contraceptive device that affords some protection against ____________________________________.

Emergency contraception is available

68. The most common types of emergency contraception are (a)____________________________, which can decrease the probability of pregnancy by about 89% and the insertion of a (b)______________________________ within five days of intercourse. This method is about 99% effective.

Sterilization renders an individual incapable of producing offspring

69. Male sterilization is accomplished by performing a (a)__________________ in which the (b)____________________ is cut.

70. Female sterilization is accomplished by performing a (a)____________________ in which the (b) ______________ are cut.

Future contraceptives may control regulatory peptides

71. One method of contraception being investigated involves the stimulation of the immune system to produce (a) __________________ which would act against hormones such as hCG. Some investigators are exploring ways to interfere with sperm (b) ____________________ or with (c) ______________________.

Abortions can be spontaneous or induced

72. The drug RU-496 is used to interrupt a pregnancy. It works by ____, but it does not activate them. This causes the endometrium to break down and uterine contractions to begin.

SEXUALLY TRANSMITTED INFECTIONS

73. (a) ______________________________________ is one the two most common STIs in the United States. About (b) ________________ (#?) people are currently infected. The second most common STI is (c)________________. This infection can lead to (d)______________________ disease in infected women. (e)______________________________ is an STI that is caused by a virus and individuals may not know that they are infected.

BUILDING WORDS

Use combinations of prefixes and suffixes to build words for the definitions that follow.

Prefixes	The Meaning
acro-	extremity, height
circum-	around, about
contra-	against, opposite, opposing
endo-	within
intra-	within
oo-	egg
partheno-	virgin
post-	behind, after
pre-	before, prior to, in advance of, early
spermato-	seed, "sperm"
vaso-	vessel

Suffixes	The Meaning
-ectomy	excision
-gen(esis)	production of
-some	body

Prefix	Suffix	Definition
________	________	1. The production of an adult organism from an egg in the absence of fertilization; virgin development.
________	________	2. The production of sperm.
________	________	3. The organelle (body) at the extreme tip of the sperm head that helps the sperm penetrate the egg.
________	-cision	4. The removal of all or part of the foreskin by cutting all the way around the penis.
________	________	5. The production of eggs or ova.
________	-metrium	6. The lining within the uterus.
________	-ovulatory	7. Pertains to a period of time after ovulation.
________	-ovulatory	8. Pertains to a period of time prior to ovulation.
vas-	________	9. Male sterilization in which the vas deferentia are partially excised.
________	-congestion	10. The engorgement of erectile tissues with blood.
________	-uterine	11. Located or occurring within the uterus.
________	-ception	12. Birth control; techniques or devices opposing conception.

MATCHING

Terms:

a. Budding
b. Clitoris
c. Corpus luteum
d. Ejaculatory duct
e. Fertilization
f. Fragmentation
g. Ovary
h. Oviduct
i. Ovulation
j. Progesterone
k. Prostate gland
l. Scrotum
m. Semen
n. Testosterone
o. Zygote

For each of these definitions, select the correct matching term from the list above.

____ 1. Fluid composed of sperm suspended in various glandular secretions that is ejaculated from the penis during orgasm.

____ 2. A hormone produced by the corpus luteum of the ovary and by the placenta; acts with estradiol to regulate menstrual cycles and to maintain pregnancy.

____ 3. A gland in male mammals that secretes an alkaline fluid that is part of the seminal fluid.

____ 4. A small, erectile structure at the anterior part of the vulva in female mammals; homologous to the male penis.

____ 5. One of the paired female gonads; responsible for producing eggs and sex hormones.

____ 6. Pocket of endocrine tissue in the ovary that is derived from the follicle cells; secretes the hormone progesterone and estrogen.

____ 7. The fusion of the male and female gametes; results in the formation of a zygote.

____ 8. A short duct that passes through the prostate gland and connects the vas deferens to the urethra.

____ 9. The release of a mature "egg" from the ovary.

____10. The external sac of skin found in most male mammals that contains the testes and their accessory organs.

____11. A form of asexual reproduction in which the adult breaks into pieces and new offspring are formed.

____12. This hormone is produced by interstitial cells.

MAKING COMPARISONS

Fill in the blanks.

Endocrine Gland	Hormone	Target Tissue	Action in Male	Action in Female
Hypothalamus	GnRH	Anterior pituitary	Stimulates release of FSH and LH	Stimulates release of FSH and LH
#1	FSH	#2	#3	#4
Anterior pituitary	LH	Gonad	#5	#6
#7	#8	General	Establishes and maintains primary & secondary sex characteristics	Not produced
#9	#10	General	Not produced	Establishes and maintains primary & secondary sex characteristics

MAKING CHOICES

Place your answer(s) in the space provided. Some questions may have more than one correct answer.

____ 1. The hormone in urine or blood that indicates pregnancy is
a. GnRH.
b. FSH.
c. hCG.
d. inhibin.
e. ABP.

____ 2. The tube(s) that transport(s) gametes away from gonads is/are the
a. vas deferens.
b. seminal vesicle.
c. sperm cord.
d. vagina.
e. oviduct.

____ 3. A human female at birth already has
a. primary oocytes in the first prophase.
b. preovulatory Graafian follicles.
c. all the formative oogonia she will ever have.
d. some secondary oocytes in meiosis II.
e. zona pellucida.

____ 4. The *basic* physiological responses that result from effective sexual stimulation are
a. vasocongestion.
b. erection.
c. vaginal lubrication.
d. muscle tension.
e. orgasm.

____ 5. Human chorionic gonadotropin (hCG)
a. is produced by embryonic membranes.
b. is secreted by the endometrium.
c. affects functioning of the corpus luteum.
d. is originally produced in the pituitary.
e. causes menstrual flow to begin.

____ 6. The hormone(s) primarily responsible for secondary sexual characteristics is/are
a. luteinizing hormone.
b. estrogen.
c. follicle-stimulating hormone.
d. testosterone.
e. human chorionic gonadotropin.

____ 7. The corpus luteum is directly or indirectly involved in
a. secreting estrogen.
b. secreting progesterone.
c. stimulating the hypothalamus.
d. responding to signals from embryonic membrane secretions.
e. maintaining a newly implanted embryo.

____ 8. Follicle-stimulating hormone (FSH) stimulates development of
a. seminiferous tubules.
b. interstitial cells.
c. sperm cells.
d. the corpus luteum.
e. ovarian follicles.

____ 9. When used properly, the most effective form of birth control is
a. condoms.
b. douching.
c. oral contraception.
d. the rhythm method.
e. spermicidal sponges.

____10. The first measurable response to effective sexual stimulation
a. is orgasm.
b. is ejaculation.
c. occurs in the excitement phase.
d. is penile erection.
e. is vaginal lubrication.

____11. The form(s) of birth control that can definitely deter contraction of sexually transmitted diseases is/are
a. condoms.
b. douching.
c. oral contraception.
d. diaphragms.
e. spermicides.

____12. Your authors consider a count of 200 million sperm in a single ejaculation as
a. clinical sterility.
b. above average.
c. below average.
d. about average.
e. impossible.

____13. A Papanicolaou test (Pap smear)
a. can detect early cancer.
b. requires an abdominal incision.
c. involves examination of cells taken from the lower portion of the uterus.
d. looks at cells from the menstrual flow.
e. examines cells scraped from the cervix.

____14. Primitive, undifferentiated stem cells that line the seminiferous tubules are called
a. sperm.
b. spermatids.
c. primary spermatocytes.
d. secondary spermatocytes.
e. spermatogonia.

____15. The external genitalia of the human female are collectively known as the
a. vagina.
b. clitoris.
c. mons pubis.
d. major and minor lips.
e. vulva.

____16. In gametogenesis in the human male, the cell that develops after the first meiotic division is the
a. sperm.
b. spermatid.
c. primary spermatocyte.
d. secondary spermatocyte.
e. spermatogonium.

____17. The portion of the ovarian follicle that remains behind after ovulation
a. develops into the corpus luteum.
b. becomes a gland.
c. is called the Graafian follicle.
d. is surrounded by the zona pellucida.
e. transforms into the endometrium.

____18. The term gonad refers to the
a. penis.
b. testes.
c. vagina.
d. ovaries.
e. gametes.

____19. Human sperm cell production occurs in the
a. epididymis.
b. sperm tube.
c. testes.
d. accessory glands.
e. seminiferous tubules.

____20. The normal, periodic flow of blood from the vagina is associated with
a. ovulation.
b. orgasm.
c. sloughing of the endometrium.
d. removal of the inner lining of the vagina.
e. miscarriages.

____21. Human female breast size is primarily a function of the amount of
a. alveoli.
b. milk glands.
c. adipose tissue (fat).
d. milk the woman can produce after giving birth.
e. colostrum.

_____22. Which of the following is/are true of the human penis?
a. It has three erectile tissue bodies.
b. Circumcision involves removal of the prepuce.
c. The glans is an expanded portion of the spongy body.
d. It contains a bone.
e. Erection results from increased blood pressure.

_____23. The term "virgin development" refers to which method of reproduction?
a. budding
b. fragmentation
c. fission
d. hermaphroditism
e. parthenogenesis

_____24. Testosterone
a. stimulates descent of the testes into the scrotum.
b. secretion is under the influence of FSH.
c. is down regulated by inhibin.
d. is produced by interstitial cells in the testes.
e. is produced by Sertoli cells in the testes.

_____25. Seminal vesicles
a. secrete fructose.
b. secrete fibrinogen.
c. secrete prostaglandins.
d. secrete an alkaline fluid.
e. secrete clotting enzymes.

_____26. Low sperm counts may be the result of
a. smoking.
b. alcohol use.
c. exposure to PCBs.
d. decreased PSA levels.
e. high levels of ABP.

_____27. Which of the following is incorrectly paired?
a. inhibin : anterior pituitary
b. androgen binding protein : Sertoli cells
c. oxytocin : stimulation of uterine contractions
d. luteinizing hormone : development of the endometrium
e. follicle stimulating hormone : Sertoli cells

VISUAL FOUNDATIONS

Color the parts of the illustrations below as indicated. Label the first meiotic division and second meiotic division.

RED	❑	sperm
PINK	❑	large chromosome
BLUE	❑	small chromosome
ORANGE	❑	primary spermatocyte
BROWN	❑	secondary spermatocyte
YELLOW	❑	spermatids

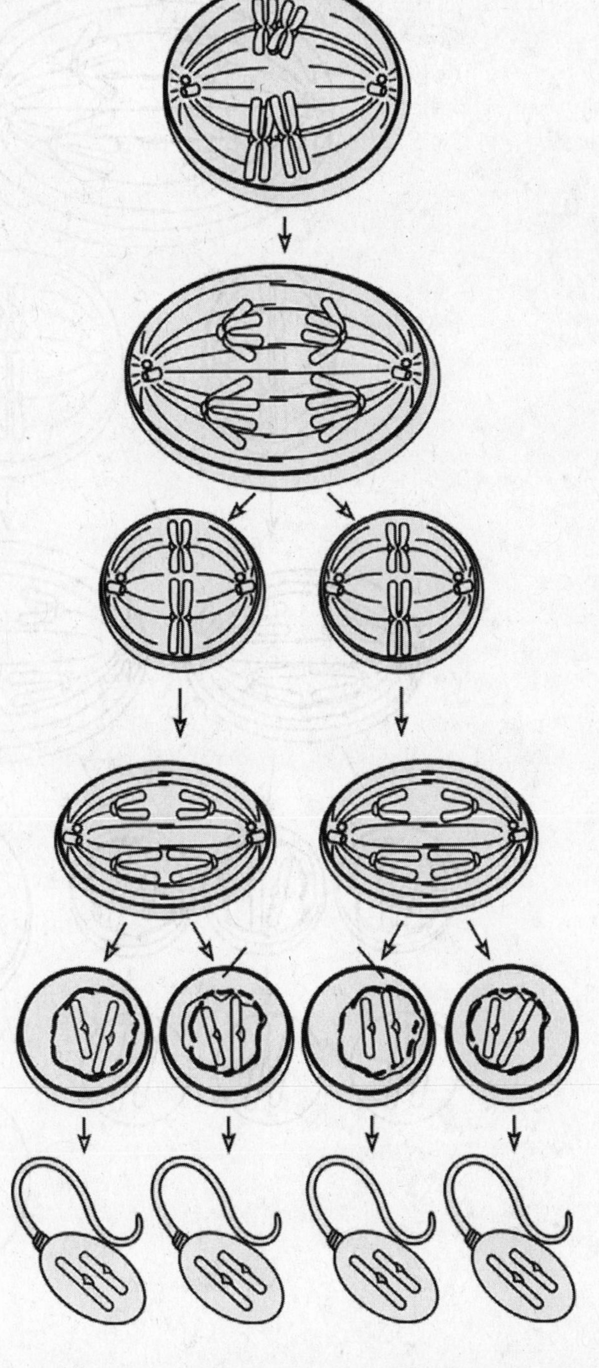

Color the parts of the illustrations below as indicated. Label first and second meiotic divisions.

RED ❑ ovum
PINK ❑ large chromosome
BLUE ❑ small chromosome
ORANGE ❑ primary oocyte
BROWN ❑ secondary oocyte
YELLOW ❑ polar body

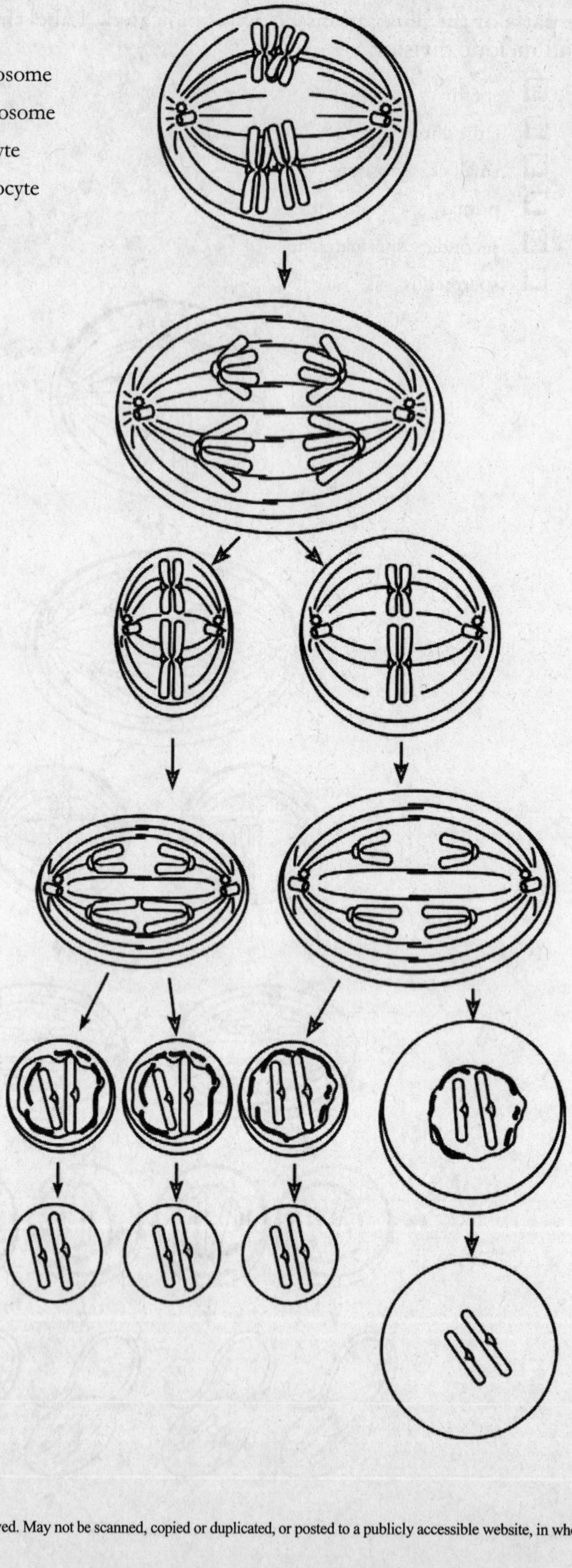

CHAPTER 51

Animal Development

Development encompasses all of the changes that take place during the life of an organism from fertilization to death. Guided by instructions encoded in the DNA of the genes, the organism develops from one cell to billions, from a formless mass of cells to an intricate, highly specialized and organized organism. Development is a balanced combination of several interrelated processes — cell division, increase in the number and size of cells, cellular movements that arrange cells into specific structures and appropriate body forms, and biochemical and structural specialization of cells to perform specific tasks. The stages of early development, which are basically similar for all animals, include fertilization, cleavage, gastrulation, and organogenesis. All terrestrial vertebrates have four extraembryonic membranes that function to protect and nourish the embryo. Environmental factors, such as nutrition, vitamin and drug intake, cigarette smoke, and disease-causing organisms, can adversely influence developmental processes. Aging is a developmental process that results in decreased functional capacities of the mature organism.

REVIEWING CONCEPTS

Fill in the blanks.

INTRODUCTION

1. In animals, growth occurs primarily by an increase in (a)________________________ and (b) ______________________.

DEVELOPMENT OF FORM

2. The process by which cells become specialized is called __________________.
3. The principle of ________________________________ states that in most cases neither cell determination nor cell differentiation entails a loss of genetic information from the nucleus.
4. (a) ____ is responsible for variations in chemistry, behavior, and structure among cells. However not all cells differentiate and some known as (b)________________________ remain in a relatively undifferentiated state and retain the ability to give rise to various types of cells.
5. Cells undergo ______________________, a process during which they become increasingly organized, shaping the intricate pattern of tissues and organs that characterizes a multicellular animal.
6. Programed cells death is called ________________________.

FERTILIZATION

7. Fertilization results when a sperm cell and an egg cell fuse to form a ____________________.

8. Fertilization restores the (a) ________________ number of chromosomes, during which time the (b) __________ of the individual is determined.
9. Fertilization involves four events. They are
   (a)____,
   (b)____,
   (c)____, and
   (d)____.

The first step in fertilization involves contact and recognition

10. A mammalian egg is enclosed by a thick, noncellular (a)________________ which is surrounded by a layer of (b)________________ cells.
11. Sea urchins have two acellular layers that interact with the sperm: a very thin (a) ________________________ and a thick glycoprotein layer called the (b) ________________.
12. Before a mammalian sperm can participate in fertilization, it must first undergo (a) ______________________, a maturation process that occurs in the female reproductive tract. During this process, sperm become increasingly motile and capable of undergoing an (b)______________________ when they encounter an egg.
13. After the acrosome reaction, the acrosome releases ______________________________ that digest a path through the eggs protective layers.
14. Mammalian fertilization requires interaction between sperm and ______________________________ in the egg's zona pellucida.

Sperm entry is regulated

15. Fertilization of the egg by more than one sperm is a condition called ____________________.
16. Two reactions prevent multiple sperm from entering an egg. During the (a) ____________________ the egg cell membrane depolarizes as the result of the opening of (b) ______________ channels. During the (c) ________________, enzymes and other proteins are released from granules into the space between the plasma membrane of the egg and the vitelline envelope.
17. In mammals a fertilization envelope does not form. The enzymes released from the granules alter the (a) ____________________ on the egg's (b) ______________________________ so that no additional sperm can bind.

Fertilization activates the egg

18. An increase in calcium ions in the egg cytoplasm stimulates the (a)______________ reaction and triggers the (b)____________________, a series of metabolic changes within the egg.
19. During activation, aerobic respiration (a) ____________________ (increases, decreases), enzymes and proteins become active and a burst of (b) ________________________ synthesis occurs.

Sperm and egg pronuclei fuse, restoring the diploid state.

20. It is believed that after the sperm enters the egg, it is guided toward the egg nucleus by a system of ________________________________.

CLEAVAGE

21. The inherent potential of the egg to form all of the different cell types of the fully differentiated individual is referred to as ______________.
22. After fertilization, the zygote undergoes (a) ________________, first forming a morula and then a (b) ______________.

The pattern of cleavage is affected by yolk

23. The invertebrates and simple chordates have (a) ________________ eggs. These eggs undergo (b) ______________________ cleavage, that is, the entire egg divides into cells of about the same size. (c)____________ cleavage of isolecithal eggs is common in deuterostomes, chordates and echinoderms, but protostomes such as annelids and mollusks undergo (d) ______________ cleavage.
24. Many vertebrate eggs are (a) ____________________, meaning they have large amounts of yolk concentrated at one end of the cell, known as the (b) __________________ pole. The more metabolically active pole is called the (c) ________________ pole.
25. The (a) __________________ cleavage that occurs in the telolecithal eggs of reptiles and birds is restricted to the (b) ______________.

Cleavage may distribute developmental determinants

26. Mosaic development is a consequence of the ________________________ of important materials in the cytoplasm of the zygote.
27. Mallals have zygotes with very homogeneous cytoplasm. They exhibit highly (a) ______________________ development in which individual cells produced by cleavage divisions are (b) ______________.
28. In some species fertilization triggers a rearrangement of the cytoplasm. The rearranged cytoplasm directly opposite the site of sperm penetration forms a region called the ________________________ which is thought to contain growth factors.

Cleavage provides building blocks for development

29. The cells that make up the blastula being to move and arrange themselves into patterns for further development. ______________________ are important in helping cells recognize one another and determining which ones adhere to form tissues.

GASTRULATION

30. During gastrulation, the ______________ becomes a three-layered embryo.
31. Embryonic layers are collectively called (a) ________________. The outermost layer is the (b) ______________ next is the (c)______________, and the innermost is the (d) ______________.

The amount of yolk affects the pattern of gastrulation

32. As gastrulation proceeds in echinoderms and amphioxus the newly formed cavity of the developing gut is called the (a) ________. This cavity opens to the exterior via the (b) ________________, which is the site of the future (c) __________ in deuterostomes.
33. In the amphibian, cells from the animal pole move down over the embryo surface and move into the interior. This spot of inward movement is called the ________________ of the blastopore.
33. In the bird, cells of the epiblast migrate toward the midline forming a thickened region called the (a) ____________________, which elongates and narrows. At the center is a narrow furrow called the (b) ________________ which is the functional equivalent of the blastopore of echinoderm, amphioxus, and amphibian embryos. However, the bird embryo contains no cavity homologous to the (c) ________________.
34. At the center of the primitive streak cells that will form the notochord are concentrated in a thickened know known as ____________________.

ORGANOGENESIS

35. The (a) ________________ induces the formation of the neural plate, from which cells migrate downward thus forming the (b)________________, flanked on each side by neural folds. When neural folds fuse, the (c) ______________ is formed. This structure gives rise to the brain and (d) ________________. When the neural folds fuse to from the neural tube, the (e) __________ cells arise from the ectoderm. These cells migrate downward to form the (f) ____________________ ganglia.
36. Blocks of mesoderm cells called ______________ form on either side of the neural tube and give rise to vertebrae, muscles, and other parts of the body axis.
37. The four chambered heart forms through the fusion of paired ____________________.
38. In fish and some amphibians the pharyngeal pouches and bronchial grooves meet to form a continuous passage from the pharynx to the outside known as (a) ______________ which function as respiratory organs. In terrestrial vertebrates pharyngeal pouches give rise to the (b) ________________________ cavity and the (c) ____________________ leading to the pharynx.

EXTRAEMBRYONIC MEMBRANES

39. Terrestrial vertebrates have four extraembryonic membranes. The outermost is the (a) ______________ In birds and reptiles the primary function of this membrane is (b) ______________. The (c) ______________ surrounds the fluid-filled space in which the embryo develops. The (d) ________________

is an outgrowth of the digestive tract. In birds and reptiles it stores (e)____________________. The fourth membrane is the (f)__________________________.

HUMAN DEVELOPMENT

40. Fertilization occurs in the (a) ______________, and within (b) ______ hours zygote has divided to become a 2-celled embryo. The (c)_______________________ dissolves at the time the embryo enters the uterus from the ovarian tube. At this time, the embryo is in the morula stage, but now begins to differentiate into a (d) _____________.
41. While the developing embryo floats in the uterus prior to implantation it is nourished by __________________________________.
42. The outer layer of embryonic cells surrounding the blastocyst, make up the _______________________ and eventually give rise to the chorion and amnion.
43. A cluster of cells, the _______________________________ projects into the cavity of the blastocyst and will give rise to the embryo proper.
44. The trophoblast cells of the blastocyst secrete enzymes that digest an opening into the uterine wall in a process called (a)_____________________________. The process begins at about the (b)__________________ day of development.

The placenta is an organ of exchange

45. Besides providing an exchange surface between mother and embryo (fetus) the placenta serves as an (a) ____________________ organ secreting (b) _______________________________ and (c)______________________________.
46. In placental mammals, the placenta develops from the (a)_______________________ of the embryo and from (b) ________________ tissue of the mother.
47. From the time the embryo begins to implant the trophoblastic cells release (a) ____ which signals the (b) _____________________________ that the pregnancy has begun.

Organ development begins during the first trimester

48. Gastrulation occurs during the _______________and _____________ weeks of human development.
49. The heart begins to develop after (a) ______________ weeks and limb buds form near the end of the (b) ____________ week.
50. At the beginning of the ___________________ month, the embryo is referred to as a fetus.
51. Sex can be determined by external observation by the end of the ____.

Development continues during the second and third trimesters

52. The fetus moves freely within the amniotic cavity during the (a)______________ trimester and during the (b)____________ month the mother becomes award of weak fetal movements.
53. During the (a) ______________ month of pregnancy the cerebrum grows rapidly and develops (b) ______________________.

More than one mechanism can lead to a multiple birth

54. (a) __________________ twins develop when two eggs have been fertilized and (b) ______________ twins develop when the inner cell mass subdivides into two separate masses.
55. A sonogram is a picture taken using __________.

Environmental factors affect the embryo

56. Drugs or other substances that interfere with morphogenesis and cause malformations are called ______________.

The neonate must adapt to its new environment

57. A neonate's first breath may be initiated by ____________________ in the blood after the umbilical cord is cut.

Aging is not a uniform process

58. ______________________ encompasses any biological change with an organism over time.

BUILDING WORDS

Use combinations of prefixes and suffixes to build words for the definitions that follow.

Prefixes	The Meaning	Suffixes	The Meaning
arch-	primitive	-age	collection of
acr(o)-	tip	-blast	embryo
blast(o)-	bud, sprout	-chord	cord
cleav-	to divide	-cyst	sac
fet-	pregnant	-enteron	gut
holo-	whole, entire	-mere	part
iso-	equal	-por(e)	opening
mero-	part, partial	-som(e)	body
neo-	new, recent	-us	person
noto-	back		
telo-	end		
tri-	three		
tropho-	nourishment		

Prefix	Suffix	Definition
____________	-lecithal	1. Having an accumulation of yolk at one end (vegetal pole) of the egg.
____________	-lecithal	2. Having a fairly equal distribution of yolk in the egg.
____________	____________	3. Any cell that is a part of the early embryo (specifically, during cleavage).
____________	____________	4. The blastula of the mammalian embryo.

________	________	5.	The opening into the archenteron of an early embryo.
________	-blastic	6.	A type of cleavage in which the entire egg divides.
________	-blastic	7.	A type of cleavage in which only the blastodisc divides; partial cleavage.
________	________	8.	The central cavity resulting from gastrulation; the primitive digestive system.
________	________	9.	The extraembryonic part of the blastocyst that chiefly nourishes the embryo or that develops into fetal membranes with nutritive functions.
________	-mester	10.	A term or period of three months.
________	-nate	11.	A newborn child.
________	________	12.	The cap at the head of a sperm cell.
________	________	13.	A rapid series of mitotic divisions with no period of growth during each cell cycle.
________	________	14.	Supporting rod of mesodermal cartilage-like cells.
________	________	15.	The time of intrauterine development between week 9 and birth.

MATCHING

Terms:

a.	Amnion	f.	Ectoderm	k.	Implantation
b.	Animal pole	g.	Endoderm	l.	Morphogenesis
c.	Blastula	h.	Fertilization	m.	Placenta
d.	Chorion	i.	Gastrula	n.	Vegetal pole
e.	Cleavage	j.	Germ layer		

For each of these definitions, select the correct matching term from the list above.

____ 1. A membrane that forms a fluid-filled sac for the protection of the developing embryo.

____ 2. The first of several cell divisions in early embryonic development that converts the zygote into a multicellular blastula.

____ 3. Any of the three embryonic tissue layers.

____ 4. The yolk pole of a vertebrate or echinoderm egg.

____ 5. Usually a spherical structure produced by cleavage of a fertilized ovum; consists of a single layer of cells surrounding a fluid-filled cavity.

____ 6. The outer germ layer that gives rise to the skin and nervous system.

____ 7. The attachment of the developing embryo to the uterus of the mother.

____ 8. The development of the form and structures of an organism and its parts.

____ 9. Early stage of embryonic development during which the embryo has three layers and is cup-shaped.

____10. The fusion of the egg and the sperm, resulting in the formation of a zygote.

____11. Embryonic contribution to the placenta.

MAKING COMPARISONS

Fill in the blanks.

Human Developmental Process	Approximate Time	Event
Fertilization	0 hour	Restores the diploid number of chromosomes
#1	2, 3 weeks	Blastocyst becomes a three-layered embryo
Early cleavage	#2	Formation of the two-cell stage
#3	#4	Activates the egg
Implantation	#5	#6
#7	0 hour	Establishes the sex of the offspring
#8	2.5 weeks	Neural plate begins to form
#9	#10	Morula is formed, reaches uterus

MAKING CHOICES

Place your answer(s) in the space provided. Some questions may have more than one correct answer.

____ 1. The liver, pancreas, trachea, and pharynx
a. arise from Hensen's node.
b. are formed prior to the primitive streak.
c. are derived from endoderm.
d. are derived from ectoderm.
e. are derived from mesoderm.

____ 2. A hollow ball of several hundred cells
a. is a morula.
b. is a blastula.
c. surrounds a fluid-filled cavity.
d. is the gastrula.
e. contains a blastocoel.

____ 3. Which of the following organ systems is the first to form during organogenesis?
a. circulatory system
b. reproductive system
c. excretory system
d. nervous system
e. muscular system

____ 4. The duration of pregnancy is referred to as the
a. parturition.
b. induction cycle.
c. gestation period.
d. neonatal period.
e. prenatal period.

____ 5. The fertilization envelope
a. promotes binding of sperm to egg.
b. facilitates egg-sperm recognition.
c. prevents polyspermy.
d. expedites entry of sperm into egg.
e. blocks entrance of sperm.

____ 6. The cavity within the embryo that is formed during gastrulation is the
a. archenteron.
b. gastrulacoel.
c. blastocoel.
d. blastopore.
e. neural tube.

____ 7. An embryo is called a fetus
a. when the brain forms.
b. when the limbs forms.
c. when its heart begins to beat.
d. after two months of development.
e. once it is firmly implanted in the endometrium.

____ 8. Embryonic cells are arranged in three distinct germ layers (endoderm, mesoderm, ectoderm) during
a. cleavage.
b. gastrulation.
c. blastulation.
d. organogenesis.
e. blastocoel formation.

____ 9. Amniotic fluid
a. replaces the yolk sac in mammals.
b. is secreted by the allantois.
c. is secreted by an embryonic membrane.
d. fills the space between the amnion and chorion.
e. fills the space between the amnion and embryo.

____10. Recognition of species compatibility between a sperm and an egg is due to the recognition protein
a. acrosin.
b. actin.
c. vitelline.
d. bindin.
e. pellucidin.

____11. The organ of exchange between a placental mammalian embryo and its mother develops from
a. chorion and amnion.
b. chorion and allantois.
c. chorion and uterine tissue.
d. amnion and uterine tissue.
e. amnion and allantois.

____12. Eggs that have a large amount of yolk concentrated at one pole
a. are isolecithal.
b. are telolecithal.
c. have a metabolically active animal pole.
d. have a vegetal pole that is metabolically active.
e. do not have food for the embryo in the egg.

____13. ______________ gives rise to skeletal tissues, muscle, and the circulatory system.
a. Ectoderm
b. Mesoderm
c. Endoderm
d. One of the germ layers
e. The blastocoel

____14. Fertilization
a. occurs in the upper portion uterus.
b. is immediately followed by the cortical reaction
c. is immediately preceded by the acrosome reaction.
d. occurs about 7 days prior to implantation.
e. restores polyploidy to the egg.

____15. Which of the following is mismatched?
a. capacitation : sperm maturation
b. blastodisc : vegetable pole
c. spiral cleavage : diagonal cytokinesis
d. notochord : mesoderm
e. primitive groove : blastopore

____16. Calcium ions
a. are involved in egg depolarization.
b. are involved in the cortical reaction.
c. are involved in capacitation.
d. stimulate morphogenesis.
e. stimulate the formation of Hensen's node.

____17. Mammalian zygotes
a. have a homogenous cytoplasm.
b. exhibit regulative development.
c. exhibit mosaic development.
d. follow rigid developmental patterns.
e. rely on surface proteins for tissue formation.

____18. The archenteron
a. opens to the outside via the blastopore.
b. opening becomes the anus in deuterstomes.
c. arises from the oral groove.
d. is lined by tissue arising from endoderm.
e. is the newly formed cavity of the developing gut.

____19. In humans, the umbilical cord
a. contains two arteries.
b. contains one vein.
c. contains two veins.
d. contains one artery.
e. blood vessels are formed from blood vessels of the allantois.

____20. Prior to fertilization a mammalian sperm cell must
a. undergo capacitation.
b. interact with the zona pellucida.
c. release analytic enzymes.
d. mature in the female reproductive tract.
e. undergo an acrosome reaction.

____21. Which words and definitions are incorrectly paired?
a. totipotent: origin of all body cells
b. morula: solid ball of 32 cells
c. blastula: 2-cell development stage
d. blastomere: cell in the early developmental stage of an embryo
e. yolk: proteins, phospholipids and fats

____22. Spiral cleavage
a. occurs in deuterostomes.
b. is typical of annelids.
c. has the plan of cytokinesis parallel to the polar axis.
d. occurs in protostomes.
e. occurs in echinoderms.

____23.The 4th stage of early development leading up to gastrulation is called
a. the zygote.
b. the blastula.
c. the gastrula .
d. the morula.
e. the blastodisc.

VISUAL FOUNDATIONS

Color the parts of the illustration below as indicated. Also label the blastocyst and umbilical cord.

RED	❑	umbilical artery	BROWN	❑	trophoblast cells
GREEN	❑	inner cell mass	TAN	❑	uterine epithelium
YELLOW	❑	yolk sac	PINK	❑	embryo
BLUE	❑	amniotic cavity and amnion	VIOLET	❑	maternal blood and blood vessel
ORANGE	❑	chorionic cavity and chorion	GREY	❑	placenta

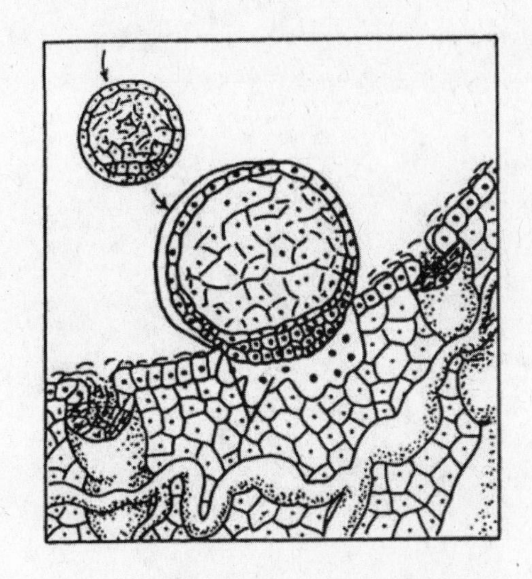

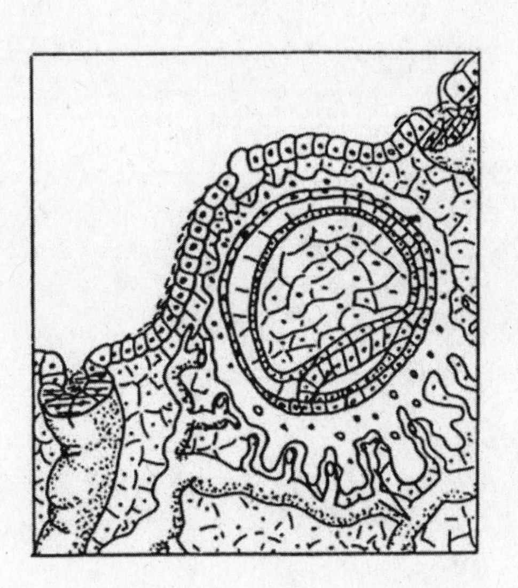

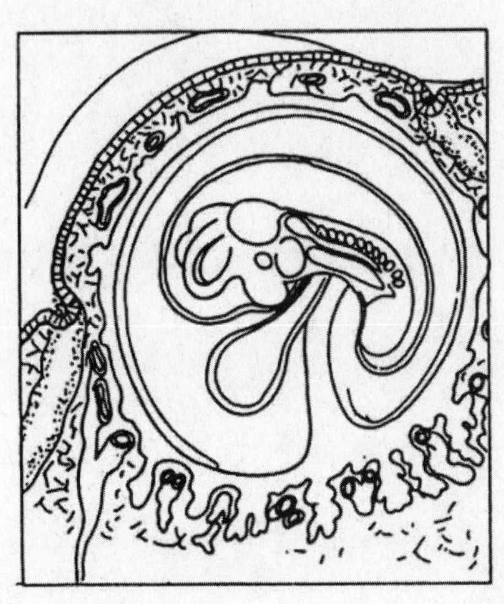

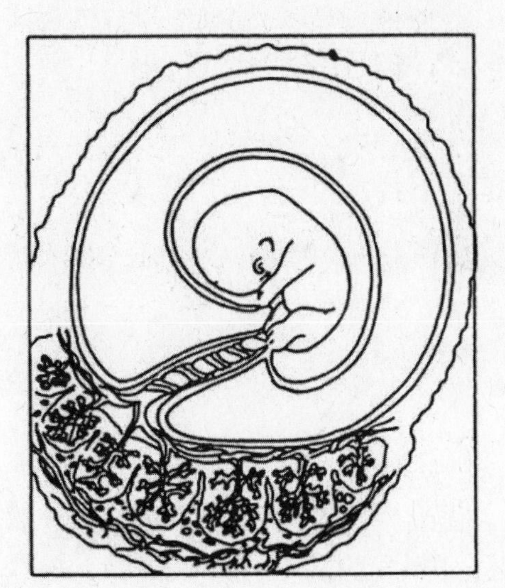

CHAPTER 52

❑

Animal Behavior

Animal behavior consists of those movements or responses an animal makes in response to signals from its environment; it tends to be adaptive and homeostatic. The behavior of an organism is as unique and characteristic as its structure and biochemistry. An organism adapts to its environment by synchronizing its behavior with cyclic change in its environment. Although behaviors are inherited, they can still be modified by experience. Behavioral ecology focuses on the interactions between animals and their environments and on the survival value of their behavior. Social behaviors, that is, interactions between two or more animals of the same species, have advantages and disadvantages. Advantages include confusing predators, repelling predators, and finding food. Disadvantages include competition for food and habitats, and increased ease of disease transmission. Some species that engage in social behaviors form societies. Some societies are loosely organized, whereas others have a complex structure. A system of communication reinforces the organization of the society. Animals communicate in a wide variety of ways; some use sound, some use scent, and some use pheromones. Some invertebrate societies exhibit elaborate and complex patterns of social interactions, such as the bees, ants, wasps, and termites. Vertebrate societies are far less rigid.

REVIEWING CONCEPTS

Fill in the blanks.

INTRODUCTION

1. Behavior refers to the responses an organism makes to ____________________ from its environment.
2. An animal's behavior is the product of ________________________ acting on its phenotype and indirectly on the genotype.
3. ____________________ is the study of behavior in natural environments from an evolutionary perspective.

BEHAVIOR AND ADAPTATION

4. An understanding of behavior requires consideration of (a)________________________ causes, the how questions, which considers immediate causes of behavior such the involvement of genetics and (b) ____________________ causes which are the why questions of behavior.

Behaviors have benefits and costs

5. (a)________________________ is an individual's reproductive success as measured by the number of viable offspring it produces, and it is one of the key benefits behavioral ecologists use when

considering (b) ______________________ to understand specific behaviors

Genes interact with environment

6. (a)________________behavior is inborn, genetically programmed behavior, whereas (b) ________________ behavior is behavior that has been modified in response to environmental experience.

Behavior depends of physiological readiness

7. Behavior is influenced primarily by two systems of the body, namely, the (a) ________________ and (b) ________________ systems.

Many behavior patterns depend on motor programs

8. Coordinated sequences of muscle actions are called ________________.
9. Behavioral patterns in which the animal shows little flexibility are elicited by a ________________.

LEARNING: CHANGING BEHAVIOR AS A RESULT OF EXPERIENCE

10. Learning is defined as a persistent change in behavior due to ______________________.
11. Animals that are poisonous display bright colors. This coloration is called ______________________.

An animal habituates to irrelevant stimuli

12. ______________________ is a type of learning in which an animal learns to ignore a repeated, irrelevant stimulus.

Imprinting occurs during an early critical period

13. Imprinting establishes a parent-offspring bond during a (a) ________ period early in development, ensuring that the offspring (b)________________ the mother.

In classical conditioning a reflex becomes associated with a new stimulus

14. In classical conditioning, an animal makes an association between an irrelevant stimulus and a normal body response. It occurs when a/an (a)______________________ stimulus becomes a substitute for a/an (b) ______________________ stimulus.

In operant conditioning spontaneous behavior is reinforced

15. In operant conditioning, the behavior of the animal is rewarded or punished after it performs a behavior discovered by chance. Repetitive rewards bring about (a) ______________________, and consistent punishment induces (b) ______________________.
16. Operant conditioning plays a role in the development of some behaviors that appear to be ______________________.

Animal cognition is controversial

17. Cognition is the process of (a) ________________ and (b)________________ knowledge. Cognition allows some animals to (c)______________________.

Play may be practice behavior

18. Play may be an example of (a) ____________________ conditioning. Hypotheses for the ultimate causes of play behavior include (b)____________________, (c)____________________, and (d)____________________.

BIOLOGICAL RESPONSES TO ENVIROMENTAL STIMULI

Biological rhythms regulate many behaviors

19. Daily cycles of activity are known as (a) ____________________. (b) ____________________ animals are most active during the day, (c) ____________________ animals are most active at night and (d)____________________ animals are most active at dawn and dusk.
20. In mammals, the master clock is located in the ____________________ in the hypothalamus.
21. The ________________ gland secretes melatonin a hormone that promotes sleep in humans.

Environmental signals trigger physiological responses that lead to migration

22. ________________ is the periodic long-distance travel from one location to another.
23. (a) ____________________ refers to travel by migrating animals in a specific direction. It involves a (b) ________________ sense.
24. (a) ____________________ requires a compass sense and a (b) ____________________ sense.

FORAGING BEHAVIOR

25. Foraging involves locating and selecting food, as well as (a)________________ and (b) ________________food.
26. ____________________ refers to an animal's ability to obtain food in the most efficient manner.

COST AND BENEFITS OF SOCIAL BEHAVIOR

27. The interaction of two or more animals, usually of the same species defines ____________________.
28. A ____________________ is an actively cooperating group of individuals belonging to the same species and often closely related.

Communication is necessary for social behavior

29. Animals use (a) ________________, (b) ________________, (c) ________________, (d) ________________, and (e) ________________ signals to transmit information to one another
30. Among nonhuman mammals, only ____________________ are known to match sounds in communicating.
31. ____________________ are chemical signals that convey information between members of a species.

32. Chemoreceptor cells in the nose of animals comprise the _______________ organ.

Dominance hierarchies establish social status

33. Once it is established, dominance hierarchy is one way to prevent _______________ within a population.
34. Dominance may be established in females and males. Establishing dominance is often strenuous and _______________.
35. In some animals, the hormone _______________ is the apparent cause of their aggressive behaviors.

Many animals defend a territory

36. Most animals have a _______________, a geographical area that they seldom leave.
37. Animals often defend a _______________, a portion of the home range which is where the raise there young and is important to their reproductive success.

Some insect societies are highly organized

38. Instructions for the honeybee society are (a) _______________ and (b) _______________.

SEXUAL SELECTION

39. For males, reproductive success is depends upon (a)_______________. For females it depends on (b) _______________, (c) _______________, and (d) _______________.

Animals of the same sex compete for mates

40. In _______________, individuals of the same sex compete for mates.

Animals select quality mates

41. In (a) _______________ females select mates on the basis of some physical trait or some resource offered to them. The physical trait typically is an indication of (b)_______________.
42. A _______________ is a small display area in which the males of some species of insects, birds, and bats compete for females.
43. _______________ ensure that the male is a member of the same species, and it provides the female with a means of evaluating the male.

Sexual selection favors polygynous mating systems

44. (a) _______________ is a mating system in which males fertilize the eggs of many females during a breeding season, whereas (b)_______________ is a mating system in which a female mates with several males.

Some animals care for their young

45. _______________ is the contribution each parent makes in producing and rearing offspring.

HELPING BEHAVIOR

Altruistic behavior can be explained by inclusive fitness

46. ______________________ behavior is when one individual behaves in a way that seems to benefit others rather than itself.
47. The ____________________________ is defined as the probability that two individuals inherit the same uncommon allele from a recent common ancestor.
48. Indirect selection, also called ____________________, is a form of natural selection that increases inclusive fitness through the breeding success of close relatives

Helping behavior may have alternative explanations

49. One example of kin selection includes ____________________, keeping watch for predators and warning of threats.

Some animals help nonrelatives

50. ______________________ is a type of behavior in which one animal helps another with no immediate benefit; however, at some later time the animal that was helped repays the debt.

CULTURE IN VERTEBRATE SOCIETIES

Some vertebrates transmit culture

51. Culture is behavior that is (a)______________________________________, learned from other members of the group, and (b)____.
52. Culture is not (a) ______________________________; it is maintained by (b) ______________________________.

Sociobiology explains human social behavior in terms of adaptation

53. Sociobiology focuses on the evolution of social behavior through ____.

BUILDING WORDS

Use combinations of prefixes and suffixes to build words for the definitions that follow.

Prefixes	The Meaning	Suffixes	The Meaning
mono-	one	-andr(y)	male
pher-	to carry		
poly-	many, much, multiple, complex	-gyn(y)	female
socio-	social, sociological, society	-(h)ormone	to excite

Prefix	Suffix	Definition
____________	-biology	1. The school of ethology that focuses on the evolution of social behavior through natural selection.
____________	____________	2. A mating system in which a female mates with many males.

__________ __________ 3. A mating system in which a male mates with many females.

__________ -gamy 4. Mating with a single partner during a breeding season.

__________ __________ 5. Chemical signal secreted into the environment.

MATCHING

Terms:

a. Altruistic
b. Biological clock
c. Circadian rhythm
d. Crepuscular
e. Diurnal
f. Ethology
g. Habituation
h. Imprinting
i. Innate
j. Migration
k. Nocturnal animal
l. Optimal foraging
m. Pheromone
n. Sign stimulus
o. Territoriality

For each of these definitions, select the correct matching term from the list above.

____ 1. Means by which activities of plants or animals are adapted to regularly-recurring changes.
____ 2. Describes animals that are most active during the day.
____ 3. Behaviors that are inherited and typical of the species.
____ 4. A substance secreted by organisms into the external environment that influences the development or behavior of other members of the same species.
____ 5. The theory that animals feed in a manner that maximizes benefits and/or minimizes costs.
____ 6. The process by which organisms become accustomed to a stimulus and cease to respond to it.
____ 7. A form of rapid learning by which a young bird or mammal forms a strong social attachment to an individual or object within a few hours after hatching or birth.
____ 8. Behavior in which one individual appears to act in such a way as to benefit others rather than itself.
____ 9. A daily cycle of activity around which the behavior of many organisms is organized.
____10. The study of animal behavior.
____11. Describes animals that are most active during the time around dusk and dawn.
____12. A simple signal that triggers a specific response.

MAKING COMPARISONS

Fill in the blanks.

Classification of Behavior	Example of the Behavior
Innate (FAP)	Egg rolling in graylag goose
#1	Wasp responding to cone arrangement to locate nest
#2	Dog salivating at sound of bell
#3	Child sitting quietly for praise
#4	Birds tolerant of human presence
#5	Chicks learning the appearance of the parent
#6	Primate stacking boxes to reach food

MAKING CHOICES

Place your answer(s) in the space provided. Some questions may have more than one correct answer.

____ 1. A behavior that resembles a simple reflex in some ways and is triggered by a sign stimulus is best described as a/an

a. releaser.
b. conditioned reflex.
c. conditioned stimulus.
d. behavioral pattern (formerly FAP).
e. unconditioned reflex.

____ 2. If your dog perks up its ears when you clap your hands, but stops perking its ears after you have repeatedly clapped your hands for a period of time, your dog has displayed a form of

a. learning.
b. operant conditioning.
c. classical conditioning.
d. sensory adaptation.
e. habituation.

____ 3. A change in your behavior derived from experience in your environment is the result of

a. a FAP.
b. learning.
c. inherited behavioral characteristics.
d. redirected behavior.
e. habituation and/or sensitization.

____ 4. Suppose you decide to perform an experiment with a type of lizard that instantly attacks any lizard with green scales on its sides. You place a live lizard with green scales on its sides in a cage next to a styrofoam block with green paint on its sides. Your experimental lizard ignores the spotted lizard, preferentially attacking the styrofoam block. The styrofoam-attacking lizard is responding to a/an

a. inducer.
b. sign stimulus.
c. conditioned reflex.
d. displacement syndrome.
e. redirected behavioral syndrome.

____ 5. Which of the following is/are used by a worker honey bee to indicate the distance of a food source?

a. round dance
b. waggle dance
c. angle relative to a north–south axis
d. angle relative to gravity
e. angle relative to sun

____ 6. The study of the behavior of animals in their natural environments from an evolutionary perspective is

a. behavioral ecology.
b. behavioral genetics.
c. social behaviorist.
d. behavioral ecology.
e. neurobiology.

____ 7. Genetically programmed behavior has been variously termed

a. innate behavior.
b. inborn behavior.
c. determinism.
d. instinct.
e. patterned behavior.

____ 8. Learning

a. utilizes past experiences.
b. occurs during a critical period.
c. involves persistent changes in behavior.
d. relies on inherited behavior.
e. is adaptive.

____ 9. You feed a group of just hatched chickens on a daily basis until they are adults. These chicks

a. are using cognition.
b. will imprint on their mother.
c. are demonstrating operant conditioning.
d. will associate you with food.
e. are demonstrating an adaptive behavior.

____10. Motor programs

a. may be mainly innate.
b. may be invoked by a specific color and pattern.
c. are subject to extinction when a specific signal is not present.
d. often have little flexibility.
e. may involve learning.

____ 11. The suprachiasmatic nucleus

a. is located in the hypothalamus.
b. receives input from the retina.
c. is connected to the spinal cord.
d. communicates with the pineal gland.
e. generates an internal 24 hour clock.

____ 12. Which of the following is not correctly paired?

a. navigation : integration of distance and time
b. circadian : 48 hours
c. lek : a nest
d. pineal gland : endocrine system
e. critical period : first month of life

____ 13. Proximate causes of behavior

a. ask why questions.
b. address costs and benefits of behavior.
c. look at immediate causes of behavior such as genetics.
d. have evolutionary explanations.
e. ask how questions.

CHAPTER 53

Introduction to Ecology: Population Ecology

This chapter focuses on the study of populations as functioning systems. A population is all the members of a particular species that live together in the same area. Populations have properties that do not exist in their component individuals or in the community of which the population is a part. These properties include birth rates, death rates, growth rates, population density, population dispersion, age structure, and survivorship. Population ecology deals with the number of individuals of a particular species that are found in an area and how and why those numbers change over time. Populations must be understood in order to manage forests, field crops, game, fishes, and other populations of economic importance. Population growth is determined by the rate of addition of new individuals through birth and immigration and the loss of individuals through emigration and death of organisms. Population growth is limited by density-dependent and density-independent factors. Each species has its own survival strategy that enables it to survive. Most organisms fall into one of two survival strategies: One emphasizes a high rate of natural increase and the other emphasizes maintenance of a population near the carrying capacity of the environment. The principles of population ecology apply to humans as well as other organisms. Environmental degradation is related to population growth and resource consumption.

REVIEWING CONCEPTS

Fill in the blanks.

INTRODUCTION

1. Interactions among organisms are called (a) ____________________, while interactions between organisms and their nonliving, physical environment are called (b) ____________________.

FEATURES OF POPULATIONS

2. Populations of organisms have properties that individual organisms do not have. Some properties of importance are (a) ______________, (b) ______________, (c) ______________, (d) ______________, (e) ______________, and (f) ______________.
3. ____________________ considers the number of individuals of a species that are found in an area along with the dynamics of the population.
4. The study of changes in populations is known as ____________________.

Density and dispersion are important features of populations

5. ______________________________ is the number of individuals of a species per unit of habitat area or volume at a given time.
6. Individuals within a population may exhibit a characteristic pattern of dispersion. For example, (a) ________________________ exists when individuals in a population are spaced throughout an area in a manner that is unrelated to the presence of others; (b)________________________occurs when individuals are more evenly spaced than would be expected from a random occupation of a given habitat; and (c) ________________________ occurs when individuals are concentrated in specific parts of the habitat.

CHANGES IN POPULATION SIZE

7. Ultimately, changes in the size of a population result from the difference between (a) ____________________ and (b)____________________.
8. The rate of change in a population is expressed as: $\Delta N / \Delta t = N(b — d)$, where ΔN is the (a) ____________________, Δt is the (b) ____________________, b is the birth rate or (c)_______________, and d is the death rate or (d) _______________.

Dispersal affects the growth rate in some populations

9. (a) _______________ occurs when individuals enter a population and thus increase its size; (b) _______________ occurs when individuals leave a population and thus decrease its size.

Each population has a characteristic intrinsic rate of increase

10. The maximum rate at which a population of a given species could increase under ideal conditions when resources are abundant and its population density is low is known as its ______________________________.
11. Generally, large species have the (a) __________________ (smallest, largest) intrinsic rates of increase.
12. ______________________________ is the accelerating population growth rate that occurs when optimal conditions allow a constant per capita growth rate.

No population can increase exponentially indefinitely

13. The largest population that can be sustained for an indefinite period by a particular environment that is not changing is that environment's ______________________________.
14. Population growth curves take a variety of forms, their shapes are primarily a function of biotic potential and limiting factors. The (a)__________________ growth curve shows the population's initial exponential increase, followed by a leveling out as the carrying capacity of the environment is approached. This curve is also called (b) ______________________ population growth.

FACTORS INFLUENCING POPULATION SIZE

Density-dependent factors regulate population size

15. The effect of density-dependent factors on population growth ________________ as population density increases.
16. Examples of density-dependent factors are (a) ________________, (b) ________________, and (c) ________________.
17. Density-dependent factors are an example of a ________________ feedback system.
18. Density-dependent limiting factors are most effective at ________ (high or low?) population densities, and therefore they tend to stabilize populations.
19. At high population densities, it is not uncommon for the (a)__________ to go up and the (b)______________ to go down.
20. Competition that occurs within a population is (a)________________ competition and competition that occurs between two different populations is (b)________________ competition.
21. In (a) ________________ competition, also called contest competition, certain dominant individuals obtain an adequate supply of the limited resource at the expense of other individuals in the population. In (b) ________________ competition, also called scramble competition, all the individuals in a population "share" the limited resource more or less equally.

Density-independent factors are generally abiotic

22. Some examples of density-independent factors that affect population size are (a) ________________, (b) ________________, and (c) ________________.

LIFE HISTORY TRAITS

23. (a) ________________ species expend their energy in a single, immense reproductive effort. (b) ________________ species reproduce during several breeding seasons throughout their lifetimes.
24. ________________ refers to a species potential capacity to produce offspring.
25. Each species has a life history strategy. (a) ________strategists are usually opportunists found in variable, temporary, or unpredictable environments where the probability of long-term survival is low. (b) _______ strategists tend to be found in relatively constant or stable environments, where they have a high competitive ability.

Life tables and survivorship curves indicate mortality and survival

26. ________________ is the probability that a given individual in a population or cohort will survive to a particular age.

27. A curve in which mortality is greatest at an early age is called a (a)_____________ survivorship curve. A curve in which the probability of survival decreases more rapidly with increasing age is called a (b) _____________ survivorship curve and a curve in which the probability of survival does not change over with age is a (c)_________________ survivorship curve.

METAPOPULATIONS

28. A population that is divided into several local populations among which individuals can disperse is known as a _________________________.

29. Good habitats, called (a) _________________________, are areas where local reproductive success is greater than local mortality. Lower-quality habitats, called (b) _____________________, are areas where local reproductive success is less than local mortality.

HUMAN POPULATIONS

30. Zero population growth is the point at which the (a)__________________ equals the (b) __________________.

Not all countries have the same growth rate

31. The science that deals with human population statistics is called _________________________.

32. Highly developed countries have (a) _______________ birth rates and (b)___________ infant mortality rates.

33. The two categories of developing countries are (a) ______________ developed and (b) ______________ developed.

34. Less developed countries have (a) ________ birth rates, (b) ________ infant mortality, (c) ________ life expectancies, and (d) ___________ average gross national income in purchasing power parity (GNI PPP).

35. The amount of time that it takes for a population to double in size is its (a) _________________. In general, the shorter this time, the (b)______(more or less?) developed the country.

36. The number of children that a couple must produce to replace themselves is known as the (a) ______________________________. This value is usually (b) ______ children in highly developed countries and (c) ______ children in developing countries.

37. The average number of children born to a woman during her lifetime is called the (a) ______________________________. On a worldwide basis, this value is currently (b) ____________ (higher or lower?) than the replacement value.

The age structure of a country helps predict future population growth

38. To predict the future growth of a population, it is important to know its age structure, which is the (a) ______________ and (b)______________ of people at each age in a population.

Environmental degradation is related to population growth and resource consumption

39. People overpopulation occurs when the environment is worsening from the effect of having too many people. Consumption overpopulation occurs when each individual in a population consumes too large a share of resources. The (a) ____________ (former or latter?) is the current problem in many developing nations, and the (b) ____________ (former or latter?) is the current problem in most affluent, highly developed nations.

BUILDING WORDS

Use combinations of prefixes and suffixes to build words for the definitions that follow.

Prefixes	The Meaning	suffixes	The Meaning
bio-	life	-ic	pertaining to
ec-	dwelling	-ology	study of
inter-	between, among		
intra-	within		
intrins-	internally		

Prefix	Suffix	Definition
____________	-specific competition	1. Competition for resources within a population.
____________	-specific competition	2. Competition for resources among populations.
____________	-sphere	3. The entire zone of air, land, and water at the surface of the earth that is occupied by living things.
____________	____________	4. The study of how living organisms and the physical environment interact.
____________	____________	5. The rate of increase in a population's size under ideal conditions.

MATCHING

Terms:

a.	Intrinsic	g.	Ecology	m.	Population
b.	Carrying capacity	h.	Emigration	n.	Population crash
c.	Contest competition	i.	Exponential	o.	Random
d.	Density-dependent	j.	Immigration	p.	Replacement-level fertility
e.	Density-independent	k.	Mortality	q.	Uniform
f.	Doubling time	l.	Natality		

For each of these definitions, select the correct matching term from the list above.

____ 1. A pattern of spacing in which individuals in a population are spaced unpredictably.

____ 2. Growth that occurs at a constant rate of increase over a period of time.

____ 3. The rate at which organisms produce offspring.

____ 4. The maximum number of organisms that an environment can support.

____ 5. Any factor, such as climate, that does not depend on the density of populations.

____ 6. The number of offspring a couple must produce in order to "replace" themselves.

____ 7. The movement of individuals out of a population.

____ 8. The maximum rate of increase of a species that occurs when all environmental conditions are optimal.

____ 9. The amount of time it takes for a population to double in size, assuming that its current rate of increase does not change.

____10. An abrupt decline in a population from a high to very low population density.

____11. The study of the interactions between living things and their environment, both physical and biotic.

____12. A group of organisms of the same species that live in the same geographical area at the same time.

____13. Dominant individuals obtain an adequate supply of resources at the expense of others in the population.

____14. The pattern of spacing when cacti in the desert are evenly spaced to maximize the available resources.

MAKING COMPARISONS

Fill in the blanks.

Examples of Environmental Factors That Affect Population Size	Is Factor Density-Dependent or Density-Independent?	Description of Factor
Predation	Density-dependent	As the density of a prey species increases, predators are more likely to encounter an individual of the prey species
Killing frost	#1	#2
Contest competition	#3	#4
Disease	#5	#6
Hurricane	#7	#8

MAKING CHOICES

Place your answer(s) in the space provided. Some questions may have more than one correct answer.

Questions 1-5 pertain to the following equation. Use the list of choices below to answer them.

$$\frac{\Delta N}{\Delta t} = N(b - d)$$

a. growth only
b. rate of growth
c. change only
d. ΔN
e. Δt
f. $b - d$
g. value "b"
h. value "d"
i. $\Delta N / \Delta t$

____ 1. Natality.

____ 2. Mortality.

____ 3. Population change.

____ 4. Rate of population change.

____ 5. The parameter derived from the equation.

____ 6. The equation dN/dt = rN is used to derive the
a. biotic potential (r_m).
b. instantaneous growth rate.
c. decline in growth due to limiting factors.
d. per capita growth rate (r).
e. growth rate at carrying capacity.

____ 7. During the exponential growth phase of bacterial growth, the population
a. increases moderately.
b. increases dramatically.
c. decreased moderately.
d. decreases dramatically.
e. does not change substantially.

____ 8. A cultivated field of wheat would most likely display
a. uniform dispersion.
b. random dispersion.
c. clumped dispersion.
d. no dispersion pattern.
e. a form of dispersion not found in nature.

____ 9. In the logistic equation, if environmental limits or resistance are sustained, "K" is the
a. growth rate.
b. birth rate.
c. mortality rate.
d. carrying capacity.
e. equivalent of the lag phase.

____10. A grove of trees that originated from one seed displays
a. uniform dispersion.
b. random dispersion.
c. clumped dispersion.
d. no dispersion pattern.
e. a form of dispersion not found in nature.

____11. Population growth is most frequently controlled by which *one* of the following?
a. environmental limits
b. predation
c. disease
d. behavioral modifications
e. resource depletion

____12. Organisms referred to as r-strategists usually
a. are mobile animals.
b. have a high "r" value.
c. inhabit variable environments.
d. survive well in the long term.
e. develop slowly.

____13. The redwood stands in California are examples of organisms that
a. use the K-strategy.
b. use the r-strategy.
c. often experience local extinction.
d. live in a relatively stable environment.
e. pioneer new habitats.

____14. The number of individuals of a species per unit of habitat area is the __________ for that species.
a. distribution curve
b. distribution pattern
c. population density
d. ΔN
e. b – d

____15. Which of the following is incorrectly paired?
a. population density : spacing of individuals over a given area
b. S-shaped growth curve : logistic population growth
c. density dependent factors : negative feedback
d. cohort : group of individuals of the same age
e. J-shaped curve : exponential population growth

____16. The carrying capacity (K)
a. may change over time.
b. affects herbivores.
c. is affected by climate.
d. has a primary effect on the growth of density independent populations.
e. affects r-selected populations the most.

____17. Species that expend their energy in a single, immense reproductive effort
a. are an iteroparous species.
b. are likely to be invertebrates.
c. are a semelparous species.
d. grow and die in a single season.
e. are exemplified by the agave plant.

____18. A population that is expanding would have an age structure diagram
a. that is widest at the top.
b. that is widest in the middle.
c. that is widest at the bottom.
d. that is even throughout.
e. that is narrowest at the top.

CHAPTER 54

❑

Community Ecology

A community consists of an association of populations of different species that live and interact in the same place at the same time. A community and its abiotic environment together compose an ecosystem. The interactions among species in a community include competition, predation, and symbiosis. Every organism has its own ecological niche, that is, its own role within the structure and function of the community. Its niche reflects the totality of its adaptations, its use of resources, and its lifestyle. Communities vary greatly in the number of species they contain. The number is often high where the number of potential ecological niches is great, where a community is not isolated or severely stressed, at the edges of adjacent communities, and in communities with long histories. A community develops gradually over time, through a series of stages, until it reaches a state of maturity; species in one stage are replaced by different species in the next stage. Most ecologists believe that the species composition of a community is due primarily to abiotic factors.

REVIEWING CONCEPTS

Fill in the blanks.

INTRODUCTION

1. An association of populations of different species living and interacting in the same place at the same time comprises a ____________________.
2. A biological community and its abiotic environment together comprise an ____________________.

COMMUNITY STRUCTURE AND FUNCTIONING

3. Communities exhibit community (a) ____________________ and community (b) ____________________,both characteristics that populations lack.
4. Species may interact in a positive way. Such interactions are referred to as ____________________.

Community interactions are complex and often not readily apparent

5. The three main types of interactions that occur among species in a community are (a) ____________________, (b)____________________, and (c) ____________________.

The niche is a species' ecological role in the community

6. The totality of adaptations by a species to its environment, its use of resources, and the lifestyle to which it is suited comprises its ____________________.

7. The potential ecological niche of a species is its (a) ________________ niche, while the lifestyle that a species actually pursues and the resources that it actually uses make up its (b) ______________ niche.
8. Any environmental resource that, because it is scarce or unfavorable, tends to restrict the ecological niche of a species is called a ____________________.

Competition is intraspecific or interspecific

9. Competition among individuals within a population is called (a)________________________ competition and when it is between members of different populations it is (b) ______________ competition.
10. According to the ________________________, two species cannot indefinitely occupy the same niche in the same community.
11. The reduction in competition for environmental resources among coexisting species as a result of each species' niche differing from the others in one or more ways is called ____________________________.
12. ____________________________ is the divergence in traits in two similar species living in the same geographic area.

Natural selection shapes the bodies and behaviors of both predator and prey

13. ______________________ refers to the interdependent evolution of two interacting species.
14. Conspicuous colors or patterns that advertise a species' unpalatability to potential predators are known as (a)________________________ or (b)______________________________.
15. ____________________coloration helps animals hide by blending into their physical surroundings.
16. ___________________________ is a defense strategy in which a defenseless species is protected from predation by its resemblance to a species that is dangerous in some way.
17. _______________ mimicry is used as a defense when both species are harmful.

Symbiosis involves a close association between species

18. The three forms of symbiosis are (a) ________________ in which both benefit, (b) ________________________in which one benefits but the other is not harmed, and (c) ______________ in which one benefits but the other is harmed.
19. When a parasite causes disease or in some cases death it is called a ________________.
20. Parasites may be either (a) _______ parasites or (b) _______parasites.

STRENGTH AND DIRECTION OF COMMUNITY INTERACTIONS

Other species of a community depended on or are greatly affected by keystone species

21. Keystone species determine the nature of an entire community, having an impact that is out of proportion to their abundance because they often affect the amount of (a) ______________, (b) ____________, or (c)______________________________.

Dominant species influence a community as a result of their greater size or abundance

22. Dominant species greatly affect a community because they are ______________________________.

Ecosystem regulation occurs from the bottom up and the top down

23. Bottom-up processes are based on ______________________.
24. A trophic level is an organism's position in a ______________________________.
25. Bottom-up processes predominate in aquatic ecosystems where (a)____________________ and (b) ________________ is limiting.
26. Top-down processes regulate ecosystem function by (a) ___________ interactions. Top-down interactions are also known as a (b)______________________________.

COMMUNITY BIODIVERSITY

27. Species diversity a measure of (a) ____________________________, the number of species within a community, and (b)______________________, a measure of each species abundance.

Ecologists seek to explain why some communities have more species than others

28. Species richness is inversely related to the ______________________________ of a community.
29. Isolated island communities are generally much ____________ (more? less?) diverse than are communities in similar environments found on continents.
30. An ______________ is a transitional zone where two or more communities meet. It contains most or all of the ecological niches of both adjacent communities as well as some that are unique.
31. The change in species composition produced at ecotones is known as the ______________________.

Species richness may promote community stability

32. Traditionally, ecologists assumed that community stability is a consequence of community ______________________.

COMMUNITY DEVELOPMENT

33. (a) ________________ succession is the change in species composition over time in a habitat that was not previously inhabited by organisms

and where no soil exists. In contrast, (b) _______________ succession is the change in species composition that takes place after some disturbance removes the existing vegetation; soil is already present.

Disturbance influences succession and species richness

34. Species richness is thought to be greatest at moderate levels of (a)_______________ of the community. Because all communities are exposed to periodic disturbances Joseph Connell proposed the (b) _______________ hypothesis to replace the idea that, over time, a climax community occurs.

Ecologists continue to study community structure

35. The cooperative view of community that stresses the interaction of the members which tend to cluster in tightly knit groups is known as the _______________ model of a community.
36. The _______________ model of communities emphasizes species individuality, with each species having its own particular abiotic living requirements.

BUILDING WORDS

Use combinations of prefixes and suffixes to build words for the definitions that follow.

Prefixes	The Meaning
apo-	away
carn-	meat
crypt-	hidden
eco-	home
herb-	plant
inter-	between, among
intra-	within
para-	along side
patho-	suffering, disease, feeling
pred-	prey upon
success-	to follow

Suffixes	The Meaning
-ation	the process of
-ic	pertaining to
-gen	production of
-ion	process of
-(i)vor(e)	to eat
-semat(ic)	mark or sign
-sit(e)	food

Prefix	Suffix	Definition
_______	_______	1. Any animal that eats flesh.
_______	-system	2. A community along with its nonliving environment.
_______	_______	3. Any disease-producing organism.
_______	-specific competition	4. Competition for resources within a population.
_______	_______	5. Any animal that eats plants.
_______	-specific competition	6. Competition for resources among populations.
_______	_______	7. The consumption of one species by another.
_______	_______	8. Coloration or markings that help an organism hide by blending in with the environment.
_______	_______	9. In a symbiotic relationship, the organism that receives nutrition from its host.

__________ __________ 10. The change in species composition in a community over time.

__________ __________ 11. Coloration that describes the display of conspicuous colors or patterns which advertise unpalatability to predators.

MATCHING

Terms:

a. Batesian mimicry
b. Commensalism
c. Detritus
d. Ecological niche
e. Ecotone
f. Edge effect
g. Mullerian mimicry
h. Mutualism
i. Parasitism
j. Pathogen
k. Primary consumer
l. Secondary
m. Succession
n. Symbiosis
o. Community

For each of these definitions, select the correct matching term from the list above.

____ 1. The sequence of changes in a plant community over time.

____ 2. An ecological succession that occurs after some disturbance destroys the existing vegetation.

____ 3. An organism that is capable of producing disease.

____ 4. Resemblance of different species, each of which is equally obnoxious to predators.

____ 5. An intimate relationship between two or more organisms of different species.

____ 6. A symbiotic relationship between individuals of different species in which one benefits from the relationship and one is harmed.

____ 7. A transition region between adjacent communities.

____ 8. A type of symbiosis in which one organism benefits and the other one is neither harmed nor helped.

____ 9. An assemblage of different species of organisms that live and interact in a defined area or habitat.

____10. The bacteria in your intestine produce vitamin K which is needed by your body for normal blood clotting.

____11. Occurs when a defenseless species is defended by resembling a dangerous species.

MAKING COMPARISONS

Fill in the blanks.

Interaction	Species 1	Effect on Species 1	Species 2	Effect on Species 2
Mutualism of Species 1 and Species 2	Coral	Beneficial	Zooxanthellae	Beneficial
#1	Silverfish	#2	Army ants	#3
#4	Orcas	#5	Salmon	#6
#7	Tracheal mites	#8	Honeybees	#9
#10	*Entamoeba histolytica*	#11	Humans	#12

Interaction	Species 1	Effect on Species 1	Species 2	Effect on Species 2
Mutualism of Species 1 and Species 2	*Rhizobium*	#13	Beans	#14
Competition between Species 1 and Species 2	*Balanus balanoides*	#15	*Chthamalus stellatus*	#16
#17	Orchids	#18	Host tree	#19

MAKING CHOICES

Place your answer(s) in the space provided. Some questions may have more than one correct answer.

____ 1. The local environment in which a species lives is its
a. ecosystem.
b. habitat.
c. fundamental niche.
d. realized niche.
e. niche.

____ 2. The actual lifestyle of a species and the resources it uses are known as the
a. ecosystem.
b. habitat.
c. fundamental niche.
d. realized niche.
e. niche.

____ 3. Resource partitioning results in
a. extinction of the least adapted species.
b. interspecific breeding.
c. reduced competition among coexisting species.
d. increased competition among coexisting species.
e. competition for niches among similar species.

____ 4. An association of different species of organisms interacting and living together is a/an
a. ecosystem.
b. community.
c. food web.
d. pyramid.
e. niche.

____ 5. The difference between an organism's potential niche and the niche it comes to occupy may result from
a. competition.
b. an inverted pyramid.
c. factors that exclude it from part of its fundamental niche.
d. overabundance of resources.
e. limited space or resources.

____ 6. The totality of an organism's adaptations, its use of resources, and the lifestyle to which it is fitted are all embodied in the term
a. ecosystem.
b. community.
c. food web.
d. pyramid.
e. niche.

____ 7. All food energy in the biosphere is ultimately provided by
a. producers.
b. nitrogen fixers.
c. phosphorus fixers.
d. consumers.
e. decomposers.

____ 8. The major roles of organisms in a community are
a. producers.
b. nitrogen fixers.
c. phosphorus fixers.
d. consumers.
e. decomposers.

____ 9. Critically important minerals and other essential substances would be permanently lost to organisms if it were not for the
a. producers.
b. nitrogen fixers.
c. phosphorus fixers.
d. consumers.
e. decomposers.

____10. The nonliving environment together with different interacting species is a/an
a. ecosystem.
b. community.
c. food web.
d. pyramid.
e. niche.

____11. Divergence of traits of two similar species living in the same geographical area is known as
a. coevolution.
b. aposematic displacement.
c. character displacement.
d. Batesian mimicry.
e. Mullerian mimicry.

____12. Both dominant and keystone species
a. are very common in the community.
b. have effects on energy in a community.
c. have large effects on a community even though they represent a relatively small biomass.
d. disproportionally effect bottom-up processes.
e. have community-wide effects.

____13. An ecological niche
a. includes an organism's habitat.
b. is difficult to precisely define.
c. represents the totality of adaptations of a species to its environment.
d. includes what a species eats.
e. is restricted by limiting resources.

____14. Predator strategies include
a. ambush.
b. pursuit.
c. generally have larger brains.
d. chemical protection.
e. attracting prey.

____15. Which of the following relationships are incorrectly paired?
a. aposematic coloration : warning coloration
b. secondary succession : no soil
c. ectoparasite : tapeworm
d. resource portioning : timing of feeding
e. cryptic coloration : camouflage

____16. Competition
a. is intraspecific if it occurs between different species.
b. is avoided by character displacement.
c. is reduced by resource partitioning in coexisting species.
d. is not always a straight-forward, direct interaction.
e. occurs when resources are limited.

____17. One species living in or on another species
a. may be endoparasites.
b. may lead to species evolution.
c. are demonstrating character displacement.
d. are comodels.
e. are symbionts.

____18. The species that is critical in determining the nature of an entire community is the
a. predator species.
b. one often present in the largest biomass.
c. scavenger species.
d. dominant species.
e. keystone species.

____19. Which of the following is not correctly paired?
a. species richness: number of species in a community.
b. top-down process : regulates ecosystems at the tropic level
c. species evenness : relative abundance of one species compared to others
d. Shannon index : represents species diversity quantitatively
e. edge effect : transitional zone between two communitie

CHAPTER 55

❑

Ecosystems and the Biosphere

This chapter examines the dynamic exchanges between communities and their physical environments. Energy cannot be recycled and reused. Energy moves through ecosystems in a linear, one-way direction from the sun to producer to consumer to decomposer. Much of this energy is converted into less useful heat as the energy moves from one organism to another. Matter, the material of which living things are composed, cycles from the living world to the abiotic physical environment and back again. All materials vital to life are continually recycled through ecosystems and so become available to new generations of organisms. Key cycles include the carbon cycle, the nitrogen cycle, the phosphorus cycle, and the water, or hydrologic, cycle. Life is possible on earth because of its unique environment. The atmosphere protects organisms from most of the harmful radiation from the sun and space, while allowing life supporting visible light and infrared radiation to penetrate. An area's climate is determined largely by temperature and precipitation. Many ecosystems contain fire-adapted organisms.

REVIEWING CONCEPTS

Fill in the blanks.

INTRODUCTION

1. An individual community *and* its abiotic components together make up an ____________________.
2. All of earth's communities combined and their interactions with the abiotic environment comprise the Earth's ____________________.

ENERGY FLOW THROUGH ECOSYSTEMS

3. Energy enters an ecosystem in the form of (a) ____________ energy, which producers can convert to useful (b) ______________ energy that is stored in the bonds of molecules. When cellular respiration breaks these molecules apart, energy becomes available to do work. As the work is accomplished, energy escapes the organisms and dissipates into the environment as (c) ________ energy where organisms cannot use it.
4. A complex of interconnected food chains in an ecosystem is a ________________.
5. Each "link" in a food chain is a trophic level, the first and most basic of which is comprised of the photosynthesizers, or (a) ____________. Once "fixed," materials and energy pass linearly through the chain from photosynthesizers to the (b)

_________________________, and then to the (c) _________________________.

6. Dead organic matter is called (a) _________________ and it is consumed by (b) _________________________.
7. _________________________ are organisms that externally break down organic molecules in remains and ingest inorganic products.

Ecological pyramids illustrate how ecosystems work

8. The ecological pyramid of (a) ______________ illustrates how many organisms there are in each trophic level; while the pyramid of (b)______________ is a quantitative estimate of the amount of living material at each level.
9. The pyramid of ______________ indicates energy content in kilocalories or kilojoules per square meter per year of the biomass of each trophic level.

Ecosystems vary in productivity

10. The rate at which energy is captured during photosynthesis is known as the _________________________ of an ecosystem.
11. The energy that remains in plant tissues after cellular respiration has occurred is called _________________________; it represents the rate at which organic matter is actually incorporated into plant tissues to produce growth.
12. The amount of net energy available for biomass production by consumer organisms is called _________________________.
13. It is estimated that humans use between ____ and ____% of annual land-based net primary productivity

Food chains and poisons in the environment

14. The buildup of a toxin in the body of an organism is known as _________________________.
15. The increase in concentration as a toxin passes through successive levels of the food web is known as _________________________.

CYCLES OF MATTER IN ECOSYSTEMS

16. The cycling of matter from one organism to another and from living organisms to the abiotic environment and back again is referred to as _________________________.
17. Four cycles that are of particular importance to organisms are the cycles for (a) ______________, (b) ______________, (c) ______________, and (d) ______________.

Carbon dioxide is the pivotal molecule in the carbon cycle

18. Carbon is incorporated into the biota, or "fixed," primarily through the process of (a) _________________________, and in the short term much of it is returned to the abiotic environment by (b)__________________. Some carbon is tied up for long periods

of time in wood and fossil fuels, which may eventually be released through combustion.

19. Carbon can be stored for millions of years in the oceans in the ________________________________ which eventually may form into limestone.
20. The oceans can absorb much CO_2. Some dissolved CO_2 is converted to ____________________________ which is acidifying ocean waters.

Bacteria and archaea are essential to the nitrogen cycle

21. The nitrogen cycle has five steps: (a) ____________________, which is the conversion of nitrogen gas to ammonia; the conversion of ammonia or ammonium to nitrate, called (b) ____________________; absorption of nitrates, ammonia, or ammonium by plant roots, called (c) __________________; the release of organic nitrogen compounds back into the abiotic environment in the form of ammonia and ammonium ions in waste products and by decomposition, a process known as (d) __________________; and finally, to complete the cycle, (e) ____________________ reduces nitrates to gaseous nitrogen.
22. Nitrogen oxides are necessary for the production of ______________________ which is a mixture of several air pollutants that injure plant tissues, irritate eyes, and cause respiratory problems.

The phosphorus cycle lacks a gaseous component

23. Erosion of (a) ________ releases inorganic phosphorus to the (b)________ where it is taken up by plant roots.

Water moves among the ocean, land, and atmosphere in the hydrologic cycle

24. The hydrological cycle results from the evaporation of water from both land and bodies of water such as the oceans and its subsequent precipitation when air is cooled. The movement of surface water from land to ocean is called ____________________.
25. The loss of water vapor from plants into the atmosphere is called ____________________.
26. Water seeps down through the soil to become __________________, which supplies much of the water to streams, rivers, ocean, soil, and plants.

ABIOTIC FACTORS IN ECOSYSTEMS

27. The two abiotic factors that probably most affect organisms in ecosystems are (a) ___________, and (b) ______________________.

The sun warms Earth

28. Without the sun's energy, the temperature on Earth would approach ____________________________.
29. Ultimately, all of the sun's energy that is absorbed by the Earth's surface and atmosphere is returned to space in the form of ____ energy. This balance prevents Earth from either heating up or cooling down.

30. The solar energy that reaches the equator strikes at an almost (a)_______________ angle which produces very warm temperatures. The solar energy near the polar regions is (b)______________________ and produces (c)______________________________________.
31. Earth's inclination on its axis of __________ degrees from the perpendicular determines the seasons.

The atmosphere contains several gases essential to organisms

32. The atmosphere is (a) _______% oxygen and (b) _______% nitrogen.
33. The atmosphere protects Earth's surface from (a) ________________ radiation. However, (b) ________________ and some (c)______________ radiation are able to penetrate it.
34. The circulation of the atmosphere is driven by differences in temperature as a result of differences in the amount of (a)__________________ at different locations on Earth's surface. Winds result from differences in atmospheric (b)____________________ and (c)____________________, Earth's (d)____________________ and (e) ______________________ of the oceans and continents. Winds generally blow from (f) ________ (high or low?) pressure areas to (g) ________ (high or low?) pressure areas, but they are deflected from these paths by Earth's rotation, a phenomenon known as the (h) __________________.

The global ocean covers most of Earth's surface

35. The ocean is separated by the continents into four "individual" oceans. They are the (a) ________________, (b) ________________, (c) ________________, and (d) ________________ oceans.
36. The ________________ Ocean covers one third of the Earth's surface and contains more than half of Earth's water.
37. Earth's rotation from west to east causes surface ocean currents to swerve to the (a) ______________ (left or right?) in the Northern Hemisphere, producing a (b) ____________________ (clockwise or counterclockwise?) gyre of water currents, and to the (c)____________ (left or right?) in the Southern Hemisphere, producing a (d)____________ (clockwise or counterclockwise?) gyre.
38. The ______________________________ is a periodic warming of the tropical eastern Pacific that alters both oceanic and atmospheric circulation patterns and results in unusual weather.

Climate profoundly affects organisms

39. ______________ is the average weather conditions, plus extremes, that occur in a given location over an extended period of time.
40. The two most important factors that determine an area's climate are (a)________________________ and (b) ____________________.

41. Unlike climate that changes slowly over thousands of years, ____________ changes rapidly.
42. (a) ____________ occurs when air heavily saturated with water vapor cools and loses its moisture-holding ability. A lack of moisture in the atmosphere produces (b) ____________ on land.
43. The dry lands on the sides of the mountains away from the prevailing wind are called ____________.
44. Local variations in climate as known as ____________.

Fires are a common disturbance in some ecosystems

45. Fires have several effects on organisms. First, they (a)____________ ; second, they (b)____________; and third, they (c)____________.

STUDYING ECOSYSTEM PROCESSES

46. ____________ is the clearance of large expanses of forest for agriculture or other uses.

BUILDING WORDS

Use combinations of prefixes and suffixes to build words for the definitions that follow.

Prefixes	The Meaning	Suffixes	The Meaning
assimil-	to bring into conformity	-ation	the process of
atmo-	air	-ic	pertaining to
bio-	life	-troph	nutrition
detrit-	worn off	-ule	small
hydro-	water	-us	thing
micro-	small		
nod-	knob		
sapr(o)-	rotten		
troph-	nutrition		

Prefix	Suffix	Definition
________	-mass	1. The total weight of all living things in a particular habitat.
________	-sphere	2. The gaseous envelope surrounding the earth; the air.
________	-logic	3. The water cycle, which includes evaporation, precipitation, and flow to the seas.
________	-climate	4. A localized variation in climate; the climate of a small area.
________	________	5. A feeding level in a food web.
________	________	6. An organism that obtains energy by breaking down organic molecules in the remains of all member of the food chain.
________	________	7. Dead organic matter.

__________ __________ 8. Oxygen-excluding swellings on the roots of legumes.

__________ __________ 9. The incorporation of ammonia and nitrate nitrogen into proteins by roots that absorb them.

MATCHING

Terms:

a.	Assimilation	f.	Food chain	k.	Nitrogen fixation
b.	Bioaccumulation	g.	Food web	l.	Rain shadow
c.	Energy pyramid	h.	Fossil fuel	m.	Tradewinds
d.	Biosphere	i.	Heterocyst	n.	Trophic level
e.	Ecosystem	j.	Nitrification	o.	Westerlies

For each of these definitions, select the correct matching term from the list above.

____ 1. The distance of an organism in a food chain from the primary producers of a community.

____ 2. Cells of certain cyanobacteria that are the site of nitrogen fixation.

____ 3. The interacting system that encompasses a community and its nonliving, physical environment.

____ 4. The system of interconnected food chains in a community.

____ 5. The ability of certain microorganisms to bring atmospheric nitrogen into chemical combination.

____ 6. An area on the downwind side of a mountain range with very little precipitation.

____ 7. All of the Earth's communities.

____ 8. A sequence of organisms through which energy is transferred from its ultimate source in a plant; each organism eats the preceding member and is eaten by the following member of the sequence.

____ 9. A graphical representation of the relative energy value at each trophic level.

____10. The oxidation of ammonium salts or ammonia into nitrites and nitrates by soil bacteria.

____11. The buildup of mercury in tuna is one example.

____12. These winds circulate from 30° N and 30° S of the equator.

MAKING COMPARISONS

Fill in the blanks.

Biogeochemical Cycle	Material Cycled	Description
Phosphorus	Phosphorus	No gas formed; dissolves in water, is taken into plants
#1	#2	As a gas, is taken into plants, converted into glucose; returned to atmosphere by respiration
#3	#4	Converted from a gas into form usable by plants, then fixed in plant tissue, then consumed by animals
#5	Water	#6

MAKING CHOICES

Place your answer(s) in the space provided. Some questions may have more than one correct answer.

____ 1. The energy that remains as biomass in plants after they have carried out cellular respiration is
a. significant only in a changing ecosystem.
b. called gross primary productivity.
c. called net primary productivity.
d. equal to the primary producer's metabolic rate.
e. evident only in a climax community.

____ 2. Compared to the overall regional climate, a microclimate
a. may be significantly different.
b. is insignificant to organisms.
c. is more important to resident organisms.
d. covers a larger area.
e. exists for a brief period.

____ 3. When it is warmest in the United States, it is coolest in
a. the Southern Hemisphere.
b. the Northern Hemisphere.
c. Brazil.
d. Europe.
e. Mexico.

____ 4. Activities performed by microorganisms involved in the nitrogen cycle include
a. liberation of nitrogen from nitrate.
b. oxidization of ammonia to nitrates.
c. production of ammonia from proteins, urea, or uric acid.
d. fixation of molecular nitrogen.
e. continuous cycling of nitrogen.

____ 5. An animal burrow in a hostile desert environment is an example of a/an
a. subterranean biome.
b. Coriolis effect.
c. macroclimate.
d. microclimate.
e. ecosystem.

____ 6. Which of the following would be positioned at the top of a typical pyramid?
a. producers
b. consumers
c. the trophic level with the greatest biomass
d. trophic level with the least numbers
e. trophic level with the least energy

____ 7. The main collector(s) of solar energy used to power life processes is/are the
a. producers.
b. atmosphere.
c. hydrosphere.
d. green leaves.
e. photosynthetic organisms.

____ 8. Most carbon is fixed ____________ and liberated ____________
a. in proteins/as organic compounds.
b. in CO_2/as complex compounds.
c. in complex compounds/as CO_2.
d. by cyanobacteria/ in oceans.
e. by plants/by plants.

____ 9. Phosphorus enters aquatic communities by means of
a. decomposers.
b. producers.
c. primary consumers.
d. secondary consumers.
e. erosion.

____10. The totality of the Earth's living inhabitants is the
a. ecosphere.
b. lithosphere.
c. biosphere.
d. biome.
e. hydrosphere.

____11. Which of the following is incorrectly paired?
a. saprotrophs : decomposers
b. detritivores : dead organic matter
c. biological magnification : increase in toxin concentration as it moves through trophic levels
d. food web : food chains
e. ecosystem energy flow : two-way

____12. Ecological pyramids may represent
a. the number of organisms at each level.
b. the rate of energy flow through an ecosystem.
c. the biomass at each level.
d. the producers, consumers, and decomposers in an ecosystem.
e. the kilocalories per square meter per year in an ecosystem.

____13. The oxygen-excluding cells of filamentous cyanobacteria that fix nitrogen are called
a. nodules.
b. nitrification cells.
c. assimilocytes.
d. anaerobe cells.
e. heterocysts.

____14. Approximatley ___% of the water a plant absorbs is transported to the leaves and is lost there by transpiration.
a. 85
b. 87
c. 90
d. 95
e. 97

____15. Air currents and the direction of wind flow are affected by
a. the Coriolis effect.
b. the Earth's rotation.
c. the position of the continents.
d. differences in atmospheric pressure.
e. upwelling.

____16. An ecosystem
a. is like a community.
b. is part of a larger ecosystem.
c.. is like a population.
d. lacks precise boundaries.
e. describes the Earth's biosphere.

____17. Consumers
a. are heterotrophs.
b. may use photosynthesis to obtain energy.
c.. occupy third or higher trophic levels.
d. are a part of Earth's biomass.
e. extract energy from inorganic molecules.

____18. The gross primary productivity of an ecosystem
a. is fixed in individual ecosystems.
b. is affected by environmental factors.
c. is the energy that remains in plant tissues after cellular respiration.
d. is the net energy available for biomass production.
e. is the rate at which energy is captured during photosynthesis.

____19. Net primary productivity would be highest in which biome?
a. a tropical rain forest
b. a boreal forest
c. agricultural land
d. an estuary
e. open ocean

____20. The buildup of toxins passing through successive levels of the food web is known as
a. bioaccumulation.
b. bioreduction.
c. persistence.
d. biological amplification.
e. biological magnification..

____21. Erosion occurs in which cycle(s)?

a. phosphorus
b. nitrogen
c. water.
d. carbon
e. oceanic

____22. Absolute zero is

a. 32° F.
b .0°F.
c. 0°K.
d. 0°C.
e. -250°C.

____23. The winds nearest the equator are called the

a. easterlies.
b. westerlies.
c. northerlies.
d. southerlies.
e. trade winds.

____24. The Coriolis effect

a. has an impact on the ocean currents.
b. has an impact on El Nino-Southern Oscillation.
c. determines the amount of moisture in the atmosphere.
d. determines the seasons.
e. is the tendency of moving air to be defected by the Earth's rotation.

VISUAL FOUNDATIONS

Color the parts of the illustration below as indicated. Label N_2 in the atmosphere, NO_3^- in the soil, and NH_3 in the soil.

RED	❑	nitrogen fixation	BLUE	❑	atmosphere	TAN	❑	animal
GREEN	❑	plants	ORANGE	❑	assimilation	PINK	❑	nitrification
YELLOW	❑	ammonification	BROWN	❑	soil	VIOLET	❑	denitrification

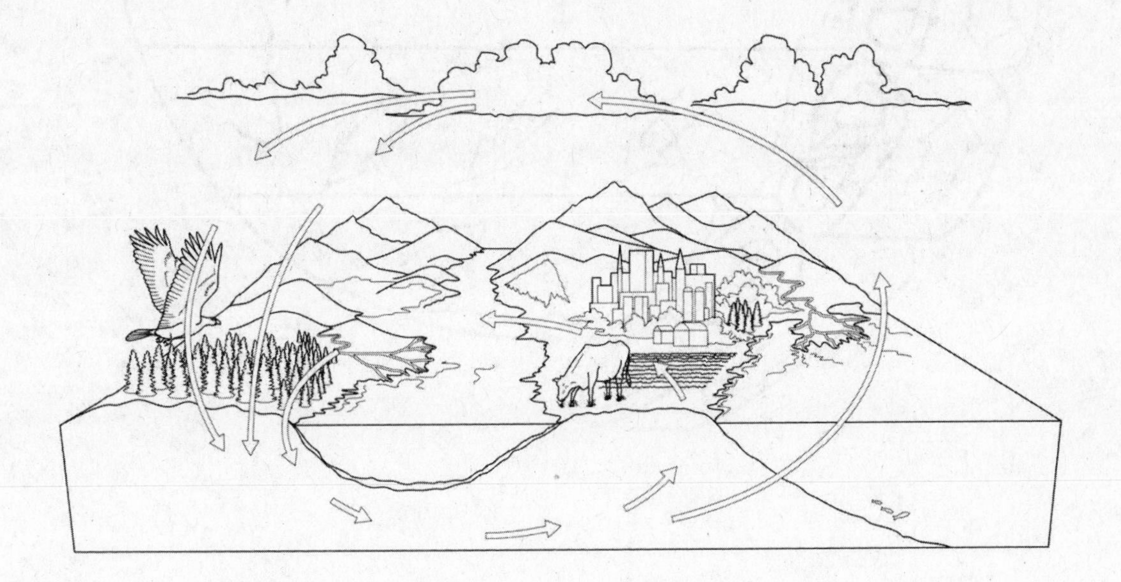

Color the parts of the illustration below as indicated.

- RED ❑ equator
- GREEN ❑ trade winds
- YELLOW ❑ warm air
- BLUE ❑ cool air
- ORANGE ❑ westerlies
- VIOLET ❑ polar easterlies

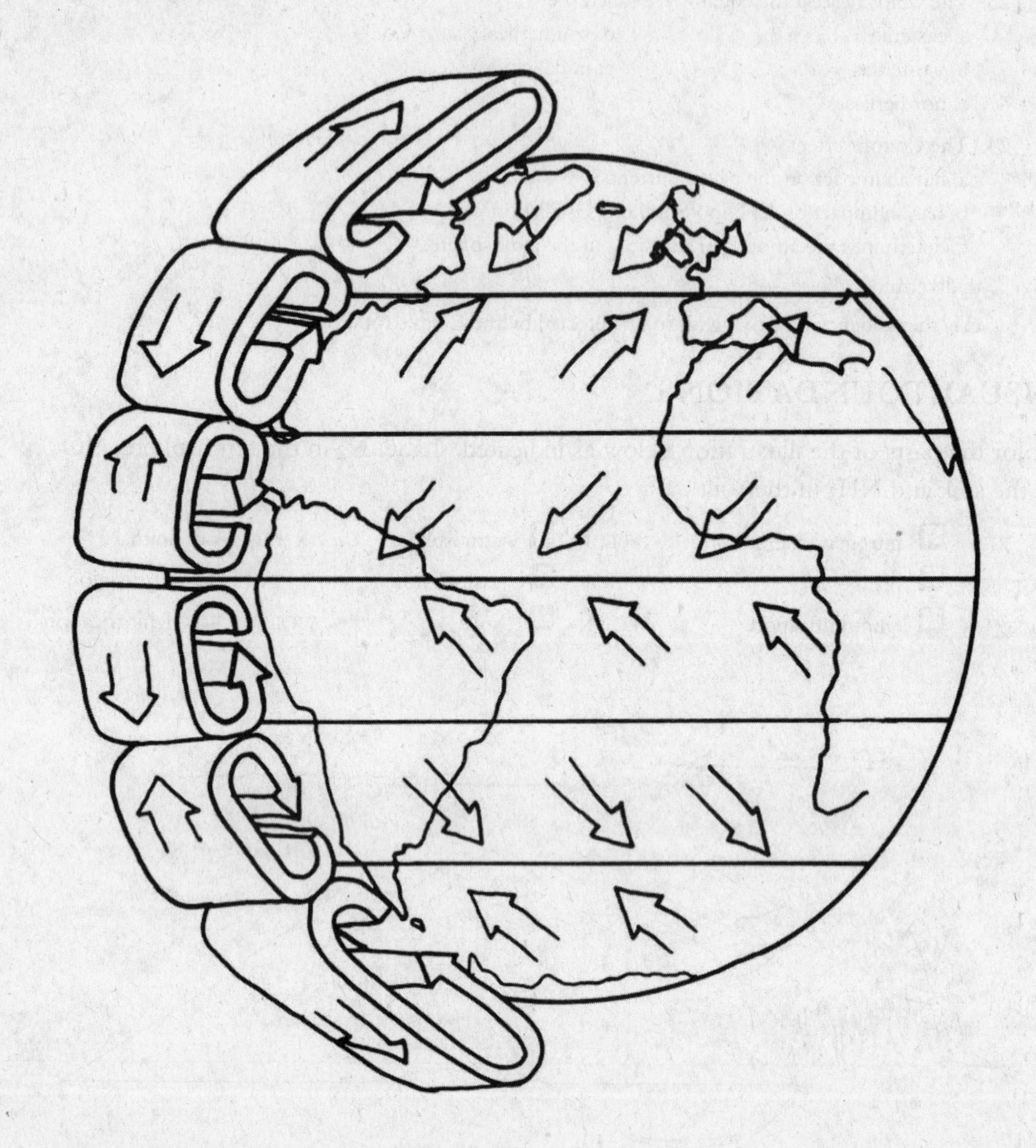

CHAPTER 56

❑

Ecology and the Geography of Life

This chapter examines the characteristics of Earth's major biomes, aquatic ecosystems, and biogeographical realms. Major terrestrial life zones (biomes) are large, relatively distinct ecosystems characterized by similar climate, soil, plants, and animals, regardless of where they occur on earth. Their boundaries are determined primarily by climate. Temperature and precipitation are the major determinants of plant and animal inhabitants. Examples of biomes include tundra; the taiga, or boreal forests; temperate forests; temperate grasslands; chaparral; deserts; tropical grasslands, or savanna; and tropical forests. In aquatic life zones, salinity, dissolved oxygen, nutrient minerals, and light, are the major determinants of plant and animal inhabitants. Freshwater ecosystems include flowing water (rivers and streams), standing water (lakes and ponds), and freshwater wetlands. Marine ecosystems include estuaries, the intertidal zone, the ocean floor, the neritic province, and the oceanic province. All terrestrial and aquatic life zones interact with one another. The study of biogeography helps biologists relate ancient communities to modern communities that exist in a given area. Biologists believe that each species originated only once, and each has a characteristic geographic distribution. The ranges of different species do not include everywhere that they could survive. The world's land areas are divided into six biogeographical realms.

REVIEWING CONCEPTS

Fill in the blanks.

INTRODUCTION

BIOMES

1. Biomes are (a) ______________ regions that have similar climate, soil, plants, and animals. (b) ____________ is the most influential factor in determining the characteristics and boundaries of a biome.

Tundra is the cold, boggy plains of the far north

2. Due to its location, tundra is the most (a) ______________ and (b) ________________ biome. It is characterized by low-growing vegetation, a short growing season, and a permanently frozen layer of subsoil called (c) ____________________.
3. Tundra may be found not only at specific latitudes, but also at high ____________________________.
4. Tundra is characterized by (a) ______________ species richness and low (b) ____________________ productivity

Boreal forest is the evergreen forest of the north

5. Another name for the boreal forest is ____________.
6. The boreal forest receives about (a) _______ cm of rain per year and its soils is typically (b) _____________ and low in minerals.

7. Most of the boreal forest is not suitable for agriculture because of its (a)____________________, and (b)____________________ soil.
8. Boreal forest trees are currently the primary source of the world's ____________________.

Temperate rain forest has cool weather, dense fog, and high precipitation

9. Coniferous temperate rainforest can be found in (a)____________________, (b)____________________, and (c)____________________.
10. Soil in temperate rain forests is nutrient-poor but rich in ____________________.
11. Temperate rain forest is one of the richest wood producers in the world, supplying us with (a) ________________and (b)________________.

Temperate deciduous forest has a canopy of broad-leaf trees

12. Temperate deciduous forest receives approximately ________ to ________ inches (cm) of rain and is dominated by broad-leaf hardwood trees.
13. Worldwide, temperate forests have been cleared and converted to ________________.

Temperate grasslands occur in areas of moderate precipitation

14. Temperate grasslands receive approximately ____ to ____ inches (cm) of rain.
15. Moist temperate grasslands with grasses up to 6.5 ft. tall are known as (a)________________prairies. Prairies in which the grass is less than 1.6 ft. tall are called (b) ________________ prairies.

Chaparral is a thicket of evergreen shrubs and small trees

16. Chaparral is an assemblage of small-leaved shrubs and trees in areas with mild winters and abundant rainfall, combined with extremely dry summers. These climatic conditions are referred to as a ________________ climate.

Deserts are arid ecosystems

17. Desserts located in temperate climates are called (a)________________ deserts and those located in subtropical or topical regions are classified as (b) ________________ deserts.
18. North America has four distinctive deserts. They are the (a)________________, (b) ________________, (c) ________________ and (d) ________________ deserts.
19. Deserts generally receive less than ________ inches of rain a year.
20. During the driest months of the year, many desert animals tunnel underground where they remain inactive; this period of dormancy is known as ____________________.

Savanna is a tropical grassland with scattered trees

21. Seasons in tropical grassland are regulated by (a) ________________ and not by (b) ________________.
22. Savannas receive between ________ and ______ inches (cm) of precipitation.
23. The best known savanna, with its large herds of grazing, hoofed animals and predators, is the ________________ savanna.
24. Desertification of savannas is occurring as a result of ________________.

There are two basic types of tropical forests

25. Tropical dry forest receive (a) ________ to ________ inches (cm) of rainfall annually and during the dry season the trees (b) __________ their leaves.
26. Tropical rain forests have about ________ to _______ inches (cm) of rainfall annually. They have many diverse species and are noted for tall foliage, many epiphytes, and leached soil.
27. Most trees of tropical rain forests are ________________________ plants.
28. Tropical rainforests support extensive ________________ communities of plants such as orchids and bromeliads.

AQUATIC ECOSYSTEMS

29. Aquatic ecosystems are classified primarily on ____________________ and not on the dominant form of vegetation as land biomes are.
30. Aquatic life is ecologically divided into the free-floating (a)____________, actively swimming (b) ________________, and the bottom dwelling (c) ______________ organisms.

Freshwater ecosystems are linked to land and marine ecosystems

31. Freshwater ecosystems include (a) ____________and (b) ________, (c) ______________________and (d) ________________ as well as (e) ______________ and (f) ________________. Freshwater ecosystems occupy about (g) ______% of the Earth's surface.
32. Rivers and streams are flowing-water ecosystems. The kinds of organisms they contain depends primarily on the ________________________.
33. Streams and rivers depend on __________ for much of their energy due to the amount of detritus that is washed into them.
34. Lakes and ponds are standing-water ecosystems. Large lakes have three zones: the (a) ________________ zone is the shallow water area along the shore; the (b) __________________ zone is the open water area away from shore that is illuminated by sunlight; and the (c)__________________ zone, which is the deeper, open water area beneath the illuminated portion. Of these, the (d)______________________ zone is the most productive.

35. Vertical thermal layering of standing water is called (a)____________________. It is usually marked in the summer by an abrupt temperature transition at some depth called the (b) ____________________. When atmospheric temperatures drop in fall, vertical layers mix, a phenomenon known as (c)____________________. In spring, surface waters sink and bottom waters return to the surface. This is called (d)____________________.
36. Freshwater wetlands may be (a) ____________, where grasslike plants dominate, or (b) ____________, where woody trees or shrubs dominate.
37. Freshwater wetlands may be small, shallow ponds called (a)____________________, or (b)____________________ that are dominated by mosses.

Estuaries occur where fresh water and salt water meet

38. Water levels in estuaries rise and fall with the (a) ____________ and salinity fluctuates with the (b) ____________, (c)____________________ and (d) ____________________.
39. Temperate estuaries usually contain ____________________, shallow wetlands in which salt-tolerant grasses dominate.
40. The ____________________ are the tropical equivalent of salt marshes.

Marine ecosystems dominate Earth's surface

41. The main marine habitats are similar to those of fresh water but are modified by tides and currents. The marine habitat near the shoreline between low and high tide is the (a) ____________________. This region has high levels of (b) ____________ and (c)____________ together with an abundance of oxygen.
42. The benthic environment is divided into zones based upon (a)____________________, (b)____________________ and (c)____________________.
43. The abyssal zone is at a depth of ___ to ___ thousand meters (miles).
44. Coral reefs are built from accumulated layers of (a)____________________. Only species of coral with (b)____________________ living within them, a symbiotic relationship, build reefs.
45. The ____________________ is open ocean that overlies the continental shelves.
46. The portion of the neritic province that contains sufficient light to support photosynthesis is called the ____________________ region.
47. The deeper region of the open ocean, beyond the reach of light, is known as the____________________.

ECOTONES

48. Within landscapes, the transition zone where two communities or biomes meet and intergrade is called an ____________________.

BIOGEOGRAPHY

49. One of the basic tenets of biogeography is that each species originated only once. The particular place where this occurred is known as the species' ____________________.

50. Localized, native species not found anywhere else in the world are called an (a) ______________ species. Some species have a nearly worldwide distribution and occur on more than one continent or throughout much of the ocean. Such species are said to be (b)______________

Land areas are divided into six biogeographic realms

51. The six biogeographical realms are the
(a)______________, (b) ______________,
(c)______________, (d) ______________,
(e)______________ and (f) ______________.

BUILDING WORDS

Use combinations of prefixes and suffixes to build words for the definitions that follow.

Prefixes	The Meaning
abys(s)-	bottomless
benth-	depth
bore-	north
chappara-	dwarf oak
estiv-	summer
inter-	between, among
littor-	seashore
phyto-	plant
thermo-	heat, hot
zoo-	animal

Suffixes	The Meaning
-al	pertaining to
-ation	the process of
-ic	pertaining to

Prefix	Suffix	Definition
________	-plankton	1. Planktonic plants; primary producers.
________	-plankton	2. Planktonic protists and animals.
________	-tidal	3. Pertains to the zone of a shoreline between high tide and low tide.
________	-cline	4. A layer of water marked by an abrupt temperature transition.
________	________	5. A zone of shallow water along the shore of a lake or pond.
________	________	6. Refers to the ocean floor or the bottom of a lake or pond.
________	________	7. Region or zone of the benthic environment of the ocean ranging from 4000 to 6000 m.
________	________	8. Another name for the evergreen forests called taiga.

__________ ______l____ 9. A thicket of evergreen shrubs and small trees.

__________ __________ 10. A period of inactivity brought on by heat and a lack of water.

MATCHING

Terms:

a. Benthos
b. Biome
c. Chaparral
d. Desert
e. Desertification
f. Ecotone
g. Endemic species
h. Estuary
i. Euphotic zone
j. Limnetic zone
k. Neritic province
l. Permafrost
m. Savanna
n. Taiga
o. Tundra

For each of these definitions, select the correct matching term from the list above.

____ 1. Northern coniferous forest biome that stretches across North America and Eurasia.
____ 2. Upper reaches of the neritic province; region of aquatic habitats lying near enough to the surface that sufficient light is present for photosynthesis to take place.
____ 3. The open water area away from the shore of a lake or pond, extending down as far as sun light penetrates.
____ 4. Organisms that live on the bottom of oceans or lakes.
____ 5. Distinctive vegetation type characteristic of certain areas with cool moist winters and long dry summers.
____ 6. Permanently frozen ground.
____ 7. Dry areas found in temperate and subtropical or tropical regions.
____ 8. The conversion of marginal savanna into desert due to severe overgrazing by domestic animals..
____ 9. A coastal body of water, partly surrounded by land, with access to the open ocean and a large supply of fresh water from rivers.
____ 10. Localized, native species that are not found anywhere else in the world.
____ 11. Transition zone where two communities or biomes meet.
____ 12. Open ocean that overlies the continental shelves.

MAKING COMPARISONS

Fill in the blanks.

Biome	Climate	Soil	Characteristic Plants and Animals
Temperate rain forest	Cool weather, dense fog, high rainfall	Relatively nutrient-poor	Large conifers, epiphytes, squirrels, deer
Tundra	#1	#2	#3
#4	Not as harsh as tundra	Acidic, mineral poor, decomposed needles	Coniferous and deciduous trees, medium to small animals and a few larger species
Temperate deciduous forest	#5	#6	#7

Biome	Climate	Soil	Characteristic Plants and Animals
#8	Moderate, uncertain rainfall	#9	Grasses, grazing animals and predators such as wolves
Chaparral	#10	Thin, infertile	#11
#12	Temperate and tropical, low rainfall	#13	Organisms adapted to water conservation, small animals
#14	Tropical areas with low or seasonal rainfall	#15	Grassy areas with scattered trees; hoofed mammals, large predators
Tropical rain forest	#16	#17	#18

MAKING CHOICES

Place your answer(s) in the space provided. Some questions may have more than one correct answer.

____ 1. One expects to find organisms with adaptations for retaining moisture and clinging securely to the substrate in the
a. salt marsh.
b. intertidal zone.
c. profundal zone.
d. littoral zone.
e. estuary.

____ 2. The communities containing the most permafrost are called
a. polar.
b. steppes.
c. boreal.
d. tundra.
e. muskegs.

____ 3. A community with two distinct soil layers abounding in trees, reptiles, and amphibians is
a. a grassland.
b. a deciduous forest.
c. a temperate rain forest.
d. temperate.
e. a chaparral.

____ 4. Crayfish are found mainly in
a. the littoral zone.
b. the limnetic zone.
c. the profundal zone.
d. shallow lakes.
e. the benthic zone.

____ 5. Small or microscopic organisms carried about by open ocean currents
a. would be found in the littoral zone.
b. would be found in the neritic provence.
c. would be found in the nektonic zone.
d. would be found in the benthic.
e. are planktonic.

____ 6. Most of midwestern North America is
a. grassland.
b. deciduous forest.
c. temperate rain forest.
d. temperate.
e. chaparral.

____7. The area(s) essentially corresponding to the waters above the continental shelf is/are the
a. littoral zone.
b. limnetic zone.
c. profundal zone.
d. neritic province.
e. benthic zone.

____8. The two most diverse of communities are the ________ and ________.
a. deciduous forests
b. tropical rain forests
c. coniferous forests
d. coral reefs
e. grasslands

____9. An aquatic zone at the juncture of a river and the sea is a/an
a. salt marsh.
b. intertidal zone.
c. profundal zone.
d. littoral zone.
e. estuary.

____10. Communities dominated by evergreens and containing acidic, mineral-poor soil is/are
a. in South America.
b. the taiga.
c. temperate coniferous forests.
d. typically impregnated with a layer of permafrost.
e. the boreal forests.

____11. The factor that determines the nature of the biogeographic realms more than any other single factor is
a. soil.
b. climate.
c. latitude.
d. types of animals.
e. altitude.

____12. Which biome is found only in the northern hemisphere?
a. chaparral
b. desert
c. tundra
d. boreal forest (taiga)
e. temperate grasslands

____13. Trees that might be found in a boreal forest include
a. spruces.
b. aspen.
c. birch.
d. firs.
e. larch.

____14. Which biome has an annual precipitation range of 30 to 50 inches (75 to 126 cm)?
a. temperate grasslands
b. savanna
c. temperate deciduous forest
d. taiga
e. temperate rain forest

____15. Which desert is located in California, Arizona, and Mexico?
a. Mojave Desert
b. Sonoran Desert
c. Death Valley Desert
d. Great Basin Desert
e. Chihuahuan Desert

____16. Desert soil is rich in
a. $MgSO_4$.
b. $CaSO_4$.
c. $CaCO_3$.
d. minerals.
e. NaCl.

____17. About ___% of the organisms living in a tropical rainforest live in the middle and upper canapoies.
a. 80
b. 85
c. 90
d. 95
e. 97

____18. Streams and rivers depend upon ____________ for much of their energy.

a. aquatic producers
b. detritus
c. plankton
d. cyanobacteria
e. land

____19. A coral reef

a. is comparable to a tropical rain forest.
b. has stinging tentacles.
c. is dependent upon a symbiotic relationship with zooxanthellae.
d. can be found in water poor in nutrients.
e. provides humans with pharmaceuticals.

VISUAL FOUNDATIONS

Color the parts of the illustration below as indicated. Label intertidal zone, benthic environment, abyssal zone, hadal zone, and pelagic environment.

GREEN ❑ neritic province
BLUE ❑ oceanic province
TAN ❑ land

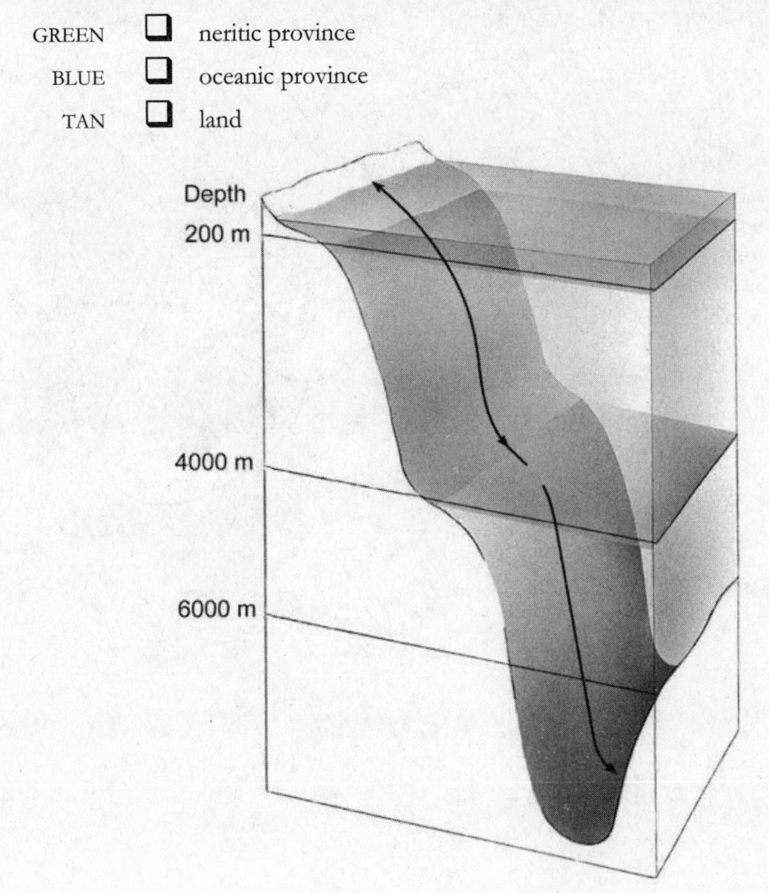

Color and label the biogeography realms of the illustration below as indicated.

GRAY	☐	Nearartic	GREEN	☐	Neotropical
BLUE	☐	Ethiopian	ORANGE	☐	Paleartic
TAN	☐	Austrailian	BROWN	☐	Oriental

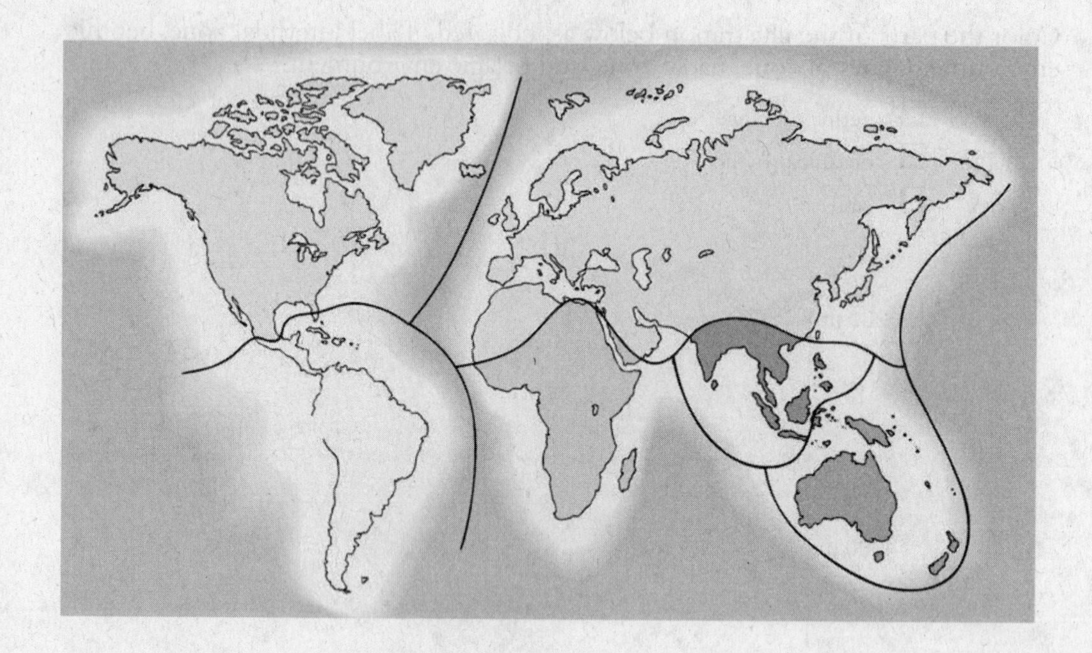

CHAPTER 57

Biological Diversity and Conservation Biology

This chapter examines some of the consequences of the growth in the human population on the biosphere. Although humans have been on earth a comparatively brief span of time, our biological success has been unparalleled. We have expanded our biological range into almost every habitat on earth, and, in the process, have placed great strain on the Earth's resources and resilience, and profoundly affected other life forms. As a direct result of our actions, many environmental concerns exist today. Species are disappearing from earth at an alarming rate. Although forests provide us with many ecological benefits, deforestation is occurring at an unprecedented rate. The production of atmospheric pollutants that trap solar heat in the atmosphere threaten to alter the earth's climate. Global warming may cause a rise in sea level, changes in precipitation patterns, death of forests, extinction of animals and plants, and problems for agriculture. It could result in the displacement of thousands or even millions of people. Ozone is disappearing from the stratosphere at an alarming rate. The ozone layer helps to shield Earth from damaging ultraviolet radiation. If the ozone were to disappear, Earth would become uninhabitable for most forms of life.

REVIEWING CONCEPTS

Fill in the blanks.

INTRODUCTION

THE BIODIVERSITY CRISIS

1. Many people are concerned with ______________________________, the ability to meet humanity's current needs without compromising the ability of future generations to meet their needs.
2. Biological diversity is the variety of living organisms considered at three levels: (a) ______________________, (b)______________________, and (c) ______________________.
3. When the number of individuals in a species is reduced to the point where extinction seems imminent, the species is said to be (a)________________, and when numbers are seriously reduced so that extinction is feared but thought to be less imminent, the species is said to be (b) ______________.

Endangered species have certain characteristics in common

4. Many threatened and endangered species share certain characteristics that make them vulnerable to extinction. Some of these characteristics are having (a) ______________________________________, requiring a large (b) ________________________, living

(c)________________________, having a
(d)__________________________________, needing specialized
(e)____________________________ and possessing specialized
(f)_________________________________.

Human activities contribute to declining biological diversity

5. Species become endangered and extinct for a variety of reasons, but the most common reasons are the (a) ______________________________, (b) ______________________________, or (c) ____ of habitat.
6. The breakup of large areas of habitat into small, isolated patches is known as (a) ____________________________. This encourages the spread of (b) _________________________ species.
7. (a) _______________________________________, the introduction of a foreign species into an area where it is not native, often upsets the balance among the organisms living in that area. A foreign species is call an (b)____________________________________ species if it causes economic or environmental harm to the area.
8. The collection of live organisms from nature is called ______________________________ harvesting.

CONSERVATION BIOLOGY

9. ______________conservation attempts to preserve biodiversity in the wild by establishing reserves (e.g., parks).
10. ___________________ conservation tries to preserve species in human-controlled settings such as zoos and seed banks.

***In situ* conservation is the best way to preserve biological diversity**

11. The single best way to protect biological diversity is to protect (a)____________________________, and (b) __________________________ habitats.
12. ____ is a subdivision of ecology studying the connections in a heterogeneous landscape consisting of multiple interacting ecosystems.
13. Strips of habitat connecting isolated habitat patches are called ____.
14. (a) _________________________________, a process in which the principles of ecology are used to return a degraded environment to one that is more functional and sustainable, is an important part of (b)____________________conservation.

***Ex situ* conservation attempts to save species on the brink of extinction**

15. Artificial insemination and host mothering ar used to increase the number of offspring particularly in endangered or threatened species. In (a) ____, sperm is collected from a male and used to impregnate a host mother. The host mother is treated with (b) ____ so she will produce more eggs. The fertilized egg is then implanted in another female of a related but less rate species.

The Endangered Species Act provides some legal protection for species and habitats

16. The Endangered Species Act authorizes the U.S. Fish and Wildlife Service to ______________________________ ______________________________.

International agreements provide some protection for species and habitats

17. The Convention on International Trade in Endangered Species of Wild Flora and Fauna protects ______________________________ ______________________________.

    It bans hunting, capturing, and selling of endangered or threatened species.

DEFORESTATION

18. Forests on mountains and hillsides help to protect valleys from flooding by ______________________________.
19. Deforestation may increase global temperature by releasing ______________________________.

    Deforestation also contributes to the loss of biodiversity.

Why are tropical rain forests disappearing?

20. There are many reasons that deforestation is occurring. However, three most notable causes of deforestation in tropical rains forests are (a) ______________________, (b) ______________________, and (c) ______________________.

Why are boreal forests disappearing?

21. Boreal forests are disappearing rapidly because they are currently the primary source of the world's ______________________________.

CLIMATE CHANGE

Greenhouse gases cause climate change

22. Gases that are accumulating in the atmosphere as a result of human activities include (a) ________________, (b) ________________, (c) ________________, (d) ________________ and (e)________________.
23. Because CO_2 and other gases trap the sun's radiation somewhat like glass does in a greenhouse, the natural trapping of heat is referred to as the ________________.
24. Greenhouse gases trap ______________________________ causing the Earth to warm.

What are the probable effects of climate change?

25. The probable effects of global warming include changes in (a)________________, and (b) ________________, as well as effects on (c) ________________, (d)________________ and (e)________________.

26. Both (a) ________________ and (b) ________________ containing substances catalyze ozone destruction.
27. The primary chemicals responsible for ozone loss in the stratosphere are a group of chlorine compounds called ________________.
28. The thinning of the ozone layer over Antarctica occurs annually between (a) ________________ and (b) ________________.
29. Excessive exposure to UV radiation is linked to human health problems, such as (a) ________________, (b)________________, and (c) ________________.
30. In 1987, the Montreal Protocol was signed by many countries to limit the production of ________________.

BUILDING WORDS

Use combinations of prefixes and suffixes to build words for the definitions that follow.

Prefixes	The Meaning	Suffixes	The Meaning
bi(o)-	life	-demic	pertaining to people or country
de-	to remove	-tic	pertaining to
en-	in		
strat(o)-	layer		

Prefix	Suffix	Definition
________	-forestation	1. The removal or destruction of all tree cover in an area.
________	-sphere	2. The layer of the atmosphere between the troposphere and the mesosphere, containing a layer of ozone that protects life by filtering out much of the sun's ultraviolet radiation.
________	________	3. Species that are found on an island and nowhere else in the world.
________	________	4. A type of pollution that occurs when a foreign species is introduced into an area where it was not found previously.

MATCHING

Terms:

a. Artificial insemination
b. Biological diversity
c. Endangered species
d. Ex situ conservation
e. Extinction
f. Genetic diversity
g. Greenhouse effect
h. Greenhouse gases
i. Host mothering
j. In situ conservations
k. Keystone species
l. Ozone
m. Slash-and-burn agriculture
n. Subsistence agriculture
o. Threatened species

For each of these definitions, select the correct matching term from the list above.

____ 1. The warming of the Earth resulting from the retention of atmospheric heat caused by the build-up of certain gases, especially carbon dioxide.
____ 2. A species whose population is low enough for it to be at risk of becoming extinct.
____ 3. The number and variety of living organisms.
____ 4. The disappearance of a species from a given habitat.
____ 5. Carbon dioxide, methane, nitrous oxide, CFCs, and ozone.

____ 6. A method of controlled breeding in zoos.
____ 7. Efforts to preserve biological diversity in the wild.
____ 8. The production of enough food to feed oneself and one's family.
____ 9. A species whose numbers are so severely reduced that it is in imminent danger of becoming extinct.
____10. The layer in the upper atmosphere that helps shield the Earth from damaging ultraviolet radiation.
____11. Species conservation in which species are conserved in human controlled settings.
____12. Species conservation in which eggs are collected from a female of a rare species, fertilized artificially and implanted in a female from a closely related species.

MAKING COMPARISONS

Fill in the blanks.

Issue	Why It Is An Issue	Cause of Problem	Solution
Declining biological diversity	Diversity contributes to a sustainable environment; lost opportunities, lost solutions to future problems	Commercial hunting, habitat destruction, efforts to eradicate or control species, commercial harvesting	*In situ* conservation, *ex situ* conservation, habitat protection
#1	Forests are needed for habitat, watershed protection, prevention of erosion	#2	#3
#4	Threatens food production, forests, biological diversity; could cause submergence of coastal areas, change precipitation patterns	#5	#6
Ozone depletion	Ozone layer protects living organisms from exposure to harmful amounts of UV radiation	#7	#8

REFLECTIONS

The subject matter covered in this chapter is of critical importance to the continuance of life on earth as we know it. This chapter points out some of the dangers humanity faces if we continue some of our habits. Biologists and other responsible citizens can make a difference. One might even say that we have a responsibility to make a difference. A few genuinely concerned individuals will assume leadership roles in bringing about the social, cultural, and political changes required to protect the environment. All of us can participate in assuring that our children and grandchildren will inherit a clean, healthy, and productive environment.

We encourage you to reflect on what your personal role could be to bring about needed reforms. What might you contribute to the welfare of the planet? What changes are needed to ensure that future generations can enjoy a reasonably good quality of life? Think about your lifestyle and aspirations. Are they compatible with a long-term investment in the future of the planet?

VISUAL FOUNDATIONS

Color the parts of the illustration below as indicated. Also label greenhouse gases in atmosphere.

- RED ❑ radiated heat redirected back to Earth
- GREEN ❑ Earth
- YELLOW ❑ solar energy
- BLUE ❑ atmosphere
- ORANGE ❑ heat escaping to space

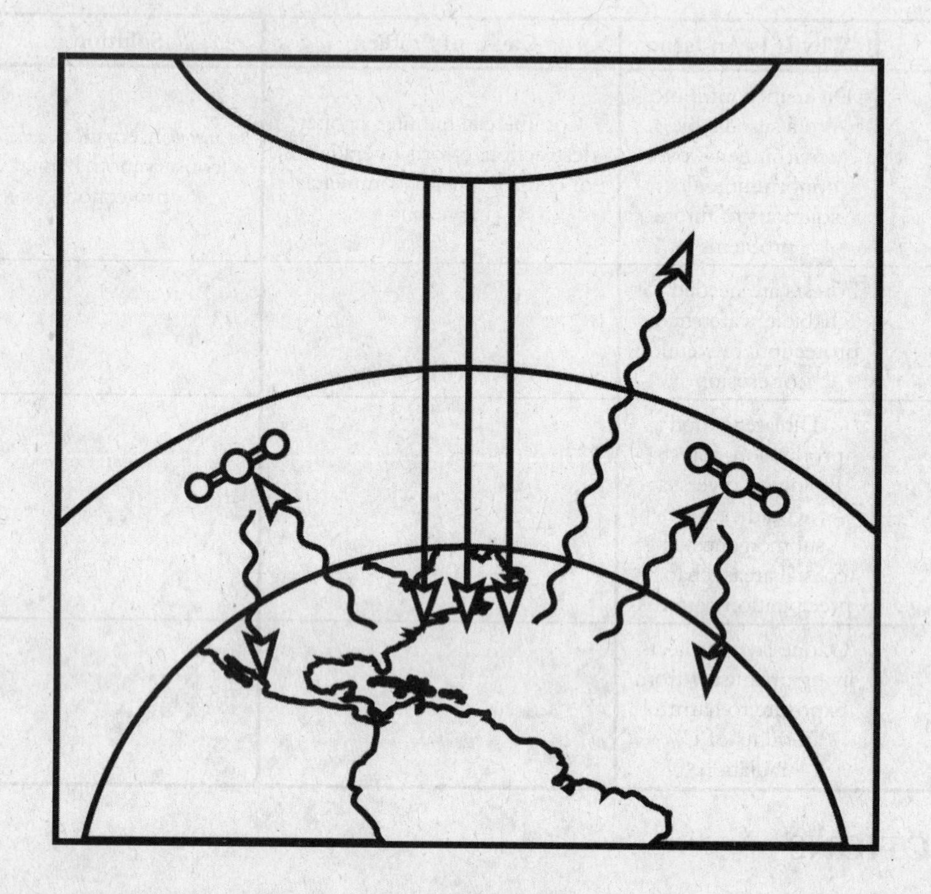

Color the parts of the illustration below as indicated. Also label normal ozone levels, and reduced ozone levels.

RED	❑	ozone molecules
GREEN	❑	oxygen molecules
BLUE	❑	troposphere
YELLOW	❑	stratosphere
ORANGE	❑	UV solar radiation

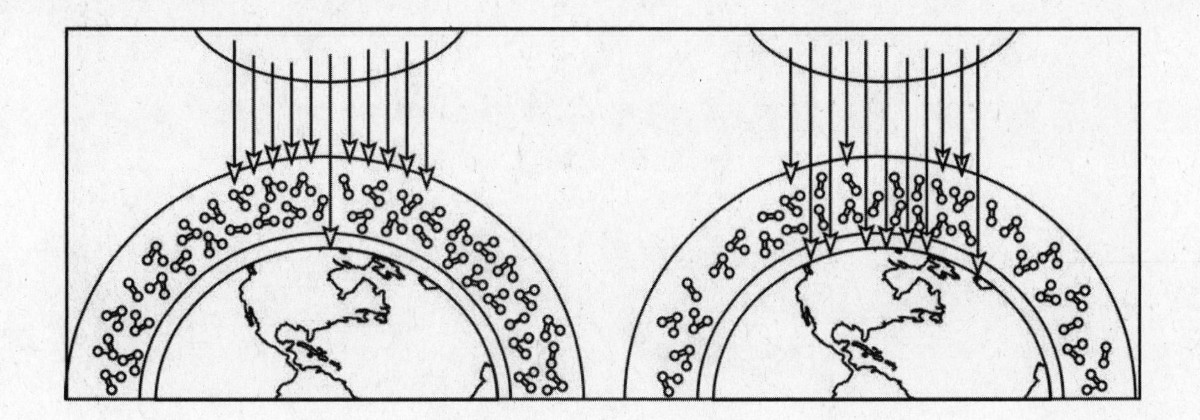

ANSWERS

Reviewing Concepts, Building Words, Matching, Making Comparisons, Making Choices, Visual Foundations

CHAPTER 1

Reviewing Concepts: **1.** (a) unspecialized cells, (b) divide, (c) more stem cells, (d) specialized types of cells, (d) repair injury **2.** (a) plants, (b) humans, (c) research animals **3.** (a) pluripotent stem cells, (b) all the tissues **4.** (a) the expanding human population, (b) decreasing biological diversity, (c) diminishing natural resources, (d) global climate change, prevention and cure of diseases **5.** (a) the interaction of biological systems, (b) the inter-relationship of structure and function, (c) information transfer, (d) energy transfer, (e) the process of evolution **6.** (a) a precise kind of organization, (b) growth and development, (c) self-regulated metabolism, (d) the ability to respond to stimuli, (e) reproduction, (f) adaptation to environmental change **7.** (a) unicellular, (b) multicellular **8.** (a) prokaryote, (b) eukaryote, (c) a nucleus **9.** (a) size, (b) number **10.** development **11.** chemical reactions **12.** (a) homoeostasis, (b) self-regulating **13.** stimuli **14.** sessile **15.** (a) light, (b) gravity, (c) water, (d) touch, (e) other stimuli **16.** Louis Pasteur **17.** (a) asexually, (b) sexually **18.** (a) adaptations, (b) structural, (c) physiological, (d) biochemical, (e) behavioral **19.** reductionism **20.** emergent properties **21.** (a) atom, (b) cell **22.** (a) muscle, (b) nervous **23.** population **24.** biosphere **25.** ecology **26.** genes **27.** nucleotides **28.** (a) Watson, (b) Crick **29.** proteins **30.** hormones **31.** neurotransmitters **32.** (a) releasing chemicals, (b) sounds, (c) a visual display **33.** (a) primary producers, (b) autotrophs **34.** metabolism **35.** cellular respiration **36.** (a) producers, (b) consumers, (c) decomposers **37.** evolution **38.** taxonomy **39.** species **40.** (a) genus, (b) species **41.** orders **42.** domain **43.** a clade **44.** (a) Bacteria, (b) Archaea, (c) Eukarya **45.** evolutionary processes **46.** (a) Charles Darwin, (b) Alfred Wallace **47.** *On the Origin of the Species by Natural Selection* **48.** (a) variation, (b) survive, (c) competition, (d) survive **49.** gene pool **50.** environmental changes (pressures) **51.** new species **52.** scientific method **53.** deductive **54.** inductive **55.** Alexander Fleming **56.** phenomenon or problem **57.** (a) reasonably consistent, (b) generate testable predictions, (c) generate test results which can be confirmed by others, (d) be falsifiable **58.** proven to be true **59.** models **60.** variable **61.** double-blind study **62.** not all cases of what is being studied can be observed **63.** conclusions **64.** theory **65.** (a) the Big Bang Theory, (b) evolution of major groups of organisms **66.** paradigm **67.** reductionists **68.** systems **69.** (a) ethically responsible, (b) educate people **70.** (a) genetic research, (b) stem cell research, (c) cloning, (d) climate change, (e) human and animal experimentation

Building Words: **1.** biology **2.** ecology **3.** asexual **4.** homeostasis **5.** ecosystem **6.** biosphere **7.** autotroph **8.** heterotroph **9.** prokaryote **10.** Photosynthesis **11.** hypothesis

Matching: **1.** k **2.** o **3.** r **4.** f **5.** p **6.** c **7.** n **8.** q **9.** a **10.** e **11.** b **12.** m

Making Comparisons: **1.** population **2.** bison, humans, wolves, elk, mountain goats, etc. that also live in Yellowstone --- any example of a number of different species living together in the same area **3.** All animal populations and physical features in Yellowstone **4.** Biosphere **5.** Tissues **6.** Atoms and molecules **7.** genes

Making Choices: **1.** e **2.** a, d **3.** b **4.** d **5.** c, e **6.** a **7.** a, b, e **8.** b, c **9.** a, c, d **10.** a, b, d, e **11.** e **12.** d **13.** c **14.** c, d **15.** c **16.** a, b, c, e **17.** a-e

Visual Foundations: **1.** d **2.** b, c ,d **3.** e

CHAPTER 2

Reviewing Concepts: **1.** (a) hydrogen, (b) helium **2.** (a) water, (b) many simple acids and bases, (c) simple salts **3.** organic compounds **4.** chemical reactions (ordinary chemical means) **5.** (a) carbon, (b) hydrogen, (c) oxygen, (d) nitrogen **6.** trace elements **7.** (a) protons, (b) neutrons, (c) electrons, (d) protons and neutrons, (e) electrons **8.** (a) protons, (b) neutrons, (c) electrons **9.** (a) number of protons, (b) $_8O$ **10.** dalton **11.** atomic mass **12.** (a) atomic mass (or numbers of neutrons), (b) atomic number (or numbers of protons) **13.** radioisotopes **14.** autoradiography **15.** electron shell **16.** valence **17.** valance electrons **18.** chemical properties **19.** two or more different elements **20.** molecule **21.** chemical **22.** structural **23.** atomic masses of its constituent atoms **24.** mole **25.** (a) mole, (b) dissolved, (c) liter **26.** (a) reactants, (b) products **27.** (a) left, (b) right **28.** (a)chemical bonds (b) valence electrons, (c) bonds **29.** (a) covalent bond, (b) filled valance shell **30.** (a) single, (b) double, (c) triple **31.** geometric shape **32.** orbital hybridization **33.** nonpolar covalent bonds **34.** polar covalent bonds **35.** (a) cations, (b) anions **36.** (a) solvent (b) solute **37.** hydration **38.** (a) an atom with a partial negative charge (b) oxygen, (c) nitrogen **39.** (a) weak, (b) strong **40.** electric charges **41.** (a) reduction, (b) oxidation, (c) electron, (d) acceptor (component that takes up electrons), (e) donor (component that donates the electrons) **42.** 70 **43.** (a) hydrogen bonding, (b) surface tension **44.** adhesion **45.** capillary action **46.** (a) hydrophilic, (b) hydrophobic **47.** hydrophobic interactions **48.** heat **49.** temperature **50.** heat of vaporization **51.** calorie (cal) **52.** specific heat **53.** (a) H^+ and an anion, (b)OH^- and a cation, **54.** (a) 0-14, (b) logarithmic, (c) 7 **55.** (a) base, (b) acid **56.** (a) changes in pH, (b) weak acid, (c) weak base **57.** salt **58.** electrolytes

Building Words: **1.** neutron **2.** equilibrium **3.** tetrahedron **4.** hydration **5.** hydrophobic **6.** nonelectrolyte **7.** hydrophilic **8.** isotopes **9.** covalent **10.** polar

Matching: **1.** n **2.** e **3.** i **4.** d **5.** h **6.** q **7.** a **8.** l **9.** g **10.** f **11.** k **12.** r **13.** b

Making Comparisons: **1.** CO_2 **2.** Ca **3.** iron (III) oxide **4.** CH_4 **5.** Cl^- **6.** potassium ion **7.** H^+ **8.** magnesium **9.** NH_4^+ **10.** H_2O **11.** $C_6H_{12}O_6$ **12.** HCO_3^- **13.** carbonic acid **14.** Sodium chloride **15.** hydroxide ion **16.** NaOH

Making Choices: **1.** e **2.** a **3.** b **4.** a **5.** e **6.** a, b **7.** e **8.** c, d **9.** a **10.** a, e **11.** d **12.** b, d, e **13.** a **14.** a, b, d **15.** a **16.** a, b, d, e **17.** a, b **18.** b, c, d, e **19.** a-e **20.** a, b, d

Visual Foundations: **1a.** Carbon-12 **1b.** Carbon-14 **2.** e **3.** See Figure 2.7 **4.** This is a polar bond because the electrons are spending more time in orbit near the oxygen atom (oxygen end of the bond) than the hydrogen atom (hydrogen end of the bond).

CHAPTER 3

Reviewing Concepts: **1.** another carbon **2.** (a) carbon, (b) hydrogen (for example CO_2) **3.** chemical groups **4.** (a) carbohydrates, (b) lipids, (c) proteins, (d) nucleic acids **5.** four **6.** (a) one, (b) two, (c) three, (d) four **7.** (a) hydrocarbons, (b) unbranched chains, (c) branched chains, (d) rings **8.** structural **9.** geometric **10.** enantiomers **11.** (a) hydroxyl, (b) polar **12.** (a) carboxyl, (b) acidic **13.** (a) amino, (b) basic **14.** Monomers **15.** (a) condensation, (b) hydrolysis reactions **16.** (a) sugars, (b) starches, (c) cellulose **17.** (a) carbon, (b) hydrogen, (c) oxygen **18.** (a) glucose, (b) ring **19.** fructose **20.** glycosidic **21.** (a) sucrose, (b) lactose **22.** simple sugars (monosaccharides) **23.** starch **24.** amylose **25.** amyloplasts **26.** glycogen **27.** (a)cellulose (b)cell walls **28.** (a) chitin, (b) cell walls **29.** (a) glycoprotein, (b) glycolipids **30.** (a) structure, (b) insoluble **31.** (a) carbon, (b) hydrogen, (c) oxygen **32.** (a) fats, (b) storage **33.** (a) alcohol, (b) unbranched hydrocarbon chain, (c) carboxyl group **34.** van der Waals interactions **35.** (a) liquid, (b) double bonds **36.** (a) hydrogenation, (b) saturated fats **37.** amphipathic **38.** (a) glycerol, (b) fatty acids, (c) phosphate, (d) organic **39.** (a) isoprene units, (b) Vitamin A **40.** (a) carbon, (b) six, (c) five **41.** cholesterol **42.** fatty acids **43.** Macromolecules, (b) amino acids, (c) versatile **44.** (a) amino, (b) carboxyl, (c) alpha carbon, (d) R group (radical) **45.** essential **46.** dipeptide **47.** polypeptide **48.** (a) condensation, (b) carboxyl, (c) amino **49.** (a)conformation, (b) function **50.** (a) primary, (b) secondary, (c) tertiary, (d) quaternary **51.** polypeptide chain **52.** (a) hydrogen (b) α-helix, (c) β-pleated sheet **53.** (a) side chains, (b) shape **54.** polypeptides **55.** molecular chaperones **56.** function **57.** denatured **58.** (a) Alzheimer's, (b) Huntington's, (c) mad cow **59.** nucleotides **60.** (a) deoxyribose, (b) ribose, (c) a purine (d) pyrimidine, (e) phosphate **61.** (a) adenine, (b) guanine, (c) cytosine, (d) thymine **62.** phospodiester linkages **63.** ATP

Building Words: **1.** isomer **2.** macromolecule **3.** monomer **4.** hydrolysis **5.** monosaccharide **6.** hexose **7.** disaccharide **8.** polysaccharide **9.** amyloplast **10.** monoacylglycerol **11.** diacylglycerol **12.** triacylglycerol **13.** amphipathic **14.** dipeptide **15.** polypeptide **16.** Pentose **17.** hydrocarbon **18.** Glycosidic **19.** Nucleotide

Matching: **1.** k **2.** c **3.** n **4.** p **5.** a **6.** q **7.** i **8.** m **9.** o **10.** g **11.** e **12.** f

Making Comparisons: **1.** phosphate **2.** organic phosphates **3.** R—CH_3 **4.** hydrocarbon; nonpolar **5.** amino **6.** amines **7.** thiols **8.** hydoxyl **9.** alcohols **10.** polar because electronegative oxygen attracts covalent electrons

Making Choices: **1.** c, e **2.** b, d **3.** c, d, e **4.** b, d **5.** c, e **6.** b, c, d **7.** a, b, c, d **8.** b, c **9.** a, c, d, e **10.** b, c, e **11.** a, b, d **12.** b, c, d, e **13.** a, b, c, d **14.** a-e **15.** a, c, e **16.** d **17.** e **18.** a, d, e **19.** a, b, c **20.** b, c, d, e **21.** d **22.** a, b **23.** c, e **24.** a-e **25.** b, c, d, e **26.** e

Visual Foundations: **1.** b, d **2.** b **3.** b **4.** c

CHAPTER 4

Reviewing Concepts: **1.** (a) cells (b) single **2.** (a) actin, (b) keratin **3.** (a) cells are the basic living units of organization and function in all organisms, (b) all cells come from other cells **4.** (a) the similarities in their structures, (b) the similarities of the molecule of which they are made **5.** homeostasis **6.** plasma membrane **7.** organelles **8.** organizing surfaces **9.** 1/1000 **10.** nanometers **11.** egg **12.** surface area-to-volume ratio **13.** (a) sizes, (b) shapes **14.** microscope **15.** (a) Hooke, (b) van Leeuwenhoek **16.** (a) magnification, (b) resolution **17.** (a) wavelengths (b) visible light **18.** staining **19.** (a) bright-field, (b) dark-field **20.** phase contrast **21.** ultrastructure **22.** (a) electrons, (b) pass, (c) TEM **23.** cell fractionation

24. (a) prokaryotic, (b) nucleoid **25.** (a) cell wall, (b) flagella **26.** (a) ribosome (b) smaller **27.** (a) membrane-bound organelles, (b) a "true" nucleus **28.** cytosol **29.** (a) soluble, (b) membrane bound or compartmentalized **30.** (a) localized, (b) in specific regions **31.** enzymes **32.** electrochemical **33.** (a) nuclear envelope, (b) nuclear pores **34.** nuclear lamina **35.** chromatin **36.** (a) nucleolus, (b) RNA **37.** (a) RNA, (b) proteins **38.** (a) endomembrane system, (b) transport vesicles **39.** endoplasmic reticulum **40.** (a) phospholipids, (b) cholesterol, (c) enzymes, (d) stored glycogen. **41.** ribosomes **42.** (a) transport vesicles (b) proteasomes **43.** cytoplasmic **44.** cisternae **45.** glycoproteins **46.** digestive enzymes **47.** (a) Golgi complex, (b) rough ER **48.** tonoplast **49.** (a) turgor pressure, (b) proteins **50.** excess water **51.** hydrogen peroxide **52.** (a) synthesis, (b) storage, (c) degredation **53.** phospholipids **54.** (a) prokaryotes, (b) endosymbiotic **55.** (a) small molecules, (b) folds, (c) cristae, (d) strictly regulates **56.** (a) apoptosis, (b) necrosis **57.** (a) electrons, (b) free radicals **58.** (a) chloroplasts, (b) chlorophyll, (c) carotenoids **59.** thylakoid **60.** (a) stroma, (b) thylakoids **61.** (a) strength, (b) shape, (c) movement **62.** (a) microfilaments, (b) microtubules **63.** (a) alpha tubulin, (b) beta tubulin **64.** (a) microtubule-associated proteins, (b) motor microtubule-associated proteins **65.** kinesin **66.** dynein **67.** (a) centrosome, (b) MTOC (microtubule-organizing center) **68.** (a) centrioles, (b) centrosome, (c) cell division **69.** (a) flagella, (b) cilia, (c) movement **70.** basal body **71.** primary cilium **72.** (a) rapidly assembling and disassembling, (b) functioning with molecular motors **73.** myosin **74.** pseudopodia **75.** intermediate filaments **76.** glycocalyx **77.** (a) carbohydrates, (b) fibrous proteins **78.** (a) cellulose, (b) polysaccharide

Building Words: **1.** chlorophyll **2.** chromoplasts **3.** cytoplasm **4.** cytoskeleton **5.** glyoxysome **6.** leukoplast **7.** lysosome **8.** microfilaments **9.** microtubules **10.** microvilli **11.** myosin **12.** peroxisome **13.** proplastid **14.** prokaryotes **15.** cytosol **16.** eukaryote

Matching: **1.** b **2.** h **3.** f **4.** m **5.** n **6.** c **7.** e **8.** l **9.** d **10.** j **11.** o

Making Comparisons: **1.** ribosomes **2.** prokaryotic, most eukaryotic **3.** cytoplasm **4.** modifies, sorts, and packages proteins **5.** cytoplasm **6.** ATP synthesis (aerobic respiration) **7.** eukaryotic **8.** nucleus (and cytoplasm during some stages of mitosis and meiosis) **9.** eukaryotic **10.** centrioles **11.** cell division, microtubule formation **12.** eukaryotic (animals primarily) **13.** cell wall **14.** encloses cell, including the plasma membrane **15.** plasma membrane **16.** boundary of cell cytoplasm **17.** prokaryotic, eukaryotic **18.** cytoplasm **19.** captures light energy (site of photosynthesis)

Making Choices: **1.** c **2.** d, e **3.** a-e **4.** c **5.** c **6.** a, b, d, e **7.** a, b, c, d **8.** a-e **9.** a, c, d, e **10.** b, c, e **11.** a, b, d, e **12.** a, d **13.** a, b, c, e **14.** e **15.** a, d **16.** a, c, d **17.** a, d **18.** a, b, c **19.** a, e **20.** a, b, e, c **21.** a-e **22.** a, d **23.** d, e **24.** a, d, e

CHAPTER 5

Reviewing Concepts: **1.** plasma membrane **2.** adhesion molecules **3.** constant motion **4.** amphipathic **5.** (a) hydrophilic, (b) hydrophobic **6.** hydrophobic core **7.** liquid crystal **8.** (a) fluid mosaic, (b) lipid bilayer, (c) proteins **9.** (a) head groups, (b) fatty acid chains **10.** membrane lipids **11.** unsaturated fats **12.** cholesterol **13.** plasma membrane **14.** (a) integral membrane proteins (b) transmembrane proteins **15.** peripheral membrane proteins **16.** (a) RER ribosomes, (b) glycoproteins, (c) Golgi complex **17.** protein **18.** (a) integrins, (b) transport, (c) receptor, (d) signal transduction **19.** (a) recognition, (b) junctions **20.** selectively permeable **21.** nonpolar (hydrophobic) **22.** (a) ions (b) large **23.** (a) carrier proteins, (b) shape change **24.** (a) channel proteins, (b) gated **25.** aquaporins

26. energy **27.** (a) simple, (b) facilitated **28.** (a) higher, (b) lower **29.** (a) size, (b) shape, (c) electrical charge **30.** osmosis **31.** osmotic pressure **32.** isotonic **33.** (a) hypertonic, (b) out, (c) shrink **34.** (a) hypotonic, (b) swell, (c) into **35.** (a) turgor pressure, (b) hypotonic, (c) plasmolysis, (d) hypertonic **36.** (a) higher, (b) lower **37.** (a) transport, (b) channel, (c) carrier **38.** a change in shape **39.** (a) concentration gradient, (b) ATP **40.** (a) membrane potential, (b) fewer **41.** negatively **42.** (a) electrical gradient, (b) electrochemical gradient **43.** (a) uniporters, (b) symporters (c) antiporters **44.** indirect active transport **45.** energy **46.** vesicle **47.** (a) phagocytosis, (b) pinocytosis **48.** receptor-mediated endocytosis **49.** (a) anchoring junctions, (b) tight junctions, (c) gap junctions, (d) plasmodesmata **50.** (a) desmosomes, (b) adhering junctions **51.** blood-brain barrier **52.** (a) communicating, (b) connexin, (c) cylinders **53.** gap junctions **54.** desmotubule

Building Words: **1.** endocytosis **2.** exocytosis **3.** hypertonic **4.** hypotonic **5.** isotonic **6.** phagocytosis **7.** pinocytosis **8.** desmosome **9.** plasmodesma **10.** peripheral

Matching: **1.** g **2.** i **3.** f **4.** b **5.** a **6.** k **7.** e **8.** m **9.** l **10.** h

Making Comparisons: **1.** cell ingests large particles such as bacteria and other food; literally means "cell eating" **2.** Facilitated diffusion **3.** endocytosis **4.** carrier-mediated transport **5.** cell junctions anchored by intermediate filaments **6.** the active transport of materials out of the cell by fusion of cytoplasmic vesicles with the plasma membrane **7.** diffusion of water across a selectively permeable membrane **8.** active transport **9.** cell takes in materials dissolved in droplets of water; literally means "cell drinking" **10.** desmosomes **11.** glycoprotein **12.** permeable **13.** aquaporins **14.** equilibrium **15.** facilitated diffusion **16.** liposome

Making Choices: **1.** a, e **2.** a, c, e **3.** c **4.** c, e **5.** b, c, d **6.** c **7.** b **8.** d **9.** b, d, e **10.** b, e **11.** b, c, d, e **12.** b, d **13.** a **14.** c, d **15.** a, b, c, e **16.** a, b, c, d **17.** a, e **18.** b, c, d, e **19.** a, b, c, d

Visual Foundations: **1.** fluid mosaic model **2.** transmembrane proteins **3.** cholesterol

CHAPTER 6

Reviewing Concepts: **1.** communicate **2.** secreting chemical signals **3.** biofilm **4.** chemical signals **5.** survive **6.** hormones **7.** cell signaling **8.** target cells **9.** signal transduction **10.** (a) through gap junctions, (b) using electrical signals, (c) temporary cell-to-cell contact, (d) chemical signals **11.** neurotransmitters **12.** endocrine glands **13.** paracrine regulation **14.** prostaglandins **15.** receptors **16.** (a) ligand, (b) hydrophilic, (c) hydrophobic **17.** enzyme **18.** (a) domains, (b) docking, (c) tail **19.** cryptochromes **20.** (a) high, (b) decreasing **21.** (a) low, (b) increasing **22.** (a) ligand-gated channels, (b) chemical, (c) electrical **23.** (a) G protein linked, (b) signal transduction **24.** (a) enzyme-linked receptors, (b) kinase **25.** (a) transcription factors, (b) small,hydrophobic **26.** plasma membrane **27.** (a) signaling pathway/cascade, (b) amplify **28.** (a) phosphorylation, (b) dephosphorylation **29.** ligand-gated chloride ion channels **30.** signal transduction **31.** (a) plasma membrane, (b) perception of sight, (c) perception of smell **32.** second messengers **33.** (a) ion, (b)small molecules **34.** cyclic AMP (cAMP) **35.** cytoplasmic **36.** (a) cAMP, (b) protein kinase A, (c) proteins **37.** (a) phosphilipase C, (b) IP_3, (c) DAG, (d)PKC, (e) calcium **38.** low **39.** (a) microtubule disassembly, (b) muscle contraction, (c) blood clotting, (d) secretion and activation of certain cells in the immune system **40.** intracellular receptors **41.** (a) intracellular signaling molecules, (b) signaling complexes **42.** (a) integrins, (b) two directions **43.** (a) ion channels open/close, (b) altered enzyme activity,

(c) gene turned on/off **44.** (a) ras proteins, (b) GTP, (c) cancer, **45.** (a) MAP kinase /ERK, (b) division, (c) differentiation **46.** signal amplification **47.** (a) signal transduction, (b) inactive **48.** (a) conserved (b) G proteins, (c) protein kinases, (d) phosphatases **49.** prokaryotes

Building Words: **1.** paracrine **2.** neurotransmitter **3.** phytochrome **4.** intracellular **5.** triphosphate **6.** diphosphate **7.** diacylglycerol **8.** intercellular **9.** interstitial **10.** substrate

Matching: **1.** h **2.** n **3.** j **4.** i **5.** f **6.** d **7.** c **8.** g **9.** p **10.** k **11.** o **12.** a **13.** b **14.** l **15.** m

Making Comparisons: **1.** phosphate group **2.** cAMP **3.** protein **4.** ATP **5.** phosphorylated protein **6.** G protein **7.** signaling molecule **8.** plasma membrane

Making Choices: **1.** c, d **2.** e **3.** b **4.** a, c, e **5.** c **6.** a, b, d **7.** e **8.** a, b, c, e **9.** a **10.** b, e **11.** c, d **12.** a, c, d, e **13.** a, b, d **14.** d **15.** d, e **16.** a, b, c, d **17.** a, c ,e **18.** a-e **19.** a, b, c, d **20.** b, c **21.** a, d **22.** c

CHAPTER 7

Reviewing Concepts: **1.** 0.02% **2.** chemical **3.** (a) work, (b) kilojoules **4.** (a) potential energy, (b) kinetic energy **5.** (a) energy, (b) transformations **6.** (a) created or destroyed, (b) transferred or converted from one form to another **7.** heat **8.** randomness or disorder **9.** (a) anabolism, (b) catabolism **10.** bond energy **11.** total bond energies **12.** entropy **13.** (a) enthalpy, (b) entropy **14.** exergonic reactions **15.** ΔG **16.** positive **17.** (a) high, (b) lower **18.** dynamic equilibrium **19.** equilibrium **20.** (a) release, (b) require input of **21.** more **22.** (a) adenine, (b) ribose, (c) three phosphates **23.** phosphorylation **24.** (a) catabolic, (b) anabolic **25.** large **26.** (a) oxidized, (b) reduced **27.** hydrogen atom **28.** nicotinamide adenine dinucleotide (NAD^+) **29.** catalase **30.** break existing bonds **31.** accelerate/speed up **32.** enzyme substrate complex **33.** active site **34.** induced fit **35.** active site **36.** ase **37.** (a) apoenzyme, (b) cofactor **38.** coenzyme **39.** (a) temperature, (b) pH, (c) ion concentration **40.** (a) product, (b) substrate **41.** metabolic pathway **42.** feedback regulation **43.** allosteric regulators **44.** reversible **45.** competitive **46.** non competitive **47.** PABA

Building Words: **1.** kilocalorie **2.** anabolism **3.** catabolism **4.** exergonic **5.** endergonic **6.** allosteric **7.** entropy **8.** catalyst **9.** substrate

Matching: **1.** m **2.** p **3.** n **4.** l **5.** q **6.** e **7.** i **8.** k **9.** o **10.** f **11.** a **12.** d **13.** g

Making Comparisons: **1.** catabolic **2.** exergonic **3.** anabolic **4.** endergonic **5.** anabolic **6.** endergonic **7.** catabolic **8.** exergonic **9.** anabolic **10.** endergonic **11.** catabolic **12.** exergonic

Making Choices: **1.** b, c, e **2.** a–e **3.** a, c **4.** a, b, e **5.** a, b, d **6.** b, c, e **7.** b, d, e **8.** b, c, d **9.** c **10.** a-e **11.** c **12.** d **13.** a, b, d **14.** a, b, d, e **15.** e **16.** a-e

CHAPTER 8

Reviewing Concepts: **1.** catabolism **2.** anabolism **3.** adenosine triphosphate (ATP) **4.** (a) cellular respiration, (b) ATP **5.** (a) aerobic, (b) anaerobic, (c) fermentation **6.** aerobic respiration **7.** (a) carbon dioxide, (b) water, (c) energy, (d) ATP) **8.** (a) glycolysis, (b) formation of acetyl coenzyme A, (c) citric acid cycle, (d) electron transport chain and chemiosmosis **9.** (a) NADH, (b) $FADH_2$ (c) oxygen, (d) water **10.** (a) dehydrogenations, (b) decarboxylations **11.** (a) oxygen, (b) aerobic, (c) anaerobic **12.** cytosol (or cytoplasm) **13.** glyceraldehyde-3-phosphate (PGAL) **14.** pyruvate **15.** (a) two, (b) substrate-level phosphorylation, (c)two **16.** (a) mitochondria, (b) cytosol, (c) oxidative decarboxylation **17.** (a) two, (b) two **18.** (a) three, (b) two, (c) one **19.** two **20.** four **21.** (a) NAD^+, (b) FAD, (c) the electron transport chain, (d) oxidative phosphorylation **22.** (a) inner mitochondrial membrane, (b) molecular oxygen **23.** (a) NADH-ubiquinone oxidoreductase, (b) succinate-ubiquinone reductase, (c) ubiquinone-cytochrome c oxidoreductase, (d) cytochrome c oxidase **24.** (a) proton, (b) intermembrane space, (c) matrix **25.** (a) protons, (b) ATP synthase, (c) exergonic **26.** chemiosmosis **27.** (a) oxidative phosphorylation, (b) substrate-level phosphorylation **28.** NADH **29.** oxygen **30.** (a) ADP, (b) phosphate **31.** fatty acids **32.** deamination **33.** (a) glycerol, (b) fatty acid **34.** beta oxidation **35.** (a) waterlogged soil, (b) stagnant ponds (water), (c) animal intestines **36.** (a) inorganic, (b) oxygen **37.** (a) organic molecule, (b) NAD+, (c) alcohol, (d) lactate **38.** (a) alcohol fermentation, (b) facultative **39.** ethyl alcohol **40.** lactate **41.** (a) two, (b) 36-38

Building Words: **1.** aerobe **2.** anaerobe **3.** dehydrogenation **4.** decarboxylation **5.** deamination **6.** glycolysis **7.** chemiosmosis **8.** fermentation

Matching: **1.** c **2.** d **3.** h **4.** i **5.** a **6.** g **7.** k **8.** l **9.** f **10.** m

Making Comparisons: **1.** fructose-6-phosphate **2.** phosphofructokinase **3.** –1 ATP **4.** dihydroxyacetone phosphate **5.** aldolase **6.** zero **7.** 3-phosphoglycerate **8.** +2 ATP **9.** 2-phosphoglycerate **10.** zero **11.** phosphoenolpyruvate **12.** pyruvate kinase

Making Choices: **1.** a, b, c **2.** b, c, d, e **3.** a, b, c, d **4.** c **5.** b, e **6.** c, e **7.** a, d **8.** c, d **9.** a, b, e **10.** a, b **11.** a-e **12.** a, b, c **13.** d **14.** c **15.** a **16.** a, e **17.** a, e **18.** b, e **19.** a, c, d

Visual Foundations: **1.** citrate synthase **2.** aconitase **3.** isocitrate dehydrogenase **4.** α-ketoglutarate dehydrogenase **5.** succinyl CoA synthetase **6.** succinate dehydrogenase **7.** fumarase **8.** malate dehydrogenase

CHAPTER 9

Reviewing Concepts: **1.** (a) carbon dioxide, (b) water, (c) light (solar), (d) chemical, (e) stored **2.** (a) wavelength, (b) photons, (c) shorter wavelengths, **3.** (a) absorbs, (b) fluorescence **4.** (a) electron acceptor, (b) reduced **5.** (a) mesophyll, (b) stroma **6.** (a) thylakoids, (b) grana, (c) chlorophyll **7.** (a) blue, (b) red **8.** (a) chlorophyll *a*, (b) light-dependent, (c) chlorophyll *b*, (d)carotenoids **9.** (a) absorption spectrum, (b) spectrophotometer **10.** (a) action spectrum, (b) accessory pigments **11.** (a) chlorophyll, (b) light energy, (c) carbon dioxide, (d) glucose, (e) oxygen **12.** (a) chemical, (b) thylakoids **13.** water **14.** (a) ATP, (b) NADPH, (c) carbon fixation **15.** (a) NADPH, (b) ATP, (c)carbohydrate, (d) stroma **16.** (a) CO_2, (b) organic **17.** (a) ADP, (b) ATP, (c) $NADP^+$, (d) NADPH **18.** chlorophyll *a* **19.** antenna complexes **20.** reaction center **21.** electron transfer reactions **22.** (a) P700, (b) P680 **23.** (a) resonance, (b) reaction center **24.** (a) ferredoxin, (b) NADP+ **25.** antenna complex **26.** photolysis **27.** (a) energy, (b) protons, (c) thylakoid, (d) ATP **28.** (a) photosystem I, (b) P700, (c) P700 **29.** (a) ATP synthase, (b) proton gradient **30.** cyclic electron transport **31.** (a) oxidized form, (b) reduced form **32.** photophosphorylation **33.** Electron transport **34.** (a) thylakoid lumen (b) ATP synthase **35.** (a) ATP, (b) NADPH, (c) carbon dioxide **36.** stroma **37.** (a) ribulose biphosphate (RuBP), (b) rubisco (c) phosphoglycerate (PGA) **38.** (a) CO_2, (b) NADPH,

(c) ATP **39.** RuBP **40.** (a) hot, dry, (b) conserve water, (c) CO_2, **41.** (a) RuBP carboxylate, (b) oxygen **42.** (a) CO_2, (b) oxaloacetate **43.** (a) Calvin cycle, (b) C_4 pathway **44.** (a) mesophyll, (b) bundle sheath **45.** (a) PEP carboxylase, (b) phosphoenolpyruvate (PEP) **46.** (a) stomata, (b) desert **47.** (a) heterotrophs, (b) autotrophs **48.** (a) photoheterotrophs, (b) chemoautotrophs **49.** (a) organic molecules (b) carbon dioxide **50.** photolysis of water

Building Words: **1.** mesophyll **2.** photosynthesis **3.** photolysis **4.** photophosphorylation **5.** chloroplast **6.** chlorophyll **7.** autotroph **8.** heterotrophy **9.** photorespiration **10.** carotenoid

Matching: **1.** q **2.** r **3.** k **4.** g **5.** c **6.** e **7.** n **8.** h **9.** m **10.** f **11.** j **12.** p **13.** s **14.** a **15.** o

Making Comparisons: **1.** anabolic **2.** catabolic **3.** site of electron transport **4.** thylakoid membrane **5.** H_2O **6.** glucose or other carbohydrate **7.** terminal hydrogen acceptor **8.** O_2

Making Choices: **1.** a **2.** b **3.** b **4.** a, b, d **5.** a, b, d, e **6.** a, c, d, e **7.** a, c, d **8.** a **9.** c **10.** c, d **11.** a, c, e **12.** c **13.** b **14.** a, c, d **15.** c **16.** b **17.** e **18.** a **19.** a, d, e **20.** a, b, c, e **21.** a, b, c, d **22.** e **23.** e **24.** a, b, c **25.** b, c, d, e **26.** a, b, c, e

CHAPTER 10

Reviewing Concepts: **1.** (a) growth, (b) repair, (c) reproduction **2.** chromatin **3.** 20,000 **4.** (a) circular, (b) chromosomes, (c) packaged **5.** large bacterial cell **6.** (a) negatively, (b) nucleosomes **7.** becoming tangled **8.** (a) H1, (b) linker DNA, (c) chromatin **9.** (a) chromosome, (b) genes **10.** (a) 46, (b) olive **11.** (a) 8, 9b) 50 **12.** cell cycle **13.** interphase **14.** (a) first gap, (b) synthesis, (c) second gap **15.** (a) mitosis, (b) meiosis, (c) cytokinesis **16.** (a) nuclear, (b) cytoplasmic **17.** first **18.** (a) S, (b) identical, (c) centromere **19.** cohesion **20.** (a) kinetochore, (b) microtubules **21.** (a) mitotic spindle, (b) microtubule-organizing center, (c) centrioles **22.** (a) nuclear envelope, (b) chromosomes **23.** (a) kinetochores, (b) spindle microtubules, (c) cell's midplane **24.** metaphase plate **25.** (a) kinetochore, (b) polar **26.** (a) sister, (b) chromosome **27.** all chromosomes have reached the poles **28.** (a) elastic, (b) contractile **29.** (a) arrival of the chromosomes at the poles, (b) interphase-like **30.** (a)the cytoplasm, (b) M **31.** (a) contractile ring, (b) cleavage furrow, (c) cell plate **32.** chromosomes **33.** binary fission **34.** origin of replication **35.** cell-cycle checkpoints **36.** (a) phosphorylating, (b) cyclin-dependent kinases (Cdks), (c) cyclins **37.** (a) M-Cdk, (b) anaphase-promoting complex (APC) **38. (a)** cytokinins, (b) steroids, (c) growth factors **39.** (a) asexual, (b) clones, **40.** (a) sexual, (b) diploid, (c) zygote **41.** size, shape, and the position of their centromeres **42.** four **43.** (a) meiosis I, (b) meiosis II **44.** synapsis **45.** (a) maternal, (b) paternal **46.** (a) crossing-over, (b) prophase, (c) genetic variability **47.** anaphase I **48.** (a) four, (b) two **49.** (a) mitosis, (b) meiosis, (c) four, (d) haploid **50.** (a) gametes, (b) gametogenesis **51.** spermatogenesis **52.** oogenesis **53.** (a) alternation, (b) sporophyte, (c) gametophyte

Building Words: **1.** chromosome **2.** interphase **3.** haploid **4.** diploid **5.** centromere **6.**gametogenesis **7.** spermatogenesis **8.** oogenesis **9.** gametophyte **10.** sporophyte **11.** metaphase **12.** anaphase **13.** prophase **14.** somatic **15.** chromatin **16.** histone **17.** mitosis **18.** polyploid

Matching: **1.** s **2.** q **3.** j **4.** f **5.** c **6.** d **7.** k **8.** o **9.** m **10.** g **11.** l **12.** n **13.** e **14.** a **15.** t **16.** r

Making Comparisons: **1.** M phase **2.** Telophase I and II **3.** does not occur **4.** anaphase I **5.** tetrads form **6.** does not occur **7.** telophase **8.** telophase I, telophase II **9.** anaphase **10.** anaphase II **11.** chromosomes line up at the midplane of the cell **12.** metaphase I, metaphase II **13.** S phase of interphase **14.** S phase of interphase

Making Choices: **1.** a, b **2.** c, d **3.** b **4.** b, c, e **5.** a, d **6.** a-d **7.** b, f **8.** b **9.** h **10.** d **11.** b **12.** c **13.** d **14.** b **15.** d **16.** g **17.** h **18.** b **19.** e **20.** a, i **21.** a, g, h **22.** c **23.** j **24.** b **25.** g **26.** d **27.** c **28.** a, d **29.** a, b, c, d **30.** a, d, e

CHAPTER 11

Reviewing Concepts: **1.** Gregor Mendel **2.** (a) pollination, (b) anthers **3.** (a) phenotype, (b) genotype **4.** the same phenotype **5.** (a) first filial, (b) F_2 or second filial **6.** (a) dominant, (b) recessive **7.** (a) genes, (b) DNA, (c) RNA or a protein product **8.** genes **9.** (a) separate, b) meiosis **10.** locus **11.** (a) variations, (b) corresponding, (c) homologous **12.** (a) heterozygous (b) homozygous **13.** Punnett square **14.** (a) genotype, (b) homozygous dominant or heterozygous **15.** homozygous recessive **16.** two loci **17.** genetic recombination (recombination) **18.** meiosis **19.** linearly arranged **20.** (a) zero, (b) one **21.** product rule **22.** product **23.** sum rule **24.** (a) past events, (b) future events **25.** exceptions **26.** (a) linked, (b) inherited **27.** crossing-over **28.** (a) recombination, (b) map unit **29.** autosomes **30.** (a) SRY, (b) testosterone **31.** (a) XY male, (b) XX female **32.** (a) males, (b) X **33.** (a) homozygous, (b) heterozygous, (c) hemizygous **34.** (a) hyperactive, (b) inactivation **35.** Barr body **36.** variegation **37.** (a) incomplete dominance, (b) intermediate **38.** codominance **39.** (a) three or more, (b) locus (c) two **40.** pleiotropy **41.** epistasis **42.** polygenic inheritance **43.** norm of reaction

Building Words: **1.** polygene **2.** homozygous **3.** heterozygous **4.** dihybrid **5.** monohybrid **6.** hemizygous **7.** genotype **8.** phenotype

Matching: **1.** h **2.** a **3.** n **4.** c **5.** p **6.** l **7.** f **8.** g **9.** i **10.** q **11.** t **12.** o

Making Comparisons: **1.** tall plant with yellow seeds (TT Yy)) **2.** tall plant with green seeds (TT yy) **3.** tall plant with yellow seeds (Tt Yy) **4.** tall plant with green seeds (Tt yy) **5.** tY **6.** tall plant with yellow seeds (Tt Yy) **7.** short plant with yellow seeds (tt YY) **8.** short plant with yellow seeds (tt Yy) **9.** ty **10.** tall plant with yellow seeds (Tt Yy) **11.** tall plant with green seeds (Tt yy) **12.** short plant with green seeds (tt yy)

Making Choices: **1.** b, e **2.** j **3.** both j **4.** RrTT **5.** l, m, n, r, s, t **6.** g **7.** c/n and d/r **8.** g **9.** b/m and e/s **10.** b, e **11.** a, b, c, e **12.** d **13.** a-d **14.** b **15.** b or d **16.** a, c, d **17.** d **18.** c, d **19.** a, e **20.** c **21.** a, b, c, d

CHAPTER 12

Reviewing Concepts: **1.** inheritance **2.** protein **3.** 4 **4.** transformation **5.** bacteriophages (phages) **6.** (a) pentose sugar (deoxyribose), (b) phosphate, (c) nitrogenous base **7.** (a) adenine, (b) guanine, (c) cytosine, (d) thymine **8.** (a) covalent, (b) phosphodiester linkage **9.** (a) phosphate, (b) hydroxyl **10.** (a) adenine, (b) thymine, (c) guanine, (d) cytosine **11.** X-ray diffraction **12.** (a) antiparallel nucleotide strands, (b) sugar-phosphate backbone, (c) bases **13.** (a) opposite, (b) antiparallel, (c) phosphate, (d) hydroxyl **14.** (a) two, (b) three **15.** complementary base pairing **16.** (a) template, (b) semiconservative **17.** density **18.** mutation **19.** (a) origin of replication, (b) helicase, (c) single-strand binding proteins

20. (a) supercoiling, (b) topoisomerase **21.** DNA polymerases **22.** (a) RNA primer, (b) DNA primase, (c) DNA polymerase, (d) subunits **23.** (a) 5, (b) 3, (c) Okazaki, (d) DNA ligase **24.** (a) bidirectional, (b) 2 **25.** 1 **26.** DNA polymerase **27.** mismatch repair **28.** (a) nuclease, (b) DNA polymerase, (c) DNA ligase **29.** (a) short, noncoding, (b) telomerase **30.** (a) cell aging, (b) apoptosis **31.** apoptosis

Building Words: **1.** virulent **2.** antiparallel **3.** semiconservative **4.** telomere **5.** transformation **6.** bacteriophage **7.** Mutation

Matching: **1.**q **2.** e **3.** m **4.** k **5.** i **6.** j **7.** b **8.** g **9.** d **10.** e **11.** n **12.** o

Making Comparisons: **1.** Found substance in heat-killed bacteria that "transformed" living bacteria **2.** Hershey-Chase **3.** found that the ratios of guanine to cytosine, purines to pyrimidines, and adenine to thymine were very close to 1 **4.** Inferred from x-ray crystallographic films of DNA patterns that nucleotide bases are stacked like rungs in a ladder **5.** Avery, MacLeod, and McCarty

Making Choices: **1.** a, c, e **2.** a, c **3.** d **4.** c **5.** d **6.** b, c **7.** b, e **8.** d **9.** b, c, e **10.** a, c, e **11** a **12.** b **13.** e **14.** d **15.** a-e **16.** d **17.** d **18.** a-c **19.** a, b, c, e **20.** a, c, d

CHAPTER 13

Reviewing Concepts: **1.** DNA bases **2.** urease **3.** (a) recessive mutant allele, (b) homologous chromosome **4.** (a) single-stranded, (b) contains ribose, (c) uracil **5.** (a) mRNA, (b) tRNA, (c) rRNA **6.** (a) tRNA, (b) ribosome **7.** codon (triplet) **8.** (a) anticodon, (b) tRNA, (c) mRNA **9.** reading frame **10.** (a) 64, (b) 61, (c) 3 **11.** evolved early in ancient life **12.** first two **13.** (a) methionine, (b) tryptophan **14.** why the third base in a codon does not always matter **15.** RNA polymerases **16.** (a) rRNA, (b) mRNA, (c) tRNA **17.** (a) 5' to 3' (b) 3' to 5' **18.** (a) upstream, (b) downstream **19.** promoter **20.** (a) elongation, (b)3' end **21.** (a) termination, (b) template DNA (c) RNA **22.** (a) leader sequence, (b) start codon **23.** trailing sequences **24.** precursor mRNA **25.** (a) 5' end, (b) poly-A tail, (c) 3' end **26.** (a) spliceosome, (b) introns, (c) exons **27.** peptide bonds **28.** aminoacyl-tRNA synthetases **29.** (a) mRNA coding, (b) amino acids **30.** (a) one, (b) three, (c) P, (d) A, (e) E **31.** (a) initiation, (b) small **32.** AUG **33.** (a) methionine, (b) large, (c) initiation **34.** (a) elongation, (b) GTP **35.** (a) peptidyl transferase, (b) ribozyme **36.** (a) translocation, (b) 5' to 3' **37.** release factor **38.** molecular chaperones **39.** polyribosome **40.** nucleotide sequence **41.** (a) silent, (b) lethal, (c) useful **42.** (a) base-substitution, (b) silent **43.** (a) missense, (b) nonsense **44.** inserted or deleted **45.** (a) movable, (b) disruption, (c) deactivation **46. (a)** retrotransposons, (b) reverse transcriptase **47.** evolve **48.** mutational hot spots **49.** (a)mutagens, (b) carcinogens **50.** snRNA **51.** (a) spliceosome, (b) intron removal **52.** (a) SRP RNA, (b) rough ER **53.** sno-RNA **54.** (a) gene expression, (b) siRNA (c) miRNA **55.** miRNAs **56.** (a) a specific RNA, (b) polypeptide **57.** reverse transcriptase

Building Words: **1.** anticodon **2.** ribosome **3.** polysome **4.** mutagen **5.** carcinogen **6.** transposon **7.** codon **8.** intron **9.** mutation

Matching: **1.** i **2.** m **3.** k **4.** p **5.** c **6.** d **7.** n **8.** e **9.** h **10.** f **11.** j **12.** o **13.** a **14.** b

Making Comparisons: **1.** 5'—ACC—3' **2.** 3'—TGG —5' **3.** 3'—CGC—5' **4.** 3'—CGC—5' **5.** 3'—ACA—5' **6.** 5'—UAA—3', 5'—UAG—3', 5'—UGA—3' **7.** no tRNA molecule binds to a stop codon **8.** 3'—TAC—5' **9.** 5'—UTC—3'

Making Choices: **1.** d **2.** c **3.** b, d **4.** c, e **5.** a, c **6.** a, c, d **7.** b, c **8.** a, c, e **9.** d **10.** a **11.** b **12.** b, c **13.** b, d **14.** d, e **15.** a, b, c, d **16.** d, e **17.** b **18.** b, d

CHAPTER 14

Reviewing Concepts: **1.** control of the amount of mRNA transcribed, (b) control the rate of translation of mRNA, (c) control of the activity of the protein product **2.** mRNA **3.** transcriptional-level control **4.** differential expression **5.** constitutive genes **6.** (a) enzyme activity, (b) number of enzyme molecules **7.** operon **8.** promoter region **9.** operator **10.** inducer **11.** on **12.** corepressor **13.** (a) repressors, (b) positive **14.** promoter **15.** posttranscriptional controls **16.** feedback inhibition **17.** operons **18.** (a) heterochromatin, (b) euchromatin (c) histones **19.** (a) methylation, (b) inactivate, (c) genomic imprinting, (d) epigenetics **20.** (a) tandemly repeated gene sequences, (b) gene amplification **21.** TATA box **22.** (a) enhancers, (b) silencers **23.** transcription factors **24.** domain **25.** Xist **26.** RNA processing **27.** alternative splicing **28.** mRNA **29.** miRNAs **30.** chemical modification **31.** kinases **32.** phosphatases **33.** (a) ubiquitin, (b) proteasome

Building Words: **1.** euchromatin **2.** heterochromatin **3.** corepressor **4.** heterodimer **5.** homodimer **6.** operon **7.** histone **8.** intron

Matching: **1.** i **2.** a **3.** l **4.** n **5.** f **6.** c **7.** j **8.** e **9.** g **10.** o **11.** d **12.** b

Making Comparisons: **1.** genes are inducible only during specific phases of life cycle **2.** molecular chaperones **3.** phenotypic expression is determined by whether an allele is inherited from the male or female parent **4.** posttranslational controls **5.** adjusts the rate of synthesis in a metabolic pathway **6.** enhancers **7.** regulate the rate of mRNA translation

Making Choices: **1.** b, e **2.** a **3.** a, d **4.** c **5.** e **6.** e **7.** b **8.** b, d **9.** a, b, c, e **10.** a, d, e **11.** a, d, e **12.** b, c, d **13.** b, e **14.** a, e **15** a-e **16.** a, c, d, e

CHAPTER 15

Reviewing Concepts: **1.** genetic engineering **2.** biotechnology **3.** restriction enzymes **4.** plasmids **5.** transformation **6.** palindromic **7.** (a) sticky ends, (b) DNA ligase **8.** (a) an origin of replication, (b) one or more restriction sites, (c) genes that confer resistance to antibiotics or let cells use a specific nutrient **9.** genomic DNA library **10.** probe **11.** hybridize **12.** reverse transcriptase **13.** (a) DNA polymerase, (b) heat **14.** Taq **15.** (a) positive pole, (b) phosphate groups, (c) size **16.** (a) Southern blot, (b) Northern blot, (c) Western blot **17.** chain termination method **18.** (a) 1000, (b) bioinformatics **19.** (a) RNA fractions, (b) cDNA, (c) reverse transcriptase **20.** (a) mRNA, (b) **21.** (a) 75, (b) non-protein-coding RNAs **22.** non-protein-coding regulatory regions **23.** comparative genomics **24.** functionally important **25.** (a) mental, (b) linguistic **26.** (a) RNAi, (b) inactivating or silencing **27.** functional genomics **28.** gene targeting **29.** homologous recombination **30.** gene therapy **31.** (a) drugs, (b) vaccines **32.** polymorphic **33.** fertilized egg, (b) ES (embryonic stem) **34.** blastocysts **35.** pharming **36.** Ti (tumor inducing) **37.** (a) seeds, (b) sexual, (c) asexually **38.** genetically modified crops **39.** risk assessment

Building Words: **1.** transgenic **2.** retrovirus **3.** bacteriophage **4.** kilobase **5.** biotechnology **6.** plasmid **7.** genome

Matching: **1.** k **2.** g **3.** b **4.** j **5.** h **6.** m **7.** i **8.** d **9.** o **10.** c **11.** a **12.** f **13.** l **14.** e

Making Comparisons: **1.** restriction enzymes **2.** a segment of DNA that is homologous to a part of a DNA sequence that is being studied **3.** cloning **4.** amplifying DNA by alternate heat denaturization and use of heat-resistant DNA polymerase to replicate DNA strands *in vitro* **5.** plasmids **6.** DNA fragments comprising the total DNA of an organism

Making Choices: **1.** a **2.** d **3.** c **4.** d **5.** c, e **6.** b **7.** a **8.** c **9.** b, e **10.** b **11.** d, e **12.** c **13.** b, c **14.** c, d, e **15.** a, c **16.** b **17.** d **18.** a, b, d, e **19.** a, b, c, d

CHAPTER 16

Reviewing Concepts: **1.** inherited variation **2.** karyotype **3.** (a) white, (b) induced to divide **4.** (a) FISH, (b) fluorescent dyes **5.** pedigree analysis **6.** (a) location, (b) functions **7.** (a) single-letter, (b) region **8.** 1.5 **9.** protein-coding **10.** genome wide associations scans **11.** 500 **12.** gene targeting **13.** polyploidy **14.** (a) aneuploidy, (b) trisomic, (c) monosomic (d) nondisjunction **15.** (a) 47, (b) trisomy, (c) nondisjunction **16.** (a) maternal age, (b) younger than 35 **17.** Klinefelter syndrome **18.** Turner syndrome **19.** XYY **20.** (a) duplication, (b) inversion, (c) deletion breakage, (d) translocation **21.** (a) 14, (b) 14/21, (c) 21 **22.** 5 **23.** fragile sites **24.** fragile X syndrome **25.** parental origin **26.** (a) epigenetic inheritance, (b) genomic imprinting **27.** inborn errors of metabolism **28.** enzyme **29.** hemoglobin **30.** ion transport **31.** mucous **32.** (a) autosomal recessive, (b) central nervous system, (c) lipid, (d) brain **33.** (a) autosomal dominant allele, (b) the central nervous system **34.** 4 **35.** factor VIII **36.** (a) cognitive abilities, (b) males **37.** (a) less than 4%, (b) 10% **38.** (a) mutant, (b) normal **39.** viral vectors **40.** (a) amniocentesis, (b) cells sloughed from the fetus **41.** chorionic villus sampling (CVS) **42.** pre-implantation genetic diagnosis **44.** preventive medicine **45.** carriers **46.** Probability **47.** (a) all, (b) all, (c) recessive **48.** consanguineous mating **49.** (a) genetic, (b) normal **50.** (a) discriminating, (b) genetic tests **51.** privacy and confidentiality

Building Words: **1.** karyotype **2.** polyploidy **3.** trisomy **4.** translocation **5.** aneuploidy

Matching: **1.** a **2.** k **3.** g **4.** i **5.** f **6.** d **7.** j **8.** b **9.** m **10.** n **11.** e

Making Comparisons: **1.** valine substitutes for glutamic acid in hemoglobin **2.** abnormal RBCs block small blood vessels. **3.** Klinefelter syndrome **4.** sex chromosome nondisjunction during meiosis **5.** absence of an enzyme that converts phenylalanine to tyrosine; instead phenylalanine is converted into toxic phenylketones that accumulate **6.** Hemophilia A **7.** X-linked recessive trait **8.** lack enzyme that breaks down a brain membrane lipid **9.** blindness, severe retardation **10.** Huntington's disease **11.** autosomal dominant allele **12.** 47 chromosomes because they have extra copies of chromosome 21 **13.** retardation, abnormalities

Making Choices: **1.** b **2.** c **3.** a, d, e **4.** b, d **5.** c, e **6.** b, c, d **7.** a, b ,c **8.** b, e **9.** b, c, e **10.** e **11.** c, e **12.** a, d **13.** d **14.** a, d, e **15.** c

Visual Foundations: **1.** c **2.** a **3.** d

CHAPTER 17

Reviewing Concepts: **1.** (a) cell differentiation, (b) development of an organism. **2.** immunoflourescence **3.** (a) determination, (b) differentiation **4.** (a) morphogenesis, (b) pattern formation **5.** nuclear equivalence **6.** transcriptional **7.** differentiated **8.** clones **9.** cell cycles **10.** transgenic organisms **11.** pluripotent **12.** placenta **13.** transcription factors

14. (a) human reproductive, (b) human therapeutic **15.** characteristics that allow for the efficient analysis of biological processes **16.** mutants **17.** (a) egg, (b) larval, (c) pupal,, (d) adult, (e) metamorphosis **18.** maternal effect **19.** (a) segmentation genes, (b) homeotic genes **20.** morphogen **21.** (a) homeobox, (b) homeodomain **22.** Hox **23.** (a) hermaphrodites, (b) genetic crosses **24.** founder cells **25.** mosaic **26.** induction **27.** (a) apoptosis, (b) caspases **28.** chimera **29.** regulative **30.** IGF **31.** SEPALLATA **32.** (a) tumor, (b) metastasis **33.** oncogenes **34.** tumor suppressor gene **35.** proto-oncogenes **36.** (a) Ras, (b) protein kinase **37.** (a) oncogenes, (b) tumor suppressor genes or anti-oncogenes, (c) growth-inhibiting factors

Building Words: **1.** morphogenesis **2.** oncogene **3.** totipotent **4.** pluripotent **5.** neoplasm **6.** morphogen

Matching: **1.** m **2.** n **3.** c **4.** l **5.** f **6.** e **7.** d **8.** b **9.** m **10.** j **11.** a **12.** i **13.** h

Making Comparisons: **1.** ectoderm **2.** epidermis **3.** ectoderm **4.** neuron **5.** germ line cells **6.** lungs **7.** endoderm **8.** mesoderm

Making Choices: **1.** c, e **2.** c, e **3.** a, c, d **4.** c **5.** a, d **6.** c, e **7.** b **8.** c **9.** d **10.** c, e **11.** a **12.** b **13.** a, b, c, d **14.** b, c, e **15.** b, d **16.** a-e

CHAPTER 18

Reviewing Concepts: **1.** gradual divergence or evolution **2.** links all fields of the life sciences into a unified body of knowledge **3.** accumulation of genetic changes within populations over time **4.** (a) individual organisms, (b) populations **5.** many generations **6.** species **7.** bioremediation **8.** scale of nature **9.** Leonardo da Vinci **10.** Jean Baptiste de Lamarck (Lamarck) **11.** Lamarck **12.** (a) H.M.S. Beagle, (b) Galapagos Islands **13.** artificial selection **14.** Thomas Malthus **15.** adaptation **16.** Alfred Russell Wallace **17.** *On the Origin of Species by Natural Selection,* or *The Preservation of Favored Races in the Struggle for Life* **18.** (a) overproduction, (b) variation, (c) limits on population growth (d) struggle for existence), (e) differential reproductive success ("survival of the fittest") **19.** (a) Darwin, (b) Mendel **20.** mutations **21.** (a) natural selection, (b) chance **22.** (a) macroevolution, (b) microevolution **23.** has evolved through time **24.** (a) rapid covering, (b) decay **25.** rain forests **26.** 195,000 **27.** index fossils **28.** (a) radioisotopes, (b) half-life **29.** biogeography **30.** continental drift **31.** (a) homologous, (b) homology **32.** convergent evolution **33.** homoplastic features **34.** vestigial structures (organs) **35.** (a) genetic code, (b) sequences of amino acids in proteins (c) sequences of nucleotides in DNA **36.** amino acid **37.** DNA sequencing **38.** developmental genes that already exist **39.** Years

Building Words: **1.** homologous **2.** biogeography **3.** homoplastic **4.** convergent

Matching: **1.** b **2.** d **3.** f **4.** m **5.** n **6.** o **7.** j **8.** e **9.** i **10.** k **11.** a **12.** h

Making Comparisons: **1.** homologous features **2.** homoplastic features **3.** the presence of nonfunctional or degenerate structures that were once more developed in ancestral species which occurs as a species adapts to a changing mode of life **4.** areas of the world that have been separated from the rest of the world for a long time have organisms unique to that area **5.** evidence that all life is related **6.** differential reproductive success **7.** fossil record **8.** convergent evolution **9.** the order of nucleotide bases in DNA is useful in determining evolutionary relationships

Making Choices: **1.** a, c, d **2.** a-e **3.** c **4.** b **5.** a **6.** b, e **7.** a **8.** e **9.** c, d **10.** c **11.** d **12.** c **13.** a, b, c, e **14.** a, b, c, e **15.** d **16.** e **17.** a, b, d **18.** a

CHAPTER 19

Reviewing Concepts: **1.** (a) genetic variability, (b) evolutionary **2.** all the alleles for all of the loci present **3.** (a) genotypic frequency, (b) phenotype frequency, (c) allele frequency **4.** allele frequencies must change over successive generations **5.**(a) 0, (b) 1 **6.** (a)$p^2 + 2pq + q^2 = 1$, (b) at genetic equilibrium **7.** (a) random mating, (b) no net mutations,(c) large population size, (d) no migration, (e) no natural selection **8.** codominant **9.** (a) natural selection, (b) visible phenotype **10.** microevolution **11.** genetically similar individuals within a population **12.** genetic fitness **13.** assortative mating **14.** (a) at the loci involved in mate choice, (b) in the entire genome **15.** mutation **16.** somatic cells **17.** usually harmful **18.** genetic drift **19.** (a) decrease, (b) increase **20.** (a) bottleneck, (b) genetic diversity **21.** a small number of individuals from a larger population establish a colony in a new area **22.** gene flow **23.** (a) traits, (b) phenotypes, (c) the environment **24.** stabilizing selection **25.** directional selection **26.** disruptive selection **27.** phenotypes **28.** (a) genetic polymorphism, (b) alleles that differ by a single nucleotide **29.** balanced polymorphism **30.** heterozygote advantage **31.** frequency-dependent selection **32.** neutral variation **33.** geographic variation

Building Words: **1.** microevolution **2.** phenotype **3.** genotype **4.** polymorphism

Matching: **1.** n **2.** k **3.** h **4.** f **5.** c **6.** b **7.** e **8.** a **9.** j **10.** m **11.** L

Making Comparisons: **1.** yes **2.** yes **3.** no **4.** yes **5.** yes **6.** yes **7.** no **8.** yes **9.** no

Making Choices: **1.** b, c **2.** a **3.** a, c **4.** c, d **5.** e **6.** c **7.** a, c, d **8.** a, c **9.** a **10.** c, e **11.** e **12.** n **13.** i **14.** y **15.** r **16.** m

CHAPTER 20

Reviewing Concepts: **1.** (a) 4-100 million, (b) 99.99 **2.** 3-4 **3.** population whose members interbred and produce fertile offspring **4.** evolution long enough for statistically significant differences in diagnostic traits to emerge **5.** DNA sequencing data of a taxonomic group has not been carefully studied **6.** gene flow between two individual species is prevented **7.** union of sperm and egg of two different species **8.** interspecific **9.** (a) temporal isolation, (b) behavioral or sexual isolation, (c) mechanical isolation, (d) gametic isolation **10.** interspecific hybrid **11.** (a) normal development, (b) aborted, (c) hybrid inviability **12.** hybrid sterility **13.** hybrid breakdown **14.** lysin **15.** (a) populations, (b) reproductively isolated, (c) gene pools **16.** geographically isolated **17.** genetic drift **18.** allele frequency **19.** (a) most common, (b) almost all **20.**small **21.** (a) a change in ploidy, (b) a change in ecology **22.** (a) plants, (b) polyploidy **23.** polyploidy, **24.** (a) autopolyploidy, (b) allopolyploidy **25.** (a) hybridization, (b) fertile **26.** (a) parasitic insects, (b) polyploidy **27.** hybrid zone **28.** (a) reinforcement, (b) less fit **29.** fusion **30.** stability **31.** ecotones **32.** punctuated equilibrium **33.** gradualism **34.** preadaptations **35.** (a) allometric growth, (b) paedomorphosis **36.** (a) adaptive zones, (b) adaptive radiation **37.** major environmental change **38.** background extinction **39.** mass **40.** (a) microevolutionary, (b) macroevolutionary

Building Words: **1.** paedomorphic **2.** polyploidy **3.** allopatric **4.** allopolyploidy **5.** macroevolution **6.** allometric **7.** species

Matching: **1.** j **2.** p **3.** t **4.** a **5.** n **6.** q **7.** m **8.** i **9.** o **10.** e **11.** k

Making Comparisons: **1.** hybrid breakdown **2.** gametes of similar species are incompatible **3.** hybrid sterility **4.** similar species have structural differences in their reproductive organs **5.** hybrid inviability **6.** temporal isolation **7.** two different species do not attempt to mate because they live in different parts of the environment **8.** prezygotic barrier

Making Choices: **1.** a **2.** e **3.** a, b, e **4.** c, d **5.** d **6.** e **7.** b, d **8.** a, c, e **9.** c, e **10.** e **11.** a, c ,e **12.** c, d **13.** b, d **14.** a, c, d, e **15.** a, b **16.** a, c, e **17.** a, b, d

CHAPTER 21

Reviewing Concepts: **1.** chemical evolution **2.** Soft-bodied marine organisms **3.** amphibians **4.** (a) amphibians, (b) reptiles **5.** 4.6 **6.** (a) little or no free oxygen, (b) energy, (c) chemical building blocks, (d) time **7.** (a) prebiotic soup, (b) sulfur world **8.** (a) Oparin, (b) Haldane **9.** (a) Miller, (b) Urey, (c) hydrogen (H2), (d) methane (CH4), (e) ammonia (NH3) **10.** protobionts **11.** microspheres **12.** metabolism first **13.** molecular reproduction **14.** (a) RNA, (b)ribozymes, (c) proteins **15.** (a) RNA, (b) double helix**16.** (a) microfossils, (b) 3.5 **17.** stromatolites **18.** (a) anaerobic, (b) prokaryotic **19.** Autotrophs **20.** cyanobacteria **21.** oxygen **22.** obligate anerobes **23.** (a) oxygen, (b) carbon dioxide **24.** 2.2 **25.** serial endosymbiosis **26.** index fossils **27.** eons **28.** (a) periods, (b) epochs **29.** Archaean **30.** Proterozoic **31.** Ediacaran **32.** (a) 542 (b) Cambrian **33.** (a) Ordovician, (b) ostracoderm **34.** Devonian **35.** (a)Carboniferous, (b) coal **36.** Permian **37.** 252 **38.** (a) Triassic, (b) Jurassic, (c) Cretaceous **39.** (a) Paleogene period, (b) Neogene period, (c) Quaternary period **40.** Paleocene **41.** Ecocene **42.**Oligocene **43.**

Building Words: **1.** autotroph **2.** protobiont **3.** heterotroph **4.** precambrian **5.** aerobe **6.** Anaerobe

Matching: **1.** n **2.** g **3.** b **4.** q **5.** l **6.** f **7.** e **8.** k **9.**p **10.** h **11.** i **12.** m

Making Comparisons: **1.** Ordovician **2.** Paleozoic **3.** Triassic **4.** Mesozoic **5.** Cretaceous **6.** Mesozoic **7.** Cambrian **8.** Paleozoic **9.** Quaternary **10.** Cenozoic **11.** Devonian **12.** Paleozoic **13.** Paleogene **14.** Cenozoic **15.** Carboniferous **16.** Paleozoic

Making Choices: **1.** d **2.** b, c **3.** d **4.** c **5.** b, c **6.** a, d **7.** a-e **8.** c, d, e **9.** a-e **10.** a, b, c, d **11.** d **12.** b, d **13.** a-e **14.** b, e **15.** a, b, d, e **16.** e **17.** b, d **18.** c **19.** b, e **20.** a,b,c,e **21.** b, c, d

CHAPTER 22

Reviewing Concepts: **1.** paleoanthropology **2.** placental mammals **3.** arboreal past **4.** (a) flexible, (b) five digits **5.** 56 **6.** opposable thumb **7.** (a) stereoscopic vision (b) integrate visual information from both eyes simultaneously **8.** (a) prosimii, (b) tarsiiformes, (c) anthropoidea **9.** (a) Africa, (b) Asia **10.** (a) of their brain, (b) cerebrum **11.** prehensile **12.** quadrupedal **13.** hominoids **14.** gibbons (*Hylobates*), orangutans (*Pongo*), gorillas (*Gorilla*), chimpanzees (*Pan*), humans (*Homo*) **15.** brachiation **16.** knuckle walking **17.** foramen magnum **18.** supraorbital ridges **19.** Africa **20.** Chad **21.** Ardipithecus **22.** Australopithecus **23.** sexual dimorphism **24.** *Australopithecus afarensis* **25.** Tanzania **26.** 2.5 **27.** (a) *Homo ergaster* (b) *H. erectus* **28.** Indonesia **29.** 1.7 **30.** *H. ergaster* **31.** Spain **32.** Neander Valley in Germany

33. dead end **34.** out of Africa **35.** (a)98, (b) 99 **36.** (a) development of hunter-gatherer societies, (b) development of agriculture, (c) the Industrial Revolution **37.** 10 **38.** 18th century **39.** urbanization

Building Words: **1.** hominoid **2.** quadrupedal **3.** bipedal **4.** arboreal **5.** supraorbital **6.** arthropoid **7.** dimorphic

Matching: **1.** e **2.** g **3.** a **4.** j **5.** l **6.** k **7.** g **8.** c **9.** d **10.** f

Making Comparisons: **1.** *Australopithecus afarensis* **2.** Neandertal **3.** *Ardipithecus ramidus* **4.** *Australopithecus africanus* **5.** 2.5 mya **6.** more modern hominid features than *Australopithecus*, fashioned primitive tools; **7.** 6-7 mya **8.** 1.7 mya **9.** larger brain than *H. habilis*, made more sophisticated tools **10.** 1.2 mya **11.** *Australopithecus anamensis*

Making Choices: **1.** c, d, e **2.** a, b, e **3.** a, e **4.** a, d, e **5.** c **6.** e **7.** b **8.** b **9.** a **10.** d, e **11.** d **12.** b, c **13.** c **14.** d **15.** b, c, d, e **16.** e **17.** b, c, d, e **18.** a, b, c, d **19.** b, c, e **20.** a, c **21.** a, c, d, e

CHAPTER 23

Reviewing Concepts: **1.** (a) 15, (b) 10 **2.** biological diversity or biodiversity **3.** 40 **4.** systematics **5.** taxonomy **6.** homologous traits **7.** binomial system **8.** (a) genus, (b) specific epithet **9.** (a) family, (b) class, (c) phylum **10.** domain **11.** taxon **12.** (a) Plantae, (b) Animalia **13.** Protista **14.** Fungi **15.** Prokaryotae **16.** (a) Bacteria, (b) Archea, (c) Eukarya **17.** DNA sequencing and analysis **18.** (a) cladogram, (b) node, (c) most recent common ancestor **19.** vertical gene transfer **20.** horizontal gene transfer or lateral gene transfer **21.** evolutionary relationships **22.** (a) hypotheses, (b) available data **23.** population **24.** convergent evolution **25.** homoplasy **26.** shared ancestral characters **27.** shared derived characters **28.** combination **29.** bar code **30.** molecular systematics **31.** homologous **32.** (a) ribosomal RNA (rRNA), (b) DNA **33.** monophyletic **34.** paraphyletic **35.** polyphyletic **36.** (a) phenetics, (b) evolutionary taxonomy, (c) phylogenetic systematic **37.** (a) outgroup, (b) ingroup **38.** derived **39.** ancestral **40.** recently a group diverged **41.** most recent common ancestor **42.** ancestor-descendent **43.** parsimony **44.** maximum likelihood **45.** (a) ecology, (b) evolution, (c) medicine

Building Words: **1.** subphylum **2.** paraphyletic **3.** Polyphyletic

Matching: **1.** Classification **2.** p **3.** e **4.** k **5.** o **6.** b **7.** j **8.** f **9.** d **10.** i

Making Comparisons: **1.** Archaea **2.** Archaea **3.** Eukarya **4.** Fungi **5.** Eukarya **6.** Animalia **7.** Eukarya **8.** Plantae

Making Choices: **1.** c **2.** d **3.** d **4.** c, e **5.** b, c, e **6.** a, c, e **7.** a **8.** b, d **9.** b **10.** e **11.** a-e

Visual Foundations: **1.** 2, 3, 4, 5, 6 **2.** 4, 5, 6 **3.** This taxon shares the most common ancestor, E. **4.** This taxon includes some but not all of the descendants of a common ancestor, A. **5.** This taxon does not share the same recent common ancestor.

CHAPTER 24

Reviewing Concepts: **1.** disease **2.** (a) evolution, (b) their own genes, (c) eykaryotic genes **3.** (a) nucleic acid, (b) a living host **4.** (a) are not composed of cells, (b) cannot carry on metabolic activities, (c) reproduce **5.** (a) 20, (b) 200 **6.** capsid **7.** (a) DNA, (b) RNA, (c) DNA-RNA **8.** (a) capsomers, (b) shape **9.** (a) helical, (b) polyhedral, (c) a combination of both shapes **10.** capsid **11.** host cell's plasma membrane **12.** (a) phospholipids, (b) proteins, (c) unique proteins **13.** host range **14.** (a) transcriptional, (b) translational **15.** polyhedral head **16.** tail **17.** infection **18.** (a) lytic, (b) lysogenic **19.** virulent **20.** (a) attachment (absorption), (b) penetration, (c) replication and synthesis, (d) assembly, (e) release **21.** (a) restriction enzymes, (b) cut up, (c) slightly modifying it **22.** (a) temperate, (b) lytic, (c) lysogenic **23.** (a) prophage, (b) lysogenic cells **24.** lysogenic conversion **25.** pathogens **26.** (a) a capsid, (b) an envelope **27.** single-stranded RNA **28.** the plant wall has been damaged **29.** bioterrorism **30.** being transmitted from one animal to another **31.** retroviruses **32.** reverse transcriptase **33.** antibiotics **34.** progressive (escape) **35.** regressive (reduction) **36.** virus-first **37.** polydnaviruses **38.** subviral **39.** (a) protective protein coat, (b) reproduction **40.** RNA **41.** (a) nucleus, (b) gene regulation **42.** transmissible spongiform encephalopathies or TSEs **43.** variability

Building Words: **1.** bacteriophage **2.** viroid **3.** capsid **4.** obligate **5.** retrovirus **6.** pathogen

Matching: **1.** h **2.** m **3.** k **4.** f **5.** d **6.** c **7.** i **8.** g **9.** j **10.** l

Making Comparisons: **1.** yes **2.** no **3.** no **4.** no **5.** yes **6.** yes **7.** yes **8.** yes **9.** yes **10.** no **11.** yes

Making Choices: **1.** a-e **2.** a, c, d **3.** a, d, e **4.** a **5.** a, b, d **6.** b, d, e **7.** a **8.** a, b, c, e **9.** a, b **10.** c **11.** d, e

CHAPTER 25

Reviewing Concepts: **1.** disease **2.** recyclers **3.** cellular **4.**(a) cocci, (b) diplococci, (c) streptococci, (d) staphylococci **5.** bacilli **6.** (a)spirillum, (b)spirochete **7.** vibrio **8.** nucleoid or nuclear area **9.** (a) plasma membrane, (b) supports, (c) shape, (d) bursting **10.** peptidoglycan **11.** (a) gram-positive, (b) gram-negative **12.** (a) gram-positive, (b) peptidoglycan **13.** cell wall **14.** other microorganisms **15.** the host's white blood cells **16.** (a) protein, (b) flagella **17.** reproduction **18.** (a) rotating flagella, (b) chemotaxis **19.** (a) swarm, (b) adhere to surfaces, (c) participate in biofilm formation **20.** plasmids **21.** (a) binary fission, (b) budding, (c) fragmentation **22.** (a) vertical, (b) horizontal **23.** (a) transformation, (b) transduction, (c) conjugation **24.** mutations **25.** autotrophs **26.** (a) chemotrophs, (b) phototrophs **27.** decomposers **28.** (a) aerobic, (b) facultative anaerobes, (c) obligate anaerobes **29.** (a) amino acids, (b) nucleic acids **30.** other organisms **31.** (a) size, (b) shape **32.** (a) Bacteria, (b) Archaea **33.** peptidoglycan **34.** eukaryotes **35.** genome sequencing **36.** extreme thermophiles **37.** methanogens **38.** extreme halophiles **39.** (a) mutualism, (b) commensalism, (c) parasitism **40.** (a) biofilms, (b) dental plaque **41.** bacteria **42.** (a) nitrogen, (b) rhizobial prokaryotes **43.** cyanobacteria **44.** antibiotics **45.** bioremidation **46.** microbiota **47.** Koch's postulates **48.** (a) adhere to a specific cell type, (b) multiply, (c) produce toxic substances **49.** R factors **50.** bioflims

Building Words: **1.** chemotaxis **2.** exotoxin **3.** endotoxin **4.** methanogen **5.** endospore **6.** pathogen **7.** anaerobe **8.** prokaryote **9.** eukaryote **10.** obligate **11.** halophile

Matching: **1.** i **2.** a **3.** h **4.** e **5.** b **6.** c **7.** j **8.** d **9.** n **10.** k **11.** l **12.** f

Making Comparisons: **1.** absent **2.** present **3.** present **4.** absent (except in mitochondria and chloroplasts) **5.** absent **6.** absent **7.** present **8.** present **9.** absent **10.** absent **11.** absent **12.** absent **13.** present

Making Choices: **1.** a, b, c, e **2.** b, d, e **3.** d, e **4.** a, b, d **5.** e **6.** a, c, d, e **7.** a-e **8.** b **9.** b, c, d, e **10.** a **11.** c **12.** a, b **13.** a, d **14.** a, c, e **15.** d **16.** a, b, e

CHAPTER 26

Reviewing Concepts: **1.** (a) Eukarya, (b) nucleus **2.** unicellular, (b) colonies **3.** coenocytes **4.** (a) pseudopodia, (b) individual cells, (c) cilia, (d) flagella **6.** 2.2 **7.** serial endosymbiosis **8.** apicomplexa **9.** (a) monophyletic, (b) polyphyletic **10.** oral groove **11.** mitochondria **12.** (a) functional mitochondria, (b) Golgi complex, (c) flagella **13.** *Giardia intestinalis* **14.** (a) termites, (b) wood-eating cockroaches **15.** (a) crystalline rod, (b) atypical mitochondria **16.** photosynthetic **17.** pellicle **18.** photosynthetic **19.** mitochondrion **20.** *Trypanosoma brucei* **21.** flattened vesicles **22.** (a) flagella, (b) cellulose **23.** (a) zooxanthellae, (b) carbohydrates, (c) coral reefs **24.** chloroplast **25.** locomotion **26.** (a) apical complex, (b) moving junction **27.** sporozoites **28.** *Plasmodium* **29.** (a) micronuclei, (b) macronucleus **30.** conjugation **31.** mycelium **32.** (a) zoosporangium, (b) zoospores **33.** oospore **34.** silica **35.** (a) radial, (b) bilateral **36.** floating plankton **37.** diatomaceous earth **38.** seaweeds **39.** (a) gametes, (b) zoospores **40.** alternation of generation **41.** unicellular **42.** (a) silica, (b) calcium carbonate **43.** nanoplankton **44.** test **45.** marine **46.** chalk **47.** index fossils **48.** (a) axopods, (b) microtubules **49.** (a) red algae, (b) green algae, (c) land plants **50.** basal holdfast **51.** coralline algae **52.** (a) agar, (b) carrageenan **53.** land plants **54.** tissues **55.** (a) mitosis, (b) cell division **56.** fragmentation **57.** motile spores with flagella **58.** posterior flagellum **59.** two flagella **60.** (a) rounded, (b) wide **61.** asymmetrical **62.** pseudopodia **63.** *Entamoeba histolytica* **64.** sporangia **65.** (a) swarm cell, (b) myxamoeba **66.** amoebas, (b) plasmodial slime molds **67.** microvilli

Building Words: **1.** protozoa **2.** pellicle **3.** pseudopodium **4.** micronucleus **5.** macronucleus **6.** conjugation **7.** sessile **8.** phagocytosis

Matching: **1.** b **2.** k **3.** g **4.** i **5.** d **6.** c **7.** f **8.** e **9.** j **10.** l **11.** m **12.** o

Making Choices: **1.** a-d **2.** c, e **3.** c, d, e **4.** a, b, d, e **5.** b **6.** c **7.** b **8.** a, c, e **9.** b **10.** a **11.** a, d **12.** a, c, d **13.** e **14.** c **15.** a, c, d, e

CHAPTER 27

Reviewing Concepts: **1.** green algae **2.** (a) *a* and *b*, (b) xanthophylls, (c) carotenes **3.** (a) starch, (b) cellulose **4.** maternal tissues **5.** carbon dioxide **6.** (a)waxy cuticle, (b)stomata **7.** gametangia **8.** alternation of generations **9.** (a) gametophyte generation, (b) sporophyte generation **10.** (a) antheridia, (b) archegonia **11.** (a) diplod zygote (fused gametes), (b) spores **12.** stoneworts (charophytes) **13.** (a) mosses and other bryophytes, (b) bryophytes, (c) seedless vascular plants, (d) gymnosperms, (e) flowering plants (angiosperms) **14.** (a) xylem, (b) phloem **15.** lignin **16.** gymnosperms **17.** (a) mosses, (b) liverworts, (c) hornworts **18.** rhizoids **19.** (a) archegonium, (b) flowing water (raindrops) **20.** sporophyte **21.** protonema **22.** (a) *Sphagnum*, (b) peat mosses **23.** thallus **24.** (a) archegonia, (b) antheridia **25.** gemmae **26.** Anthocerophyta **27.** indeterminate growth **28.** photoperiodism **29.** monophyletic **30.** bryophytes **31.** (a) microphyll, (b) megaphyll, (c) megaphyll **32.** (a) ferns, (b) club mosses

33. coal deposits **34.** (a) rhizome, (b) fronds **35.** (a) sporangia, (b) sori, (c) prothallus **36.** (a) upright stems, (b) dichotomous **37.** fungus **38.** Australia **39.** silica **40.** strobilus **41.** (a) homospory, (b) heterospory, (c) microspores, (d) megaspores **42.** apical meristem **43.** apical cell **44.** megafossils

Building Words: **1.** xanthophyll **2.** gametangium **3.** archegonium **4.** gametophyte **5.** sporophyte **6.** microphyll **7.** megaphyll **8.** sporangium **9.** homospory **10.** heterospory **11.** megaspore **12.** microspore **13.** bryophyte **14.** bryology **15.** thallus **16.** strobilus

Matching: **1.** g **2.** d **3.** j **4.** l **5.** b **6.** h **7.** a **8.** c **9.** p **10.** k **11.** e **12.** f

Making Comparisons: **1.** vascular **2.** sporophyte **3.** vascular **4.** sporophyte **5.** vascular **6.** sporophyte **7.** vascular **8.** naked seeds **9.** nonvascular **10.** seedless, reproduce by spores **11.** gametophyte **12.** seedless, reproduce by spores **3.** seedless, reproduce by spores **14.** sporophyte

Making Choices: **1.** a, c, e **2.** a-e **3.** a, c, d **4.** b, d **5.** b, c, e **6.** c, d **7.** b, d **8.** a, e **9.** a, c **10.** a, c, d **11.** a, e **12.** b, d **13.** a-e **14.** d **15.** b **16.** a, c, d, e

CHAPTER 28

Reviewing Concepts: **1.** seeds **2.** (a) more developed, (b) an abundant food supply, (c) a multicellular seed coat **3.** pinon pine **4.** (a) sporophyte, (b) gametophyte **5.** (a) ovules, (b) seed **6.** (a) gymnosperms, (b) ovary wall, (c) angiosperms **7.** (a) xylem, (b) phloem **8.** (a) Coniferophyta, (b) Ginkgophyta, (c) Cycadophyta, (d) Gnetophyta **9.** resin **10.** monoecious **11.** Taxol **12.** (a) *Pinus*, (b) sporophytes **13.** (a) sporophylls, (b) microsporocytes, (c) male gametophytes (pollen grains) **14.** (a) seed cones, (b) megasporangia, (c) megaspores, (d) female gametophyte **15.** pollen tube **16.** (a) external water, (b) air **17.** on land **18.** Cycadophyta **19.** dioecious **20.** Ginkgophyta **21.** completely exposed **22.** Gnetophyta **23.** vessel elements **24.** *Welwitschia* **25.** (a) vascular, (b) sexually **26.** (a) vessel elements, (b) sieve tube elements **27.** Anthophyta **28.** palms, grasses, orchids, irises, onions, lilies **29.** oaks, roses, mustards, cacti, blueberries, sunflowers **30.** (a) three, (b) one, (c) endosperm **31.** (a) four or five, (b) two, (c) endosperm **32.** (a) sepals, (b) petals, (c) stamens, (d) carpels; (e) stamens; (f) carpels; (g)perfect, (h) complete **33.** (a) sepals, (b) leaflike **34.** (a) calyx, (b) corolla **35.** (a)anther, (b) carpels, (c) ovary **36.** (a)megaspores, (b)embryo sac **37.** (a)microspores, (b)male gametophyte (pollen grain) **38.** (a) zygote, (b) endosperm tissue **39.** fruit **40.** (a) cross-fertilization, (b) genetic variation **41.** (a) food, (b) water storage **42.** (a) leaves, (b) stamens **43.** leaves or leaflike structures **44.** (a) progymnosperms, (b) seed ferns **45.** gymnosperms **46.**Conifers **47.** basal angiosperms **48.** (a) eudicots, (b) monocots

Building Words: **1.** gymnosperm **2.** angiosperm **3.** monoecious **4.** dioecious **5.** fertilization **6.** endosperm

Matching: **1.** a **2.** e **3.** f **4.** i **5.** k **6.** m **7.** h **8.** l **9.** b **10.** c **11.** n

Making Comparisons: **1.** herbaceous or woody **2.** endosperm **3.** cotyledons **4.** mostly broader than in monocots **5.** 1 cotyledon **6.** floral parts in multiples of 4 or 5 **7.** parallel venation **8.** netted venation

Making Choices: **1.** b, c **2.** a **3.** b **4.** b **5.** a, c, d, e **6.** c, d, e **7.** a, e **8.** b, e **9.** a-e **10.** b, c **11.** d **12.** a, d **13.** a-e **14.** d **15.** c

CHAPTER 29

Reviewing Concepts: **1.** mycologists **2.** decomposers **3.** symbiotic associations **4.** eukaryotes **5.** (a) 2 to 9, (b) less **6.** heterotrophs **7.** extracellularly **8.** lipid droplets or glycogen **9.** a cell wall **10.** (a) chitin (b) microbial, (c) Arthropoda **11.** yeasts **12.** hyphae **13.** (a) mycelium (b) molds **14.** septa (cross walls) **15.** spores **16.** (a) fruiting body, (b) spores) **17.** (a) buds, (b) spores **18.** haploid **19.** (a) monokaryotic, (b) dikaryotic **20.** pheromones **21.** cellulose **22.** flagellated protist **23.** Animals **24.** (a) Chytridiomycota, (b) Zygomycota, (c) Ascomycota, (d) Basidiomycota, (e) Glomeromycota **25.** monophyletic **26.** (a) Ascomycota, (b) Basidiomycota, (c) septa, (d) dikaryotic **27.** sexual stage **28.** (a) thallus, (b) rhizoids **29.** flagellated **30.** earliest **31.** decomposers **32.** *Rhizopus stolonifer* **33.** (a) heterothallic, (b) + (plus), (c) – (minus) **34.** (a) opportunistic (b) compromised immune systems **35.** (a) mitochondria, (b) simpler **36.** mycorrhizae **37.** arbuscules **38.** mutualistic **39.** (a) asci, (b) conidia, (c) conidiophores **40.** homothallic **41.** ascocarp **42.** budding **43.** (a) basidium, (b) basidiospores **44.** (a) button, (b) basidiocarp **45.** fairy ring **46.** (a) carbon dioxide, (b) minerals **47.** (a) cellulose (b) lignin **48.** (a) endomycorrhizal, (b) ectomycorhizae **49.** (a) green alga, (b) cyanobacterium, (c) both (d) ascomycete **50.** (a) crustose, (b) foliose, (c) fruticose **51.** soredia **52.** lignin **53.** (a) yeasts (*Saccharomyces*), (b) fruit sugars, (c) sugar from starch in grains, (d) carbon dioxide **54.** (a) *Penicillium*, (b) *Aspergillus* **55.** *Amanita* **56.** psilocybin **57.** (a) *Penicillium notatum*, (b) penicillin **58.** ergot **59.** (a) coal tars, (b) petroleum **60.** (a) pathogens, (b) insect pests **61.** aflatoxins **62.** cutinase **63.** haustoria

Building Words: **1.** coenocytic **2.** monokaryotic **3.** heterothallic **4.** conidiophore **5.** homothallic **6.** mycelium **7.** plasmogamy **8.** pheromone **9.** rhizoid **10.** basidiocarp

Matching: **1.** j **2.** k **3.** b **4.** c **5.** o **6.** l **7.** e **8.** d **9.** h **10.** f **11.** m

Making Comparisons: **1.** Zygomycota **2.** Ascomycota **3.** Basidiomycota **4.** Glomeromycota

Making Choices: **1.** a, c, d, e **2.** c, e **3.** a-e **4.** e **5.** a-e **6.** a, c, e **7.** c, d, e **8.** a, d **9.** b **10.** b, d **11.** a-e **12.** c **13.** a, c, d **14.** a-e **15.** a, e **16.** a, b, c **17.** b, d, e **18.** a, b, d **19.** a, b, c, e **20.** a-e **21.** e **22.** a-e **23.** d

CHAPTER 30

Reviewing Concepts: **1.** 99 **2.** monophyletic **3.** collagen **4.** heterotrophs **5.** (a) diploid, (b) zygote **6.** cleavage **7.** (a) fluid, (b) salt **8.** (a) movements, (b) currents **9.** hypotonic **10.** (a) water, (b) salts **11.** less constant **12.** scorpion-like arthropods **13.** dehydrate (dry out) **14.** (a) internal fertilization, (b) shelled egg **15.** choanoflagellate **16.** parsimony **17.** Cambrian radiation or Cambrian explosion **18.** Proterozoic **19.** *Hox* **20.** (a) similarities, (b) morphology **21.** (a) bilateral symmetry, (b) cephalization **22.** (a) anterior, (b) posterior, (c) dorsal, (d) ventral **23.** (a) medial, (b)lateral, (c) cephalic, (d) caudal **24.** (a) germ layers, (b) ectoderm, (c) mesoderm, (d) endoderm, (e) triploblastic **25.** acoelomates **26.** (a) pseudocoelomates, (b) coelomates **27.** (a) mouth, (b) anus **28.** (a) radial cleavage, (b) spiral cleavage **29.** metozoa **30.** eumetazoa **31.** (a) Lophotrochozoa, (b) Ecydsozoa **32.** segmentation **33.** bilateral animals

Building Words: **1.** pseudocoelom **2.** ectoderm **3.** mesoderm **4.** protostome **5.** deuterostome **6.** schizocoely **7.** enterocoely **8.** blastula **9.** gastrodermis **10.** cleavage

Matching: **1.** m **2.** j **3.** a **4.** e **5.** h **6.** c **7.** p **8.** n **9.** o **10.** i

Making Comparisons: **1.** radial **2.** spiral **3.** indeterminate **4.** determinate **5.** anus **6.** mouth **7.** enterocoelous (enterocoely) **8.** schizocoelous (shizocoely)
Making Choices: **1.** c, e **2.** b, c, d **3.** c **4.** d, e **5.** a **6.** b **7.** a, d, e **8.** e **9.** d **10.** a, c, d, e **11.** a, b, d, e **12.** a-e **13.** b, c, d, e **14.** b, c

CHAPTER 31

Reviewing Concepts: **1.** 99 **2.** (a) capture food, (b) escape from predators, (c) reproduce **3.** (a) protosomtes, (b) deuterostomes **4.** (a) Parazoa, (b) asymmetrical, (c) simple **5.** form true tissues **6.** choanoflagellates **7.** (a) asconoid, (b) syconoid, (c) leuconoid **8.** (a) spongocoel, (b) osculum **9.** (a) mesohyl, (b) spicules **10.** (a) stimuli, (b) nerve **11.** (a) calcerous, (b) glass, (c) demosponges **12.** (a) polyp, (b) medusa, (c) radial **13.** cnidocytes **14.** nematocysts **15.** (a) ectoderm (epidermis), (b) endoderm (gastrodermis), (c) mesoglea **16.** (a) Hydrozoa, (b) Scyphozoa, (c) Anthozoa, (d) Cubozoa **17.** (a) freshwater, (b) sexually **18.** (a) medusa, (b) *Cyanea* **19.** sea turtles **20.** planula **21.** *zooxanthellae* **22.** biradial **23.** (a) cnidocytes, (b) mouth, (c) anal pores **24.** (a) convergent, (b) similar environments **25.** (a) protostomes, (b) lophophore, (c) trochophore **26.** (a) bilateral symmetry, (b) three, (c) true, (d) tube-within-a-tube **27.** (a) platyhelminthes (b) Turbellaria, (c) Trematoda, (d) Monogenea, (e) Cestoda **28.** ganglia **29.** protonephridia **30.** (a) auricles, (b) pharynx **31.** (a) free-living flatworms, (b) hooks and suckers **32.** scolex **33.** proglottid **34.** (a) everted from the anterior end of the body, (b) derived **35.** mantle **36.** (a) radula, (b) hemocoel, (c) cephalopods **37.** (a) trochophore, (b) veliger **38.** (a) eight, (b) reduced, (c) eyes, (d) tentacles **39.** insects **40.** torsion **32.** (a) shell, (b) calcium carbonate (mother-of-pearl) **42.** (a) tentacles, (b) ten, (c) eight **43.** "head-foot" **44.** horny beak **45.** (a) Polychaeta, (b) Oligochaeta, (c) Hirudinida **46.** (a) skin, (b) gills, (c) metanephridia **47.** (a) septa, (b) setae **48.** (a) marine, (b) parapodia **49.** (a) crop, (b) gizzard **50.** (a) hemoglobin, (b) metanephridia, (c) moist skin **51.** (a) Brachiopoda, (b) Phoronida, (c) Bryozoa **52.** (a) brain, (b) sense organs, (c) flame cells **53.** increase in cell size **54.** (a) nematodes, (b) arthropods, (c) molt **55.** (a) decomposers, (b) parasites, (c) predators of smaller organisms **56.** *Ascaris* **57.** 80 **58.** (a) exoskeleton, (b) chitin, (c) protein **59.** compound **60.** tracheae (tracheal tubes) **61.** trilobites **62.** (a) Chilopoda, (b) one, (c) predators, (d) poison **63.** (a) Diplopoda, (b) two, (c) herbivores **64.** Chelicerata **65.** (a) cephalothorax, (b) six, (c) four **66.** book lungs **67.** spinnerets **68.** (a) Crustacea, (b) mandibles, (c) maxillae, (d) biramous, (e) two **69.** barnacles **70.** Decapoda **71.** (a) maxillae, (b) maxillipeds, (c) chelipeds, (d) walking legs **72.** (a) reproductive, (b) swimmeretes **73.** (a) Insecta, (b) Hexapoda **74.** (a) three, (b) one or two, (c) one, (d) Malpighian tubules

Building Words: **1.** exoskeleton **2.** bivalve **3.** biramous **4.** uniramous **5.** trilobite **6.** cephalothorax **7.** hexapod **8.** arthropod **9.** spongocoel **10.** choanocytes **11.** hydrostatic **12.** auricle(s)

Matching: **1.** s **2.** j **3.** q **4.** f **5.** d **6.** c **7.** l **8.** e **9.** r **10.** i **11.** u **12.** n **13.** t

Making Comparisons: **1.** hydras, jellyfish, coral **2.** biradial symmetry; diploblastic; gastrovascular cavity with mouth and anal pores **3.** Platyhelminthes **4.** biradial symmetry; triploblastic; simple organ systems; gastrovascular cavity with one opening, bilateral symetry , acoelomate **5.** Mollusca **6.** soft body with dorsal shell; muscular foot; mantle covers visceral mass; most with radula **7.** some marine worms, earthworms, leeches **8.** segmented body; most with setae **9.** Rotifera **10.** crown of cilia; constant cell number **11.** cylindrical, threadlike body **12.** centipedes, crabs, lobsters, spider, insects **13.** segmented body; jointed appendages; exoskeleton; some with compound eyes; insects with tracheal tubes **14.** Echinodermata **15.** bilateral larva, pentaradial adult; triploblastic; organ systems; complete digestive tube, coelom

Making Choices: **1.** a, b, d **2.** c **3.** c, e **4.** e **5.** a **6.** b **7.** a, d, e **8.** c, e **9.** d **10.** a, c, e **11.** c **12.** b, c, e **13.** a, b, d, e **14.** d **15.** b, d, e **16.** a, d **17.** a, b, d **18.** b, c, e **19.** a, c, e **20.** b, d, e **21.** d **22.** b, d **23.** a, b **24.** b, c

CHAPTER 32

Reviewing Concepts: **1.** filter **2.** (a) echinoderms, (b) chordates **3.** Proterozoic **4.** (a) shared derived, (b) radial, (c) indeterminate, (d) spiral, (e) determinate **5.** (a) coelom, (b) bilateral, (c) pentaradial **6.** (a) endoskeleton (internal skeleton), (b) pedicellariae **7.** water vascular **8.** Early Cambrian **9.** (a) Crinoidea, (b) upper **10.** (a) Asteroidea, (b) central disk **11.** tube feet **12.** (a) carnivores (active predators), (b) scavengers **13.** Ophiuroidea **14.** (a) locomotion, (b) to collect and handle food **15.** (a) Echinoidea, (b) arms, (c) test, (d) spines **16.** (a) Holothuroidea, (b) tube feet **17.** Evisceration **18.** Chordata **19.** (a) notochord, (b) dorsal, tubular nerve cord, (c) postanal tail, (d) endostyle **20.** tadpoles **21.** (a) tunic, (b) carbohydrate **22.** *Brachiostoma* **23.** filtering particles from water **24.** elongated **25.** (a) vertebral column, (b) cranium, (c) cephalization **26.** marine **27.** (a) 6, (b) 3, (c) *Tetrapoda*, (d) *Amniota* **28.** *ostracoderms* **29.** (a) Myxini, (b) vertebrae, (c) notochord **30.** lampreys **31.** fins **32.** (a) gill arch skeleton, (b) active predators **33.** (a) Chondrichthyes, (b) sharks, (c) skates, (d) rays **34.** placoid scales **35.** (a) lateral line organs, (b) electroreceptors **36.** (a) oviparous, (b) ovoviviparous, (c) viviparous **37.** tooth **38.** (a) ray-finned, (b) sarcopterygians **39.** swim bladders **40.** Hox **41.** (a) coelocanths, (b) lungfishes, (c) tetrapods **42.** lung fish **43.** lobed fins of fishes **44.** (a) *Tiktaalik,* (b) movable neck, (c) ribs that supported lungs **45.** (a) Caudata, (b) Anura, (c) Gymnophiona **46.** paedomorphosis **47.** (a) skin, (b) three **48.** amniotic egg **49.** keratin **50.** (a) two, (b) one **51.** endothermic **52.** protective leathery shell **53.** (a) two, (b) partially, (c) completely divided **54.** ectotherms **55.** (a) Testudines, (b) Squamata, (c) Sphenodonta, (d) Crocodilia, (e) Aves **56.** keratin beak **57.** overlap **58.** pit organ **59.** iguanas **60.** (a) tapered snout, (b) large fourth tooth **61.** asymmetrical **62.** endothermy **63.** the two clavicles fuse **64.** attachment of flight muscles **65.** (a) lay eggs, (b) scales **66.** (a) four, (b) high metabolic **67.** uric acid **68.** (a) crop, (b) proventriculus, (c) gizzard **69.** (a) hair, (b) mammary glands, (c) a pair of temporal openings in the skull, (d) differentiation of teeth, (e) three middle ear bones **70.** (a) viviparous, (b) placenta **71.** (a) therapsids, (b) Triassic **72.** (a) arboreal, (b) nocturnal **73.** (a) Protheria, (b) Theria, (c) Metatheria, (d) Eutheria

Building Words: **1.** agnathan **2.** Anura **3.** Apodal **4.** Chondrichthyes **5.** tetrapod **6.** caudata **7.** echinoderm **8.** endoskeleton **9.** notochord **10.** ampulla **11.** cephalization **12.** endotherm

Matching: **1.** k **2.** i **3.** l **4.** a **5.** o **6.** q **7.** n **8.** r **9.** p **10.** f **11.** c

Making Comparisons: **1.** sharks, rays, skates, chimeras **2.** cartilage **3.** jawed, gills, marine and fresh water, placoid scales, well-developed sense organs **4.** Actinopterygii **5.** bone **6.** salamanders, frogs and toads, caecilians **7.** three chambered heart **8.** Reptilia (order Aves) **9.** four chambered heart **10.** light hollow bone with air spaces **11.** Mammalia **12.** bone

13. mostly tetrapods, possess hair and mammary glands, endothermic, highly developed nervous system

Making Choices: **1.** d **2.** d **3.** b, c **4.** c **5.** a, b **6.** c, e **7.** a, b, c, d **8.** a-e **9.** a, b **10.** b, e **11.** b, c, d **12.** c, e **13.** a, d **14.** d **15.** c, e **16.** d, e **17.** a, d **18.** a-e **19.** a, d, e **20.** a-e **21.** d, e **22.** e **23.** b, d, e

CHAPTER 33

Reviewing Concepts: **1.** (a) annual, (b) biennial, (c) perennial **2.** (a) deciduous, (b) evergreens **3.** (a) root, (b) shoot **4.** (a) tissue, (b) simple tissues, (c) complex tissues **5.** organs **6.** (a) parenchyma, (b) collenchyma, (c) sclerenchyma **7.** (a) photosynthesis, (b) storage, (c) secretion **8.** differentiate **9.** support **10.** (a) sclereids, (b) fibers **11.** (a) hemicelluloses, (b) pectins **12.** lignin **13.** (a) water, (b) dissolved minerals, (c) parenchyma cells or xylem parenchyma, (d) fibers **14.** (a) apoptosis, (b) hollow **15.** (a) food, (b) sieve tube elements **16.** plasmodesmata **17.** (a) epidermis, (b) periderm, (c) cuticle, (d) stomata **18.** trichomes **19.** (a) protection, (b) cork **20.** meristems **21.** grow **22.** (a) primary, (b) secondary **23.** (a) area of cell division, (b) area of cell elongation, (c) area of cell maturation **24.** (a) leaf primordia, (b) bud primordia **25.** (a) vascular cambium, (b) cork cambium **26.** bark **27.** preprophase band **28.** (a) increase in number, (b) increase in size **29.** differential gene expression **30.** different concentrations of signaling molecules

Building Words: **1.** trichome **2.** biennial **3.** epidermis **4.** stoma **5.** deciduous **6.** cuticle

Matching: **1.** o **2.** m **3.** a **4.** n **5.** c **6.** k **7.** b **8.** h **9.** f **10.** g

Making Comparisons: **1.** ground tissue **2.** stems and leaves **3.** sclerenchyma **4.** structural support **5.** xylem **6.** vascular tissue **7.** conducts sugar in solution **8.** extends throughout plant body **9.** dermal tissue **10.** protection of plant body, controls gas exchange and water loss on stems, leaves **11.** covers body of herbaceous plants **12.** periderm **13.** protection of plant body

Making Choices: **1.** d **2.** a **3.** e **4.** a, d **5.** d **6.** c, d, e **7.** c **8.** a, b, e **9.** c **10.** a, b, c, e **11.** b **12.** d **13.** b **14.** a-e **15.** a, b **16.** b, d, e **17.** a, c, d, e

CHAPTER 34

Reviewing Concepts: **1.** (a) light energy, (b) oxygen, (c) carbon dioxide **2.** water loss **3.** (a) blade, (b) petiole, (c) stipules **4.** (a) simple, (b) compound **5.** (a) alternate leaf arrangement, (b) opposite leaf arrangement, (c) whorled leaf arrangement **6.** (a) parallel, (b) monocots, (c) netted, (d) eudicots **7.** (a) epidermis, (b) chloroplasts **8.** (a) cuticle, (b) cutin **9.** subsidiary cells **10.** (a) mesophyll, (b) palisade mesophyll, (c) spongy mesophyll **11.** (a) xylem, (b) phloem, (c) bundle sheath **12.** (a) transparent, (b) photosynthesis, (c) mesophyll **13.** (a) carbon dioxide, (b) oxygen **14.** (a) rigidity, (b) flaccid (limp) **15.** (a) pigment, (b) light **16.** (a) yellow, (b) blue light **17.** (a) malic acid, (b) starch **18.** (a) proton pumps, (b) H^+ **19.** (a) K^+ , (b) chloride, (c) malate (d) solute **20.** (a) sucrose, (b) starch **21.** (a) light , (b) darkness (c) CO_2 concentration in the leaf, (d) water stress, (e) a circadian rhythm **22.** soil **23.** (a) prevent plant from overheating, (b) distribute essential minerals throughout plant **24.** sweating **25.** guttation **26.** shedding leaves **27.** ethylene **28.** (a) carotenoids, (b) anthocyanins **29.** (a) abscission zone, (b) fibers **30.** middle lamella **31.** spines **32.** tendrils **33.** (a) passive, (b) active

Building Words: **1.** abscission **2.** circadian **3.** mesophyll **4.** transpiration **5.** guttation **6.** petiole **7.** trichomes

Matching: **1.** b **2.** j **3.** g **4.** c **5.** h **6.** f **7.** d **8.** m **9.** i **10.** a

Making Comparisons: **1.** allows for efficient capture of sunlight **2.** helps reduce or control water loss, enabling plants to survive the dry terrestrial environment **3.** stomata **4.** allows light to penetrate to photosynthetic tissue **5.** air space in mesophyll tissue **6.** bundle sheaths and bundle sheath extensions **7.** transports water and minerals from roots **8.** phloem in veins

Making Choices: **1.** a **2.** d **3.** b, c ,e **4.** c **5.** b, c, e **6.** d **7.** a, d, e **8.** a, b, e **9.** a-e **10.** a, b, c **11.** c **12.** a, e **13.** e **14.** a-e **15.** b, c, e **16.** c, e **17.** a, e **18.** b **19.** a, c, d **20.** a-e **21.** b, d, e

CHAPTER 35

Reviewing Concepts: **1.** (a) roots, (b) leaves, (c) stems **2.** (a) support, (b) conduction (internal transport), (c) produce new stem tissue **3.** primary **4.** herbaceous eudicot **5.** vascular bundles **6.** (a) xylem, (b) phloem, (c) vascular cambium **7.** (a) xylem, (b) phloem **8.** lateral meristems **9.** vascular cambium **10.** cork cambium **11.** periderm **12.** (a) xylem (wood), (b) phloem (inner bark) **13.** radially (at right angles to the normal division) **14.** crushed **15.** (a) rays, (b) parenchyma cells **16.** (a) periderm, (b) epidermis **17.** bark **18.** (a) cork cells , (b) cork parenchyma **19.** (a) bud scales, (b) leaves **20.** bud scale scars **21.** leaf scars **22.** (a) heartwood, (b) sapwood **23.** (a) hard, (b) soft **24.** (a) fibers, (b) vessel elements **25.** (a) springwood, (b) late summerwood **26.** natural physical process **27.** (a) pushed, (b) pulled, (c) pulled **28.** (a) water potential, (b) less, (c) higher (less negative), (d) lower (more negative) **29.** negative **30.** transpiration **31.** (a) cohesion, (b) adhesion **32.** (a) mineral ions, (b) decrease **33.** (a) source, (b) sink **34.** (a) decrease, (b) osmosis, (c) increase

Building Words: **1.** translocation **2.** periderm **3.** internode **4.** adhesion **5.** cohesion **6.** osmosis **7.** apical

Matching: **1.** a **2.** g **3.** h **4.** b **5.** c **6.** l **7.** m **8.** j **9.** i **10.** d **11.** k **12.**

Making Comparisons: **1.** secondary phloem **2.** conducts dissolved sugar **3.** produced by cork cambium **4.** storage **5.** cork cells **6.** periderm **7.** vascular cambium **8.** produces secondary xylem and secondary phloem **9.** a lateral meristem, usually arises from parenchyma **10.** produces periderm (secondary growth)

Making Choices: **1.** c **2.** d **3.** a, b, c, d **4.** a, d **5.** a, b, d **6.** b, d, e **7.** a-e **8.** a, b, d **9.** a, c, e **10.** b **11.** e **12.** a, c, d, e **13.** e **14.** b, d, e **15.** a **16.** e

CHAPTER 36

Reviewing Concepts: **1.** (a) anchoring, (b) water, (c) minerals, (d) storage **2.** (a) taproot, (b) fibrous **3.** (a) taproot, (b) fibrous **4.** adventitious **5.** (a) root cap, (b) apical meristem **6.** root hairs **7.** (a) parenchyma, (b) collenchyma **8.** (a) plasmodesmata, (b) symplast **9.** (a) endodermis, (b) Casparian strip **10.** aquaporins **11.** (a) pericyle, (b) xylem, (c) phloem **12.** (a) cortex, (b) endodermis, (c) pericycle **13.** vascular cambium **14.** roots **15.** (a) periderm, (b) cork parenchyma **16.** nodes **17.** (a) prop roots, (b) buttress roots **18.** (a) epiphytes, (b) aerial **19.** mycorrhizae **20.** cell signaling **21.** rock **22.** (a) inorganic matter, (b) organic matter, (c) water, (d) air **23.** (a) sand, (b) silt, (c) clay **24.** (a) protons (H^+), (b) cation exchange **25.** (a) sand, (b) silt, (c) clay, (d) air, (e) water **26.** humus **27.** air **28.** (a) leaching, (b) illuviation **29.** castings **30.** acid precipitation **31.** (a) 90, (b) 60 **32.** (a) macronutrients, (b) micronutrients **33.** (a) as free K^+, (b) turgidity, (c) stomata **34.** hydroponics **35.** (a) nitrogen, (b) phosphorus, (c) potassium **36.** soil erosion **37.** salinization

Building Words: **1.** hydroponics **2.** macronutrient **3.** epiphyte **4.** endodermis **5.** Mycorrhiza
6. nodule **7.** adventitious **8.** humus

Matching: **1.** i **2.** g **3.** d **4.** a **5.** c **6.** k **7.** j **8.** e **9.** m **10.** l **11.** n **12.** o **13.** h

Making Comparisons: **1.** area of cell division that causes an increase in length of the root **2.** root hairs **3.** protects root **4.** storage **5.** endodermis **6.** pericycle **7.** xylem **8.** conducts dissolved sugars

Making Choices: **1.** a, c, e **2.** b, e **3.** c **4.** b, d **5.** a, b ,d **6.** b **7.** a, b, d, e **8.** c **9.** c **10.** c **11.** c **12.** e **13.** c **14.** a, e **15.** a, b, d **16.** a, d **17.** e **18.** a, c, d **19.** b

CHAPTER 37

Reviewing Concepts: **1.** (a) fertilization, (b) different **2.** (a) asexually, (b) similar **3.** (a) alternation of generations, (b) gametophyte generation, (c) sporophyte generation **4.** flowering **5.** (a) sepals, (b) petals, (c) stamens, (d) carpels **6.** (a) calyx, (b) corolla **7.** ovules **8.** (a) microsporocytes, (b) microspores, (c) pollen grain, (d) sperm cells **9.** (a) anther, (b) stigma **10.** (a) inbreeding, (b) outcrossing or outbreeding **11.** self-incompatability **12.** (a) showy petals, (b) scent **13.** (a) blue, (b) yellow, (c) red **14.** *bee's purple* **15.** (a) red, (b) orange, (c) yellow, (d) visible light **16.** (a) at night, (b) dull white, (c) fermented fruit **17.** female bee **18.** insect **19.** flowers **20.** (a) style, (b) ovule **21.** endosperm **22.** (a) embryo, (b) nutrients **23.** suspensor **24.** (a) radicle, (b) cotyledons **25.** plumule **26.** (a) simple, (b) aggregate, (c) multiple, (d) accessory **27.** (a) simple, (b) berries, (c) drupes **28.** (a) follicle, (b) legumes, (c) capsules **29.** (a) aggregate, (b) multiple **30.** (a) accessory, (b) receptacle, (c) floral tube **31.** (a) wind, (b) animals, (c) water, (d) explosive dehiscence **32.** imbibition **33.** abscisic acid **34.** scarification **35.** coleoptile **36.** (a) rhizomes, (b) food or starch, (c) tubers, (d) bulbs, (e) corms, (f) stolons **37.** suckers **38.** dispersed **39.** sexual reproduction

Building Words: **1.** endosperm **2.** hypocotyl **3.** coevolution **4.** plumule **5.** dormancy **6.** germinate **7.** coleoptile **8.** rhizome **9.** angiosperm **10.** sporophyte

Matching: **1.** d **2.** g **3.** l **4.** b **5.** a **6.** k **7.** j **8.** m **9.** c **10.** f **11.** i **12.** e

Making Comparisons: **1.** dry: does not open; nut **2.** stony wall does not split open at maturity **3.** single seeded; fully fused to fruit wall **4.** simple **5.** fleshy: berry **6.** accessory **7.** multiple **8.** formed from the ovaries of many flowers that fuse together and enlarge after fertilization **9.** aggregate **10.** many separate ovaries from a single flower **11.** simple **12.** fleshy, drupe **13.** hard, stony pit; single seeded

Making Choices: **1.** c **2.** e **3.** a **4.** b **5.** b, d, e **6.** d **7.** b, d, e **8.** b, c, d **9.** e **10.** d **11.** d **12.** e **13.** b, d, e **14.** a, c **15.** c **16.** a, b, d, e **17.** c, d, e **18.** a, c **19.** b, d, e **20.** a-e **21.** b

CHAPTER 38

Reviewing Concepts: **1.** hormones **2.** gene expression **3.** directional response **4.** (a) phototropic, (b) blue, (c) kinases **5.** photoreceptor **6.** (a) gravitropism, (b) thigmotropism **7.** (a) auxins, (b) gibberellins, (c) cytokinins, (d) ethylene, (e) abscisic acid, (f) brassinosteroids **8.** (a) enzyme-linked receptors, (b) enzymatic **9.** coleoptile **10.** indoleacetic acid (IAA) **11.** (a) polar, (b) shoot apical meristem, (c) root **12.** apical dominance **13.** fruit **14.** fungus **15.** (a) flowering, (b) seed germination **16.** (a) adenine cell, (b) division, (c) differentiation **17.** senescence **18.** thigmomorphogenesis **19.** (a) ethylene, (b) auxin **20.** dormancy **21.** environmental stress **22.** (a) gibberellins, (b) cytokinins **23.** dwarf **24.** florigen **25.** photoperiodism **26.** long-night plants **27.** (a) short-day (long-night), (b) long-day (short-night), (c) intermediate-day, (d) day-neutral **28.** (a) blue-green, (b) photoperiodism **29.** shade avoidance **30.** (a) red, (b) Pr, (c) Pfr **31.** (a) potassium, (b) turgor **32.** transcription factor **33.** the time of day **34.** (a) innate, (b) hypersensitive **35.** phloem **36.** (a) prostaglandins, (b) herbivores (plant-eating insects) **37.** (a) willow bark, (b)

Building Words: **1.** phototropism **2.** gravitropism **3.** phytochrome **4.** photoperiodism **5.** senescence **6.** promoter **7.** tropism **8.** amyloplast **9.** auxin **10.** dormancy

Matching: **1.** b **2.** a **3.** m **4.** d **5.** c **6.** f **7.** o **8.** e **9.** p **10.** l **11.** h **12.** k

Making Comparisons: **1.** auxin **2.** cytokinin **3.** gibberellin **4.** ethylene **5.** gibberellin and cytokinin **6.** abscisic acid

Making Choices: **1.** a, c **2.** b **3.** a-e **4.** c **5.** a **6.** a, d **7.** a **8.** b **9.** e **10.** a-e **11.** d **12.** a, c, d, e **13.** e **14.** a, c **15.** d **16.** b, c **17.** a-e **18.** d **19.** a-e **20.** a

CHAPTER 39

Reviewing Concepts: **1.** ratio of its surface area to its volume. **2.** (a) number, (b) size **3.** complex organs systems **4.** (a) tissue, (b) organs, (c) organ systems **5.** (a) continuous, (b) tightly fitted, (c) basement membrane **6.** (a) protection, (b) absorption, (c) secretion, (d) sensation **7.** (a) squamous, (b) cuboidal, (c) columnar **8.** (a) simple, (b) stratified **9.** pseudostratified **10.** (a) exocrine glands, (b) endocrine glands **11.** (a) mucous, (b) serous **12.** (a) intercellular substance, (b) microscopic fibers **13.** intercellular substance **14.** (a) collagen, (b) elastin, (c) collagen and glycoprotein **15.** (a) fibers, (b) protein, (c) carbohydrate **16.** macrophages **17.** (a) loose connective tissue, (b) move **18.** collagen **19.** cartilage **20.** skeleton **21.** (a) chondrocytes, (b) collagen fibers, (c) lacunae **22.** (a) osteocytes, (b) lacunae, (c) vascularized **23.** canaliculi **24.** (a) osteons, (b) lamellae, (c) Haversian canal **25.** plasma **26.** transport oxygen **27.** defend the body against disease-causing microorganisms **28.** platelets **29.** muscle fiber **30.** (a) actin, (b) myosin, (c) myofibrils **31.** (a) skeletal, (b) smooth, (c) cardiac **32.** (a) neurons, (b) glial **33.** (a) cell body, (b) dendrite(s), (c) axon(s) **34.** synapses **35.** nerve **36.** (a) integumentary, (b) skeletal, (c) muscular, (d) digestive, (e) cardiovascular, (f) immune (lymphatic), (g) respiratory, (h) urininary, (i) nervous, (j) endocrine, (k) reproductive **37.** homeostasis **38.** stressors **39.** regulators **40.** (a) sensor, (b) intergrator (control center) **41.** opposite (negative) **42.** intensifies **43.** lower **44.** six **45.** hypothalamus **46.** (a) torpor, (b) hibernation, (c) estivation

Building Words: **1.** multicellular **2.** pseudostratified **3.** fibroblast **4.** intercellular **5.** macrophage **6.** chondrocyte **7.** osteocyte **8.** myofibril **9.** homeostasis **10.** dendrite **11.** synapsis **12.** niche **13.** interstitial

Matching: **1.** q **2.** b **3.** o **4.** d **5.** m **6.** g **7.** c **8.** j **9.** i **10.** e **11.** p **12.** s **13.** a

Making Comparisons: **1.** simple squamous epithelium **2.** air sacs of lungs, lining of blood vessels **3.** epithelial tissue **4.** some respiratory passages, ducts of many glands **5.** secretion, movement of mucus, protection **6.** connective tissue **7.** food storage, protection of some organs, insulation **8.** bone **9.** support and protection of internal organs, calcium reserve, site of skeletal muscle attachments **10.** connective tissue **11.** within heart and blood vessels of the circulatory system **12.** transport of oxygen, nutrients, waste product and other materials **13.** walls of the heart **14.** skeletal muscle **15.** nervous tissue

Making Choices: **1** a **2.** a, b, c, e **3.** d **4.** a, b, e **5.** c **6.** b, d **7.** a, c, e **8.** b, e **9.** d **10.** b **11.** c **12.** c, e **13.** a, c, d **14.** a, c **15.** c **16.** b **17.** d **18.** a, c, d **19.** c **20.** c, d, e **21.** a, c, d, e **22.** d, e **23.** a-e **24.** a, c, d, e **25.** a, b, e **26.** a, c, e **27.** b, d, e

CHAPTER 40

Reviewing Concepts: **1.** MRSA **2.** protection **3.** skeleton **4.** (a) environment, (b) lifestyle **5.** cuticle **6.** (a) lubricants, (b) adhesives **7.** (a) skin, (b) structures that develop from the skin. **8.** (a) feathers, (b) hair **9.** (a) claws (nails), (b) hair, (c) sweat glands, (d) oil glands, (e) sensory receptors **10.** (a) epidermis, (b) strata **11.** stratum basale **12.** (a) keratin, (b) strength, (c) flexibility, (d) a diffusion barrier **13.** (a) connective tissue, (b) collagen, (c) subcutaneous **14.** (a) thicken, (b) melanin, (c) DNA, (d) cancer **15.** mechanical forces **16.** fluid-filled body compartments **17.** muscle contraction **18.** (a) hydrostatic skeleton, (b) longitudinally, (c) circularly, (d) outer, (e) inner **19.** septa **20.** exoskeleton **21.** (a) chitin, (b) molting (ecdysis) **22.** (a) endoskeletons, (b) chordates **23.** calcium salts **24.** transmits muscle forces **25.** (a) axial, (b) appendicular **26.** (a) skull, (b) vertebral column, (c) rib cage **27.** (a) pectoral girdle, (b) pelvic girdle, (c) limbs **28.** (a) periostium, (b) tendons, (c) ligaments **29.** (a) epiphyses, (b) diaphysis, (c) metaphysis, (d) epiphyseal line, (e) compact bone **30.** (a) osteoblasts, (b) osteocytes, (c) osteoclasts **31.** (a) joints, (b) immovable joints, (c) slightly movable joints, (d) freely movable joints **32.** (a) capsule, (b) synovial fluid **33.** (a) actin, (b) myosin **34.** (a) striated (skeletal), (b) smooth **35.** asynchronous **36.** tendons **37.** antagonistically **38.** (a) agonist (b) antagonist **39.** an elongated cell **40.** (a) myofibrils, (b) filaments (myofilaments) **41.** (a) actin myofilaments, (b) myosin myofilaments **42.** sarcomere **43.** (a) acetylcholine (neurotransmitter), (b) action potential **44.** (a) sacroplasmic reticulum, (b) troponin, (c) actin, (d) mysosin-binding **45.** center **46.** ATP **47.** rigor mortis **48.** creatine phosphate **49.** glycogen **50.** (a) slow-oxidative, (b) aerobic respiration, (c) myoglobin **51.** fast-oxidative **52.** (a) fast-glycolytic, (b) glycolysis **53.** motor unit **54.** simple twitch **55.** muscle tone (state of tonus) **56.** (a) smooth, (b) cardiac

Building Words: **1.** epidermis **2.** periosteum **3.** endochondral **4.** osteoblast **5.** osteoclast **6.** myofilament **7.** keratin **8.** chitin **9.** axial **10.** sarcomere 11. sarcolemma **12.** tendon

Matching: **1.** i **2.** f **3.** l **4.** o **5.** g **6.** b **7.** e **8.** p **9.** j **10.** a **11.** m **12.** k **13.** h **14.** q

Making Comparisons: **1.** chitin **2.** exoskeleton **3.** external covering jointed for movement, does not grow so animal must periodically molt **4.** echinoderms **5.** calcium salts **6.** cartilage or bone **7.** endoskeleton **8.** internal, composed of living tissue and grows with the animal **9.** fast **10.** fast **11.** slow **12.** fast **13.** aerobic respiration **14.** aerobic respiration **15.** low **16.** high

Making Choices: **1.** c **2.** a, b, c **3.** d **4.** c **5.** a **6.** e **7.** c, d **8.** a **9.** d **10.** c, e **11.** a **12.** a, b, d **13.** b, c **14.** a, c **15.** b, e **16.** b, e **17.** d **18.** a **19.** a, c, e **20.** b, c **21.** a, b, c, e **22.** b, c, d **23.** a, c **24.** b, d, e **25.** c **26.** b, c, d **27.** a-e **28.** a, b, e **29.** a, b, c

CHAPTER 41

Reviewing Concepts: **1.** stimuli **2.** nervous system **3.** neurogenesis **4.** reception **5.** (a) afferent or sensory neuron, (b) interneuron, (c) efferent neuron, (d) motor neurons **6.** effectors **7.** (a) dendrites, (b) cell body, (c) axon **8.** (a) terminal branches, (b) synaptic terminals, (c) neurotransmitter **9.** (a) Schwann cells, (b) myelin sheath, (c) nodes of Ranvier **10.** neurogenesis **11.** (a) nerve, (b) tract (pathway) **12.** (a) ganglion, (b) nucleus **13.** neuroglia **14.** (a) astrocytes, (b) oligodendrocytes, (c) ependymal cells, (d) microglia **15.** membrane potential **16.** (a) resting potential, (b) -70 mV **17.** (a) differences in the concentrations of specific ions inside and outside of the cell, (b) selective permeability of the plasma membrane to these ions 18. (a) passive ion channels, (b) voltage-activated channels, (c) chemically activated ion channels **19.** potassium (K^+) **20.** (a) positive, (b) negative) **21.** (a) inside, (b) outside **22.** (a) sodium-potassium pumps, (b) 3, (c) 2 **23.** (a) excitatory, (b) transmitting an impulse **24.** (a) inhibitory, (b) decreases **25.** graded potential **26.** (a) threshold level, (b) action potential (nerve impulse) **27.** (a) neurons, (b) muscle cells **28.** voltage-activated Na^+ channels **29.** (a) voltage-activated sodium channels, (b) sodium, (c) depolarization **30.** positive **31.** (a) sodium channel inactivation gates, (b) voltage-activated potassium channels, (c) repolarization **32.** repolarization **33.** absolute refractory **34.** all-or-none **35.** wave of depolarization **36.** continuous **37.** (a) salutatory, (b) node of Ranvier, (c) 50 **38.** synapse **39.** (a) presynaptic, (b) postsynaptic **40.** (a) electrical synapses, (b) chemical synapses, (c) neurotransmitter molecules **41.** (a)acetylcholine, (b) brain, (c) autonomic nervous system **42.** adrenergic **43.** (a) norepinephrine, (b) epinephrine, (c) dopamine **44.** (a) serotonin, (b) histamine **45.** (a) endorphins, (b) enkephalins **46.** (a) endorphins, (b) enkephalins **47.** nitric oxide **48.** (a) synaptic vesicles, (b) calcium (Ca^{++}) **49.** ligand-gated **50.** (a) excitatory postsynaptic potential (EPSP), (b) inhibitory postsynaptic potential (IPSP) **51.** graded potentials **52.** (a) summation, (b) temporal summation, (c) spatial summation **53.** (a) converging, (b) diverging

Building Words: **1.** interneuron **2.** neuroglia **3.** neurotransmitter **4.** multipolar **5.** postsynaptic **6.** presynaptic **7.** hyperpolarized **8.** synapse

Matching: **1.** d **2.** j **3.** f **4.** b **5.** i **6.** k **7.** a **8.** h **9.** n **10.** g **11.** m **12.** o
Making Comparisons: **1.** -70mV **2.** stable **3.** sodium-potassium pump active, sodium and potassium diffusion channels open, voltage-activated sodium and potassium channels closed **4.** threshold potential **5.** voltage-activated sodium ion channels open **6.** varies, but is about +35mV **7.** wave of depolarization (rise: depolarization; fall: repolarization) **8.** depolarization **9.** neurotransmitter-receptor combination opens sodium ion channels **10.** hyperpolarization

Making Choices: **1.** b, c, e **2.** b **3.** c, d, e **4.** c, d **5.** b **6.** a, d **7.** c **8.** b, c **9.** d **10.** c, e **11.** a **12.** e **13.** e **14.** a, d **15.** a, b, d, e **16.** a, b, c **17.** c, e **18.** b, d **19.** b, d **20.** b **21.** a **22.** a **23.** a-e **24.** e

CHAPTER 42

Reviewing Concepts: **1.** (a) nerve net, (b) cnidarians, (c) central control organ **2.** nerve ring **3.** (a) ladder-type, (b) cerebral ganglia **4.** (a) ganglia, (b) brain **5.** functional regions **6.** (a) sense, (b) intelligent **7.** (a) increased number of nerve cell, (b) concentration of nerve cells, (c) specialization of function, (d) increased number of interneurons and more complex synaptic contacts, (e) cephalization or the formation of a head **8.** (a) 100 billion, (b) learning ability and behavior **9.** (a) central nervous (CNS), (b) peripheral nervous (PNS) **10.** (a) somatic, (b) autonomic, (c) sympathetic (d) parasympathetic **11.** (a) neural tube, (b) brain, (c) spinal cord **12.** (a) cerebellum, (b) pons, (c) metencephalon, (d) medulla, (e) myelencephalon **13.** (a) medulla, (b) pons, (c) midbrain **14.** (a) muscle tone, (b) posture, (c) equilibrium **15.** association **16.** (a) superior colliculi, (b) inferior colliculi, (c) red nucleus **17.** (a) thalamus, (b) hypothalamus, (c) cerebrum, (d) olfactory bulbs **18.** (a) hemispheres, (b) lateral ventricle **19.** (a) white, (b) myelinated, (c) gray **20.** neocortex **21.** (a) convolutions (gyri), (b) sulci, (c) fissures **22.** (a) dura matter, (b) arachnoid, (c) pia mater, (d) choroid plexus **23.** second lumbar **24.** (a) white matter, (b) tracts, (c) central canal **25.** reflexes **26.** (a) sensory, (b) motor, (c) association **27.** (a) frontal, (b) parietal, (c) central sulcus, (d) temporal, (e) occipital **28.** basal ganglia **29.** (a) dopamine, (b) inhibition, (c) excitation **30.** dopamine **31.** (a) pineal gland, (b) melatonin **32.** sleep-wake cycle **33.** brain stem and thalamus **34.** (a) electroencephalogram (EEG), (b) alpha, (b) deta, (c) theta, (d) delta **35.** (a) rapid eye movement (REM), (b) non-REM, (c) REM **36.** cerebrum **37.** amygdala **38.** reward **39.** (a) encoding, (b) storing, (c) retrieving, (d) implicit memory, (e) declarative memory (explicit memory) **40.** synaptic plasticity **41.** hypocampus **42.** (a) long-term potentiation, (b) long-term depression **43.** (a) gene expression, (b) protein synthesis **44.** temporal lobe **45.** Broca's area **46.** (a) sensory receptors, (b) sensory neurons, (c) motor neurons **47.** (a) cranial, (b) spinal **48.** (a) sympathetic, (b) parasympathetic **49.** (a) preganglionic neuron, (b) postganglionic neuron **50.** 25 **51.** alcohol **52.** psychological dependence **53.** tolerance **54.** mesolimbic dopamine

Building Words: **1.** hypothalamus **2.** postganglionic **3.** preganglionic **4.** paravertebral **5.** sensory **6.** limbic **7.** ventricle **8.** cerebrum

Matching: **1.** g **2.** d **3.** j **4.** f **5.** h **6.** a **7.** e **8.** k **9.** b **10.** i **11.** n **12.** o

Making Comparisons: **1.** pons **2.** midbrain **3.** center for visual and auditory reflexes **4.** diencephalon **5.** relay center for motor and sensory information **6.** hypothalamus **7.** diencephalon **8.** metencephalon **9.** cerebrum **10.** reticular activating system **11.** limbic system

Making Choices: **1.** a **2.** a **3.** b **4.** a-e **5.** a, d, e **6.** b **7.** a, b **8.** d **9.** a, c **10.** b **11.** b **12.** a, b **13.** b **14.** c **15.** a-e **16.** a-e **17.** c, e **18.** b, e **19.** b, d **20.** a, c, d **21.** a, d, e **22.** e **23.** b **24.** e **25.** a **26.** a, d, e

CHAPTER 43

Reviewing Concepts: **1.** echolocation **2.** reception **3.** (a) transduced (converted), (b) electrical, (c) transduction **4.** (a) receptor potential, (b) graded response **5.** (a) depolarizes, (b) action potential, (c) sensory **6.** sensory adaptation **7.** (a) receptor, (b) brain **8.** (a) total number of sensory neurons transmitting the signal, (b) the specific neurons transmitting action [potential and their targets], (c) the total number of action potentials transmitted by a given neuron, (d) the frequency of the action potentials transmitted by a given fiber **9.** interneurons **10.** (a) sensory perception, (b) brain **11.** (a) exteroceptors, (b) interoceptors **12.** (a) thermoreceptors, (b) electromagnetic receptors, (c) nociceptors, (d) chemoreceptors, (e) photoreceptors **13.** (a) pit vipers, (b) boas, (c) boas **14.** (a) skin, (b) tongue **15.** hypothalamus **16.** (a) sharks, (b) rays, (c) bony fishes **17.** stun **18.** (a) mechanical,

(b) thermal, (c) polymodal **19.** (a) spinal cord, (b) glutamate, (c) substance P **20.** (a) endorphins, (b) enkephalins **21.** change shape **22.** (a) hair, (b) bristle **23.** Merkel cells **24.** (a) Meissner corpuscles, (b) Ruffini endings, (c) Pacinian corpuscles **25.** (a) muscle spindles, (b) Golgi tendon organs, (c) joint receptors **26.** (a) sensory hair cells, (b) statoliths **27.** kinocilium **28.** microvilli **29.** vibrations **30.** (a) canal, (b) sensory hair cells, (c) cupula **31.** (a) saccule, (b) utricle, (c) gelatinous cupula, (d) otoliths, (e) statocysts **32.** (a) angular acceleration, (b) endolymph, (c) crista (e), (d) otoliths **33.** (a) cochlea, (b) mechanoreceptors **34.** (a) tympanic membrane, (b) malleus, (c) incus, (d) stapes, (e) oval window **35.** (a) Organ of Corti, (b) tectorial, (c) cochlear nerve **36.** eustachian tube **37.** (a) frequency, (b) amplitude **38.** (a) gustation (taste), (b) olfaction (smell) **39.** (a) sweet, (b) sour, (c) salty, (d) bitter, (e) umami (glutamate), (f) fatty acid **40.** olfaction **41.** olfactory epithelium **42.** limbic system **43.** small volatile molecules **44.** rhodopsins **45.** eyespots (ocelli) 46. lens **47.** ommatidia **48.** (a) sclera, (b) shape (rigidity) (c) cornea **49.** vitreous body **50.** aqueous humor **51.** iris **52.** photoreceptors **53.** (a) fovea, (b) cones, (c) sharpest vision **54.** (a) photoreceptors, (b) bipolar cells, (c) ganglion cells, (d) horizontal cells, (e) amacrine cells **55.** (a) opsin, (b) retinal **56.** dark adaptation **57.** (a) blue, (b) green, (c) red, (d) range of wavelengths **58.** (a) hypothalamus, (b) optic chiasm

Building Words: **1.** proprioceptor **2.** otolith **3.** endolymph **4.** chemoreceptor **5.** thermoreceptor **6.** interoceptor **7.** tactile **8.** statoctyst **9.** olfactory **10.** optic chiasm **11.** optic

Matching: **1.** l **2.** b **3.** j **4.** i **5.** n **6.** k **7.** o **8.** d **9.** a **10.** c **11.** g **12.** m

Making Comparisons: **1.** electroreceptor **2.** exteroceptor **3.** pressure waves (sound) **4.** mechanoreceptor **5.** interoceptor **6.** receptor in the human hypothalamus **7.** muscle contraction **8.** interoceptor **9.** thermoreceptor **10.** pit organ of a viper **11.** light energy **12.** photoreceptor **13.** chemoreceptor **14.** mammalian taste buds **15.** chemoreceptor **16.** exteroceptor

Making Choices: **1.** c **2.** d **3.** a **4.** a **5.** c **6.** b **7.** c **8.** b, d **9.** b, d **10.** c **11.** d **12.** e **13.** a, d **14.** b, d, e **15.** a-e **16.** a, c, d, e **17.** c, d **18.** b, d, e **19.** a, d, e **20.** b, e **21.** d, e

CHAPTER 44

Reviewing Concepts:

1. cholesterol **2.** (a) sponges, (b) cnidarians, (c) ctenophores, (d) flatworms, (e) nematodes **3.** (a) circulatory, (b) digestive **4.** fluid **5.** (a) blood (circulatory fluid), (b) heart, (c) spaces, (d) blood vessels **6.** (a) open-ended, (b) sinuses (hemocoel) **7.** (a) hemolymph, (b) interstitial fluid **8.** (a) arthropods, (b) most mollusks **9.** three **10.** (a) hemocyanin, (b) copper **11.** (a) nutrients, (b) hormones **12.** (a) annelids, (b) some mollusks (cephalopods), (c) echinoderms **13.** (a) dorsal, (b) ventral, (c) five **14.** plasma **15.** ventrally **16.** (a)nutrients, (b) oxygen, (c) metabolic wastes, (d) hormones **17.** (a) plasma, (b) red blood cells, (c) white blood cells, (d) platelets **18.** (a) interstitial, (b) intracellular **19.** (a) fibrinogen, (b) globulins, (c) albumin **20.** (a) red bone marrow, (b) hemoglobin, (c) 120 **21.**)a) anemia, (b) loss of blood (hemorrhage), (c) increased breakdown of RBCs **22.** leukocytes **23.** (a) granular, (b) agranular **24.** (a) monocytes, (b) macrophages **25.** (a) neutrophils, (b) basophils, (c) eosinophiles **26.** (a) lymphocytes, (b) monocytes **27.** thrombocyte **28.** platelets **29.** (a) arteries, (b) veins **30.** capillaries **31.** arterioles **32.** (a) ventricles, (b) atria **33.** (a) one, (b) one **34.** (a) three, (b) two, (c) one, (d) sinus venosus **35.** (a) birds, (b) crododillians, (c) mammals **36.** higher **37.** (a) 5, (b) 20 **38.** (a) pericardium, (b) fluid-filled **39.** (a)interventricular, (b)interatrial septum

40. (a) four, (b) atrioventricular valves, (c) tricuspid valve, (d) mitral valve, (e) semilunar valves **41.** (a) sinoatrial (SA) node, (b) atrioventricular (AV) node, (c) atrioventricular (AV) bundle **42.** (a) systole, (b) diastole **43.** (a) lub, (b) AV valves **44.** (a) dub, (b) semilunar valves **45.** (a) nervous, (b) endocrine **46.** Starling's law of the heart **47.** (a) cardiac output (CO), (b) stroke volume **48.** five liters/minute **49.** (a) hypertension, (b) increase, (c) salt **50.** diameter of arterioles **51.** low-resistance **52.** baroreceptors **52.** (a)pressure changes, (b)cardiac and vasomotor centers **53.** (a) pulmonary circuit, (b) systemic circuit **54.** (a) pulmonary circuit, (b) systemic circuit **55.** (a) rich, (b) left **56.** (a) aorta, (b) brain, (c) shoulder area, (d) legs **57.** (a) superior vena cava, (b) inferior vena cava **58.** circulatory system **59.** (a) interstitial fluids, (b) an immune response, (c) absorbs fats **60.** lymph **61.** lymph nodes **62.** (a) tonsils, (b) thymus gland, (c) spleen **63.** (a) subclavian veins, (b) thoracic duct, (c) right lymphatic duct **64.** (a) net filtration, (b) hydrostatic, (c) osmotic, (d) osmotic **65.** (a) atherosclerosis, (b) lipid (fat) **66.** (a) ischemia, (b) angina pectoris **67.** thrombosis **68.** (a) angioplasty, (b) stent

Building Words: **1.** hemocoel **2.** hemocyanin **3.** erythrocyte **4.** leukocyte **5.** neutrophil **6.** eosinophil **7.** basophil **8.** leukemia **9.** vasoconstriction **10.** vasodilation **11.** pericardium **12.** semilunar **13.** baroreceptor **14.** sinus **15.** ventricle **16.** leukemia

Matching: **1.** i **2.** j **3.** h **4.** l **5.** d **6.** m **7.** b **8.** o **9.** a **10.** q **11.** k **12.** e

Making Comparisons: **1.** lipoproteins **2.** plasma **3.** cell component **4.** transport of oxygen and carbon dioxide **5.** blood clotting **6.** cell component **7.** defense; differentiate to form phagocytic macrophages in tissue **8.** albumins and globulins **9.** plasma **10.** defense, principle phagocytic cell in blood

Making Choices: **1.** d **2.** a **3.** d **4.** c, e **5.** a **6.** b, c **7.** d **8.** a, b, d **9.** a, c, e **10.** b, d, e **11.** b **12.** a, d, e **13.** c, e **14.** a, b, c, e **15.** a, c **16.** c **17.** d **18.** a, d, e **19.** a, e **20.** b, d, e **21.** a, b, d, e **22.** b **23.** e **24.** b, c, e **25.** e **26.** b **27.** c **28.** a, d **29.** b, c **30.** b, e **31.** c, e **32.** a

CHAPTER 45

Reviewing Concepts: **1.** pathogens **2.** immunology **3.** signal **4.** recognizing foreign or dangerous macromolecules **5.**(a) innate, (b) adaptive (acquired) **6.** antigen **7.** antibodies **8.** outer covering **9.** pattern recognition **10.** phagocytes **11.** inflammatory **12.** natural killer cells **13.** antimicrobial **14.** nonspecific **15.** jawed vertebrates **16.** (a) outer covering, (b) pathogens **17.** defensins **18.** lysozyme **19.** (a) phagocytes, (b) natural killer cells, (c) dendritic cells **20.** macrophages **21.** natural killer (NK) **22.** (a) dendritic, (b) interferons **23.** (a) peptides, (b) proteins **24.** (a) autocrine, (b) paracrine, (c) hormones **25.** type I **26.** type II **27.** tumor necrosis factor (TNF) **28.** interleukins **29.** chemokines **30.** (a) lysis of viruses, bacteria and other cells, (b) coating pathogens so that can be more easily phagocytized, (c) attracting white blood cells to sites of infection, (d) binding specific receptors on cells of the immune system, stimulating specific actions **31.** (a) vasodilation, (b) increased capillary permeability, (c) increased phagocytosis **32.** (a) acquired, (b) highly specific **33.** (a) lymphocytes, (b) antigen-presenting cells **34.** (a) T-cells, (b) B-cells **35.** bone marrow **36.** bone marrow **37.** plasma cells **38.** (a) thymus, (b) immunocompetent **39.** (a)cytotoxic (CD_8), (b)helper (CD_4) **40.** (a) macrophages, (b) dendritic cells, (c) B-cells, (d) foreign antigens, (e) their own surface proteins **41.** (a) major histocompatibility complex (MHC) or human leukocyte antigen compolex (HLA), (b) "fingerprint" **42.** (a) most nucleated cells, (b) antigen presenting cells **43.** properly presented **44.** NK **45.** (a) T_H, (b) B, (b) clone **46.** lymph nodes **47.** (a) immunoglobulins, (b) Ig, (c) antigenic determinants (epitopes), (d) binding sites

48. (a) IgG, (b) IgA, (c) IgD, (d) IgE, (e) IgM **49.** (a) IgG, (b) IgA, (c) IgD, (d) IgE **50.** (a) pathogens, (b) toxins, (c) phagocytosis of pathogens **51.** clonal selection **52.** (a) identical, (b) a single cell **53.** antigenic determinant **54.** (a) hybridoma, (b) monoclonal antibodies **55.** (a) survival, (b) apoptosis **56.** (a) primary immune, (b) IgM **57.** (a) secondary immune, (b) IgG **58.** (a) actively, (b) artificially **59.** temporary **60.** (a) gamma globulin, (b) memory **61.** IgA **62.** (a) molecular structure, (b) source **63.** angiogenesis **64.** protein malnutrition **65.** opportunistic **66.** thymus **67.** (a) body fluids, (b) mucosa **68.** (a) T_H , (b) reverse-transcribed **69.** (a) tuberculosis, (c) lymphomas, (c) Kaposi's sarcoma **70.** dementia complex **71.** self-tolerance **72.** autoreactive **73.** multiple sclerosis **74.** antigen D **75.** erythroblastosis fetalis **76.** IgE **77.** allergic rhinitis **78.** (a) mast, (b) contract and constrict the airways **79.** systemic anapnylaxis **80.** MHC antigens (HLA)

Building Words: **1.** antibody **2.** antihistamine **3.** lymphocyte **4.** monoclonal **5.** autoimmune **6.** lysozyme **7.** phagocyte **8.** complement **9.** pathogen

Matching: **1.** b **2.** r **3.** f **4.** q **5.** g **6.** j **7.** p **8.** o **9.** l **10.** n **11.** e **12.** d **13.** i

Making Comparisons: **1.** plasma cells **2.** memory B cells **3.** B-cells and macrophages **4.** helper T cells **5.** cytotoxic T cells **6.** memory T cells **7.** natural killer cells **8.** macrophages **9.** neutrophils

Making Choices: **1.** a, d **2.** a, c, d **3.** b **4.** b **5.** b, c, e **6.** a-e **7.** a, c, e **8.** b, c, d **9.** a, b, c, e **10.** a, b, c, d **11.** b, c, e **12.** a-e **13.** c **14.** a-e **15.** a, e **16.** a **17.** c **18.** a, b, c, e **19.** a, b, c, e **20.** d, e **21.** a, b, c, e **22.** c, e **23.** a-e **24.** a, c, d, e

CHAPTER 46

Reviewing Concepts: **1.** organismic respiration **2.** carbon dioxide **3.** (a) more, (b) faster **4.** 20 **5.** (a) moist, (b) thin walls **6.** (a) body surface, (b) tracheal tubes, (c) gills, (d) lungs **7.** ventilation **8.** (a) surrounding water, (b) fluid secretions **9.** (a) tracheal tubes (tracheae), (b) spiracles, (c) muscles **10.** tracheoles **11.** (a) dermal, (b) water, (c) coelomic fluid **12.** (a) cilia, (b) mantle **13.** filaments **14.** (a) countercurrent exchange system, (b) higher, (c) 80 **15.** (a) body surface, (b) body cavity **16.** book lungs **17.** swim bladder **18.** across the body surface **19.** long period of activity **20.** respiratory system **21.** (a) one direction, (b) two, (c) posterior air sacs, (d) into the lungs, (e) anterior air sacs **22.** cross-current **23.** moist, ciliated **24.** (a) pharynx, (b) larynx, (c) trachea **25.** epiglottis **26.** (a) thoracic, (b) three, (c) two **27.** (a) pleural membrane, (b) pleural cavity, (c) alveoli **28.** (a) increases, (b) contracting, (c) decreases, (d) relaxing **29.** (a) external intercostal, (b) increasing **30.** pulmonary surfactant **31.** (a) tidal volume, (b) 500, (c) residual volume, (d) vital capacity **32.** pressure (tension) **33.** Dalton's law of partial pressures **34.** (a) partial pressure, (b) surface area **35.** (a) 100 mm Hg, (b) 40 mm Hg, (c) into the tissues **36.** hemocyanins **37.** (a) hemoglobin, (b) myoglobin **38.** (a) iron-porphyrin, (b) globin **39.** (a) iron, (b) heme, (c) oxyhemoglobin (HbO_2) **40.** (a) acidic, (b) Bohr effect **41.** (a) carbon dioxide in the plasma, (b) bound to hemoglobin, (c) is carried as HCO_3^- in the plasma, (d) carbonic anhydrase **42.** respiratory acidosis **43.** (a)brain stem, (b) medulla, (c) pons **44.** (a) medulla, (b) walls of the aorta, (c) walls of the carotid arteries, (d) breathing rate **45.** (a) faster, (b) slower **46.** oxygen **47.** hypoxia **48.** 4 **49.** blood pressure **50.** (a) 21, (b) partial pressure **51.** decompression sickness **52.** (a) a slowing of heart rate, (b) increased blood pressure, (c) shunting of blood to internal organs and decreases in flow to extremities, (d) a decrease in metabolic rate **53.** bronchial constriction

Building Words: **1.** hyperventilation **2.** hypoxia **3.** ventilation **4.** oxyhemoglobin **5.** ventilation **6.** diffusion **7.** dermal **8.** alveolus **9.** spiracle **10.** operculum **11.** diaphragm

Matching: **1.** n **2.** a **3.** f **4.** l **5.** c **6.** j **7.** b **8.** h **9.** k **10.** g **11.** e **12.** m

Making Comparisons: **1.** gills **2.** some, tracheal tubes; some book lungs **3.** lungs **4.** lungs **5.** body surface **6.** tracheal tube **7.** dermal gills **8.** gills **9.** lungs

Making Choices: **1.** e **2.** c, d **3.** a-e **4.** b, c, e **5.** a, e **6.** a, b, e **7.** a, b, d **8.** a **9.** a, b, c, d **10.** e **11.** b, c, e **12.** a, d **13.** a, b, d, e **14.** a-e **15.** c **16.** a **17.** a, d, e **18.** d **19.** a **20.** b, d **21.** a, b ,c, e **22.** a, c, d **23.** a-e **24.** a-e

CHAPTER 47

Reviewing Concepts: **1.** heterotrophs (consumers) **2.** nutrients **3.** (a) carbohydrates, (b) lipids, (c) proteins, (d) vitamins, (e) minerals **4.** (a) selection, (b) acquisition, (c) ingestion **5.** (a) ingestion, (b) digestion, (c) absorption, (d) egestion (elimination) **6.** (a) herbivores, (b) primary consumers **7.** (a) cellulose, (b) symbiotic **8.** ruminants **9.** (a) secondary (b) predators, (c) canine **10.** omnivores **11.** filter feeders **12.** deposit **13.** (a) cnidarians, (b) flatworms **14.** (a) mouth, (b) anus **15.** peristalsis **16.** (a) pharynx, (b) esophagus, (c) small intestine **17.** (a) mucosa, (b) submucosa, (c) visceral peritoneum **18.** (a) incisors, (b) canines, (c) molars, (d) premolars, (e) enamel, (f) dentin, (g) pulp cavity **19.** salivary amylase **20.** bolus **21.** epiglottis **22.** peristalsis **23.** rugae **24.** (a) gastric glands, (b) parietal, (c) chief, (d) pepsin **25.** *Helicobacter pylori* **26.** 3-4 **27.** (a) duodenum, (b) jejunum, (c) ileum, (d) duodenum **28.** (a) villi, (b) microvilli **29.** mechanical digestion of fats **30.** (a) salts, (b) pigments, (c) cholesterol, (d) lecithin (phosphatidylcholine) **31.** (a) gall bladder, (b) concentrated **32.** (a) trypsin, (b) chymotrypsin (c) pancreatic lipase, (d) pancreatic amylase **33.** (a) two glucose molecules, (b) amino acids **34.** (a) emulsifies, (b) monoacylglycerols, (c) diacylglycerols, (d) fatty acids, (e) glycerol **35.** (a) subsrance P, (b) encephalin **36.** endocrine cells **37.** villi **38.** (a) liver, (b) hepatic portal vein **39.** (a) chylomicrons, (b) lacteal **40.** ileocecal valve **41.** (a) cecum, (b) ascending colon, (c) transverse colon, (d) descending colon, (e) sigmoid colon, (f) rectum, (g) anus **42.** (a) K, (b) B **43.** (a) excretion, (b) elimination **44.** (a) 45 to 65, (b) 20 to 35, (c) 10 to 35 **45.** (a) Calories, (b) kilocalories **46.** (a) starch, (b) cellulose **47.** complex carbohydrates **48.** (a) glycogen, (b) fatty acids, (c) glycerol, (d) fat deposits **49.** insulin **50.** (a) cellulose, (b) indigestible carbohydrates **51.** triacylglycerols **52.** (a) linoleic acid, (b) linolenic acid, (c) arachidonic acid **53.** 300 **54.** (a) lipoproteins, (b) high-density lipoproteins (HDLs), (c) low-density lipoproteins (LDLs) **55.** (a) saturated fats, (b) cholesterol, (c) unsaturated fats, (d) cholesterol **56.** (a) omega-3, (b) trans **57.** (a) 20, (b) 9, (c) essential amino acids **58.** (a) fish, (b) meat, (c) nuts, (d) eggs, (e) milk **59.** (a) deaminate, (b) ammonia **60.** (a) fat-soluble, (b) water-soluble **61.** fat soluble **62.** (a) sodium, (b) chloride, (c) potassium, (d) calcium, (e) phosphorus, (f) magnesium, (g) manganese, (h) sulfur **63.** trace elements **64.** free radicals **65.** (a) C, (b) E, (c) A **66.** (a) fruits, (b) vegetables, (c) fats **67.** (a) basal metabolic rate (BMR), (b) total metabolic rate **68.** (a) weight, (b) height **69.** (a) hypothalamus, (b) neuropeptide Y (NPY), (c) melanocortins **70.** (a) ghrelin, (b) NPY, (c) YY, (d) NPY **71.** leptin **72.** (a) 33, (b) 33 **73.** set point **74.** essential amino acids **75.** (a) kwashiorkor, (b) marasmus

Building Words: **1.** herbivore **2.** carnivore **3.** omnivore **4.** submucosa **5.** peritonitis **6.** epiglottis **7.** microvilli **8.** chylomicron **9.** Intrinsic **10.** parietal **11.** jejunum **12.** cecum **13.** vitamin(e)

Matching: **1.** h **2.** e **3.** j **4.** i **5.** o **6.** l **7.** a **8.** c **9.** k **10.** n **11.** g **12.** f

Making Comparisons: **1.** splits maltose into 2 glucose molecules **2.** small intestine **3.** pepsin **4.** stomach **5.** trypsin and chymotrypsin **6.** pancreas **7.** RNA to free nucleotides **8.** pancreas **9.** pancreatic lipase **10.** dipeptidase **11.** duodenum **12.** lactose to glucose and galactose

Making Choices: **1.** a, b, d **2.** c, d, e **3.** d **4.** a, d **5.** a **6.** a, b **7.** a, b, c, e **8.** a, b, c **9.** d, e **10.** a-d **11.** b **12.** c **13.** a, d, e **14.** d **15.** c, e **16.** e **17.** c, d **18.** a, d **19.** a, b, e **20.** b, e **21.** a-e **22.** a, c **23.** c, d **24.** a, c, e **25.** b, c, d, e **26.** c, d, e **27.** c, d **28.** a, d, e **29.** b, d, e **30.** a, c, e **31.** a-e **32.** a **33.** a, e **34.** a, d **35.** a-e **36.** a, b, d **37.** c **38.** a, b, c, d **39.** a, c b, d, e **40.** b, d

CHAPTER 48

Reviewing Concepts: **1.** water **2.** (a) osmoregulation and excretion **3.** electrolytes **4.** osmole **5.** osmoregulation **6.** (a) water, (b) carbon dioxide, (c) nitrogenous wastes **7.** (a) ammonia, (b) urea, (c) uric acid **8.** (a) uric acid, (b) ammonia, (c) break-down of nucleotides **9.** urea **10.** (a) isosmotic (at osmotic equilibrium), (b) osmoconformers **11.** osmoregulators **12.** (a) nephridiopores, (b) protonephridia, (c) metanephridia **13.** (a) gut wall, (b) hemocoel, (c) hemolymph **14.** (a) diffusion, (b) active transport **15.** (a) kidney, (b) skin, (c) lungs, (d) gills **16.** hypertonic **17.** gills **18.** (a) ammonia, (b) urea **19.** dilute urine **20.** hypo-osmotic (hypotonic) **21.** drink sea water **22.** (a) urea, (b) hypertonic **23.** concentrated **24.** (a) uric acid, (b) urea **25.** (a) renal cortex (cortex), (b) renal medulla (medulla) **26.** (a) ureters, (b) bladder, (c) urethra **27.** (a) Bowman's capsule, (b) renal tubule, (c) Bowman's capsule, (d) loop of Henle, (e) collecting duct **28.** (a) cortical, (b) juxtamedullary **29.** (a) filtration, (b) reabsorption, (c) tubular secretion **30.** Bowman's capsule **31.** (a) blood hydrostatic pressure, (b) large surface area for filtration, (c) permeability of the glomerular capillaries **32.** (a) podocytes, (b) filtration slits, (c) filtration membrane **33.** (a) 99, (b) 1.5 **34.** proximal convoluted tubule **35.** tubular transport maximum **36.** (a) becomes irregular, (b) aldosterone **37.** (a) 300, (b) 1200 **38.** (a) descending, (b) water moving out of the filtrate (loop membrane permeable to water), (c) ascending, (d) salt diffusing out of thin ascending limb and then being transported out in the thick ascending limb, (e) counterflow (countercurrent mechanism) **39.** vasa recta **40.** adjusted filtrate **41.** urinalysis **42.** (a) antidiuretic hormone (ADH), (b) posterior lobe of the pituitary gland, (c) hypothalamus **43.** hypothalamus **44.** diabetes insipidus **45.** (a) aldosterone, (b) distal tubules, (c) collecting ducts **46.** (a) juxtaglomerular apparatus, (b) angiotensinogen, (c) angiotensin I, (d) angiotensin I, (b) angiotensin II **47.** (a) heart atrial cells, (b) sodium, (c) blood pressure, (d) dialates **48.** antagonistically

Building Words: **1.** protonephridium **2.** podocyte **3.** juxtamedullary **4.** glomerulus **5.** electrolyte **6.** urethra

Matching: **1.** e **2.** a **3.** f **4.** c **5.** o **6.** j **7.** m **8.** l **9.** i **10.** b **11.** d

Making Comparisons: **1.** kidney **2.** Malpighian tubules **3.** metanephridia **4.** protonephridia

Making Choices: **1.** a, c, e **2.** b **3.** a, b **4.** a, d, e **5.** c **6.** c, d **7.** b, d **8.** d **9.** c **10.** e **11.** c, e **12.** c **13.** a, c, d **14.** b, d **15.** a, d, e **16.** e **17.** b, c, d **18.** a, c **19.** e **20.** d **21.** a-e **22.** a, c, d **23.** a, b, c **24.** b, d **25.** a **26.** a, b, d **27.** d **28.** a, e **29.** b, c **30.** b, c, e **31.** a, c, d, e

CHAPTER 49

Reviewing Concepts: **1.** (a) growth and development, (b) metabolism, (c) fluid balance and concentrations of specific ions and chemical compounds, (d) reproduction, (e) response to stress **2.** "orphan" **3.** (a) ducts, (b) into the surrounding interstitial fluid or blood **4.** 10 **5.** (a) digestive tract, (b) heart, (c) kidneys **6.** (a) the blood, (b) bind to a specific receptor on a target cell **7.** homeostasis **8.** slowly **9.** longer **10.** (a) negative feedback systems, (b) homeostasis **11.** (a) fatty acid derivatives, (b) steroids, (c) amino acid derivatives, (d) peptides or proteins **12.** neuropeptides (neurohormones) **13.** (a) classical endocrine, (b) neuroendocrine, (c) autocrine, (d) paracrine **14.** plasma proteins **15.** (a) neuroendocrine (b) axons, (c) interstitial fluid **16.** (a) autocrine, (b) paracrine **17.** Histamine **18.** (a) nitric oxide, (b) prostaglandins, (c) smooth muscle **19.** (a) receptor up-regulation, (b) receptor down-regulation **20.** (a) in the cytoplasm or in the nucleus, (b) lipid soluble (pass through the membrane) **21.** (a) gene activation, (b) protein synthesis **22.** are not lipid soluble **23.** (a) G protein, (b) enzyme **24.** signal transduction **25.** (a) adenylyl cyclase, (b) protein kinases **26.** (a) inositol trisphosphate (IP_3), (b) diacylglyerol (DAG) **27.** second messengers **28.** tyrosine kinase **29.** neurons **30.** (a) brain hormone (BH), (b) prothoracic, (c) molting hormone (MH), (d) ecdysone **31.** (a) hypersecretion, (b) hyposecretion **32.** pituitary stalk **33.** master gland **34.** (a) antidiuretic hormone (ADH) , (b) oxytocin, (c) posterior pituitary **35.** (a) releasing, (b) inhibiting, (c) portal **36.** tropic **32.** (a) anterior pituitary, (b) neurons, (c) suppresses appetite **38.** (a) somatotropin, (b) anabolic hormone **39.** (a) gigantism, (b) acromegaly **40.** (a) thyroxine, (b) iodine, (c) triiodothyronine, (d) iodine **41.** (a) anterior pituitary, (b) thyroid-stimulating hormone (TSH) **42.** cretinism **43.** Graves disease **44.** (a) bones, (b) kidney tubules, (c) thyroid, (d) calcitonin **45.** (a) islets of Langerhans, (b) beta, (c) insulin, (d) alpha, (e) glucagon **46.** (a) liver, (b) muscle, (c) fat **47.** (a) fat, (b) protein **48.** (a) liver, (b) gluconeogenesis **49.** (a) autoimmune disease, (b) beta cells, (c) overweight, (d) metabolic, (e) insulin resistance **50.** (a) fat, (b) protein, (c) carbohydrate, (d) electrolyte **51.** tubular transport maximum **52.** (a) epinephrine, (b) norepinephrine, (c) catecholamines **53.** (a) steroid, (b)sex hormone precursors, (c) mineralocorticoids, (d) glucocorticoids, (e) testosterone, (f) estradiol **54.** aldosterone **55.** cortisol (hydrocortisone) **56.** (a) corticotropin-releasing factor, (b) adrenocorticotropic hormone (ACTH) **57.** Addison's **58.** (a) pineal, (b) melatonin **59.** thymosin **60.** atrial natriuretic factor

Building Words: **1.** neurohormone **2.** neurosecretory **3.** hypersecretion **4.** hyposecretion **5.** hyperthyroidism **6.** hypoglycemia **7.** hyperglycemia **8.** acromegaly **9.** homeostasis **10.** steroid **11.** glucogon

Matching: **1.** i **2.** d **3.** l **4.** a **5.** g **6.** c **7.** e **8.** n **9.** m **10.** k

Making Comparisons: **1.** steroid **2.** maintain sodium and potassium balance; increase sodium reabsorption and potassium excretion **3.** amino acid derivative **4.** stimulates metabolic rate; essential to normal growth and development **5.** ADH **6.** glucocorticoids (cortisol) **7.** steroid **8.** prostaglandins **9.** fatty acid derivative **10.** stimulates secretion of adrenal cortical hormones **11.** amino acid derivative **12.** help body cope with stress; increase heart rate, blood pressure, metabolic rate; raise blood sugar (glucose) level **13.** oxytocin **14.** peptide

Making Choices: **1.** c **2.** a, c, e **3.** b **4.** a, d, e **5.** b **6.** a, d **7.** a, d **8.** b, c **9.** b, c **10.** a, c, d **11.** d **12.** a, d **13.** d **14.** b, d **15.** a, b, c, e **16.** a, c, d **17.** a, b. c. d **18.** a, b, c, e **19.** d **20.** c **21.** a, b, c **22.** b, d **23.** a, d **24.** b

CHAPTER 50

Reviewing Concepts: **1.** (a) male, (b) female, (c) sperm, (e) eggs **2.** genetically identical **3.** budding **4.** fragmentation **5.** unfertilized egg **6.** zygote **7.** (a) external fertilization, (b) internal fertilization **8.** hermaphroditism **9.** asexual **10.** (a) fitness, (b) harmful mutations, (c) beneficial mutations **11.** (a) spermatogenesis, (b) spermatogonium, (c) primary spermatocyte, (d) meiotic, (e) two spermatids **12.** (a) acrosome, (b) penetrate the egg **13.** (a) Sertoli, (b) tight junctions **14.** immune response **15.** (a) scrotum, (b) seminiferous **16.** inguinal canal **17.** (a) epididymis, (b) vas deferens, (c) ejaculatory duct, (d) urethra **18.** 200 **19.** (a) seminal vesicles, (b) prostaglandins, (c) prostate gland **20.** bulbourethral **21.** (a) shaft, (b) glans, (c) prepuce (foreskin), (d) circumcision **22.** (a)cavernous bodies, (b) spongy body (c) blood **23.** (a) androgen, (b) steroids, (c) estradiol **24.** interstitial **25.** (a) follicle-stimulating hormone (FSH), (b) luteinizing hormone (LH) **26.** (a) testosterone **27.** (a) androgen-binding protein, (b) inhibin **28.** (a) oogenesis, (b) oogonia, (c) primary oocytes **29.** (a) follicle, (b) FSH, (c) polar bodies, (d) secondary oocyte **31.** (a) uterine, (b) fallopian **31.** (a) corpus luteum, (b) progesterone **32.** wall of the oviduct **33.** (a) endometrium (epithelium), (b) myometrium **34.** (a) endometrium, (b) menstruation **35.** cervix **36.** (a) perineal, (b) birth canal **37.** vulva **38.** (a) clitoris, (b) penis **39.** alveoli **40.** (a) lactation , (b)prolactin, (c) oxytocin **41.** uterus **42.** estrogen **43.** 50 44. (a) preovulatory, (b) ovulation, (c) postovulatory **45.** (a) gonadotropin releasing hormone (GnRH), (b) follicle-stimulating (FSH), (c) luteinizing (LH) **46.** (a) FSH, (b) GnRH **47.** ovulation **48.** (a) leuteal, (b) estrogen, (c) progesterone **49.** menstruation **50.** (a) hypothalamus, (b) ova, (c) infertile **51.** reabsorbed **52.** conception **53.** (a) 48, (b) 24 **54.** (a) estrogen, (b) prostaglandins **55.** enzymes **56.** (a) human chorionic gonadotropin (hCG), (b) estrogen, (c) progesterone **57.** placenta **58.** (a) stretching of uterine muscle, (b) estrogen, (c) oxytocin **59.** (a) the cervix, (b) amniotic fluid, (c) positive feedback. **60.** born **61.** (a) placenta, (b) fetal membranes **62.** (a) vasocongestion, (b) increased muscle tension **63.** (a) excitement phase, (b) plateau phase, (c) orgasm, (d) resolution **64.** contraception **65.** (a) estrogen, (b) progestin (progesterone) **66.** (a) reversible, (b) 90 million **67.** sexually transmitted infections (STIs) **68.** (a) progestin pills, (b) copper IUD **69.** (a) vasectomy, (b) vas deferens **70.** (a) tubal sterilization, (b) oviducts **71.** (a) antibodies, (b) motility, (c) fertilization **72.** binding progesterone receptors in the uterus **73.** (a) human papillomavirus or HPV, (b) 20 million, (c) *chlamydia* (d) pelvic inflammatory, (e) genital herpes

Building Words: **1.** parthenogenesis **2.** spermatogenesis **3.** acrosome **4.** circumcision **5.** oogenesis **6.** endometrium **7.** postovulatory **8.** preovulatory **9.** vasectomy **10.** vasocongestion **11.** intrauterine **12.** contraception

Matching: **1.** m **2.** j **3.** k **4.** b **5.** g **6.** c **7.** e **8.** d **9.** i **10.** l **11.** f **12.** n

Making Comparisons: **1.** anterior pituitary **2.** gonad **3.** stimulates development of seminiferous tubules, stimulates spermatogenesis **4.** stimulates follicle development, secretion of estrogen **5.** stimulates interstitial cells to secrete testosterone **6.** stimulates ovulation, corpus luteum development **7.** testes **8.** testosterone **9.** ovaries **10.** estrogen (estradiol)

Making Choices: **1.** c **2.** a, e **3.** a, c **4.** a, d **5.** a, c **6.** b, d **7.** a, b, d, e **8.** a, c, e **9.** c **10.** c, d, e **11.** a **12.** d **13.** a, c, e **14.** e **15.** e **16.** c **17.** a, b **18.** b, d **19.** c, e **20.** c **21.** c **22.** a, b, c, e **23.** e **24.** a, d **25.** a, b, c **26.** a, b, c **27.** d

CHAPTER 51

Reviewing Concepts: **1.** (a) cell number, (b) cell size **2.** cell differentiation **3.** nuclear equivalence **4.** (a) differential gene expression, (b) stem cells **5.** morphogenesis **6.** apoptosis **7.** zygote **8.** (a) diploid, (b) sex **9.** (a) sperm contacts the egg and recognition occurs, (b) the sperm or sperm nucleus enters the egg, (c) the egg becomes activated, and developmental changes begin, (d) the sperm and egg nuclei fuse **10.** (a) zona pellucida, (b) granulosa **11.** (a) vitelline envelope, (b) jelly coat **12.** (a) capacitation, (b) acrosome reaction **13.** proteolytic enzymes **14.** species-specific glycoproteins **15.** polyspermy **16.** (a) fast block, (b) calcium, (c) slow block (cortical reaction) **17.** (a) sperm receptors, (b) zona pelluc**22.** ida **18.** (a) cortical, (b) activation program **19.** (a)increases, (b) protein **20.** microtubules **21.** totipotent **22.** (a), cleavage, (b) blastula **23.** (a) isolecithal, (b) holoblastic, (c) radial, (d) spiral **24.** (a) telolecithal, (b) vegetal, (c) animal **25.** (a) meroblastic, (b) blastodisc **26.** unequal distribution **27.** (a) regulative, (b) equivalent **28.** gray crescent **29.** surface proteins **30.** blastula **31.** (a) germ layers, (b) ectoderm, (c) mesoderm, (d) endoderm **32.** (a) archenteron, (b) blastopore, (c) anus **33.** dorsal lip **33.** (a) primitive streak, (b) primitive groove, (c) archenteron **34.** Hensen's node **35.** (a) notochord, (b) neural groove, (c) neural tube, (d) spinal cord, (e) neural crest, (f) dorsal root **36.** somites **37.** blood vessels **38.** (a) gill slits, (b) middle-ear, (c) eustachian tube **39.** (a) chorion, (b) gas exchange, (c) amnion, (d) allantois, (e) nitrogenous wastes, (f) yolk sac **40.** (a) oviduct, (b) 24, (c) zona pellucida, (d) blastocyst **41.** secretions from uterine glands **42.** trophoblast **43.** inner cell mass **44.** (a) implantation, (b) seventh **45.** (a) endocrine, (b) estrogens, (c) progesterone **46.** (a) chorion, (b) uterine **47.** (a) human chorionic gonadotropin (hCG), (b) corpus luteum **48.** second and third **49.** (a) 3.5, (b) 4 **50.** third **51.** first trimester **52.** (a) second, (b) fifth **53.** (a) seventh, (b) convolutions **54.** (a) dizygotic, (b) monozygotic **55.** ultrasound **56.** teratogens **57.** carbon dioxide **58.** development

Building Words: **1.** telolecithal **2.** isolecithal **3.** blastomere **4.** blastocyst **5.** blastopore **6.** holoblastic **7.** meroblastic **8.** archenteron **9.** trophoblast **10.** trimester **11.** neonate **12.** acrosome **13.** cleavage **14.** notochord **15.** fetus

Matching: **1.** a **2.** e **3.** j **4.** n **5.** c **6.** f **7.** k **8.** l **9.** i **10.** h **11.** d

Making Comparisons: **1.** gastrulation **2.** 24 hours **3.** fertilization **4.** 0 hour **5.** 7 days **6.** blastocyst begins to implant **7.** fertilization **8.** organogenesis **9.** late cleavage **10.** 3 days

Making Choices: **1.** c **2.** b, c, e **3.** d **4.** c **5.** c, e **6.** a **7.** d **8.** b **9.** c, e **10.** d **11.** c **12.** b, c **13.** b, d **14.** c, d **15.** b **16.** a, b **17.** a, b, e **18.** a, b, d, e **19.** a, b, e **20.** a, b. d. e **21.** c **22.** b, d **23.** b

CHAPTER 52

Reviewing Concepts: **1.** stimuli **2.** natural selection **3.** behavioral ecology **4.** (a) proximate, (b) ultimate **5.** (a) direct fitness, cost-benefit analysis **6.** (a) innate, (b) learned **7.** (a) nervous, (b) endocrine **8.** motor programs **9.** sign stimulus or releaser **10.** experience **11.** aposematic or warning coloration **12.** habituation **13.** (a) critical, (b) recognizes **14.** (a) conditioned, (b) unconditioned **15.** (a) positive reinforcement, (b) negative reinforcement **16.** genetically programmed **17.** (a) gaining, (b) using, (c) problem solve **18.** (a) operant, (b) exercise, (c) learning to coordinate movements, (d) learning social skills **19.** (a) circadian rhythms, (b) diurnal, (c) nocturnal, (d) crepsular **20.** suprachiasmatic nucleus (SCN) **21.** pineal **22.** migration **23.** (a) directional orientation, (b) compass **24.** (a) navigation, (b) map **25.** (a) gathering, (b) capturing **26.** optimal foraging **27.** social behavior **28.** society

29. (a) electrical, (b) tactile, (c) visual, (d) auditory, (e) chemical. **30.** Bottlenose dolphins **31.** pheromones **32.** vomeronasal **33.** fighting **34.** dangerous **35.** testosterone **36.** home range **37.** territory **38.** (a) inherited, (b) preprogrammed **39.** (a) how many females he impregnates, (b) how many eggs she can produce, (c) the quality of the sperm that fertilize them, (d) survival of her offspring to reproductive age **40.** intrasexual selection **41.** (a) intersexual selection, (b) genetic quality (good health) **42.** lek **43.** courtship rituals **44.** (a) polygyny, (b) polyandry **45.** parental investment **46.** altruistic **47.** coefficient of relatedness **48.** kin selection **49.** sentinel behavior **50.** reciprocal altruism **51.** (a) common to a population, (b) transmitted from one generation to the next **52** (a). inherited, (b) social learning **53.** natural selection

Building Words: **1.** sociobiology **2.** polyandry **3.** polygyny **4.** monogamy **5.** pheromone

Matching: **1.** b **2.** e **3.** i **4.** m **5.** l **6.** g **7.** h **8.** a **9.** c **10.** f **11.** d **12.** n

Making Comparisons: **1.** learned **2.** classical conditioning **3.** operant conditioning **4.** habituation **5.** imprinting **6.** insight learning

Making Choices: **1.** d **2.** e **3.** b **4.** b **5.** a, b **6.** a **7.** a, b, d **8.** a, c, e **9.** d, e **10.** a, b, d, e **11.** a, b, d, e **12.** b, c, e **13.** c, e

CHAPTER 53

Reviewing Concepts: **1.** (a) biotic factors, (b) abiotic factors **2.** (a) population density, (b) population dispersion, (c) birth and death rates, (d) growth rates, (e) survivorship, (f) age structure **3.** population ecology **4.** population dynamics **5.** population density **6.** (a) random dispersion, (b) uniform dispersion, (c) clumped (aggregated) dispersion **7.** (a) natality, (b) mortality **8.** (a) change in number of individuals in the population, (b) change in time, (c) natality, (d) mortality **9.** (a) immigration, (b) emigration **10.** intrinsic rate of increase (r_{max}) **11.** smallest **12.** exponential population growth **13.** carrying capacity **14.** (a) S-shaped, (b) logistic **15.** increases **16.** (a) predation, (b) disease, (c) competition **17.** negative **18.** high **19.** (a) death rate, (b) birth rate **20.** (a) intraspecific, (b) interspecific **21.** (a) interference, (b) exploitation **22.** (a) frost, (b) blizzards, (c) hurricanes, (other weather related events and natural phenomenon) **23.** (a) semelparous, (b) iteroparous **24.** fecundity **25.** (a) *r*, (b) *K* **26.** survivorship **27.** (a) type III, (b) type I, (c) type II **28.** metapopulation **29.** (a)source habitats, (b)sink habitats **30.** (a) birth rate, (b) death rate **31.** demographics **32.** (a) low, (b) low **33.** (a) moderately, (b) less **34.** (a) high, (b) high, (c) lowest, (d) lowest **35.** (a) doubling time, (b) less **36.** (a) replacement-level fertility, (b) 2.1, (c) 2.7 **37.** (a) total fertility rate, (b) lower **38.** (a) number, (b) proportion **39.** (a) former, (b) latter

Building Words: **1.** intraspecific competition **2.** interspecific competition **3.** biosphere **4.** ecology **5.** intrinsic

Matching: **1.** o **2.** i **3.** l **4.** b **5.** e **6.** p **7.** h **8.** a **9.** f **10.** n **11.** g **12.** m **13.** c **14.** q

Making Comparisons: **1.** density-independent **2.** environmental factor that affects the size of a population but is not influenced by changes in population density **3.** density-dependent **4.** dominant individuals obtain adequate supply of a limited resource at expense of others **5.** density-dependent **6.** as density increases, encounters between population members increase as does opportunity to spread disease **7.** density-independent **8.** random weather event that reduces vulnerable population

Making Choices: **1.** g **2.** h **3.** d **4.** f **5.** b **6.** b **7.** b **8.** a, e **9.** d **10.** c **11.** a **12.** b, c **13.** a, d **14.** c **15.** a **16.** a, b, c **17.** b, c, d, e **18.** c, e

CHAPTER 54

Reviewing Concepts: **1.** community **2.** ecosystem **3.** (a) structure, (b) functioning **4.** facilitation **5.** (a) competition, (b) predation, (c) symbiosis **6.** niche **7.** (a) fundamental, (b) realized **8.** limiting resource **9.** (a) intraspecific, (b) interspecific **10.** competitive exclusion principle **11.** resource partitioning **12.** character displacement **13.** coevolution **14.** (a) aposematic coloration, (b) warning coloration **15.** cryptic **16.** Batesian mimicry **17.** Mullerian **18.** (a) mutualism, (b) commensalism, (c) parasitism **19.** pathogen **20.** (a) ecto, (b) endo **21.** (a) food, (b) water, (c) other resources **22.** very common **23.** food webs **24.** in a food web **25.** (a) nitrogen, (b) phosphorous **26.** (a) trophic, (b) trophic cascade **27.** (a) species richness, (b) species evenness **28.** geographic isolation **29.** less **30.** ecotone **31.** edge effect **32.** complexity **33.** (a) primary, (b) secondary **34.** (a) disturbance, (b) intermediate disturbance **35.** organismic **36.** individualistic

Building Words: **1.** carnivore **2.** ecosystem **3.** pathogen **4.** intraspecific competition **5.** herbivore **6.** interspecific competition **7.** predation **8.** cryptic **9.** parasite **10.** succession **11.** aposematic

Matching: **1.** m **2.** l **3.** j **4.** g **5.** n **6.** i **7.** e **8.** b **9.** o **10.** h **11.** a

Making Comparisons: **1.** commensalisms of species 1 and species 2 **2.** beneficial **3.** no effect **4.** predation of species 2 by species 1 **5.** beneficial **6.** adverse **7.** parasitism of species 2 by species 1 **8.** beneficial **9.** adverse **10.** parasitism of species 2 by species 1 **11.** beneficial **12.** adverse **13.** beneficial **14.** beneficial **15.** adverse **16.** adverse **17.** commensalism of species 1 and species 2 **18.** beneficial **19.** no effect

Making Choices: **1.** b **2.** d **3.** c **4.** b **5.** a, c, e **6.** e **7.** a **8.** a, d, e **9.** e **10.** a **11.** c **12.** b, e **13.** a-e **14.** a, b, c, e **15.** b, c **16.** b, c, d, e **17.** a, b, e **18.** e **19.** e

CHAPTER 55

Reviewing Concepts: **1.** ecosystem **2.** biosphere **3.** (a) radiant, (b) chemical, (c) heat **4.** food web **5.** (a) producers, (b) primary consumers (herbivores), (c) secondary consumers (carnivores or omnivores) **6.** (a) detritus, (b) detritivores **7.** decomposers (saprotrophs) **8.** (a) numbers, (b) biomass **9.** energy **10.** gross primary productivity **11.** net primary productivity **12.** secondary productivity **13.** 32, 40 **14.** bioaccumulation **15.** biological magnification **16.** a biogeochemical cycle **17.** (a) carbon, (b) nitrogen, (c) phosphorus, (d) water **18.** (a) photosynthesis, (b) respiration **19.** shells of marine organisms **20.** carbonic acid (H_2CO_3) **21.** (a) nitrogen fixation, (b) nitrification, (c) assimilation, (d) ammonification, (e) denitrification **22.** photochemical smog **23.** (a) rocks, (b) soil **24.** runoff **25.** transpiration **26.** groundwater **27.** (a) water, (b) temperature **28.** absolute zero **29.** long-wave infrared **30.** (a) vertical, (b) less concentrated, (c) lower temperatures **31.** 23.5 **32.** (a) 21, (b) 78 **33.** (a) high-energy, (b) visible light, (c) infra-red **34.** (a) solar energy, (b) temperature, (c) pressure, (d) rotation, (e) uneven heating, (f) high, (g) low, (h) Coriolis Effect **35.** (a) Pacific, (b) Atlantic, (c) Indian, (d) Artic **36.** Pacific **37.** (a) right, (b) clockwise, (c) left,

(d) counterclockwise **38.** El Nino-Southern Oscillation (ENSO) **39.** climate **40.** (a) temperature, (b) precipitation **41.** weather **42.** (a) precipitation, (b) deserts **43.** rain shadows **44.** microclimates **45.** (a) free minerals locked in dry organic matter, (b) remove plant cover and expose the soil (thereby stimulating the germination and establishment of seeds requiring bare soil, as well as encouraging the growth of shade-intolerant plants), (c) removes plant cover (increase soil erosion, leaving the soil more vulnerable to wind and water) **46.** deforestation

Building Words: 1. biomass **2.** atmosphere **3.** hydrologic cycle **4.** microclimate **5.** trophic **6.** saprotroph **7.** detritus **8.** nodules **9.** assimilation

Matching: 1. n **2.** i **3.** e **4.** g **5.** k **6.** l **7.** d **8.** f **9.** c **10.** j **11.** b 12. m

Making Comparisons: 1. carbon **2.** carbon dioxide **3.** nitrogen **4.** nitrogen **5.** hydrologic **6.** enters atmosphere by transpiration, evaporation; exits as precipitation

Making Choices: 1. c **2.** a, c **3.** a, c **4.** a-e **5.** d **6.** b, d, e **7.** a, d, e **8.** c, e **9.** e **10.** c **11.** e **12.** a, c, d, e **13.** e **14.** e **15.** a, b, d **16.** a, b, d, e **17.** a, c, d **18.** b, e **19.** a **20.** e **21.** a, c, d **22.** c **23.** a, b, e

CHAPTER 56

Reviewing Concepts: 1. (a) terrestrial, (b) climate **2.** (a) northern, (b) southern, (c) permafrost **3.** elevations **4.** (a) low, (b) primary **5.** taiga **6.** 50, (b) acidic **7.** (a) short growing season, (b) mineral-poor **8.** industrial wood and wood fiber **9.** (a) northwestern North America, (b) southeastern Australia, (c) southwestern South America **10.** organic content **11.** (a) lumber, (b) pulpwood **12.** 75-126 **13.** agricultural use **14.** 10-30 inches (25-75 cm) **15.** (a) tall grass, (b) shortgrass **16.** Mediterranean **17.** (a) cold, (b) warm **18.** (a) Great Basin, (b) Mojave, (c) Chihuahuan, (d) Sonoran **19.** 10 **20.** estivation **21.** (a) precipitation (b) temperature **22.** 34 to 60 (85 to 150 cm) **23.** African **24.** over grazing by domestic animals **25.** (a) 60 to 80 (150 to 200 cm) (b) shed **26.** 80-180 inches (200-450 cm) **27.** evergreen flowering **28.** epiphyte **29.** abiotic factors **30.** (a) plankton, (b) nekton, (c) benthos **31.** (a) streams, (b) rivers, (c) ponds, (d) lakes, (e) marshes, (f) swamps, (g) 2 **32.** strength of the current **33.** land **34.** (a) littoral, (b) limnetic, (c) profundal, (d) littoral **35.** (a) thermal stratification, (b) thermocline, (c) fall turnover, (d) spring turnover **36.** (a) marshes, (b) swamps **37.** (a) prairie potholes, (b) peat moss bogs **38.** (a) tides, (b) tides, (c) time of year, (d) precipitation **39.** salt marshes **40.** mangrove forests **41.** (a) intertidal zone, (b) light, (c) nutrients **42.** (a) distance from land, (b) light availability, (c) depth **43.** 4-6 meters (2.5-3.7 miles) **44.** (a) calcium carbonate, (b) zooxanthellae **45.** neritic province **46.** euphotic **47.** oceanic province **48.** ecotone **49.** center of origin **50.** (a) endemic, (b) cosmopolitan **51.** (a) Palearctic, (b) Nearctic, (c) Neotropical, (d) Ethiopian, (e) Oriental, (f) Australian

Building Words: 1. phytoplankton **2.** zooplankton **3.** intertidal **4.** thermocline **5.** littoral **6.** benthic **7.** abyssal **8.** boreal **9.** chapparal **10.** estivation

Matching: 1. n **2.** i **3.** j **4.** a **5.** c **6.** l **7.** d **8.** e **9.** h **10.** g **11.** f **12.** k

Making Comparisons: 1. long, harsh winters, short summers, little rain **2.** permafrost **3.** low-growing mosses, lichens, grasses, and sedges adapted to extreme cold and short growing season; voles, weasels, arctic foxes, grey wolves, snowshoe hares, snowy owls, oxen, lemmings, caribou, etc. (few plant and animal species) **4.** taiga **5.** seasonality, high rainfall **6.** rich topsoil, clay in lower layer **7.** broad-leaf trees that lose leaves seasonally, variety of large mammals **8.** temperate grassland **9.** deep, mineral-rich soil **10.** wet, mild winters; dry summers **11.** thickets of small-leafed shrubs and trees, fire adaptation; mule deer **12.** desert **13.** low in organic material, often high mineral content **14.** savanna **15.** low in essential mineral nutrients **16.** high rainfall evenly distributed throughout year **17.** mineral-poor **18.** high species diversity, distinct stories of vegetation, epiphytes in organic material, often high mineral content

Making Choices: **1.** b **2.** d **3.** b, d **4.** d, e **5.** b, e **6.** a, d **7.** d **8.** b, d **9.** e **10.** b, e **11.** b **12.** d **13.** a-e **14.** c **15.** b **16.** b-e **17.** c **18.** b, e **19.** a-e

CHAPTER 57

Reviewing Concepts: **1.** environmental sustainability **2.** (a) genetic diversity, (b) species diversity, (c) ecosystem diversity **3.** (a) endangered, (b) threatened **4.** (a) an extremely small range, (b) territory, (c) on islands, (d) low reproductive success, (e) breeding areas, (f) feeding habits **5.** (a) destruction, (b) fragmentation, (c) degredation **6.** (a) habitat fragmentation, (b) invasive **7.** (a) biotic pollution, (b) invasive **8.** commercial **9.** *In situ* **10.** *Ex situ* **11.** (a) animal, (b) plant **12.** landscape ecology **13.** habitat corridors **14.** (a) restoration ecology, (b) *in situ* **15.** (a) artificial insemination, (b) fertility drugs **16.** protect endangered and threatened species in the U.S. and abroad **17.** endangered animals and plants considered valuable in international wildlife trade **18.** trapping and absorbing precipitation **19.** carbon stored in trees into the atmosphere as carbon dioxide **20.** (a) subsistence agriculture, (b) commercial logging, (c) cattle ranching **21.** industrial wood and wood fiber **22.** CO_2, CH_4, O_3, N_2O, hydrochlorofluorocarbons (HCFCs) **23.** greenhouse effect **24.** Infrared radiation **25.** (a) sea level, (b) precipitation patterns, (c) biological diversity, (d) human health, (e) agriculture **26.** (a) chlorine, (b) bromine **27.** chlorofluorocarbons (CFCs) **28.** September and November **29.** (a) cataracts, (b) skin cancer, (c) weakened immune system **30.** CFCs

Building Words: **1.** deforestation **2.** stratosphere **3.** endemic **4.** biotic

Matching: **1.** g **2.** o **3.** b **4.** e **5.** h **6.** a **7.** j **8.** n **9.** c **10.** l **11.** d **12.** i

Making Comparisons: **1.** deforestation **2.** permanent removal of forests for agriculture, commercial logging, ranching, use for fuel **3.** reduce current practices **4.** global warming **5.** atmosphere pollutants that trap solar heat **6.** prevent build-up of gases, mitigate effects, adapt to the effects **7.** chemical destruction of stratospheric ozone **8.** reduce use of chlorofluorocarbons and similar chlorine containing compounds